Gijsbertus de With

**Structure, Deformation, and
Integrity of Materials**

Related Titles

Baltes, H., Brand, O., Fedder, G. K., Hierold, C., Korvink,
J. G., Tabata, O., Löhe, D., Haußelt, J. (eds.)

Microengineering of Metals and Ceramics

Part I: Design, Tooling, and Injection Molding
2005
ISBN 3-527-31208-0

Zehetbauer, M., Valiev, R. Z. (eds.)

Nanomaterials by Severe Plastic Deformation
2004
ISBN 3-527-30659-5

Trebin, H.-R. (ed.)

Quasicrystals

Structure and Physical Properties
2003
ISBN 3-527-40399-X

Riedel, R. (ed.)

Handbook of Ceramic Hard Materials
2000
ISBN 3-527-29972-6

Meyers, M. A., Armstrong, R. W., Kirchner, H. O. K. (eds.)

Mechanics and Materials

Fundamentals and Linkages
1999
ISBN 0-471-24317-5

Gijsbertus de With

Structure, Deformation, and Integrity of Materials

Volume II: Plasticity, Visco-elasticity, and Fracture

WILEY-VCH

WILEY-VCH Verlag GmbH & Co. KGaA

The author

Prof. Dr. Gijsbertus de With
Eindhoven University of Technology
Department of Chemical Engineering
Den Dolech 2
5600 MB Eindhoven
The Netherlands

G.d.With@tue.nl

Library of Congress Card No.:
applied for

British Library Cataloguing-in-Publication Data
A catalogue record for this book is available from the British Library.

Bibliographic information published by Die Deutsche Bibliothek
Die Deutsche Bibliothek lists this publication in the Deutsche Nationalbibliografie; detailed bibliographic data is available in the Internet at <http://dnb.ddb.de>.

Printed in the Federal Republic of Germany

Printed on acid-free paper

Printing Strauss GmbH, Mörlenbach

Bookbinding Litges & Dopf GmbH, Heppenheim

ISBN-13: 978-3-527-31426-3

ISBN-10: 3-527-31426-1

no solider wisdom than that which is acquired in struggling against trouble.

Simon L. Altmann, *Icons and Symmetries*, Clarendon Press, Oxford, 1992.

Preface

For many processes and applications of materials a basic knowledge of their mechanical behaviour is a must. This is obviously not only true for materials with a primarily structural function but also for those materials for which the primary function is an electrical, dielectrical, magnetic or optical one and which are frequently (and wrongly) known as functional materials. Although many books on the mechanical behaviour of materials exist, a few drawbacks are generally present. On the one hand, the treatment is either too limited or too extensive and on the other hand the emphasis is typically too much on one type of materials, that is either on metals, on polymers or on inorganics. Moreover, the relation between the behaviour at the atomic, microstructural and macroscopic level is generally poorly developed. In this book a basic but as far as possible self-contained and integrated treatment of the mechanical behaviour of materials and their simplest applications is presented. We try to avoid the drawbacks mentioned by giving an approximately equal weight to the three material categories at a sound basic level. This does imply that not all topics can be treated and a certain initial acquaintance with materials science is probably an advantage for the reader. Meanwhile we try not to forget the need for somewhat more advanced discussion on several topics. Hopefully the proper balance is implemented by the two-level presentation: the 'basic' sections for all students and the 'advanced', or more properly 'intermediate', sections labelled with an asterisk for those who wish to deepen their knowledge.

A particular feature of this book is the attempt to give a basic but balanced presentation of the various aspects relevant at the micro- (atomic or molecular), meso- (microstructural or morphological) and macro-scale (bulk material properties and behaviour) for polymers, metals and inorganics. Another, also quite important aspect is that, wherever useful, the thermodynamic aspects are emphasised. We realise that this approach is not customary but we are convinced that this will make access to the more advanced literature easier. To that purpose we present in the Overview (part I) an introduction and an outline of constitutive behaviour. Part II describes the Fundamentals. It contains some mathematical preliminaries and the essentials of the continuum theory of kinematics, kinetics and thermodynamics. Also a summary of atomic and structural tools is given in this part. The latter has been incorporated to be somewhat self-contained. The remaining chapters discuss several topics in more detail and have been divided into various parts, namely Elasticity (part III), Plasticity (part IV), Visco-elasticity (part V) and Fracture (part VI). Of course, it is quite impossible to deal with every aspect and therefore we have limited ourselves, apart from the essentials of each of these topics, mainly to similarities and differences between the type of materials and their thermodynamic and structural background.

The whole of topics presented is conveniently described as mechanics of materials: it describes the thermomechanical behaviour of materials itself with applications to elementary structures and processes. With respect to the latter aspect the book by A.H. Cottrell, *The mechanical properties of matter*, has been an enlightening example. Unfortunately, in the past the term 'mechanics of materials', sometimes also called 'strength of materials', was claimed to denote the description of the deformation of beams, plates and other structures, given the constitutive behaviour of the material. The material aspect thus only appears on the phenomenological level. This area would rightfully have been called 'mechanics of structures' or 'structural mechanics'. The somewhat different description 'mechanical behaviour of materials' is not entirely adequate for this book since that title suggests that the treatment is essentially mechanical. Moreover, the application of the mechanical behaviour to simple structures and the explanation of the behaviour in structural terms, which is considered as essential in our approach, are not incorporated. The title reflects our final choice! For brevity I generally refer to the field as *thermomechanics*.

Since I realised that my style of writing is compact, I introduced in several chapters some panels. These panels do not interrupt the line of the discussion but can be read as an aside, which put a certain topic in perspective, either from a pragmatic, a context or a historical viewpoint. With respect to history I restricted myself to two or so panels per chapter with a short biography of eminent scientists. Most of the information is taken from the books written by Timoshenko[*], Struik[†], Nye[‡], Love[§], Tanner and Walters[**] and Cahn[††]. Another useful source was the book by Hoddeson et al.[‡‡]. The choice of the short biographies is arbitrary and it is likely that other authors will make another choice. It must be said that history is not always kind to people and certain topics or subjects are not known by the name of their first discoverer. I refrained from critical remarks in this respect. For those interested in these aspects the very detailed works[§§] of Clifford Ambrose Truesdell[***] might be useful. Contrary to many textbooks, reference is made to the literature, generally to original presentations, other textbooks and reviews. On the other hand, only incidental reference is made to experimental methods.

After having treated the phenomenological equations, the applications of the theory, and the structural aspects of elasticity, plasticity, visco-elasticity and fracture, in the last chapter, I tried to provide a personal view and perspective on the whole of thermomechanics. I hope that the remarks made will be useful to many, although I am quite sure that it does not cover these areas to the satisfaction of everybody.

The essential ingredients of these notes were already contained in a course on the mechanical behaviour of materials at the Department of Chemical Engineering and Chemistry at Eindhoven University of Technology, which I took over some 10 years ago. The overall set-up as given here has been evolved in the last few years in which hopefully both the balance in topics and their presentation is improved. I am obliged to my students and instructors who have followed and used this course and provided

---

[*] Timoshenko, S.P. (1953), *History of strength of materials*, McGraw-Hill, New York (see also Dover, 1983).
[†] Struik, D.J. (1948), *A concise history of mathematics*, Dover, 1948.
[‡] Nye, M.J. (1996), *Before big science*, Harvard University Press, Cambridge, MA.
[§] Love, A.E.H. (1927), *A treatise on the mathematical theory of elasticity*, 4th ed., Cambridge University Press, Cambridge (see also Dover, 1944).
[**] Tanner, R.I. and Walters, K. (1998), *Rheology: an historical perspective*, Elsevier, Amsterdam.
[††] Cahn, R.W. (2001), *The coming of materials science*, Pergamon, Amsterdam.
[‡‡] Hoddeson, L., Braun, E., Teichmann, J. and Weart, S. (1992), *Out of the crystal maze*, Oxford University Press, New York.
[§§] Truesdell, C.A. (1968), *Essays in the history of mechanics*, Springer, Berlin, and Truesdell, C.A. (1980), *The tragicomical history of thermodynamics 1822-1854*, Springer, Berlin.
[***] Clifford Ambrose Truesdell (1919-2000). American scientist, who played a highly instrumental role in the development of so-called rational mechanics and thermodynamics. His main criticism on the development of the thermodynamics is the continuous and complete mixing up of constitutive behaviour and basic laws as compared to the more or less separation of these aspects in mechanics.

many useful remarks. In particular I want to thank my colleague Dr. Paul. G. Th. van der Varst for the careful reading of and commenting on many parts of the manuscript and many discussions on almost all of the topics covered, which I always enjoyed and which made clearer to me a great number of aspects. Hopefully this led to an improvement in the presentation. It has been said before that authors do not finish their manuscript but abandon it. After the experience of writing this book I recognise that sentiment. My greatest indebtedness is to my wife who I 'abandoned' for many hours and days. Without her patience the book would never have been finished.

Obviously, the border between various classical disciplines is fading out nowadays. It is therefore hoped that these notes are not only useful for the original target audience, chemists and chemical engineers, but also for materials scientists, mechanical engineers, physicists and the like. Finally, I fear, the text will not be free of errors. They are my responsibility. Any comments, corrections or indications of omissions will be appreciated.

G. de With, July 2005

Acknowledgements

Figures

Figures 1.01, 1.08, 8.06, 15.04 and 15.06 are reprinted by permission of the Addison-Wesley Publishing Company, San Francisco, CA.

Figure 11.15 is reprinted by permission of the American Physical Society, College Park, MD.

Figure 22.14 is reprinted by permission of ASM Int., Metals Park, Ohio.

Figures 12.15 and 12.16 are reprinted by permission of the ASTM, Philadelpia.

Figures 14.11 and 14.12 are reprinted by permission of the Akademie-Verlag, Berlin.

Figures 8.21, 11.13, 12.07, 12.22, 23.06, E-3, E-4, E-5, E-7, E-8, E-9, E-10, E-11, E-12 and E-13 are reprinted by permission of the Cambridge University Press, Cambridge.

Figures 1.09, 8.01, 17.09 and 24.14 are reprinted by permission of Chapman & Hall, now CRC Press, Baco Raton.

Figure 22.17 is reprinted by permission of CIRP, Paris.

Figures 1.03, 1.05, 1.11, 1.12, 1.13, 1.14, 8.07, 8.33, 8.34, 15.02, 15.03, 15.13, 15.19, 15.22, 15.23b, 15.32, 15.33, 15.35, 15.36, 16.09, 16.17, 20.05, 20.06, 22.21, 23.07, 23.08 are reprinted by permission of Elsevier, Oxford.

Figure 21.04 is reprinted by permission of Marcel Dekker Inc., now CRC Press, Baco Raton.

Figures: 11.08 and 25-A are reprinted by permission of the Materials Research Society, Warrendale, PA.

Figures 1.15, 11.04, D-2, D-3, 15.09, 15.24, 15.31 and 16.14 are reprinted by permission of McGraw-Hill, New York.

Figures 14.17, 20.01, 20.02 and 20.03 are reprinted by permission of the McMillan Company, Basingstoke, Hampshire.

Figures 7.01, 7.02, 8.12, 8.13, 8.27, 8.28, 11.05 11.10, 11.11, 14.03, 15.28 and 15.30 are reprinted by permission of Oxford University Press, Oxford.

Figures 19.03, 19.04 and 19.05 are reprinted by permission of Prentice-Hall, now Pearson Education, Upper Saddle River, NJ.

Figures 24.16 and 24.17 are reprinted by permission of Scientific American, New York.

Figures 8.17, 20.07, 20.08, 20.09, 20.10, 20.11, 20.12, 20.13, 21.01, 21.11, 21.12, 21.13, E-14 and E-15 are reprinted by permission of Springer, Berlin.

Figure 2.12 are reprinted by permission of Syracuse University Press, Syracuse., New York.

Figure 22.24 is reprinted by permission of the American Ceramic Society, Westerville, Ohio.

Figures 22.15, 22.16, 22.18 and 22.19 are reprinted by permission of NIST, Gaithersburg, MD.

Figures 2.05, 7.03, 8.15, 8.16, 8.18, 11.16, 15.18, 15.20, 15.23a, 16.02, 16.12, 17.01, 17.04, 23.04 and 23.14 are reprinted by permission of Wiley, Chichester, UK.

All portraits have been reproduced from various websites by permission of the copyright holders. Wiley-VCH and the author have attempted to trace the copyright holders of all material reproduced in this publication and apologise to copyright holders if permission to publish in this form has not been obtained.

Cover

Scales: an artist's impression of the length scales aspects in thermomechanics. Martijn de With, 2005.

Contents Volume I

Contents Volume II

Preface
Acknowledgements
Contents
List of important symbols and abbreviations

Part IV: Plasticity
13. Continuum plasticity

Index

List of important symbols and abbreviations

Φ	dissipation function, Airy stress function, wave function, potential energy	Ψ	wave function
		Ω	external potential energy
Ξ	grand partition function		

α	constant	λ	Lamé constant, stretch
α_{ij}	thermal expansion tensor	μ	Lamé constant, shear modulus
β	constant, kT	π	(second) Piola-Kirchoff stress
γ	(engineering) strain, shear strain	ρ	density, radius of curvature
δ_{ij}	Kronecker delta	σ	(true) stress
ε	strain, small scalar, energy	σ_{ij}	Cauchy stress tensor
ε_{ij}	strain tensor	τ	shear stress
φ	specific dissipation function	v	Poisson's ratio
ϕ	(pair) potential energy	ω	frequency

C	right Cauchy-Green tensor	**Q**	generalised force
D	left Cauchy-Green tensor	**O**	zero tensor
E	Euler strain tensor	**0**	zero vector
F	deformation gradient	**I**	unit tensor
L	Lagrange strain tensor	**1**	unit vector

a	acceleration, generalised displacement	**q**	torque
		r	direct lattice vector, material co-ordinate
b	body force, Burgers vector		
d	rate of deformation	**s**	(first) Piola-Kirchhoff stress tensor, shear stress vector
e	unit vector		
f	force	**t**	stress vector, traction
g	reciprocal lattice vector	**u**	displacement
m, n	outer normal vector	**v**	velocity
l	angular momentum, dislocation line	**x**	spatial co-ordinate
p	linear momentum, generalised momentum		

A	area, fatigue parameter, generalised force	I	moment of inertia
		$J_{(i)}$	invariant
$A_{(i)}$	basic invariant of A_{ij}	K	bulk modulus, reaction constant
A_I	principal value of A_{ij}	L	power, length, Lagrange function
C	constant		
C_{ij}	elastic stiffness constants	M	moment, orientation factor
C_{ijkl}	elastic stiffness constants	N_A	Avogadro's number
E	Young's modulus, energy	P	porosity, probability, power
F	Helmholtz energy, fatigue limit, force	Q	partition function, charge
		Q_i	component of generalised force
G	Gibbs energy, shear modulus, strain energy release rate	R	gas constant, radius, fatigue parameter, fracture energy
H	enthalpy, Hamilton function	S	entropy, strength

S_{ij}	elastic compliance constants	W	work, strain energy
S_{ijkl}	elastic compliance constants	Y	uniaxial yield strength
T	kinetic energy, temperature	Z	section modulus, density of states, co-ordination number, partition functio
U	(internal) energy		
V	potential energy, volume		
a	generalised displacement	n	Mie constant, material constant
a_i	component of acceleration	p	pressure, plastic constraint factor
b	length of Burgers vector		
b_i	component of body force	p_i	component of (generalised) momentum
c	constant, inverse spring constant (compliance)	q_i	component of (generalised) co-ordinate
d_{ij}	rate of deformation		
e	strain	s	specific entropy, (engineering) stress
e_{ijk}	alternator		
f	(volume) fraction, specific Helmholtz energy, force	t	time
		t_i	component of traction
g	specific Gibbs energy	u,v,w	displacement in x,y,z direction
k	Boltzmann's constant, yield strength in shear, spring constant (stiffness)	v	volume
		w	strain energy density
		z	single particle partition function
l	length		
m	Weibull modulus, Mie constant, Schmid factor, mass		

BCC	body centered cubic	PMPE	principle of minimum potential energy
CRSS	critically resolved shear stress		
FCC	face centred cubic	PVP	principle of virtual power
FEM	finite element method	PVW	principle of virtual work
HCP	hexagonal closed packed	SDG	small displacement gradient
LEFM	linear elastic fracture mechanics	SFE	stacking fault energy
		SIF	stress intensity factor
PCVP	principle of complementary virtual power	SSY	small scale yielding

$\cong$	approximately equal	(hkl)	specific plane
$\equiv$	identical	$\{hkl\}$	set of planes
$\sim$	proportional to	$[hkl]$	specific direction
$\Leftrightarrow$	corresponds with	$\langle hkl \rangle$	set of directions

13

Continuum plasticity

After dealing with the basic aspects of elasticity and the elastic aspects of structures in the previous chapters, now plasticity is discussed with the continuum point of view in mind. First, we focus on the criteria for the onset of flow. The effect of pressure, strain rate and temperature on the yield criterion is briefly discussed. Work hardening and the flow behaviour using the conventional presentation are addressed thereafter. Finally, we reconsider the matter from a thermodynamic point of view.

13.1   General considerations

We have seen in Chapter 2 that after reaching a certain stress the behaviour of the material becomes non-linear and dissipative. We have to distinguish between the criterion for the onset of flow, i.e. the *yield criterion* and the flow behaviour itself, i.e. the *flow rule* (Maugin, 1992; Lemaitre and Chaboche, 1990).

With respect to the yield criterion three aspects have to be discussed: yield strength, hardening and strain rate dependence.

For many solids it is assumed that there exists a certain threshold in load below which no flow can occur. To characterise this behaviour the *yield strength Y* was introduced as a measure of the capability of the material to withstand irreversible flow (as indicated by Y_1 in Fig. 13.1). It is conventional to address the behaviour above the threshold as *plasticity* and the deformation of the material is called *plastic* (or *ductile*) *deformation*. At room temperature most metals and polymers show ductile deformation while most inorganic materials do not.

The yield strength of a material that has been deformed before is usually higher as compared to the value for a non-deformed material. Unloading a material at a certain strain and re-loading therefore leads to a higher yield strength as for the 'virgin' material (as indicated by Y_1' in Fig. 13.1). This behaviour is addressed as *hardening* and characterised by the *hardening modulus* $h = d\sigma/d\varepsilon$ (for the range where $\sigma > Y$, see

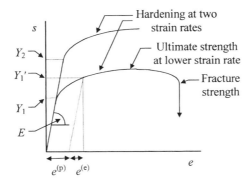

Fig. 13.1: The stress-strain relationship for a tensile test using engineering strain *e* and stress *s* showing increased yield strength at higher strain and strain rate.

Chapter 2). For some metals the hardening is quite extensive, e.g. Cu, while for others it is limited, e.g. Al or stainless steel. It should be noted though that some materials show *softening*, i.e. a decreasing yield strength with increasing strain after a certain threshold (in strain, stress or energy) has been reached. Certain coarse-grained inorganic materials, e.g. refractory ceramics, and certain composite materials, e.g. fibre-reinforced polymers, provide examples.

In principle the yield strength of a material is strain rate dependent. Generally the larger the strain rates, the larger the yield strength (as indicated by Y_2 in Fig. 13.1). The limiting situation of the latter, in which no elastic effects are present, at least in shear loading, is *viscous flow*, where the stress is proportional to the strain rate. If elastic effects are present, the behaviour is called *visco-elastic*. If both visco-elasticity and plasticity are present, the behaviour is addressed as *visco-elasto-plastic*. We limit us here to *elasto-plastic* behaviour, meaning that the deformation below the threshold is purely elastic and that the flow is assumed to be purely plastic, independent of strain rate. In Section 13.2 a simple approach is given. Some elaborations are made in Sections 13.3-13.5. Hardening is briefly addressed in Section 13.6.

The description of the flow behaviour is more complex than that of the yield behaviour and relates the strain rate after yield to the applied stress. The conventional description is presented in Section 13.7. A thermodynamically-based approach to both yield and flow behaviour is presented in Section 13.8.

13.2   A simple approach

We have seen that for a bar loaded in uniaxial tension in first approximation a yield strength Y exists. If the tensile stress σ in the bar has reached this value, flow will occur. Here a one-to-one correspondence between the field quantity σ, a tensor (component), and the material quantity Y, a scalar, exists. The condition is denoted as the *yield criterion* and we can express this as (Derby et al., 1992; Lubliner 1990)

$$X = \sigma - Y = 0$$

For $X < 0$, the behaviour is elastic while $X = 0$ expresses the yield condition. We have also seen that the stress distribution generally can be complex and needs the use of the stress tensor or its matrix representation. In order to be able to make still a simple comparison between a (tensor) field quantity and a (scalar) material property, one generally introduces the *yield surface* or *limit surface X*, described by

$$X(\sigma_{ij}) = f(\sigma_{ij}) - K = 0 \tag{13.1}$$

where $f(\sigma_{ij})$ is a function of the stress components σ_{ij} and K a material constant,

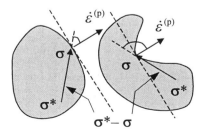

Fig. 13.2: A convex (left) and non-convex surface (right) in stress space.

related to the uniaxial yield strength Y. In general, the function $X = f - K = 0$ can be envisaged as a five-dimensional surface embedded in a six-dimensional space. It is conventional to define X in such a way that $X < 0$ when the material is deforming only elastically. This renders the material constant K positive. Yield then occurs for $X = 0$ while $X > 0$ cannot occur. The yield surface has various properties, some general, valid for (almost) all materials, and some more specific, valid only for a restricted set of materials.

In general one assumes that any yield surface should be convex[a]. Essentially this property implies that any cross-section of the surface is also convex (Fig. 13.2). It holds that such a cross-section shows no re-entrant angles. The basic arguments why the yield surface is convex can be found in stability considerations (although often it is stated that the yield surface should be convex on thermodynamic grounds).

Justification 13.1*

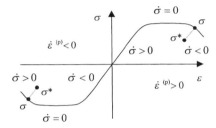

Fig. 13.3: The stress-strain curve from a tensile test showing hardening, ideal plastic and softening behaviour.

The basic arguments[b] for convexity are extensions of the stability arguments for a tensile test as given by Drucker[c]. From Fig. 13.3 it is clear that:

$\dot{\sigma}\dot{\varepsilon}^{(p)} > 0$ for hardening,

$\dot{\sigma}\dot{\varepsilon}^{(p)} = 0$ for ideal plastic and

$\dot{\sigma}\dot{\varepsilon}^{(p)} < 0$ for softening behaviour.

Obviously, for stability upon further loading the material has to show hardening. The above relations remain valid for the products $d\sigma d\varepsilon^{(p)}$ representing the plastic work per unit volume. Suppose now that we have a body initially in equilibrium with stress state σ^*. Moreover suppose that an external agency, different from the one that causes the state initial σ^*, slowly imposes a set of self-equilibrating forces and then slowly removes them. A stable work hardening material is now defined by *Drucker's postulate* as a material for which the plastic work done by the external agency during

---

[a] A function $f(x)$ is convex if for every value of the parameter λ within the range $0 \le \lambda \le 1$ it holds that $f(\lambda x + (1-\lambda)x) \le \lambda f(x) + (1-\lambda)f(x)$.
[b] Drucker, D.C. (1951), page 487 in Proc. 1st U.S. Natl. Cong. Appl. Mech, ASME, New York.
[c] Daniel C. Drucker (1918-2001). American scientist who introduced the concept of material stability, now known as "Drucker's Stability Postulate," which provided a unified approach for the derivation of stress-strain relations for plastic behavior of metals. His theorems led directly to limit design; a technique to predict the load carrying capacity of engineering structures.

application of the extra stresses is positive and the net total work during the cycle of adding and removing the stresses is non-negative. Since $d\sigma d\varepsilon^{(e)} > 0$ and $d\sigma d\varepsilon^{(p)} \geq 0$ for a stable material, we have $d\sigma d\varepsilon > 0$. This definition, known as *Drucker's inequality*, is extended to three-dimensional states and then reads

$$d\sigma d\varepsilon = d\sigma_{ij} d\varepsilon_{ij} > 0 \quad \text{and} \quad d\sigma d\varepsilon^{(p)} = d\sigma_{ij} d\varepsilon_{ij}{}^{(p)} \geq 0$$

and is valid for both hardening and ideal-plastic materials. In fact if σ^* represents an elastic state or a state at the yield surface far removed from the state σ on the yield surface, the external agency might bring the system elastically to state σ, with small increment $d\sigma$ produce an increment in plastic strain $d\varepsilon^{(p)}$ and finally bring the system back elastically to the state σ^*. Drucker's postulate implies that

$$(\sigma - \sigma^*).d\dot{\varepsilon}^{(p)} \geq 0$$

a statement often referred to as the *postulate of maximum plastic dissipation*. It is valid for hardening and ideal-plastic materials but also for softening materials, as can be seen from Fig. 13.3. This postulate is therefore of wider significance than Drucker's postulate. If we suppose that the yield surface is smooth everywhere, so that a well-defined tangent hyper-plane and normal direction exist, it will be clear from Fig. 13.2 that in order to satisfy the inequality for all states σ^*, the state σ^* must be on the inward side of the tangent plane since for a sharp angle between σ^* and $d\varepsilon^{(p)}$ the inner product becomes negative. The yield surface is therefore *convex*. Moreover, $d\varepsilon^{(p)}$ must be directed along the outward normal, a consequence referred to as the *normality rule*.

Let us now specialise to an isotropically yielding body. In that case the yield surface expression X and therefore f should be independent of the orientation of the axes system. If we express the components of the stress matrix in a new axes system by $\tilde{\sigma}_{ij} = R_{ip}\sigma_{pq}R_{qj}$, where the R_{ij} denote the elements of the rotation matrix, it thus must hold that (Section 6.8)

$$\tilde{f}(\tilde{\sigma}_{ij}) = f(\tilde{\sigma}_{ij}) = f(R_{ip}\sigma_{pq}R_{qj}) = f(\sigma_{ij}) \tag{13.2}$$

It can be shown that this is only possible if the function f is a function of the invariants $J_{(1)}, J_{(2)}$ and $J_{(3)}$ of the stress matrix only, i.e. if

$$f = f(J_{(1)}, J_{(2)}, J_{(3)}) \tag{13.3}$$

Alternatively using the decomposition of the stress tensor in a hydrostatic (isotropic) and deviatoric part, i.e.

$$\sigma = \sigma' + \tfrac{1}{3}\text{tr}(\sigma)\mathbf{I} = \sigma' - p\mathbf{I} \quad \text{or} \quad \sigma_{ij} = \sigma_{ij}' + \frac{1}{3}\sigma_{kk}\delta_{ij} = \sigma_{ij}' - p\delta_{ij} \tag{13.4}$$

we can write[d]

---

[d] The representation $X' = f^2 - K^2 = 0$ is equivalent and sometimes easier to use.

$$X = f\!\left(J_{(1)}, J_{(2)}', J_{(3)}'\right) - K = 0$$

where $J_{(2)}'$ and $J_{(3)}'$ denote the second and third invariants of the deviatoric stress tensor and $J_{(1)}$ being the first invariant of the complete stress tensor is equal to three times the hydrostatic pressure. Experimentally it appears that plastic deformation is largely independent of the hydrostatic pressure $p = -\tfrac{1}{3}\sigma_{kk}$, at least for metals and inorganic materials. Since the first invariant is given by $J_{(1)} = \sigma_{kk}$, it follows that the dependence on $J_{(1)}$ can be neglected, resulting in

$$X = f\!\left(J_{(2)}', J_{(3)}'\right) - K = 0 \tag{13.5}$$

Another representation of the yield surface is in terms of the principal stresses σ_{I}, σ_{II} and σ_{III} and reads

$$X = F\!\left(\sigma_{\mathrm{I}}, \sigma_{\mathrm{II}}, \sigma_{\mathrm{III}}\right) - K = 0 \tag{13.6}$$

Analysing $J_{(2)}'$ and $J_{(3)}'$ Eq. (13.5) shows that the function F in fact can only be a function of the difference of the principal stresses. Moreover, because the order of the principal stresses is not fixed, the function F must be symmetric in σ_{I}, σ_{II} and σ_{III}. To elucidate a bit, let us consider the *principal axes space*[e]. This is a Cartesian space with the principal values labelling the co-ordinate axes (Fig. 13.4). A plane perpendicular to the hydrostatic axis contains only deviatoric stresses and is often addressed as a *Π-plane*. Since at any position in a mechanically loaded body or structure the stress tensor can be brought at principal axes, the state of stress at any point can be represented by a point in principal axes space. Hydrostatic stresses, characterised by $\sigma_{ij} = -p\delta_{ij}$ or equivalently with $\sigma_{\mathrm{I}} = \sigma_{\mathrm{II}} = \sigma_{\mathrm{III}} = -p$, are all lying on an axis that makes equal angles to the co-ordinate axes. This axis is called the *hydrostatic axis*. The hydrostatic axis must be a three-fold symmetry axis of the yield surface, since the function F should be symmetrical in σ_{I}, σ_{II} and σ_{III}. Similarly, the planes $\sigma_{\mathrm{I}} = \sigma_{\mathrm{II}}$, $\sigma_{\mathrm{II}} = \sigma_{\mathrm{III}}$ and $\sigma_{\mathrm{III}} = \sigma_{\mathrm{I}}$ must be symmetry planes of the function F. Any cross-section of the yield surface perpendicular to the hydrostatic axis thus must have a three-fold axis of symmetry. If we assume that no *Bauschinger effect*[f] is present, i.e. the material shows the same response in tension and compression, we must also have

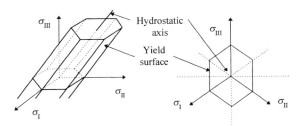

Fig. 13.4: Principal axes space, a general yield surface and the projection along the hydrostatic axis. Here straight lines represent the surfaces between the symmetry lines but they may take any convex shape. For clarity only the mirror line perpendicular to σ_{III} is indicated.

---

[e] Also referred to as Haigh-Westergaard or Westergaard space, after their first users.
[f] Johann Bauschinger (1833-1893). German scientist, well known for his experimental contributions to mechanics amongst which there are the presence of a yield point, the effect named after him, elastic recovery and the existence of a fatigue limit.

$$F\left(\sigma_{I},\sigma_{II},\sigma_{III}\right) = F\left(-\sigma_{I},-\sigma_{II},-\sigma_{III}\right) \qquad (13.7)$$

implying mirror symmetry with respect to lines perpendicular to the axes σ_I, σ_{II} and σ_{III}. Thus each equivalent sector with an opening angle of $\pi/3$ itself is symmetrical about its bisector (as indicated for σ_{III} in Fig. 13.4), leading to a six-fold axis. Since J_3' changes sign when the stresses are reversed, it follows that f should be even in J_3' or not depending on it at all. In summary, the yield surface for an isotropic material without Bauschinger effect shows an overall six-fold symmetry axis with each sector symmetrical about its bisector. The yield locus on a Π-plane shows the same symmetry.

The function f (or F) is also addressed as an *equivalent stress* σ^* since it is a scalar measure of the applied stress field σ_{ij}. In this way we can say that, similar to the tensile loaded bar, plastic deformation occurs if the equivalent stress σ^* (a field quantity) reaches the *yield strength* Y (a material quantity as determined, e.g. from a tensile test). In order to match with the uniaxial tensile test the additional requirement

▶ $\sigma^* = \sigma = Y$ $\qquad (13.8)$

where σ is the applied stress in the tensile test, should hold. Until now we considered only properties of the function f, i.e. the equivalent stress σ^*, but did not specify it. Obviously, one expects the specific expression for σ^* to depend on the material one considers. Fortunately it turns out that in many cases materials yield according to the Tresca or to the von Mises criterion and each criterion entails a certain definition of the equivalent stress σ^*. We distinguish them by using a subscript T and vM, respectively.

Tresca's criterion

In a tensile test plastic deformation occurs when the applied stress, here denoted by σ, reaches the uniaxial yield strength Y. The occurrence of slip bands in the tensile specimen suggests that it is not the tensile stress σ that is responsible for the deformation but the shear stress, here denoted by τ. The simplest yield criterion is thus that the maximum shear stress τ_{max} reaches some material property k, the *yield strength in shear*, or $\tau_{max} = k$. The maximum shear stress is easily obtained from the principal stresses σ_I, σ_{II} and σ_{III}. If the principal stresses are ordered with decreasing value, $\sigma_I \geq \sigma_{II} \geq \sigma_{III}$, the maximum shear stress τ_{max} is given by

$$\tau_{max} = \tfrac{1}{2}\left(\sigma_I - \sigma_{III}\right) \qquad \text{or in general} \qquad \tau_{max} = \tfrac{1}{2}\left(\sigma_{max} - \sigma_{min}\right) \qquad (13.9)$$

For a tensile test $\sigma_I = \sigma$ and $\sigma_{II} = \sigma_{III} = 0$, so that $\tau_{max} = \tfrac{1}{2}(\sigma_I - \sigma_{III}) = \tfrac{1}{2}\sigma$. Similarly, the yield strength in shear k is related to the yield strength in tension Y by $k = Y/2$. The yield criterion thus becomes

▶ $\sigma_T{}^* \equiv 2\tau_{max} = Y$ $\qquad (13.10)$

This criterion is known as *Tresca's criterion*[g] introduced in 1869 (elaborated by St. Venant in 1871) and the Tresca equivalent stress is defined by $\sigma_T^* = (\sigma_I - \sigma_{III})$.

---

[g] H. Tresca (1814-1885). French engineer who was the first to formulate laws of plastic flow in solids, in the 1870s. He carried out a number of experiments showing the effects of multiaxial stress systems on plastic flow and creep. Many of his ideas, which form the foundations of the mathematical theory of plasticity, are used in advanced creep analyses.

Although the expression looks simple, it is not convenient for more complex calculations. This is due to the fact that the order of principal stresses can vary from position to position in the body. This can be remedied by writing

$$F = [(\sigma_{II}-\sigma_{III})^2 - 4k^2]\,[(\sigma_{III}-\sigma_{I})^2 - 4k^2]\,[(\sigma_{I}-\sigma_{II})^2 - 4k^2] = 0$$

which is of the form of Eq. (13.6). When expressed in terms of invariants, the expression becomes somewhat complex

$$f = 4J_{(2)}^{\;3} - 27J_{(3)}^{\;2} - 36k^2 J_{(2)}^{\;2} + 96k^4 J_{(2)} - 64k^6 = 0 \qquad \text{or}$$

$$f = \tfrac{1}{2}J_{(2)}'^{\,3} - 3J_{(3)}'^{\,2} - 9k^2 J_{(2)}'^{\,2} + 48k^4 J_{(2)}' - 64k^6 = 0 \qquad (13.11)$$

The latter is of the form of Eq. (13.5). It will be obvious that the value of Y or k has to be found through experiments by using a tensile or a shear test, e.g. using the torsion of a circular shaft.

von Mises' criterion

Another hypothesis is that plastic deformation occurs when the strain energy density associated with shape modification exceeds a certain critical value. In Chapter 9 we have noticed that the stress tensor can be decomposed in an isotropic and deviatoric part given by

$$\sigma_{kk} = (3\lambda + 2\mu)\varepsilon_{kk} \qquad \text{and} \qquad \sigma_{ij}' = 2\mu\varepsilon_{ij}' \qquad (13.12)$$

where λ and μ are Lamé's constants. Moreover, we also noticed that the stress (or complementary strain) energy density $w^{(c)}$ of an isotropic body can be written as the sum of a term $w_{\text{sha}}^{(c)}$ only dependent on the shape change and a term $w_{\text{vol}}^{(c)}$ only dependent on the volume change

$$w^{(c)} = \int \varepsilon_{ij}(\sigma_{kl})\,d\sigma_{ij} = \tfrac{1}{2}\left(\frac{\sigma_{(2)}'}{2G} + \frac{\sigma_{(1)}^2}{9K}\right) = w_{\text{sha}}^{(c)} + w_{\text{vol}}^{(c)} \qquad (13.13)$$

Here $\sigma_{(1)} = \sigma_{kk}$ and $\sigma_{(2)} = \sigma_{ij}\sigma_{ij}$ denote the basic invariants of the stress tensor σ_{ij}, G and K the shear and bulk modulus, respectively, and the prime indicates, as usual, the deviatoric nature. Because $\sigma_{(1)}' = 0$ we have[h]

$$J_{(2)}' = \tfrac{1}{2}(\sigma_{(2)}' - \sigma_{(1)}'^2) = \tfrac{1}{2}\sigma_{(2)}'$$

The shape stress energy density thus becomes

$$w_{\text{sha}}^{(c)} = \frac{J_{(2)}'}{2G} \qquad (13.14)$$

Since for elastic materials the shear modulus G is a constant, a critical value of the shape stress energy density $w_{\text{sha}}^{(c)}$ is reached when $J_{(2)}'$ reaches a critical value. Note that this hypothesis implies that the yield surface does not depend on $J_{(3)}'$. This criterion corresponds to an equivalent stress σ^*. We arrive at the expression for the

---

[h] See Chapter 3 for details on the invariants. For convenience we repeat here the expressions for $J_{(2)}' = \tfrac{1}{2}\sigma_{ij}'\sigma_{ij}'$ in terms of the components of the stress tensor σ_{ij} and in terms of the principal values of the stress tensor σ_I, σ_{II} and σ_{III}: $J_2' = [(\sigma_{11}-\sigma_{22})^2+(\sigma_{22}-\sigma_{33})^2+(\sigma_{33}-\sigma_{11})^2]/6 + \sigma_{12}^2+\sigma_{23}^2+\sigma_{31}^2 = [(\sigma_I-\sigma_{II})^2+(\sigma_I-\sigma_{III})^2+(\sigma_{II}-\sigma_{I})^2]/6$

equivalent stress by considering a tensile specimen for which the only stress component is $\sigma_{11} = \sigma$. Hence $J_{(2)}' = \sigma^2/3$ and consequently the yield criterion becomes[i]

$$\sigma_{vM}^{*2} - Y^2 = 3J_{(2)}' - Y^2 = 0 \tag{13.15}$$

or equivalently

$$\blacktriangleright \qquad \sigma_{vM}^{*} \equiv \sqrt{3J_{(2)}'} = \sqrt{\frac{3}{2}\sigma_{ij}'\sigma_{ij}'} = Y \tag{13.16}$$

The *von Mises* equivalent stress introduced in 1913 (indepentdently of Huber's work in 1904) is therefore defined by $\sigma_{vM}^{*} = (3J_{(2)}')^{1/2}$. In fact von Mises proposed Eq. (13.16) for mathematical convenience and Hencky gave the physical interpretation in terms of elastic distorsion energy in 1924.

In case the loading is pure shear, the only stress components are, say, $\sigma_{12} = \sigma_{21} = \tau$ and the equivalent stress becomes

$$\sigma_{vM}^{*} = \sqrt{3\tau^2} = \tau\sqrt{3} \tag{13.17}$$

Comparing the von Mises criterion with the Tresca criterion we note that for pure shear the critical values of the equivalent stresses for the two criteria are

$$\blacktriangleright \qquad \sigma_{vM}^{*} = \tau\sqrt{3} \qquad \text{and} \qquad \sigma_{T}^{*} = 2\tau$$

respectively. The two criteria thus predict a slightly different yield strength in shear when the yield strength in tension is given (or vice versa). This provides a way to distinguish the two criteria experimentally using two tests. For example, starting with a tensile test leading to the onset of yield at $\sigma = Y$, the von Mises criterion predicts that yield in a shear test occurs whenever $\tau = Y/\sqrt{3}$. If, however, the material behaves according to the Tresca criterion, one predicts yield to happen at $\tau = Y/2$. Alternatively, starting with a shear test leading to yield at $\tau = k$, the von Mises criterion predicts that in a tensile test yield starts at $\sigma = k\sqrt{3}$ whereas according to the Tresca criterion yield is predicted to occur at $\sigma = 2k$.

Finally, similar to an equivalent stress σ^*, an equivalent strain increment $d\varepsilon^*$ can be defined. This is useful in order to express the plastic work density $w_{pla} = \int \sigma_{ij} d\varepsilon_{ij}^{(p)}$ in terms of the scalar quantities equivalent stress σ^* and equivalent plastic strain increment $d\varepsilon^*$. This increment is so chosen that $w_{pla} = \int \sigma^* d\varepsilon^* = \int \sigma_{ij} d\varepsilon_{ij}^{(p)}$ for uniaxial tension, a statement sometimes known as the *principle of equivalent dissipation rate*. This leads to

$$d\varepsilon^* = \sqrt{\frac{2}{3} d\varepsilon_{ij}^{(p)} d\varepsilon_{ij}^{(p)}} \tag{13.18}$$

Anisotropic materials: Hill's criterion*

So far we have been discussing only isotropic materials. However, anisotropy is important for single crystals as well as oriented materials. Hill (1950) had given a frequently used generalisation of the von Mises criterion for orthorhombic materials. Referring to the three orthogonal axes his criterion take the form

---

[i] Employing the form $X' = f^2 - K^2 = 0$.

$$F(\sigma_{yy} - \sigma_{zz})^2 + G(\sigma_{zz} - \sigma_{xx})^2 + H(\sigma_{xx} - \sigma_{yy})^2 + 2L\sigma_{yz}^2 + 2M\sigma_{zx}^2 + 2N\sigma_{xy}^2 = 1$$

where F, G, H, L, M and N are parameters. The criterion reduces to the von Mises criterion for vanishingly small anisotropy, shows no Bauschinger effect and is independent of the mean pressure. If F, G, H, M and N are small the criterion reduces to $2L\sigma_{yz}^2$, equivalent to the Tresca criterion.

Richard von Mises (1883-1953)
Born in Lemberg, Austria (now Lvov, Ukraine) and educated in mathematics, physics and engineering at the Technische Hochschule, Vienna where he was awarded a doctorate in 1907. In 1908 he was awarded his habilitation from Brünn (now Brno), becoming qualified to lecture on engineering and machine construction. He was professor of applied mathematics at Strasburg from 1909 until 1918. After the war von Mises was appointed to a new chair of hydrodynamics and aerodynamics at the Technische Hochschule in Dresden. Appointed in 1919 he soon moved again to the University of Berlin to become the director of the new Institute of Applied Mathematics. In 1921 he founded the journal *Zeitschrift für Angewandte Mathematik und Mechanik* and he became its first editor. His Institute rapidly became a centre for research in probability, statistics, numerical solutions of differential equations, elasticity and aerodynamics. In 1933, however, Hitler came to power and all civil servants who were not of Aryan descent were removed. Von Mises accepted a chair in Turkey but in 1939 he left Turkey for the United States. He became professor at Harvard University and in 1944 he was appointed Gordon-McKay Professor of Aerodynamics and Applied Mathematics there. Von Mises worked on fluid mechanics, aerodynamics, aeronautics, statistics and probability theory. He introduced the equivalent stress named after him. His studies of wing theory for aircraft led him to investigate turbulence. Much of his work involved numerical methods and this led him to develop new techniques in numerical analysis. His most famous, and at the same time most controversial, work was in probability theory. Von Mises was an excellent lecturer. His interest in philosophy was only one of von Mises' interests outside the realm of mathematics. Another was the fact that he was an international authority on the Austrian poet Rainer Maria Rilke (1875-1926).

Problem 13.1

Consider an isotropic material with an arbitrary stress-strain curve. This material is tested in uniaxial tension, hence $\sigma_{ij} = 0$ except $\sigma_{11} = \sigma$.

a) Show that beyond the yield point the strain increments are given by $d\varepsilon_{11} = d\sigma/h$ and $d\varepsilon_{22} = d\varepsilon_{33} = -d\sigma/2h$, where h is the hardening modulus (the derivative of the flow curve at the relevant strain).

b) Show now that the definition $d\varepsilon^* = \sqrt{\dfrac{2}{3}d\varepsilon_{ij}^{(p)}d\varepsilon_{ij}^{(p)}}$ leads to $w_{pla} = \int\sigma^* \, d\varepsilon^* =$

$\int\sigma_{ij}d\varepsilon_{ij}^{(p)}$.

Problem 13.2

Show that the (total) equivalent strain in a shear test is given by $\varepsilon^* = \gamma/\sqrt{3}$, where γ denotes the engineering shear strain.

Problem 13.3

In the accompanying figures the results of an uniaxial compressive test and a pure shear test for a hardened epoxy resin are represented. Determine whether the material satisfies von Mises yield criterion ('behaves like a von Mises material') or satisfies the Tresca yield criterion ('behaves like a Tresca material').

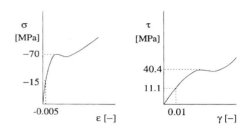

Problem 13.4

A uniaxial test is applied to a brass wire with a length $l = 400.00$ mm and a diameter $D = 15.00$ mm is loaded up to failure. At failure the length increase $\Delta l = 2.00$ mm. Linear hardening is assumed. The material data are:

Young's modulus $E = 110$ GPa,   Shear modulus $G = 42.3$ GPa
Yield strength $Y = 220$ MPa,   Ultimate tensile strength $UTS = 310$ MPa

a) Show that the total strain ε, the elastic strain $\varepsilon^{(e)}$ and the plastic strain $\varepsilon^{(p)}$ are $\varepsilon = \varepsilon^{(e)} + \varepsilon^{(p)} = 2.0\times10^{-3} + 3.0\times10^{-3} = 5.0\times10^{-3}$.
b) Show that the hardening modulus $h = 30$ GPa.
c) Show that for the direction perpendicular to the tensile direction the total strain ε', the elastic strain $\varepsilon^{(e)'}$ and the plastic strain $\varepsilon^{(p)'}$ are $\varepsilon' = \varepsilon^{(e)'} + \varepsilon^{(p)'} = -0.6\times10^{-3} - 1.5\times10^{-3} = -2.1\times10^{-3}$.
d) Show that the plastic deformation in loading can be described with the elastic equations but using a value for Poisson's ratio $\nu = \frac{1}{2}$.
e) Show that the diameter of the wire after failure $D_f = 14.98$ mm.

13.3   Graphical representation

Essentially two possibilities exist to represent the yield criterion graphically. These are a representation in principal axes space and a representation using Mohr's circles.

Representation in principal axes space

Let us again consider the *principal axes space*, the Cartesian space with the principal values labelling the co-ordinate axes. As discussed, the yield criterion $\sigma^* = Y$ can be represented as a closed surface in principal axes space. A stress state is

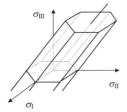 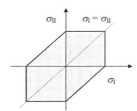

Fig. 13.5: Graphical representation of the Tresca yield criterion and the hexagonal cross-section in the σ_I-σ_{II} plane.

characterised by a point within or on the yield surface. If the point is within the yield surface, the material is elastic. As soon as the stress state reaches the yield surface, plastic deformation occurs. A stress state can never be outside the yield surface since the applied stress cannot exceed the yield strength. If during deformation the material hardens, the yield strength Y increases and consequently the yield surface expands and the space available for the stress state increases. Let us now consider the two yield criteria discussed.

The Tresca yield criterion was given by $\sigma_I - \sigma_{III} = 2k$ where $\sigma_I \geq \sigma_{II} \geq \sigma_{III}$. If we interpret $\sigma_I - \sigma_{III}$ as the difference between the largest and the smallest principal value in any quadrant, the yield criterion is represented in principal axes space by a hexagonal cylinder with the hydrostatic axis as the cylinder axis (Fig. 13.5). If one of the principal values, say σ_{III}, is zero, the remaining two principal values are in the cross-section between the cylinder and the σ_I-σ_{II} plane. This cross-section is also given in Fig. 13.5.

The von Mises yield criterion was given by $\sigma_{vM}{}^* = (3J_{(2)}')^{1/2}$ and represents in principal axes space a cylinder with the hydrostatic axis as the cylinder axis and with radius $k\sqrt{2}$ (Fig. 13.6). If in this case one of the principal values, say again σ_{III}, is zero, the cross-section with the plane $\sigma_{III} = 0$ is an ellipse.

It appears that in general von Mises criterion satisfies the experimental data somewhat better than the Tresca citerion. Since this criterion is also numerically somewhat more convenient, it is often used as a first approximation. However, it should also be stated that the difference between the two criteria is small. The most extreme difference is found for pure shear with a shear stress τ where the equivalent stresses are $\sigma_{vM}{}^* = \tau\sqrt{3}$ and $\sigma_T{}^* = 2\tau$, respectively. Using the Tresca criterion with $\sigma_I - \sigma_{III} = mY$, where m is a suitably chosen empirical number in the range $1 < m < 2/\sqrt{3} = 1.155$, the maximum error for a given stress state is never more than 8%.

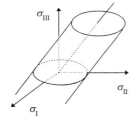 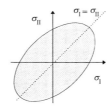

Fig. 13.6: Graphical representation of the von Mises yield criterion and the elliptical cross-section with the σ_I-σ_{II} plane.

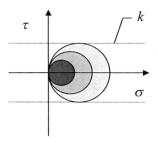

Fig. 13.7: Mohr's circle for the uniaxial tensile specimen with increasing tensile (and thus shear) stress showing that for Tresca's criterion yield occurs when $\tau = k$.

Representation by Mohr's circles

Both the Tresca and von Mises criteria indicate that yielding occurs if a certain critical shear stress is reached. For the Tresca criterion this is straightforward $\tau_{max} = k$ ($= \frac{1}{2}Y$), as indicated. For the von Mises criterion one can show that the critical shear stress involved is the shear stress on the octahedral planes[j], the so-called *octahedral shear stress* τ_{oct}. Expressed in the octahedral shear stress the von Mises yield condition reads[k] $\tau_{oct} = k\sqrt{2/3}$ ($= Y\sqrt{2}/3$). Let us plot this in a Mohr diagram for uniaxial tension (Fig. 13.7). The principal stresses are $\sigma_I = \sigma$, $\sigma_{II} = \sigma_{III} = 0$. The shear stress τ_{max} is $\tau_{max} = \sigma/2$. For the Tresca case it holds that when τ_{max} reaches the critical shear stress k, yielding occurs. Similarly for the von Mises case $\tau_{oct} = (\frac{2}{3}J_{(2)}')^{1/2}$ and yielding occurs when $\tau_{oct} = k\sqrt{\frac{2}{3}}$.

Problem 13.5

In the accompanying figure the stress components on an infinitesimal cube on the visible faces are given in MPa. The material properties are Young's modulus $E = 5$ GPa, Poisson's ratio $v = 0.25$ and yield strength $Y = 55$ MPa. Show that the Tresca equivalent stress $\sigma_T^* = 60.8$ MPa. Sketch the cross-section of the yield surface perpendicular to the hydrostatic axis.

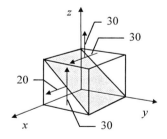

Problem 13.6

Show that for the stresses as given in Problem 13.5 the equivalent von Mises stress is given by $\sigma_{vM}^* = 58.3$ MPa.

---

[j] The octahedral plane is the plane making equal angles with the three principal directions of the stress tensor. With respect to co-ordinate base vectors along the principal directions the normal on the octahedral plane has equal components in all directions and they are $1/\sqrt{3}$ or $-1/\sqrt{3}$.

[k] Nadai, A. (1937), Appl. Phys. **8**, 205.

Problem 13.7

Show that:
a) the normal stress on the octahedral shear plane is given by $\sigma^{(n)} = \frac{1}{3}J_1 = \sigma_m$,
b) the shear stress on the octahedral shear plane is given by $\tau_{oct}^2 = \frac{2}{3} J_{(2)}'$ and
c) the yield criterion reads $Y^2 - 9\tau_{oct}^2/2 = 0$ or $\tau_{oct} = \sqrt{\frac{2}{3}}k = \frac{1}{3}\sqrt{2}Y$.

Problem 13.8

From a stress measurement using strain gauges one deduces that on a particular position for a thin plate made of a non-hardening aluminium alloy, the stresses are

$$\sigma_{11} = 70 \text{ MPa} \qquad \sigma_{22} = 120 \text{ MPa} \qquad \text{and} \qquad \sigma_{12} = 60 \text{ MPa}$$

respectively. Determine whether yielding will occur according to the von Mises and Tresca criteria, given that the uniaxial yield strength of the material $Y = 150$ MPa.

13.4 Pressure dependence

The yield criteria as outlined in the previous sections are pressure independent. For metals and inorganic materials this generally is a good approximation. However, for polymers and granular materials (soil) this is not true and the yield strength becomes dependent on the normal stress on the shear plane. The simplest way to describe this effect is to include *Coulomb friction* on the shear plane, which implies that the maximum allowable shear stress k becomes a function of the normal stress $\sigma^{(n)}$ to the shear plane:

$$k = k_0 - \mu\sigma^{(n)} \tag{13.19}$$

Here k_0 and μ are material constants, denoting the *yield strength* (cohesion) under zero normal stress and the *friction coefficient*, respectively. In Fig. 13.8 this criterion is shown in the Mohr circle representation. A number of consequences is immediately clear:

- Uniaxial yielding occurs at a lower value in tension then in compression (Bauschinger effect).
- Yielding does not occur on a plane at $\theta = \pi/4$ with the tensile direction but at a smaller angle. One can show that $\theta = \pi/4 - \phi/2$, where $\phi = \arctan \mu$ is the *angle of internal friction*. Experiments confirm the predictions roughly.

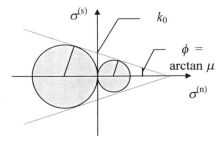

Fig. 13.8: The Coulomb-Mohr criterion using the normal stress in the Mohr circle representation.

Table 13.1: Yield for some polymers.

Polymer	k_0 (MPa)	μ
PMMA	47.4	0.16 (0.02)
PS	40.0	0.25 (0.05)
PET	31.0	0.09 (0.02)
PVC	42.0	0.11 (0.02)
Epoxy	49.0	0.09 (0.02)
HDPE	17.4	<0.05

Temperature 22 °C, Strain rate 0.002 s^{-1}.

The limiting behaviour of Eq. (13.19) occurs either when $\mu = 0$ or $k_0 = 0$. For $\mu = 0$ the behaviour reduces to pure Tresca yield, as discussed before. If one assumes that $k_0 = 0$ there is no cohesion and Eq. (13.19) represents a granular material like dry sand. In that case the angle $\phi = \arctan \mu$ describes the *angle of repose*, the steepest angle a sand hill can withstand without slipping.

There are, however, some objections to the use of the criterion in this form. In particular, the criterion is not expressed in invariant quantities, necessary to obtain a relationship independent of the choice of co-ordinate axes. One way to improve is to replace the normal stress $\sigma^{(n)}$ by the pressure p so that a pressure dependent Tresca criterion results, usually referred to as the *Coulomb-Mohr criterion*.

Another way is to use the von Mises criterion and 'correct' it in the same way, i.e.

$$\tau_{oct} = \sqrt{2/3}\,(k_0 + \mu p)$$

This is known as the *Drucker-Prager criterion*[1]. All three formulations can describe the experimental results approximately equally well. Data for the yield strength k_0 and the friction coefficient μ for some polymers are provided in Table 13.1. A typical value for the friction coefficient is 0.15.

The above description provides a reasonable framework for the so-called *yielding* of polymers. Shear bands nearly parallel to the plane with the highest shear stress show the morphology of yielding similar to the case of plastic metals (Fig. 13.9). If there is a considerable tensile component in the stress field, polymers can also fail by *crazing*. Crazes are regions of highly localised deformation similar to that developed on a macroscopic scale at large tensile strains in polymers that yield. The craze contains fibrils of highly oriented molecules separated by porous regions. Although the fibrils are strong, the overall strength is reduced by the presence of pores. Stress

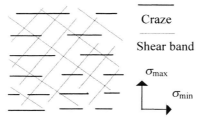

Fig. 13.9: Schematic of the morphology of crazes and shear bands. The arrows indicate the principal axes directions.

---

[1] Drucker, D.C. and Prager, W. (1952), Q. Appl. Math. **10**, 157.

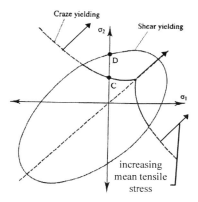

Fig. 13.10: Crazing and yielding.

concentration at the boundary between these regions serves to propagate the craze in a direction perpendicular to the largest tensile stress (Fig. 13.9). Although many crazes can develop, the overall behaviour becomes more brittle as fracture starts to intervene. Crazing is favoured by higher tensile stresses and lower temperature. Of course, yielding and crazing can occur simultaneously. It appears that crazing can be described well[m] by

$$\sigma_{max} - \sigma_{min} \geq A(T) + \frac{B(T)}{\sigma_m} \tag{13.20}$$

where σ_{max} and σ_{min} are the maximum and minimum principal stress, respectively, $A(T)$ and $B(T)$ are temperature dependent material constants and $\sigma_m = \sigma_{kk}/3$ the mean tensile stress (Fig. 13.10). Since crazing and yielding are competitive, the craze criterion and yield criterion intersect, primarily in the first quadrant of the σ_I-σ_{II} plane. With increasing mean tensile stress, the crazing critical line shifts outward along the $\sigma_I = \sigma_{II}$ line.

Problem 13.9

Derive graphically the relation $\theta = \pi/4 - \phi/2$ for the criterion as given by Eq. (13.19).

Problem 13.10

Show that for the Drucker-Prager criterion the yield criterion for uniaxial tension becomes $Y = k_0\sqrt{3}/(1+\mu/\sqrt{3})$.

13.5 Rate and temperature dependence

So far we have neglected the dependence of the plastic deformation on strain rate and temperature. Careful measurements show that for all materials there is a strain rate effect for the initial yield strength (and, to lesser extent, also for the hardening

---

[m] Sternstein, S.S. and Onghin, L. (1969), Polymer Reprints **10**, 1117.

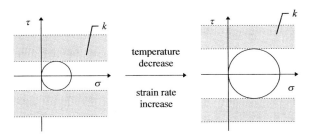

Fig. 13.11: Strain rate and temperature influence on the yield stress.

modulus). This indicates that some visco-elastic effects play a role. Evidently the yield strength is also temperature dependent. The yield strength can increase considerably with decreasing temperature and increasing strain rate and to a certain extent increased strain rate and decreased temperature are interchangeable. This implies that the yield strength is not such a precisely defined property as tacitly assumed in the previous discussions. This can be indicated in the Mohr circle representation by replacing the yield line with a yield region (Fig. 13.11), the borders of which are not clearly defined. In Chapters 16 and 17 simple models dealing with the temperature and strain rate dependence of the yield strength for inorganic and polymer materials, respectively, are discussed. Although a strain rate effect is present, nevertheless often modelling is done using rate independent plasticity but using a yield strength appropriate to the strain rate as used in practice.

13.6   Hardening*

In many cases the elastic-perfect plastic behaviour as just outlined is not followed. The material hardens as the deformation proceeds and the yield strength increases with increasing deformation. In the representation in principal axes space (Fig. 13.12) this implies either an inflating yield surface (*isotropic hardening*) or a shift of the yield surface (*kinematic hardening*). In the case of kinematic hardening also the *Bauschinger effect* occurs: the yield stress in tension and compression become different. In Mohr's circle representation hardening simply implies a shift of the k-line along the τ-axis. It should be mentioned that there might be also a change in shape of the yield surface during plastic deformation. Although we do not discuss it here for some materials this effect may be considerable (Maugin, 1992).

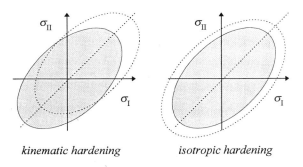

    kinematic hardening   *isotropic hardening*

Fig. 13.12: Hardening as observed in the σ_I-σ_{II} plane for a von Mises material.

The yield condition for isotropic hardening can be written as

$$\blacktriangleright \qquad X = f(\sigma_{ij}) - g(\beta) = 0 \qquad\qquad (13.21)$$

where $g(\beta)$ is a function of β, a positive parameter characterising the accumulated plastic strain. In fact β is an internal variable. One possible scalar hardening measure β is the work of the plastic strains $\varepsilon^{(p)} = \varepsilon_{ij}^{(p)}$, defined by

$$\beta \equiv w_{\text{pla}} = \int \sigma_{ij}\, d\varepsilon_{ij}^{(p)} = \int_0^t \sigma_{ij}\, \dot{\varepsilon}_{ij}^{(p)}\, dt = \int \sigma^* d\varepsilon^* \qquad\qquad (13.22)$$

Another possibility is the length of the path of plastic strains as defined by Odqvist in 1933

$$\beta \equiv \varepsilon^* = \int \left(\frac{2}{3} d\varepsilon^{(p)} : d\varepsilon^{(p)} \right)^{1/2} = \int_0^t \left(\frac{2}{3} \dot{\varepsilon}_{ij}^{(p)} \dot{\varepsilon}_{ij}^{(p)} \right)^{1/2} dt = \int d\varepsilon^* \qquad\qquad (13.23)$$

Of course, other options exist. In particular the parameter β may be a tensor $\boldsymbol{\beta} = \beta_{ij}$.

The yield condition for kinematic hardening can be written as

$$\blacktriangleright \qquad X = f(\sigma_{ij} - \alpha_{ij}) - K = 0 \qquad\qquad (13.24)$$

where α_{ij} are parameters which vary with the plastic stress and K a material parameter. A simple example is

$$\alpha_{ij} = C\varepsilon_{ij}^{(p)} \qquad\qquad (13.25)$$

where C denotes another material constant. As stated, this formulation allows in principle for the Bauschinger effect. Finally, a combination of isotropic and kinematic hardening can be represented by

$$\blacktriangleright \qquad X = f(\sigma_{ij} - \alpha_{ij}) - g(\beta_{ij}) = 0 \qquad\qquad (13.26)$$

A simple example would be

$$X = \sqrt{J_2(\sigma_{ij}' - B_0\beta\delta_{ij})} - K_0\beta = 0 \qquad\qquad (13.27)$$

a material satisfying von Mises yield criterion with a shift proportional to the parameter β as well as a yield strength proportional to the parameter β, where β is given by Eqs. (13.22) or (13.23) and B_0 and K_0 parameters.

In Chapter 2 some consequences of hardening have been already discussed, in particular the effect of necking and its relation to the stress-strain curve. Finally, it should be noted that a hardened material often becomes anisotropic, also in the elastic sense.

13.7 Incremental equations*

So far we have addressed mainly the onset phase of plastic deformation, i.e. the yield criterion. We obtained the yield function $X = f(J_2', J_3') - K = 0$ where, limiting ourselves to the von Mises criterion, $f = f(J_2') = (3J_2')^{1/2}$ and the material constant K was identified as $K = Y = 2k$ with Y the uniaxial yield strength and k the yield strength in shear, possibly dependent on the strain history, i.e. dependent on β. However, once plastic flow sets in, the development of strain with increasing stress has to be considered and therefore we need a *flow rule*.

According to von Mises there is a close connection between the yield condition and the flow rule. This connection is known as the *theory of the plastic potential* and was introduced by von Mises[n] in 1928. Here we follow closely the treatment of Hill (1950). If we assume that the shape of the yield locus is preserved during deformation, this implies that the function f is homogeneous in σ_{ij} but not a function of the strain history, i.e. of β. Generally therefore for a *neutral change* in stress state, i.e. a change where the stress state remains at the yield locus, we have for isotropic materials

$$df = \frac{\partial f}{\partial \sigma_{ij}} d\sigma_{ij} = \frac{\partial f}{\partial J_2'} dJ_2' + \frac{\partial f}{\partial J_3'} dJ_3' = 0 \tag{13.28}$$

and for a neutral change in stress state we may write for the strain increments

$$d\varepsilon_{ij}^{(p)} = G_{ij} df \tag{13.29}$$

where G_{ij} is symmetric, dependent on σ_{ij} and β but not on $d\sigma_{ij}'$. The functions G_{ij} must satisfy two criteria. First, since we consider incompressible materials we require that $G_{ii} = 0$ and, second, since the material is isotropic we require that the principal axes of the strain increment $d\varepsilon_{ij}^{(p)}$ and the stress σ_{ij} coincide. These conditions can be met by assuming that

$$G_{ij} = h \frac{\partial g}{\partial \sigma_{ij}} \tag{13.30}$$

where h and g are scalar functions of J_2' and J_3'. The function g is known as the *plastic potential*. If the material has to remain isotropic, the function g should be homogeneous in σ_{ij} but not a function of the strain history, i.e. β. Using the same considerations as for the function f in the yield criterion, it can be shown that the function g has an overall six-fold symmetry axis with each sector symmetrical about its bisector and that it is an even function of J_3'.

Not all functions h and g are allowed since any new locus must pass through the stress point and therefore they have to satisfy certain conditions. To analyse these conditions, we write

$$f(J_2', J_3') = M(w_{\text{pla}}) = M\left(\int \sigma_{ij} \, d\varepsilon_{ij}^{(p)} \right) = M\left(\int \sigma * d\varepsilon * \right) \tag{13.31}$$

where we used $M = M(\beta) = M(w_{\text{pla}})$. Differentiation leads to

$$df = M'(\sigma_{ij} d\varepsilon_{ij}^{(p)}) = M'\sigma * d\varepsilon * \tag{13.32}$$

where M' is the differential of M with respect to its argument. Combining Eqs. (13.29), (13.30) and (13.32) and multiplying by σ_{ij} yields

$$\sigma_{ij} d\varepsilon_{ij}^{(p)} = \sigma_{ij} h \frac{\partial g}{\partial \sigma_{ij}} df = h\sigma_{ij} \frac{\partial g}{\partial \sigma_{ij}} M'\sigma_{ij} d\varepsilon_{ij}^{(p)} = hngM'\sigma_{ij} d\varepsilon_{ij}^{(p)} \tag{13.33}$$

---

[n] Von Mises, R. (1928), Z. Ang. Math. Mech. **8**, 161.

where the last step is made via the Euler equation for homogeneous functions, $\sigma_{ij}\partial g/\partial\sigma_{ij} = ng$ with n the degree of g. The condition for h and g to satisfy thus reads

$$hngM' = 1$$

and the expression for the strain increment becomes

$$d\varepsilon_{ij}^{(p)} = h\frac{\partial g}{\partial\sigma_{ij}}df = \frac{1}{ngM'}\frac{\partial g}{\partial\sigma_{ij}}df \qquad (df \geq 0) \tag{13.34}$$

So for the increments in strain we have to consider, apart from the material properties contained in M', the function f describing the yield locus and the function g describing the flow. In Justification 13.1 we indicated already that $d\varepsilon_{ij}^{(p)}$ is perpendicular to the function f so that we may assume that $g = f$ or, in other words, that the yield function f acts itself as the plastic potential g. In this case the flow rule is associated with the yield condition and one speaks of an *associated* (or *associative*) *flow rule* and the material behaviour is sometimes called *standard*. A flow rule where $\partial g/\partial\sigma_{ij}$ is not proportional to $\partial f/\partial\sigma_{ij}$ is obviously called *nonassociated*.

Restricting ourselves to standard material behaviour with a yield condition given by the von Mises criterion, we have $g = J_2' = \frac{1}{2}\sigma_{ij}'\sigma_{ij}'$ and thus $\partial g/\partial\sigma_{ij} = \sigma_{ij}'$ and $n = 2$. Hence the strain increment becomes

$$d\varepsilon_{ij}^{(p)} = \frac{\sigma_{ij}'}{2J_2'}\frac{df}{M'} \quad \text{with} \quad f = M\left(\int \sigma_{ij}\,d\varepsilon_{ij}^{(p)}\right) = M\left(\int \sigma*d\varepsilon*\right) \tag{13.35}$$

Now using the von Mises criterion $f = \sigma* = (J_2')^{1/2}$ (where we omitted the subscript vM for $\sigma*$) we obtain $M' = (\sigma*)^{-1}d\sigma*/d\varepsilon*$ and for the strain increment

$$d\varepsilon_{ij}^{(p)} = \frac{3\sigma_{ij}'}{2\sigma*^2}\frac{d\sigma*}{M'} \tag{13.36}$$

Since $\sigma* = M(\int\sigma*d\varepsilon*)$, $\sigma*$ is a function of $\int d\varepsilon* = \varepsilon*$ and the criterion $\sigma* = M(w_{pla})$ is equivalent to the criterion $\sigma* = N(\varepsilon*)$. This results in $M' = N'/\sigma*$ and hence the strain increment can also be written as

$$d\varepsilon_{ij}^{(p)} = \frac{3\sigma_{ij}'}{2\sigma*}\frac{d\sigma*}{N'} \tag{13.37}$$

However, from $\sigma* = N(\varepsilon*)$ it also follows that $d\sigma* = N'd\varepsilon*$ and therefore Eq. (13.37) is equivalent to

$$\frac{d\varepsilon_{ij}^{(p)}}{d\varepsilon*} = \frac{3\sigma_{ij}'}{2\sigma*} \quad \text{and} \quad \sigma* = N\left(\int d\varepsilon*\right) \tag{13.38}$$

and therefore for the ideal plastic case we can write

$$d\varepsilon_{ij}^{(p)} = \frac{3d\varepsilon*}{2\sigma*}\sigma_{ij}' \tag{13.39}$$

Summarising so far the theory of the plastic potential postulates that the yield function is convex (which has been rationalised by Drucker in 1951) and that the deformation rate d_{ij} is given by the flow rule

$$d_{ij} = \dot\lambda \frac{\partial X}{\partial \sigma_{ij}}$$

with X the yield function, in this connection better known as the *plastic potential*. The undetermined multiplier $\dot\lambda$ is dependent on the state of deformation and obeys

$$\lambda = 0 \quad \text{for} \quad X < 0 \quad \text{as well as} \quad X = 0 \text{ and } \dot X < 0 \quad \text{and}$$

$$\lambda \geq 0 \quad \text{for} \quad X = 0 \text{ and } \dot X = 0$$

Equivalently we may write

$$d\varepsilon_{ij}^{(p)} = \frac{\partial X}{\partial \sigma_{ij}} d\lambda$$

Choosing the von Mises yield function

$$X = f(J_2') - K \quad \text{with} \quad f = \sigma^* = \sqrt{3J_2'} = \sqrt{\frac{3}{2}\sigma_{ij}'\sigma_{ij}'} \quad \text{leads to}$$

$$d\varepsilon_{ij}^{(p)} = \sigma_{ij}' d\lambda \quad \text{where} \quad \sigma_{ij}' \equiv \sigma_{ij} - \sigma_m \delta_{ij} \quad \text{with} \quad \sigma_m = \frac{1}{3}\sigma_{kk}$$

In fact the flow rule $d\varepsilon_{ij}^{(p)} = \sigma_{ij}'$ $d\lambda$ implies that the material remains isotropic during plastic deformation and thus that the principal axes of stress and plastic strain *increment* coincide when we increase the strain. Multiplying $d\varepsilon_{ij}^{(p)}$ by itself we obtain

$$d\varepsilon_{ij}^{(p)} d\varepsilon_{ij}^{(p)} = \sigma_{ij}'\sigma_{ij}' (d\lambda)^2 \quad \text{or} \quad (3/2)\, d\varepsilon^{*2} = (2/3)\sigma^{*2}\,(d\lambda)^2$$

since the equivalent strain increment is given by

$$d\varepsilon^* = \sqrt{\frac{2}{3} d\varepsilon_{ij}^{(p)} d\varepsilon_{ij}^{(p)}}$$

and we obtain $d\lambda = (3/2)d\varepsilon^*/\sigma^*$. Consequently, we have

▶ $$d\varepsilon_{ij}^{(p)} = \frac{3d\varepsilon^*}{2\sigma^*}\sigma_{ij}'$$ (13.40)

These equations were first presented by Levy[0] in 1871 and von Mises in 1913 and yield the strain increment at a certain stress state, if one accepts von Mises equivalent stress. These equations neglect the elastic contributions. However, they can be included easily. If we write the elastic equations in terms of a hydrostatic part $\sigma_{kk}\delta_{ij}/3$ and deviatoric part σ_{ij}' they read

$$\varepsilon_{ij}^{(e)} = \frac{1+v}{E}\sigma_{ij}' + \frac{1-2v}{E}\frac{\sigma_{kk}}{3}\delta_{ij}$$

where, as usual, E and v denote Young's modulus and Poisson's ratio, respectively. In differential form these equations become

▶ $$d\varepsilon_{ij}^{(e)} = \frac{1+v}{E}d\sigma_{ij}' + \frac{1-2v}{E}\frac{d\sigma_{kk}}{3}\delta_{ij}$$ (13.41)

---

[0] Maurice Lévy (1838-1910). French scientist, well known for his contributions to problems in elasticity, in particular plate theory.

To obtain the total strain increment we add the elastic and plastic strain increments leading to

$$\blacktriangleright \qquad \mathrm{d}\varepsilon_{ij} = \mathrm{d}\varepsilon_{ij}^{(e)} + \mathrm{d}\varepsilon_{ij}^{(p)} \qquad\qquad (13.42)$$

The latter equations have first been given by Prandtl[P] in 1925 and Reuss in 1930 and describe the overall deformation of a material in an incremental way. The final deformation is dependent on the path via which that deformation is reached. In general these equations have to be solved numerically, a brief discussion of which is given in the next chapter.

13.8   The thermodynamic approach*

So far, plasticity has been described by, admittedly plausible but, ad-hoc arguments. In this section we start anew with a different approach towards plasticity and follow closely the treatment as presented by Ziegler (1983). The main objective is to obtain a strongly thermodynamically based formulation. For a more advanced treatment, it is convenient to introduce the plastic behaviour as a limiting behaviour of the *visco-plastic* body, i.e. a material with a threshold stress and strain rate dependence but without an elastic part. In a later stage the elastic part can then be added to obtain the *elasto-plastic* body.

Conventional treatment

The material we consider for the moment is the (incompressible) viscous liquid. This material is characterised by a stress tensor σ_{ij} which consists of a quasi-conservative part $\sigma_{ij}^{(q)}$ and a dissipative part $\sigma_{ij}^{(d)}$ given by

$$\sigma_{ij} = \sigma_{ij}^{(q)} + \sigma_{ij}^{(d)} = -p\delta_{ij} + 2\mu' d_{ij} \qquad\qquad (13.43)$$

where p is the hydrostatic pressure.

For simple shear $d_{12} = d_{21}$ is the only component of the rate of deformation tensor which is non-vanishing while $d_{(2)} = d_{12}^2$ is the only non-zero invariant in that case. The only non-zero dissipative stress is $\sigma_{12}^{(d)}$. Hence, since in principle μ' is a function of the invariants $d_{(2)}$ and $d_{(3)}$, the following stress-strain relation is obtained

$$\sigma_{12}^{(d)} = 2\mu'\big(d_{(2)}, d_{(3)}\big)d_{12} \qquad\qquad (13.44)$$

For this particular case, because $d_{(3)} = 0$ for the considered rate of deformation tensor, the expression reduces to

$$\sigma_{12}^{(d)} = 2\mu'(d_{12}^2)d_{12}$$

A qualitative sketch of the behaviour is given in Fig. 13.13. One possible representation of the smooth curve is given by the relationship

$$\sigma_{12}^{(d)} = 2ad_{12} + \frac{2k}{\pi}\arctan\frac{d_{12}}{b} \equiv 2\mu'(d_{12}^2)d_{12} \qquad\qquad (13.45)$$

[P] Ludwig Prandtl (1875-1953). German scientist, in engineering mechanics well known for the membrane analogy of torsion but also for his major contributions to aerodynamics. Among his pupils were Theodore von Kármán, A. Nadai, and W. Prager.

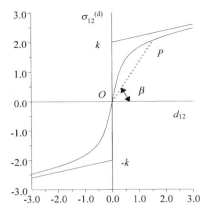

Fig. 13.13: Shear deformation of a visco-plastic body.

where a, b and k are arbitrary positive parameters so that

$$2\mu'\left(d_{12}^2\right)=2a+\frac{2k}{\pi}\frac{\arctan\left(d_{12}/b\right)}{d_{12}} \qquad (13.46)$$

For an arbitrary point P this is the tangent of the angle β between the line OP and the d_{12}-axis. Obviously for a small deformation rate d_{12} the viscosity μ' is high while at higher shear rates the value for μ' drops. As limiting cases we distinguish $k \rightarrow 0$ and $b \rightarrow 0$. In case the parameter $k = 0$, the material is a Newtonian liquid with a viscosity μ' = a and represented by a straight line through the origin. In case $b \rightarrow 0$, the curve approaches the broken line in the graph consisting of the part along the axis

$$\left|\sigma_{12}^{(d)}\right|\le k \qquad \text{for} \qquad d_{12}=0 \qquad (13.47)$$

and the other two parts

$$\sigma_{12}^{(d)}=\left(2a+\frac{k}{|d_{12}|}\right)d_{12}=\left(2a+\frac{k}{\sqrt{d_{12}^2}}\right)d_{12} \qquad \text{for} \qquad d_{12}\neq 0 \qquad (13.48)$$

The viscosity for this limiting case is given by

$$2\mu'\left(d_{12}^2\right)=2a+\frac{k}{\sqrt{d_{12}^2}} \qquad \text{for} \qquad d_{12}\neq 0 \qquad (13.49)$$

Finally, we recall the power of dissipation, described in Chapter 5, and given by

$$l^{(d)}=\frac{1}{\rho}\sigma_{ij}^{(d)}d_{ij}=\frac{1}{\rho}\left(\sigma_{12}^{(d)}d_{12}+\sigma_{21}^{(d)}d_{21}\right)=\frac{2}{\rho}\left(2a+\frac{k}{\sqrt{d_{12}^2}}\right)d_{12}^2 \qquad (13.50)$$

This expression is plotted in Fig. 13.14. Note the sharp corner at the origin caused by the term $2kd_{12}^2/(\rho\sqrt{d_{12}^2}) = 2k/|d_{12}|/\rho$. For a Newtonian fluid, $k = 0$ and the curve becomes a parabola.

For the material considered here the dissipative shear stress $\sigma_{12}^{(d)}$ is identical to the deviatoric (or total) shear stress, usually indicated by σ_{12}'. Fig. 13.13 shows that

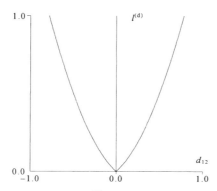

Fig. 13.14: The dissipation power $\rho \cdot l^{(d)}$ for a visco-plastic body with $k = a = 0.25$.

for $|\sigma_{12}'| \le k$ the material is rigid and above the *yield strength* k it flows with a shear rate with the sign of σ_{12}' and proportional to $|\sigma_{12}'| - k$.

The material described can be referred to as *visco-plastic*. Generalising the results for simple shear to arbitrary deformations, the relevant constitutive equations for an isotropic fluid without bulk viscosity are

$$\sigma_{ij}^{(q)} = -p\delta_{ij} \qquad \text{and} \qquad \sigma_{ij}^{(d)} = 2\mu'\big(d_{(2)}, d_{(3)}\big)d_{ij} \tag{13.51}$$

where p is the pressure and the (shear viscosity) function $2\mu'(d_{(2)}, d_{(3)})$ is to be chosen in such a way that it reduces to Eq. (13.49) for the case of simple shear. The simplest possible choice is that μ' depends on the invariant $d_{(2)}$ alone. In simple shear $d_{(2)} = 2d_{12}^2$ so that the generalisation of Eq. (13.49) for general stress states is given by

$$2\mu(d_{(2)}) = 2a + \frac{k}{\sqrt{d_{(2)}/2}} \qquad \text{for} \qquad d_{ij} \neq 0 \tag{13.52}$$

and the constitutive relation reads

$$\sigma_{ij}' = 2\mu'(d_{(2)})d_{ij} = \left(2a + \frac{k}{\sqrt{d_{(2)}/2}}\right)d_{ij} \qquad \text{for} \qquad d_{ij} \neq 0 \tag{13.53}$$

For $d_{ij} \to 0$ it follows that $\sigma_{(2)}' \to 2k^2$, so that the generalisation of Eq. (13.47) is given by

$$\sigma_{(2)}' \le 2k^2 \qquad \text{for} \qquad d_{ij} = 0 \tag{13.54}$$

According to the above equations, the stress deviator is limited, but undetermined, for $d_{ij} = 0$. As soon as $d_{ij} \neq 0$, the stress deviator contains two parts: a Newtonian part characterised by a viscosity a and a threshold value k commonly known as the yield strength in simple shear. Put differently, the material remains rigid if inequality (13.54) is satisfied. The stress for which the equality $\sigma_{(2)}' = 2k^2$ holds is denoted as the *yield limit*. For $\sigma_{(2)}' > 2k^2$, the material deforms according to Eq. (13.53) with a rate of dissipation

$$l^{(d)} = \frac{1}{\rho}\sigma_{ij}^{(d)}d_{ij} = \frac{1}{\rho}\left(2a + \frac{k}{\sqrt{d_{(2)}/2}}\right)d_{(2)} \tag{13.55}$$

The equality $\sigma_{(2)}' = 2k^2$ is also known as the *von Mises condition*. Since $\sigma_{(2)}'$ is the second basic invariant of the stress deviator one has $(3\sigma_{(2)}'/2)^{1/2} = \sigma_{vM}*$ where the quantity on the right-hand side is the von Mises equivalent stress from Section 11.2.

The use of the orthogonality theorem

So far we have been reiterating essentially previously obtained results, albeit in a somewhat different form. Let us now consider the orthogonality condition and its consequences. In the present connection we assume incompressibility, therefore $d_{(1)} = 0$, and accounting for this constraint via the Lagrange multiplier γ the orthogonality condition reads

$$\sigma_{ij}^{(d)} = \lambda \frac{\partial}{\partial d_{ij}}\left(\Phi - \gamma d_{(1)}\right) \tag{13.56}$$

with Φ the dissipation function. The value of λ can be obtained in the usual way from

$$\sigma_{kl}^{(d)} d_{kl} = \lambda \frac{\partial}{\partial d_{kl}}\left(\Phi - \gamma d_{(1)}\right) d_{kl} = \Phi \tag{13.57}$$

which leads, since $\partial d_{(1)}/\partial d_{kl} = \delta_{kl}$, to

$$\sigma_{ij}^{(d)} = \lambda \frac{\partial}{\partial d_{ij}}\left(\Phi - \gamma \delta_{ij}\right) \quad \text{and} \quad \lambda = \Phi\left(\frac{\partial \Phi}{\partial d_{kl}} d_{kl}\right)^{-1} \tag{13.58}$$

For a homogeneous dissipation function Φ of degree r we have seen that, according to Euler's equation for homogeneous functions,

$$\frac{\partial \Phi}{\partial d_{ij}} d_{ij} = r\Phi \quad \text{and therefore} \quad \lambda = \frac{1}{r} \tag{13.59}$$

The Lagrange multiplier γ can be obtained from $\sigma_{(1)}^{(d)} = 0$.

The next step is to find an expression for Φ. For an educated guess we use the dissipation function of the Newtonian liquid. Since in that case the total stress can be written as

$$\sigma_{ij} = \sigma_{ij}^{(q)} + \sigma_{ij}^{(d)} \quad \text{with} \quad \sigma_{ij}^{(q)} = -p\delta_{ij} \quad \text{and} \quad \sigma_{ij}^{(d)} = \lambda' d_{kk} + 2\mu' d_{ij} \tag{13.60}$$

where $\lambda' + 2\mu'/3$ and μ' are the bulk and shear viscosity coefficients, respectively, the dissipation function per unit volume is given by

$$\rho\varphi = \sigma_{ij}^{(d)} d_{ij} = \lambda' d_{ii} d_{jj} + 2\mu' d_{ij} d_{ij} = \lambda' d_{(1)}^2 + 2\mu' d_{(2)} \tag{13.61}$$

For $\lambda' + 2\mu'/3 = 0$, i.e. the bulk viscosity is zero, or for an incompressible liquid this reduces to

$$\rho\varphi = \sigma_{ij}^{(d)} d_{ij} = 2\mu' d_{ij}' d_{ij}' = 2\mu' d_{(2)}' \tag{13.62}$$

The corresponding stress reads

$$\sigma_{ij}^{(d)} = 2\mu'\left(d_{ij} - \frac{1}{3}d_{kk}\delta_{ij}\right) = 2\mu' d_{ij}' \tag{13.63}$$

Now we generalise Eq. (13.62) to

$$\Phi = 2\mu'd_{(2)}^{n} \qquad \text{with} \qquad n \geq \tfrac{1}{2} \qquad (13.64)$$

which is homogeneous of degree $2n$ in d_{ij}. Inserting in Eq. (13.58) and using Eq. (13.59) yields the total dissipative stress

$$\sigma_{ij}^{(d)} = 2\mu'd_{(2)}^{n-1}d_{ij} - \frac{\gamma}{2n}\delta_{ij} \qquad (13.65)$$

where the first term on the left-hand side represents the deviatoric part, here denoted by σ_{ij}' . We thus have

$$\sigma_{ij}' = 2\mu'd_{(2)}^{n-1}d_{ij} \qquad (13.66)$$

and the basic invariant $\sigma_{(2)}'$ becomes

$$\sigma_{(2)}' = \sigma_{ij}'\sigma_{ij}' = 4(\mu')^{2}d_{(2)}^{2n-1} \qquad (13.67)$$

and therefore we can obtain the inversion

$$d_{ij} = \left(2\mu'\right)^{-1/(2n-1)}\left(\sigma_{(2)}'\right)^{(1-n)/(2n-1)}\sigma_{ij}' \qquad (13.68)$$

For $n = 1$ this expression reduces to the dissipation function of the Newtonian fluid, as it should. For $n = \tfrac{1}{2}$ the stress becomes

$$\sigma_{ij}' = 2\mu'd_{(2)}^{-1/2}d_{ij} \qquad (13.69)$$

but the inversion cannot be obtained in this case. Physically this makes sense since it represents the perfect-plastic case. Setting $2\mu' = k\sqrt{2}$ we regain the von Mises yield condition.

If we express Eq. (13.66) in the principal axes system it takes for the first principal value the form

$$\sigma_{1}' = 2\mu d_{(2)}^{n-1}d_{1} \qquad (13.70)$$

and similar expressions for the principal values II and III. For the case of uniaxial stress $\sigma_{I} = \sigma$ and $\sigma_{II} = \sigma_{III} = 0$, so that $\sigma_{(1)} = \sigma$, $\sigma_{I}' = \tfrac{2}{3}\sigma$ and $\sigma_{II}' = \sigma_{III}' = -\tfrac{1}{3}\sigma$. In view of the incompressibility we also have $d_{II} = d_{III} = -\tfrac{1}{2}d_{I}$ and hence $d_{(2)} = 3d_{1}^{2}/2$. Finally we obtain

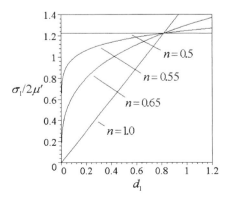

Fig. 13.15: Plot of $\sigma_{I}/2\mu'$ versus d_{I} representing power law behaviour.

$$\sigma_1 = 2\mu'\left(\frac{3}{2}\right)^n \left(d_1^2\right)^{n-1} d_1 \tag{13.71}$$

which is plotted in Fig. 13.15 for various values of n. This plot is very similar to the one given in Fig. 13.13 but it used somewhat simpler mathematics.

Hardening can be introduced by using the dissipation function $\Phi = \Phi(d_{(2)})$, here taken as being of the von Mises type $\Phi = k(2d_{(2)})^{1/2}$, in combination with elastic potential which is of the form $F = \mu\varepsilon_{(2)}$, because we still assume incompressibility, i.e. $d_{(1)} = 0$. A quasi-conservative stress results given by

$$\sigma_{ij}^{(q)} = \frac{\partial F}{\partial \varepsilon_{ij}} = 2\mu\varepsilon_{ij} \tag{13.72}$$

The total deviatoric stress is the sum of $\sigma_{ij}^{(q)}$ and $\sigma_{ij}^{(d)}$ and becomes

$$\sigma_{ij}' = 2\mu\varepsilon_{ij} + k(\tfrac{1}{2}d_{(2)})^{-1/2} d_{ij} \tag{13.73}$$

This expression can be rearranged to

$$(\sigma' - 2\mu\varepsilon)_{(2)} = (\sigma_{ij}' - 2\mu\varepsilon_{ij})(\sigma_{ij}' - 2\mu\varepsilon_{ij}) = 2k^2 \tag{13.74}$$

which stills describes the von Mises yield circle on the Π-plane surface with radius $k\sqrt{2}$ but now with a shifted origin. This represents *kinematic hardening*.

To describe an elasto-ideal plastic material we may use

$$F = \mu(\varepsilon - \alpha)_{(2)} \qquad\qquad \text{and} \qquad\qquad \Phi = k(2\dot{\alpha}_{(2)})^{1/2} \tag{13.75}$$

where α_{ij} represents an internal deviatoric strain. The quantities $\varepsilon_{ij} - \alpha_{ij}$ and α_{ij} denote the elastic and plastic strain, respectively. This leads to the external deviatoric stresses

$$\sigma_{ij}^{(q)} = \frac{\partial F}{\partial \varepsilon_{ij}} = 2\mu(\varepsilon_{ij} - \alpha_{ij}) \qquad \text{and} \qquad \sigma_{ij}^{(d)} = \frac{\partial \Phi}{\partial d_{ij}} = 0 \tag{13.76}$$

and the internal deviatoric stresses

$$\beta_{ij}^{(q)} = \frac{\partial F}{\partial \alpha_{ij}} = -2\mu(\varepsilon_{ij} - \alpha_{ij}) \qquad \text{and} \qquad \beta_{ij}^{(d)} = \frac{\partial \Phi}{\partial \dot{\alpha}_{ij}} = k(\tfrac{1}{2}\dot{\alpha}_{(2)})^{-1/2} \dot{\alpha}_{ij} \tag{13.77}$$

As usual there is also an indeterminate hydrostatic stress on account of the incompressibility. Since $\sigma_{ij} = \sigma_{ij}^{(q)} + \sigma_{ij}^{(d)}$ and $\beta_{ij}^{(d)} = -\beta_{ij}^{(q)}$ the external stress deviator is determined by

$$\sigma_{ij}' = 2\mu(\varepsilon_{ij} - \alpha_{ij}) = k(\tfrac{1}{2}\dot{\alpha}_{(2)})^{-1/2} \dot{\alpha}_{ij} \tag{13.78}$$

and therefore depends in the same way on $\dot{\alpha}_{ij}$ as it depends on d_{ij} in a perfect plastic material. Taking the second invariant we obtain

$$\sigma_{(2)}' = 2k^2 \tag{13.79}$$

From this expression it becomes clear that the yield surface is still a circular cylinder. Thus for this model, as long as no plastic flow has occurred, $\alpha_{ij} = 0$, therefore we have $\sigma_{ij}' = 2\mu\varepsilon_{ij}$ and the behaviour is elastic. When $\sigma_{(2)}' = 2k^2$ the material becomes plastified and obeys the normality rule due to Eq. (13.78). During plastic flow we have $\varepsilon_{ij} = (2\mu)^{-1}\sigma_{ij}' + \alpha_{ij}$ and during unloading $(\sigma_{(2)}' < 2k^2)$, α_{ij} remains constant.

Next we just note that an incompressible *hardening elasto-plastic material* can be described by

$$F = \mu(\varepsilon-\alpha)_{(2)} + \mu'\alpha_{(2)} \qquad \text{and} \qquad \Phi = k(2\dot{\alpha}_{(2)})^{1/2} \tag{13.80}$$

and that, although leading to somewhat more extensive expressions, the extension to compressible materials is not difficult.

As a application, let us note that in Chapter 6 we stated that the orthogonality theorem in the space of strain rates can be inverted to one in force space but that the normality rule in stress space does only hold if the dissipation function is a function of the strain rates only. This applies in the conventional treatment of the plasticity of metals, as discussed before. However, in powder compacts and soils, of which the flow behaviour depends on the pressure, the inversion cannot be made and the theory of the plastic potential is longer valid. As an example consider an incompressible von Mises material with a yield strength dependent on the pressure where the dissipation function is given by

$$\Phi = A\left(B - \sigma_{(1)}/\sqrt{3}\right)d_{(2)}^{1/2} \tag{13.81}$$

with A and B constants. Since it must hold that $\Phi \geq 0$, the constant $A > 0$ and for positive B, σ_1 is restricted by $\sigma_1 < B\sqrt{3}$. Further Φ is homogeneous in d_{ij}. Using the orthogonality theorem once more we have for the stress

$$\sigma_{ij}^{(d)} = \frac{\partial}{\partial d_{ij}}\left(\Phi - \gamma\delta_{ij}\right) = A\left(B - \sigma_{(1)}/\sqrt{3}\right)d_{(2)}^{1/2}d_{ij} \tag{13.82}$$

and therefore for the yield surface

$$\sigma_{(2)}' = A^2\left(B - \sigma_{(1)}/\sqrt{3}\right)^2 \tag{13.83}$$

In principal axis space the yield surface is a semi-cone along the hydrostatic axis and parameters A and B as indicated in Fig. 13.16. From this it follows that the yield strength in hydrostatic tension is $\sigma_{(1)} = B\sqrt{3}$ while for simple shear it becomes $k = AB/\sqrt{2}$. In uniaxial tension we have $\sigma_{(1)} = \sigma$, $\sigma_1' = \frac{2}{3}\sigma = \frac{2}{3}\sigma_{(1)}$ and $\sigma_{II}' = \sigma_{III}' = -\frac{1}{3}\sigma = -\frac{1}{3}\sigma_{(1)}$. From Eq. (13.83) it follows that the yield strength in tension and compression are given by

$$\sigma_I^+ = \frac{AB\sqrt{3}}{\sqrt{2}+A} \qquad \text{and} \qquad \sigma_I^- = -\frac{AB\sqrt{3}}{\sqrt{2}-A}$$

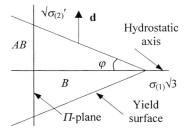

Fig. 13.16: Cross-section of the yield surface for pressure dependent behaviour showing the semi-cone with angle $\varphi = \tan A$.

respectively. The yield strength in tension σ_1^+ always exist and, as required, is smaller than $B\sqrt{3}$. The yield strength in compression σ_1^- on the other hand approximates $-\infty$ as $B \to 0$. Ideal plastic behaviour is regained if we let $B \to \infty$ and $A \to 0$ in such a way that $AB \to k\sqrt{2}$. In fact this dissipation represents the Drucker-Prager model of pressure dependent yield (Section 13.4). Since, except for a small elastic phase which has been neglected here, $d_{(1)} = 0$, the rate of deformation $\mathbf{d}$ is parallel to the deviatoric plane and thus normality does not hold. Abandoning the incompressibility condition $d_{(1)} = 0$ leads to a more complex yield surface expression but does not restore normality. Isotropic hardening can be included by making A and/or B depend on the strain history. For kinematic hardening A and B must be kept constant but an elastic potential, e.g. $F = \mu\varepsilon_{(2)}$, has to be added, as before. This would lead to a yield surface

$$(\sigma' - 2\mu\varepsilon)_{(2)} = A^2\left(B - \sigma_{(1)}/\sqrt{3}\right)^2 \tag{13.84}$$

and consequently to the stress deviator

$$\sigma_{ij}' = 2\mu\varepsilon_{ij} + A\left(B - \sigma_{(1)}/\sqrt{3}\right)d_{(2)}^{1/2}d_{ij}$$

Further combinations are possible, of course.

Finally, let us mention that the orthogonality theorem can also be applied at the strength-of-materials level using any combination of generalised strains and stresses that describes the problem at hand, facilitating the solution considerably. Recall that the global dissipation $L^{(d)}$ is given by

$$L^{(d)} = \int \sigma_{ij} d_{ij}\, dV$$

Now assume that the material is slowly loaded. For sufficiently small loads all local stresses are below the yield strength but with increasing load, locally the yield strength will be reached. As long as the material is only plastified partly, the rigidity of the remaining part will prevent global flow until the plastified zone becomes large enough. Denoting the corresponding stress field by σ_{ij} and an arbitrary stress state below or at the yield surface by σ_{ij}^* we may write

$$L^{(d)} = \int (\sigma_{ij} - \sigma_{ij}^*)d_{ij}\, dV \geq 0$$

since the elastic deformation does not contribute to the dissipation. This has brought as back to the principle of maximum dissipation rate, as introduced in Section 13.1.

In conclusion we see that the basic features of plasticity are well described by the internal variable approach and together with the orthogonality theorem provides a convenient framework for models about continuum plasticity. In particular, it includes in a natural way the theory of the plastic potential and its exceptions.

13.9   Bibliography

Derby, B., Hills, D., Ruiz, C. (1992), *Materials for engineering*, Longman Scientific & Technical, Harlow, UK.

Hill, R. (1950), *The mathematical theory of plasticity*, Oxford University Press, London.

Lemaitre, J. and Chaboche, J.-L. (1990), *Mechanics of solid materials*, Cambridge University Press, Cambridge.

Lubliner, J. (1990), *Plasticity theory*, McMillan, New York.

Maugin, G.A. (1992), *The thermomechanics of plasticity and fracture*, Cambridge University Press, Cambridge.

Ziegler, H. (1983), *An introduction to thermomechanics*, 2nd ed., North-Holland, Amsterdam.

14

Applications of plasticity theory

In this chapter some applications of plasticity are discussed with the continuum point of view in mind. After discussing materials testing a brief discussion is given of simplified models of plasticity as used for some processes. Finally elements of modelling plasticity in structures are discussed.

14.1   Materials testing

One of the most direct applications of the theory of plasticity is in materials testing. In Chapter 2 we already discussed the tensile test. Here we focus on hardness testing since it directly provides information on several materials properties.

Plasticity theory applications

Plastic deformation is used in many industrial processes and for these plasticity theory is extensively used. Rolling, wire drawing and deep drawing of metals are just three of these processes. In these cases not only estimates for the required forces in the process are made but also calculations of the resulting residual stress. For these calculations a reliable stress-strain relationship under multi-axial conditions of the material at hand is required. Plasticity is also widely used as a safe guard mechanism to prevent catastrophic failure. Typically, a design based on an elastic calculation is made and for overloads one relies on local plastic deformation so that the structure does not fail as a whole.

Hardness

Although a tensile test can provide a great deal of information on the behaviour of materials, the test is not as simple as it looks with respect to the performance. Moreover, it needs a special specimen. Therefore, alternative tests are used and by far the most frequently used one is the hardness test. Hardness is resistance to indentation. It is determined by pressing a hard indenter into a flat surface of the material and measuring the size of the resulting indentation. From this measurement an estimate of the yield strength can be made. Apart from the simplicity and low cost of the test, it is also essentially non-destructive, which means that the test can be carried out in situ on ready (semi-)products but also in the different stages of the forming process. Often also other properties than the plastic behaviour may be estimated from it. On the other hand, hardness is a complex quantity. It is possible to convert one hardness number to

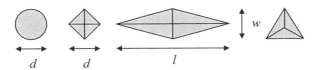

Fig. 14.1: Schematic view of Brinell, Vickers, Knoop and Berkovich indentations.

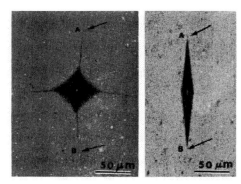

Fig. 14.2: Real Vickers (left, 40 N) and Knoop (right, 26 N) indentations on Si$_3$N$_4$ ceramics. Note the cracks at the end of the Vickers indentation. The pin-cushion shape of the Vickers indentation is due to the sinking-in of the material along the flat faces of the indenter. The converse, a barrel-shaped indentation, can occur for piling-up along the flat faces.

another. However, since hardness is a complex quantity, conversion is essentially based on empirical data.

One has to distinguish between blunt and sharp indenters. In both cases elastic and plastic deformation interact to a major extent. Sharp indenters generate yield (irreversible deformation) right from the start and produce a geometrically self-similar contact when used to indent a homogeneous material. For a cone with a semi-apical angle ($90°-\beta$), the contact radius a is related to the depth of penetration h by $h/a = \tan \beta =$ constant. In principle the mean indentation pressure, $\bar{p} = F/\pi a^2$, where F denotes the load, and the mean strain $\varepsilon \sim h/a$ are thus independent of the indentation depth h. Several types of sharp indenters are in use. Here the Vickers, Knoop and Berkovich indenters are described. Blunt indenters initially only produce an elastic deformation. Moreover, they do not produce geometrically self-similar contacts. We limit ourselves to Brinell and Rockwell indentations. The outline of the shape of the various indentations is shown in Fig. 14.1 while in Fig. 14.2 some real indentations are shown.

Vickers, Knoop and Berkovich hardness

For the Vickers[a] test a square-base diamond pyramid is used (Fig. 14.3). The included angle between opposite faces of the pyramid θ is 136°. The indentation size d (in this case the diagonal of the square impression) is measured and the hardness H_V is calculated from the actual contact area

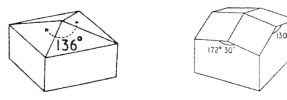

Fig. 14.3: Vickers (left) and Knoop (right) hardness indenters.

---

[a] Devised in the 1920s at Vickers Ltd., U.K., also known as the diamond pyramid hardness test.

$$H_V = \frac{2F\sin(\theta/2)}{d^2} = \frac{1.854F}{d^2} \qquad (14.1)$$

Since the projected area is equal to $d^2/2$, the effective radius of an impression a is $(d^2/2\pi)^{1/2}$. The mean indentation pressure $\bar{p}$ is thus equal to

$$\bar{p} = \frac{F}{\pi a^2} = \frac{2F}{d^2} = 1.079 H_V \qquad (14.2)$$

For a microhardness test, which always enables 'damage free' testing, the load used is typically in the range from 0.1 to 10 N and is often indicated between brackets, e.g. H_V (0.2), which means the indentation at a load of 0.2 kg or 2 N. Frequently the hardness is expressed in units kg/mm² though GPa is also employed.

For the Knoop test a rhombic pyramid with edge angles of 172° 30' and 130° is used (Fig. 14.3). The ratio of the long diagonal l to the short diagonal w is $l/w = 7.117$. The Knoop hardness test is therefore especially suitable for determining anisotropy in a sample, e.g. in a single crystal or in a textured polycrystal. The Knoop hardness H_K is calculated as the force F per projected area and is given by

$$H_K = \frac{2F}{lw} = \frac{14.23F}{l^2} \qquad (14.3)$$

For the Berkovich test a triangle-base diamond pyramid is used. For the indentation size the depth of the triangular indent is used and the hardness is again calculated according to $H = F/A$ where A is the contact area. Since three planes always meet at a point, the fabrication of a Berkovich indenter is much simpler than that of a Vickers indenter. As a result it is easier to obtain sharp indenter tips with only a limited amount of rounding. An example of a Berkovich indentation[b] is shown in Fig. 14.4.

Brinell and Rockwell hardness

In a Brinell[c] test a steel (or hard metal) ball with diameter D, usually 10 mm, is brought in contact with the material and loaded for a certain time. The Brinell hardness H_B is expressed as the load F divided by the actual surface area of the impression. This quantity is given by

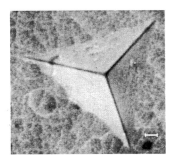

Fig. 14.4: A Berkovich indentation in a TiC coating on steel. The scale bar represents 1 μm.

[b] van der Varst, P.G.Th. and de With, G. (2001), Thin Solid Films **384**, 85.
[c] Johann August Brinell (1849-1925). Swedish engineer well known for the hardness test but who did also pioneering work concerning phase transformations in steel.

$$H_{\mathrm{B}} = \frac{2F}{\pi D(D - \sqrt{D^2 - d^2})} = \frac{F}{\pi D t} \qquad (14.4)$$

where d and t are the diameter and depth of the indentation, respectively. The Brinell hardness will vary with load unless F/D^2 is kept constant for various loads. In 1908 Meyer[d] suggested that a more rational definition of hardness would be based on the projected area. The mean indentation pressure $\bar{p}$ or Meyer hardness is given by

$$\bar{p} = \frac{4F}{\pi d^2} \qquad (14.5)$$

Meyer also proposed an empirical relation between the load and size of the indentation. This relation is called Meyer's law

$$F = kd^{n'} \qquad (14.6)$$

where $n' \cong n+2$ is a material parameter related to the strain hardening exponent n of the material and k a parameter dependent on the size of the ball and the material.

In Rockwell hardness testing a $120°$ diamond cone with slightly rounded point or a 1/16 (or 1/8) inch steel ball is used. The depth of the indentation is measured in arbitrary units of 8×10^{-5} inch. The scale runs from 0 to 100 units. Since the results are obviously dependent on the load and type of indenter, it is necessary to specify the combination used. A letter in front of the number does this. The A scale (diamond indenter, 60 kg load) is usable from soft materials like brass to hard materials like hard metals. The B scale (1/16 inch steel ball, 100 kg load) is generally used for relatively soft materials. Hardened steels are conventionally tested by the C scale (diamond indenter, 150 kg load) and range from R_C 20 to R_C 70. The general acceptance of the Rockwell tests by metallurgists, in spite of its arbitrariness, is due to its freedom of personal errors and small indentation size so that heat-treated metal parts can be tested without introducing damage.

Nano-indentation

A relatively recent development in indentation testing is *recording* or *nano-indentation*. This is basically an indentation test in which during indentation the load and indentation depth are recorded. Generally relatively low loads (< 1 N) are used leading to shallow indents (upto about a micrometre), hence the name nano-indentation. Often a Berkovich indenter is used, although other shapes are also used.

Fig. 14.5 shows a typical force-displacement curve obtained by this test. At least three different characteristics can be distilled from such a graph. First, assuming only elastic response during unloading, the unloading slope S at maximum load is related to Young's modulus E of the material[e]. The relevant relationship reads

$$E = \frac{\sqrt{\pi}}{2} \frac{S}{\sqrt{A}}$$

---

[d] Meyer, E. (1908), Z. Ver. Deut. Ing. **52**, 645.
[e] In fact one should read the reduced modulus E_{red} given by $1/E_{\mathrm{red}} = (1-v_{\mathrm{ind}}^2)/E_{\mathrm{ind}} + (1-v^2)/E$ where the subscript 'ind' refers to the indenter.

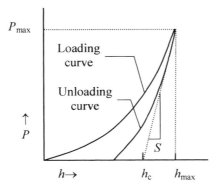

Fig. 14.5: A typical nano-indentation curve in which the unloading slope S is indicated.

where A the projected area of contact[f] between specimen and indenter at maximum load. Second, from the final indentation depth after complete unloading, the conventional hardness can be calculated. Third, from the area under the loading-unloading curve the energy dissipated during the indentation process can be obtained. The method is applied to monolithic materials (Fischer-Cripps, 2000) as well as coatings[g] and appears to be promising for further research.

Estimating the flow curve

For an estimate of the total flow curve, we need a measure for the stress as well as the strain. Since hardness is obtained from the size of the permanent impression – determined by the yield behaviour of the material – it follows that there must be a relation between hardness and flow behaviour. It has been shown by Tabor[h] that for a spherical indentation the true strain ε is proportional to d/D and can be expressed as

$$\varepsilon \cong 0.22 \frac{d}{D} \tag{14.7}$$

Moreover, he showed from an approximate elastic-plastic analysis of indentation by a flat cylindrical punch, that for materials where the ratio yield strength/elastic modulus ratio Y/E is less than about 0.01, the mean indentation pressure $\bar{p}$ is proportional to the momentary yield strength Y and can be given as[i]

$$\bar{p} \cong 2.8Y \tag{14.8}$$

It also appears that for power law strain hardening materials, the 0.2% offset uniaxial yield strength Y can be determined with reasonable precision by[j]

---

[f] For an ideal Berkovich indenter, i.e. with no tip rounding, the function A, generally denoted as *area function*, reads $A = ah_c^2$ with $a = 24.5$ and h_c the contact depth. Since tip rounding is unavoidable, the expression has to be corrected and a frequently used expression is $A = ah_c^2 + bh_c$ with b a parameter to be determined experimentally.

[g] Malzbender, J., den Toonder, J., Balkenende, R. and de With, G. (2002), Mater. Res. Eng. **R36**, 47.

[h] Tabor, D. (1951), *The hardness of metals*, Clarendon, Oxford.

[i] Tabor quotes 7 values with as average 2.7. He further uses 2.8, while in the literature 3 is often used.

[j] Cahoon, J.B., Broughton, W.H. and Kutzak, A.R. (1971), Metall. Trans. **2**, 1979.

$$Y_{0.2} = \frac{H_v b^{n'-2}}{3} \quad \text{with} \quad b = 0.1 \tag{14.9}$$

where n' can be estimated using Eq. (14.6).

A better approximation for $\bar{p}$, given by a more detailed semi-empirical analysis[k], which is appropriate for materials with an Y/E ratio larger than 0.01, yields

$$\frac{\bar{p}}{Y} = C + K \frac{3}{3-\lambda} \ln\left(\frac{3}{\lambda + 3\mu + \lambda\mu}\right) \quad \text{where}$$

$$\lambda = \frac{(1-2v)Y}{E}, \quad \mu = \frac{(1+v)Y}{E}, \quad C = 2/3 \quad \text{and} \quad K = 2/3 \tag{14.10}$$

This result is based on the spherical cavity model in plasticity theory (Appendix E). Marsh used the constants C and K as parameters in a fitting on a wide range of materials and using $C \cong 0.28$ and $K \cong 0.60$ could describe the experimental data well. Taking $v = 0.5$ in Eq. (14.10) results in the relation[l]

$$\frac{\bar{p}}{Y} = a + b \ln\frac{E}{Y} \quad a = 0.40, \quad b = 0.66 \tag{14.11}$$

Using $C \cong 0.28$ and $K \cong 0.60$ instead of $C = 2/3$ and $K = 2/3$ results in

$$\frac{\bar{p}}{Y} = a' + b' \ln\frac{E}{Y} \quad a' = 0.04, \quad b' = 0.6 \tag{14.12}$$

which, although approximate, is often accurate enough.

The complete flow curve can now be obtained from a series of spherical indentations. For Y one can take one of the Eqs. (14.8) or (14.10) with $C \cong 0.28$ and $K \cong 0.60$ or Eq. (14.12). The choice for ε is limited to Eq. (14.7). Plotting Y versus ε yields the flow curve.

It is also possible to estimate the strength S of metals from the hardness number. Tabor showed that the ultimate strength is given by[m]

$$S \cong \left(\frac{H_v}{2.9}\right)\left[1 - (n'-2)\right]\left(\frac{12.5(n'-2)}{1-(n'-2)}\right)^{n'-2} \tag{14.13}$$

This relationship has been verified experimentally and yields reasonable estimates.

Digression: Empirical relations

Some empirical relations like Meyers's law, Eq. (14.6), or the relation between the yield strength and the Vickers hardness, Eq. (14.9), may cause enormous confusion. The basic problem is that they are in fact incorrect from a methodological point of view because they are not invariant for a change in basic units. Consider for example Eq. (14.6) and assume that we have a set of data $d_1, d_2, \ldots, d_N$ and $F_1, F_2, \ldots, F_N$ given in units of mm and N, respectively. Let the proportionality factor k for these data be indicated by k_1. Using the same data but now in units of μm, one finds for the proportionality factor $k_2 = k_1 \times 10^{-3n'} \cong 0.317 \times k_1 \times 10^{-6}$ if $n' = 2.5$. So, if the value of k_1

[k] Marsh, D.M. (1964), Proc. Roy. Soc. London, **A279**, 420.
[l] Johnson, K.L. (1970), J. Mech. Phys. Solids **18**, 115.
[m] Tabor, D. (1951), *The hardness of metals*, Clarendon, Oxford.

is given and the original units for which the value was determined are not known, one cannot use the relation safely. Therefore, Meyer's law should be interpreted as follows. Suppose you have a data set as given above. If we now plot $q_i = F_i/F_N$ ($i = 1$, ..., $N–1$) versus $r_i = d_i/d_N$ ($i = 1$, ..., $N–1$) one finds

$$q_i = r_i^{n'}$$

The proportionality constant in this equation is numerically 1 and does not change its value if the units for F and d are changed.

14.2 Plasticity in processes

In this section we discuss briefly the use of plasticity in the shaping of materials. Plastic deformation is utilised in the fabrication of many materials and components. Although originating from the processing of metals, plasticity is also used for polymers and inorganics in the 'green', that is, non-sintered, state. Modelling of plastic processes for metals has always been economically important since it can provide estimates for the forces and energy involved in such a process. The forces are important since they relate to the initial investments for the equipment one has to make whereas the energy costs contribute the production costs. Nowadays also the modelling of plastic processes in polymer materials and components is done. In this case prevention of plastic failure is the most important reason. As indicated in the previous chapter pressure dependent yield criteria may be required. Also in particulate mechanics plasticity concepts are used, for example for the prediction of the density distribution during compaction of powder compacts, again with pressure dependent yield criteria. As will become clear, apart from the material behaviour, the friction of the material to be deformed with the processing equipment material is highly relevant.

Before the widespread use of the finite element method, modelling of plastic processes was always cumbersome. Although the processes are complex, the use of FEM has provided a means to analyse more complicated processes as before and with considerably higher accuracy. Nevertheless many problems remain, amongst which the availability of proper input data, both for the materials and the boundary conditions, is just one. Although FEM is used nowadays routinely, simple models are still useful because of the clarity of the various contributions.

As an example of a simple model of plastic processing we discuss the forging in plane strain of a plate of constant thickness h, width $2a$ and length l between two dies. The configuration is sketched in Fig. 14.6. We consider an infinitesimal volume element of length dx loaded with a normal pressure p. In this element (taken for the moment of unit length) the equilibrium of forces in the lateral direction, taken as x, is

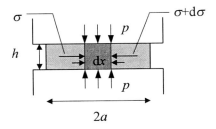

Fig. 14.6: Schematic of the forging of a slab.

$$\sigma h - (\sigma + d\sigma)h - 2\tau dx = 0 \qquad \text{or} \qquad \frac{d\sigma}{dx} = -\frac{2\tau}{h} \tag{14.14}$$

where σ is the stress in the x-direction and τ the (frictional) shear force at the interface of roll and slab. The von Mises yield criterion $\sigma_1 - \sigma_{III} = 2k = 2\tau_{max}$ for this particular configuration becomes

$$\sigma_1 - \sigma_{III} = 2k = p - \sigma \tag{14.15}$$

where p is the normal stress exerted by the die on the slab. In these expressions both the compressive stresses σ and p are taken positive. Assuming ideal plastic behaviour the yield strength in shear k is constant and therefore we have $dp/dx = d\sigma/dx$. Substituting this equation in Eq. (14.14) the governing differential equation becomes

$$\frac{dp}{dx} = -\frac{2\tau}{h} \tag{14.16}$$

Assuming further for the dependence of τ on p *Coulomb friction*, i.e. $\tau = \mu p$ where μ is the friction coefficient, the final equation to solve becomes

$$\frac{dp}{dx} = -\frac{2\mu p}{h}$$

The solution is obtained via separation of variables and integration leading to

$$\frac{dp}{p} = -\frac{2\mu}{h} dx \qquad \text{or} \qquad \ln p = -\frac{2\mu x}{h} + \ln C$$

The constant of integration C can be evaluated by using the boundary condition $\sigma(x = a) = 0$. Therefore from the yield condition (14.15) $p = 2k$ and

$$\ln C = \ln 2k + 2\mu a/h$$

Together this results in

$$p = 2k \exp\left\{\frac{2[\mu(a-x)]}{h}\right\} \cong 2k\left[1 - \frac{2\mu(a-x)}{h}\right] \qquad \text{and} \tag{14.17}$$

$$\sigma = 2k\left\{1 - \exp\left[\frac{2\mu(a-x)}{h}\right]\right\} \cong 2k\left[\frac{2\mu(a-x)}{h}\right]$$

where the second step can be made if μ is small. The mean die pressure becomes

$$\bar{p} = \int_0^a \frac{p}{a} dx = 2k \frac{\exp(2\mu a/h) - 1}{2\mu a/h} \tag{14.18}$$

while the total force is $F = 2al\bar{p}$ with l the length of the slab parallel to the die and $2a$ the width of the slab. Another way to write Eq. (14.17) is

$$p = 2k \exp\left[\left(\frac{\mu L}{h}\right)\left(1 - \frac{2x}{L}\right)\right]$$

with $L = 2a$ the contact length. This expression shows that with increasing ratio L/h the resistance to deformation rapidly increases. Both p and σ have a maximum in the middle of the plate, illustrating the existence of a *friction hill*. In this case the neutral line is in the centre of the plate. With increasing inward motion of the dies, the material at the neutral surface is stationary but flows for the remainder outward in the horizontal direction. In more complex processes the position of the neutral line (or plane) may not be so easy to establish.

If we use instead of $\tau = \mu p$, a *constant friction* factor m in $\tau = mk$, we obtain

$$p = 2k\frac{m}{h}(a - x) + 2k \qquad (14.19)$$

and in this case the pressure distribution is linear with x. For sticking friction we have $m = 1$ and the mean forging load becomes

$$\bar{p} = 2k\left(\frac{a}{2h} + 1\right) \qquad (14.20)$$

In practice the friction conditions are often intermediate between full sticking and slipping so that there may be sliding friction at the edges of the plate, where the pressure is lower, but sticking fraction closer to the centre, where the pressure increases to a point where $\tau = k$. This transition occurs at the position

$$x_t = a - \frac{h}{2\mu}\ln\frac{1}{2\mu} \qquad (14.21)$$

Although relatively simple, this model introduces the same considerations as for more complex processes, in particular the balance between material and friction behaviour, and shows that a non-uniform deformation field arises. In the next sections we discuss first a simple model for wire drawing and rolling, highlighting only the most salient features. After that we provide for both processes a somewhat more complex model.

Rolling

Rolling is one of the most important deformation processes for metals. The intention is to reduce the thickness of a sheet between two rolls to a desired value. In Fig. 14.7 a schematic view of the rolling process illustrates the important characteristics. A sheet of width b and initial thickness h_i is fed with a velocity v_i between the rolls and leaves the rolls with a thickness h_f and a velocity v_f. The angle between the entrance of the sheet and the centre line of the rolls is the *bite angle α*. In first approximation the width does not increase during the deformation process. Due to the incompressibility we have at all times during the rolling process $bh_iv_i = bhv = bh_fv_f$, where v is the velocity at thickness h intermediate between the initial and final thickness. Therefore the exit velocity must be larger than the entrance velocity and in

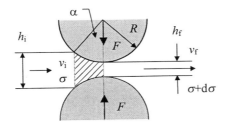

Fig. 14.7: Schematic of the rolling process. The hatched area indicates the region where deformation takes place.

fact the velocity increases steadily from entrance to exit. Only at point the velocity of the roll is equal to the velocity of the sheet and this point is called the *no-slip point*. Between the entrance and the no-slip point the sheet moves lower than the roll and the friction helps to draw the sheet into the rolls. Between the no-slip point and the exit the sheet moves faster than the roll, the friction reversed and opposes the delivery of the sheet from the rolls. A back tension is sometimes applied at the entrance side and/or a front tension at the exit side, usually in the case of a much more complex roll geometry with several rolls.

In a simple model the arc of the rolls in contact with the sheet can be taken as straight in view of the large diameter of the rolls with respect to the sheet thickness. In first approximation therefore the rolling process resembles the compression of a slab with average height $h = (h_i + h_f)/2$ in plane strain (Fig. 14.7) with the contact length $L \cong (R\Delta h)^{1/2}$ and R the roll radius. The total rolling load F is, according to Eq. (14.18),

$$\frac{F}{2k} = \frac{\overline{p}lL}{2k} = \frac{\exp(\mu L/h) - 1}{\mu L/h} l\sqrt{R\Delta h}$$

and we see that the total load increases more rapidly than with $R^{1/2}$, dependent on the friction conditions. The load also increases with decreasing thickness of the plate due to the exponential term in $\mu L/h$. Eventually a thickness is reached where the load applied cannot any longer deform the plate.

*Rolling extended**

A more complete theory of rolling is initiated by von Karman[n]. It is based on a model using the von Mises yield criterion and the following assumptions:
- The arc of contact is circular and deformation of the rolls does not occur.
- The coefficient of friction is constant at all contact points.
- There is no lateral spread and the problem is one in plane strain.
- Plane vertical sections remain plane.
- The roll velocity is constant.
- The elastic deformation of the sheet can be neglected.

The configuration is sketched in Fig. 14.8 indicating the roll radius R. At the interface the contact point is characterised by θ while the radial pressure is indicated by p and the tangential shear stress by τ. The horizontal stress σ is assumed to be uniformly distributed over the vertical faces of the element. In Fig. 14.8 the resolution in horizontal and vertical components is indicated as well.

A summation over the horizontal forces of the element results in

$$(\sigma + d\sigma)(h + dh) + 2\mu pR \cos\theta \, d\theta - \sigma h - 2pR \sin\theta \, d\theta = 0 \quad \text{or}$$

$$\frac{d\sigma h}{d\theta} = 2pR(\sin \theta \pm \mu \cos \theta) \tag{14.22}$$

often referred to as the *von Karman equation*. The positive sign applies between the exit plane and no-slip point while the negative sign applies between the entrance plane and the no-slip point. This change of sign occurs since the direction of the friction reverses at the no-slip point. The forces in the vertical direction are balanced by the roll pressure p and taking the equilibrium results in

---

[n] Von Karman, T. (1925), Z. Angew. Math. Phys. **5**, 139.

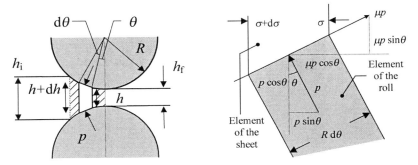

Fig. 14.8: Details of the rolling geometry.

$$n = p(1 \mp \mu \tan\theta) \tag{14.23}$$

where n is the normal pressure and p the radial pressure. The relationship between the normal pressure and the horizontal compressive stress σ is given by the von Mises criterion for plane strain reading

$$\sigma_I - \sigma_{III} = 2k \quad \text{or} \quad n - \sigma = 2k \tag{14.24}$$

where k is the yield strength in shear and n is the greater of the two compressive stresses. In agreement with the literature we take the compressive stresses positive.

The rolling process is described by the solutions of Eq. (14.22) in conjunction with Eqs. (14.23) and (14.24), which is a considerable mathematical problem so that normally various approximations are made. Orowan[o] provided a rather complete solution including the change in flow stress with θ due to strain hardening. The final equations have to be integrated numerically although the expression for the roll pressure[p] can be obtained analytically. The effect of roll flattening[q] can be taken into account as well.

Some approximations have been introduced[r] therefore by restricting the analysis to low friction and bite angles less than 6° so that $\sin\theta \cong \theta$ and $\cos\theta \cong 1$. In this case Eq. (14.22) reduces to

$$\frac{\mathrm{d}\sigma h}{\mathrm{d}\theta} = 2pR(\theta \pm \mu) \tag{14.25}$$

If it is further assumed that $p \cong n$, Eq. (14.24) can be written as $\sigma = p - 2k$. Substituting this expression in Eq. (14.25) and integrating result in the following expressions:

Entrance to no-slip point $\qquad p = \dfrac{2kh}{h_i}\left(1 - \dfrac{\sigma_i}{2k_i}\right)\exp\left[\mu(H_i - H)\right]$

No-slip point to exit $\qquad p = \dfrac{2kh}{h_f}\left(1 - \dfrac{\sigma_f}{2k_f}\right)\exp(\mu H) \qquad$ with

[o] Orowan, E. (1943), Proc. Inst. Mech. Eng. London **150**, 140.
[p] Cook, M. and Larke, E.C. (1947), J. Inst. Met. **74**, 55.
[q] Hockett, J.E. (1960), Trans. Am. Soc. Met. **52**, 675.
[r] Bland, D.R. and Ford, H. (1948), Proc. Inst. Mech. Eng. London **159**, 144.

$$H = 2\left(\frac{R}{h_{\mathrm{f}}}\right)^{1/2} \tan^{-1}\left[\left(\frac{R}{h_{\mathrm{f}}}\right)^{1/2}\theta\right]$$

Here the subscripts i and f refer to the values evaluated at the entrance and exit plane, respectively. In the analysis a back tension σ_i, applied at the entrance plane, and a front tension σ_f, applied at the exit plane, are included. The total load F is given by

$$F = Rl \int_0^\alpha n\,\mathrm{d}\theta$$

where, as before, l is the width of the sheet and α the bite angle. It has been shown by full numerical solutions[s] of the von Karman equations that, while a reasonable prediction of the rolling load can be obtained, the simplified models are incapable of providing an accurate prediction of the torque required.

Wire drawing

The goal of wire drawing for metals is to make the diameter of a wire smaller. This is done by pulling a wire through a die. The die contains a tapered orifice and is made of a wear-resistant material, e.g. a diamond or WC-Co composite (hard metal). During the drawing process also the microstructure of the wire changes but we will not discuss this aspect here. For the moment we assume that the metal behaves like a rigid-ideally plastic material. Furthermore, we will invoke the principle of virtual work on the scale of the process, i.e. globally. We assume that the stress state is uniaxial and that the cross-section of the wire is cylindrical.

In Fig. 14.9 a schematic view of the process is provided. Consider a part of the wire of length l_0 before it passes through the die. The initial cross-section is A_0 while the final cross-section is A. This corresponds to a change in length from l_0 to l. Since the volume is conserved, we have $l/l_0 = A_0/A$. The pulling stress F/A can never be larger than the yield strength Y. The work $\mathrm{d}W$ done in stretching the wire an infinitesimal amount $\mathrm{d}l$ is given by $\mathrm{d}W = F\,\mathrm{d}l = YA\,\mathrm{d}l$. The energy per unit volume u required for deformation is

$$u = \int_{l_0}^l \mathrm{d}u = \int_{l_0}^l \sigma\,\mathrm{d}\varepsilon = \int_{l_0}^l \frac{Y}{l}\mathrm{d}l = Y\ln\frac{l}{l_0} \qquad (14.26)$$

On the other hand the work done by the pulling force F in drawing a wire of length l is

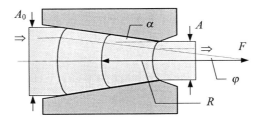

Fig. 14.9: Schematic view of the wire drawing process and the assumed deformation field.

[s] Alexander, J.M. (1972), Proc. Roy. Soc. London **A326**, 535.

given by $W = Fl$. This work must be equal to the energy per unit volume for the deformation times the total volume and thus we have

$$Fl = (Al)Y\ln\frac{l}{l_0} \qquad \text{or} \qquad F = AY\ln\frac{l}{l_0} = AY\ln\frac{A_0}{A} \qquad (14.27)$$

From this equation we see that, since the drawing stress $\sigma = F/A$ may not exceed Y, it holds that the factor $\ln(A_0/A) \le 1$. This implies that a maximum reduction in area of 63% in one pass can be obtained without failure.

In this simple model we neglected the actual complex deformation field in the die and the effect of friction. Both effects increase the stress necessary and in practice the above estimate is upper bound for the reduction for ideally plastic materials. In practice area reductions of about 50% can be obtained. Although a limit to the area reduction is obtained, the above analysis provides no estimate for the optimum drawing angle α. To that purpose a more complex model taking into account the friction and the simplest conceivable deformation field is required.

Wire drawing extended*

A systematic way to improve a model of any structure is by guessing a deformation field dependent on one or more parameters, calculating the (Helmholtz) energy and minimising this energy with respect to the parameters. The approach we follow is essential the one presented by Geleji[t]. Since the Helmholtz energy is given by the work W and $dW = Fdl$, we can consider directly the forces involved. The force required for the work of deformation can be thought to consist of three contributions

$$F = F_{\text{def}} + F_{\text{wal}} + F_{\text{int}} \qquad (14.28)$$

where F_{def} is the force required purely for the shape deformation, F_{wal} the force required to overcome the friction at the wall of the die and F_{int} the force for the internal friction in the metal. We discuss these terms in the next paragraph using the geometry and symbols as indicated in Fig. 14.10.

First, for the calculation of F_{def} we use an average normal stress s at the die wall to deform the material. The infinitesimal force dF_{def} for the deformation of a cross-section with thickness dz is given by

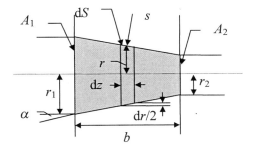

Fig. 14.10: Details on the wire drawing geometry.

[t] Geleji, A. (1960), *Bildsame Formung der Metalle in Rechnung und Versuch*, Akademie-Verlag, Berlin.

$$dF_{\text{def}} = (s\sin\alpha)dS \tag{14.29}$$

where dS denotes the area of the circumference of the cross-section. Some consideration[u] shows that the factor $\sin\alpha \, dS$ is given by

$$\sin\alpha \, dS = dA = d(\pi r^2) = 2\pi r \, dr \tag{14.30}$$

Here r and A denote the radius and the cross-sectional area of the die, respectively. The total deformation force is then

$$F_{\text{def}} = 2\pi s \int_{r_1}^{r_2} r \, dr = \pi s (r_1^2 - r_2^2) = s(A_1 - A_2) \tag{14.31}$$

where the labels 1 and 2 denote the entrance and outlet side of the die, respectively.

Second, the friction force F_{wal} depends on the friction coefficient μ and from Fig. 14.10 this force is obtained as

$$F_{\text{wal}} = \mu s Q \tag{14.32}$$

where Q is the area of contact, given by

$$Q = \frac{(r_1 + r_2)\pi b}{2\cos\alpha} = \frac{A_1 - A_2}{\sin\alpha} \tag{14.33}$$

Finally we need the force necessary to overcome the internal friction of the material F_{int}. In fact this is the troublesome term and we need to make an assumption on the deformation behaviour of the material in the die. Here we assume the simplest possible deformation field in which the deformation that occurs is conform the shape of the die. Referring to Fig. 14.9 this implies that the broken line indicated by $\Rightarrow$, deforms by an angle φ at the entrance and the outlet side of the die. Since this angle is small, it also represents the engineering shear strain γ. The equivalent plastic strain, defined by $\varepsilon^* = (2\varepsilon_{ij}\varepsilon_{ij}/3)^{1/2}$, is thus $\varepsilon^* = \gamma/\sqrt{3}$. Assuming von Mises behaviour the equivalent stress σ^* is given by k, the yield stress in shear[v]. Hence the work done per unit volume is given by $w_{\text{int}} = 2\cdot\frac{1}{2}\cdot\sigma^*\varepsilon^* = k\varphi/\sqrt{3}$. For the total work of deformation in a die of unit length we obtain[w]

$$W_{\text{int}} = \int w_{\text{int}} dV = \int \frac{k\varphi}{\sqrt{3}} dV = 2\pi R^2 \frac{k}{\sqrt{3}} \int_0^\alpha \varphi^2 d\varphi = 2R\pi^2 \frac{k}{\sqrt{3}} \frac{\alpha^3}{3} \tag{14.34}$$

Since the area A_1 for $\alpha \ll 1$ is given approximately by $A_1 \cong \pi(R\alpha)^2$ we can also write

$$W_{\text{int}} = \frac{2k}{3\sqrt{3}}\alpha V \tag{14.35}$$

where V denotes the volume of the die. An identical contribution arises from the outlet side of the die so that the total internal friction work is given by

[u] The (curved) area S of a truncated cone with radii r_1 and r_2, opening angle 2α and length b is given by $S = \pi b(r_1+r_2)/\cos\alpha = \pi(r_1^2-r_2^2)/\sin\alpha$.
[v] In case a work-hardening material is considered frequently an average yield strength is taken, defined by $k = (k_{\text{ini}} + k_{\text{fin}})/2$, where k_{ini} and k_{fin} denote the initial yield strength and final yield strength, respectively.
[w] The volume of a segment of a thin spherical shell with radius R, opening angle $\varphi \ll 1$ and thickness h is given by $V = \pi(R\varphi)^2 h$ so that $dV = \pi R^2\varphi h \, d\varphi$.

$$W_{int} = \frac{4k}{3\sqrt{3}} \alpha V = \left(\frac{4k}{3\sqrt{3}} \alpha A_2 \right) l_2 \equiv F_{int} l_2 \qquad (14.36)$$

where F_{int} denotes the internal friction force and l_2 the length of the wire after drawing.

The total force F is thus

$$F = F_{def} + F_{wal} + F_{int} = s(A_1 - A_2) + \mu s \frac{(A_1 - A_2)}{\sin \alpha} + \frac{4k}{3\sqrt{3}} \alpha A_2 \qquad (14.37)$$

What remains to be calculated is the connection between s and k. This is the topic of the next paragraph.

Since the die has cylinder symmetry, it holds that two principal strains and two principal stresses are equal, i.e. $\varepsilon_I = \varepsilon_{II}$ and $\sigma_I = \sigma_{II}$. For this symmetry it is immaterial whether the von Mises or Tresca yield condition is used. Both read $\sigma_I - \sigma_{III} = k = 2\tau_{max}$. Because $s = \sigma_I$, exactly at the interface between die and wire and to a high degree of approximation in the bulk of the wire, we may write

$$s - \sigma_{III} = k \qquad (14.38)$$

The average value of σ_{III} in the die is

$$\overline{\sigma}_{III} = \frac{F}{2A_2} \qquad (14.39)$$

since $\sigma_{III} = 0$ at the entrance side and $\sigma_{III} = F/A_2$ at the outlet side of the die. Replacing σ_{III} with $\overline{\sigma}_{III}$ in Eq. (14.38) we have

$$F = 2A_2(s - k) \qquad (14.40)$$

Substitution in Eq. (14.37) and solving for s results in

$$s = \left[k \left(1 + \frac{2}{3\sqrt{3}} \alpha \right) \right] \Big/ \left[1 - \frac{(A_1 - A_2)}{2A_2} \left(1 + \frac{\mu}{\sin \alpha} \right) \right] \qquad (14.41)$$

Taking into account the fact that $\alpha \ll 1$, we may approximate the above further by

$$s = k \Big/ \left[1 - \frac{(A_1 - A_2)}{2A_2} \left(1 + \frac{\mu}{\alpha} \right) \right] \qquad (14.42)$$

Similarly $\sin \alpha$ may be replaced by α in Eq. (14.37).

Using Eq. (14.37) and Eq. (14.42) the total force for drawing as a function of die angle for a certain friction coefficient μ can be calculated. The friction coefficient ranges from 0.01 to 0.15, a typical value being 0.05. A comparison with some experimental results is given in Fig. 14.11. Agreement within about 20% can be observed. In view of the crudeness of the model the drawing force is well predicted. Moreover, the dependence on the die angle α is also reasonably predicted.

From Eq. (14.37) and Eq. (14.42) an estimate can be made for the optimum drawing angle α, that is the angle corresponding to the minimum force. In fact such an estimate corresponds to minimisation of the work W (since $W = Fl_2$) which is the most important contribution to the Helmholtz energy. The above procedure is thus an illustration of the variational theorems of solid mechanics. Fig. 14.11 shows that such

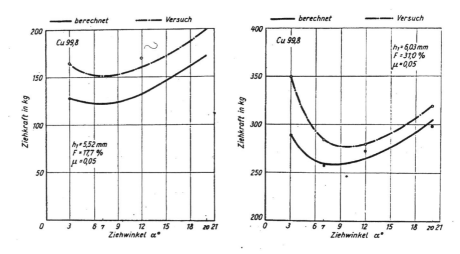

Fig. 14.11: Drawing force as a function of die angle for a cross-section reduction of 17.7% and 31%, respectively, for Cu wire with diameter 16.5 mm and friction coefficient $\mu = 0.05$.

an approach can result in useful results. Experimental results for the minimum drawing force for steel wire as a function of die angle are presented in Fig. 14.12. The optimum drawing angle can be described empirically by

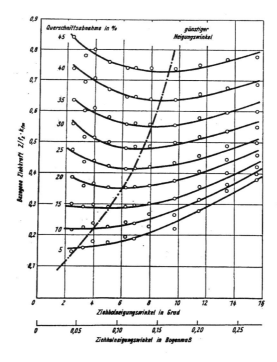

Fig. 14.12: Experimental values of the minimum drawing force for steel wire as a function of die angle for $\mu = 0.05$.

$$\alpha = 53.5 \sqrt{\mu \left(\frac{A_1}{A_2} - 1 \right)} \qquad (14.43)$$

where α is given in degrees. It can be seen from in Fig. 14.12 that the rather shallow curves are obtained so that a small variation in angle will result in not too large differences in drawing force.

Finally, it should noted that various other models for wire drawing have been proposed, all with more or less the same basic idea. The model as presented here shows clearly the various contributions to the Helmholtz energy and the possibility to improve on the model, in particular by admitting more complex deformation fields, the introduction of a position dependent normal stress and a more detailed treatment of work hardening by admitting a position dependent yield strength. Also in this case full numerical solutions have been obtained for which we refer to the literature.

Problem 14.1

Derive the results for slab compression under conditions of constant friction.

Problem 14.2

Derive an expression for the minimum plate thickness that can be reached during slab compression.

Problem 14.3

For small drawing angle α the force is given by Eq. (14.37). Using Eq. (14.42) for the average normal stress s, derive that the optimum drawing angle is

$$\alpha = \frac{-\mu \pm \sqrt{\mu^2 + 6\sqrt{3}H}}{2H} \qquad (14.44)$$

where H is given by $H = (A_1+A_2)/(A_1-A_2)$. Discuss this expression and compare the numerical result with the experimental values as shown in Fig. 14.12.

14.3 Plasticity in structures*

We have seen in Chapter 13 that after reaching the critical yield stress or yield strength, the behaviour of the material becomes non-linear and dissipative. The linear equations of elasticity no longer apply and the incremental equations of plasticity have to be solved. Generally yield is not desirable since the dimensions of the structure change. However, plasticity has also a significant impact on the safety of structures. Yield before fracture is desirable since in that case the structure deforms as a whole. Moreover, the prediction of yielding of a structure, though generally not trivial, is connected with less uncertainties than the prediction of fracture. For most structures the displacements that can be tolerated are limited. This has led to considerations with respect to the stiffness of a structure by taking into account in the design the geometry of the structure as well as the material's properties. For metallic structures the plasticity of the material is often considered as an important aspect to prevent catastrophic failure in case excessive loads are present.

The general incremental equations are generally difficult to solve analytically and it is useful to obtain information on the problem at hand without embarking on an incremental analysis. There are two general methods for calculating the collapse loads of structures: the method of *virtual power* (in this context also denoted as *static* or *stress-equilibrium* method) and the method of *complementary virtual power* (also known as *kinematic* or *plastic-work* method). The first method requires that all stresses are in equilibrium (statically admissible) without worrying about the associated strains and strain rates and that the stress nowhere exceeds the yield strength. The second method requires that the strains (strain rates) are compatible (kinematically admissible) without worrying about the associated stresses. For exact solutions both requirements are of course fulfilled. However, for complex structures an exact solution is often impossible and we may use both methods as the basis for *limit analysis*. If we do not know the actual mode of deformation we may make a guess for it, as simple as convenient but always requiring geometric compatibility. The collapse load for this guessed deformation provides an *upper limit* to the true collapse load. On the other hand, we may also guess a distribution of stresses that will lead to deformation, again as simple as convenient provided they are always in equilibrium and do not exceed the yield strength. The resulting estimate provides a *lower limit*. We do not discuss these approaches here but refer to the literature where also formal proofs can be found, e.g. Hill (1950), Martin (1975) and Lubliner (1990). Here we discuss only a bend beam as a simple example of how plasticity influences the deformation behaviour of structures (see e.g. De4by et al., 1992).

Plastic bending of a bar

For a beam loaded in bending we have seen that the maximum stress σ is at the outside of the beam (at position y_{max}) and given by $\sigma = My_{max}/I$, where I denotes the moment of inertia and M the moment applied to the beam. Of course, for plastic deformation the general relations

$$\int \sigma \, dA = 0 \quad \text{and} \quad \int \sigma y \, dA = M \tag{14.45}$$

remain true.

To be specific, let us take a beam of length l loaded in 3-point bending with central force F. The maximum stress $\sigma_{3pb} = My/I = 3Fl/2bh^2$. Here the moment M is given by $M = (F/2)(l/2)$ and the moment of inertia by $I = bh^3/12$, where b and h are the width and height of the beam, respectively. If we assume that the material of the beam

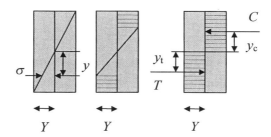

Fig. 14.13: Stress distribution in a cross-section of a beam of an elastic-perfectly plastic material. The hatched area indicates the plasticised region.

behaves like an isotropic elastic-perfectly plastic material, the maximum stress at the outer surfaces cannot exceed the yield strength Y. The corresponding moment acting on the beam is the *yield moment* $M_{yie} = YI/\tfrac{1}{2}h = YZ_{ela}$, where $Z_{ela} = 2I/h$ is the (elastic) section modulus (Fig. 14.13, left). During further loading the plastic zone will spread in the depth of the beam (Fig. 14.13, middle) until the stress has reached over the whole cross-section of the beam the yield stress Y (Fig. 14.13, right). The moment that corresponds to this situation is the *plastic moment* M_{pla}. The beam cannot withstand any further loading and will deform in uncontrolled way. The determination of the plastic moment is obviously of great importance. If we consider only situations where the beam is symmetric, like the rectangular beam, the neutral line for elastic bending and plastic bending are at the same position. The plastic moment is obtained by integrating the second part of Eq. (14.45), which is equivalent to taking the moments of the tensile and compressive forces about the neutral axis. If A denotes the area of the cross-section of the beam, the total tensile force T and total compressive force C in the lower and upper half of the beam (Fig. 14.13) are $T = YA/2$ and $C = YA/2$, respectively. Therefore,

$$M_{pla} = Ty_t + Cy_c = \frac{YA(y_t + y_c)}{2} = YZ_{pla} \tag{14.46}$$

where y_t and y_c denote the distances from the neutral axes to the centroids c_t and c_c of the areas loaded in tension and compression, respectively. Analogous to the elastic case, the quantity Z_{pla} is denoted as the *plastic section modulus*.

The ratio f of the plastic moment M_{pla} to the yield moment M_{yie} is for ideally plastic materials a function of the geometry of the beam only and often referred to as the *shape factor*:

$$f = \frac{M_{pla}}{M_{yie}} = \frac{Z_{pla}}{Z_{ela}} \tag{14.47}$$

For a rectangular beam of width b and height h the plastic section modulus Z_{pla} becomes

$$Z_{pla} = \frac{bh}{2}\left(\frac{h}{4} + \frac{h}{4}\right) = \frac{bh^2}{4} \tag{14.48}$$

while the elastic section modulus Z_{ela} was given by

$$Z_{ela} = \frac{bh^2}{6} \tag{14.49}$$

Therefore the shape factor $f = 3/2$. Thus, a rectangular beam is fully 'plasticised' at a load 3/2 times the load where the first yielding starts. Similar calculations can be done for I-section beams, which are left as an exercise though. In that case the shape factor f is typically 1.1 to 1.2.

For a moment larger than M_{yie} but less than M_{pla}, a region of contained plastic deformation will exist in the central part of the beam. At the lower part of the beam the size of that zone is indicated by l_{pla}. With increasing moment the plastic zone spreads laterally but also in the depth of the beam towards the neutral line. As soon as the plastic zone reaches the neutral line, i.e. when $M = M_{pla}$, unrestrained plastic flow may take place. In this case the beam behaves like two rigid bars connected by a *plastic hinge* that permits the two bars to rotate relative to each other under the action

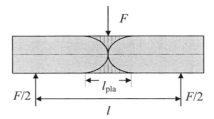

Fig. 14.14: Plastic beam acting as a plastic hinge.

of the moment M_{pla} (Fig. 14.14) The length of the plastic zone can be calculated from the fact that the moment at the edge of the zone is M_{yie}. We thus have

$$M_{\text{yie}} = \frac{F}{2}\left(\frac{l - l_{\text{pla}}}{2}\right) \tag{14.50}$$

Moreover, the maximum moment M_{pla} is equal to $Fl/4$ so that $F = 4M_{\text{pla}}/l$. Substitution and solving for l_{pla} results in

$$l_{\text{pla}} = l\left(1 - \frac{M_{\text{yie}}}{M_{\text{pla}}}\right) = l\left(1 - \frac{1}{f}\right) \tag{14.51}$$

For a rectangular beam $f = 3/2$ so that $l_{\text{pla}} = l/3$. For an I-section beam $f \cong 1.1$ and thus $l_{\text{pla}} = 0.09l$. The plastic zone is thus much smaller for an I-section beam. The increased efficiency in terms of material when using I-sections beams in comparison to rectangular beams is thus paid by a smaller 'safety range' as indicated by the value of the shape factor f and plastic zone size l_{pla}.

More complex structures and loading can be analysed in a more or less analogous way. Timoshenko and Gere (1973) provide an elementary introduction. Also the behaviour for cracked structures can be analysed. Here the critical region in the structure is at the position of the crack. The stress at which a structure becomes fully 'plasticised' is known as the *plastic collapse stress*. Of course, it can be determined using full-scale plasticity calculations. However, in the spirit of this section, an approximate method can be expected. Indeed the plastic collapse stress σ_{col} can be determined by limit analysis and is generally given[x] for simple structures, such as beams and plates, by

$$\sigma_{\text{col}} = \gamma Y (1 - a/W)^n$$

where W is the width of the structure at the position of the crack, γ a dimensionless constant and $n = 1$ for centre and double-edged cracked plates loaded in tension and $n = 2$ for bending. For example, for a plate of width W with a central crack of length a, the constant $\gamma = 1$ so that we obtain $\sigma_{\text{col}} = Y(1 - a/W)$. This approach will be used later in the two-criteria approach to describe the ductile-brittle transition.

---

[x] Green, A.P. and Hundy, B.B. (1956), J. Mech. Phys. Sol. **4**, 128 and Ford, H. and Lianis, G. (1975), Z. Ang. Math. Phys. **8**, 360.

Problem 14.4

Show that for an I-beam as indicated in the accompanying figure the shape factor f is given by

$$f = bt_{\text{fla}}\left(h - t_{\text{fla}}\right) + t_{\text{web}}\left(\frac{h}{2} - t_{\text{fla}}\right)^2$$

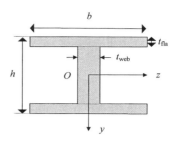

in which the first term represents the contributions of the flanges and the second term from the web.

Problem 14.5

Calculate the length of the plastic zone at the edge of the I-beam as shown in the previous problem. Assume $h = b$ and $t = 0.15h$.

Problem 14.6

Show that the stress distribution for a beam with width b and height h of which the material satisfies power-law behaviour, $\sigma = k\varepsilon^n$ where $0 \le n \le 1$, is given by

$$\sigma = \frac{2^{n+1}(n+2)M}{bh^2}\left(\frac{y}{h}\right)^n$$

Show that for $n = 1$ the behaviour reduces to that of an elastic material. Plot the stress distribution over the height of the beam for various values of n.

Problem 14.7

Consider the torsion of a circular shaft of radius R by a torque T. The only stress component is $\sigma_{rz} = \tau = Tr/I_p = Tr/\pi R^4$, where I_p denotes the polar moment of inertia. Accepting the yield condition $\sigma_2' = 2k^2$, show that:
a) The surface becomes plastified at a torque $T_0 = 2M/\pi R^3$ where M is the applied moment.
b) The shaft is fully plastified (and thus collapses) when $T = 4T_0/3$.

Problem 14.8

Consider a tube of inner radius R_i and outer radius R_o, which is internally pressurised. With increasing pressure the material will first deform elastically and thereafter plastically. Ideal plastic behaviour obeying Tresca's criterion is assumed.
a) At what position does the yielding start?
b) Show that the pressure p where yielding starts is given by

$$p_i = [(R_o^2 - R_i^2)/2R_o^2]Y$$

c) Show that the pressure p_f where the tube is fully plasticised is given by

$$p_i = Y \ln(R_o/R_i)$$

14.4   Slip-line field theory*

One of the methods to solve plastic problems is the slip-line method. This method supposes plane strain conditions and a rigid-perfectly plastic material with yield strength in shear k. In spite of these limitations it is a useful method both for educational and practical purposes. We first discuss the equivalence of the von Mises and Tresca criteria for plane strain conditions and then show that in plane strain the stress distribution always can be separated in a pure hydrostatic and pure shear part. After that we discuss briefly slip-line theory itself.

Let us consider the yield criterion first. In plane strain the stresses at each point for plastic conditions ($\nu = \frac{1}{2}$) are determined by two principal stresses, say σ_I and σ_{III}. The third principal stress is given by $\sigma_{II} = \frac{1}{2}(\sigma_I + \sigma_{III})$. For the mean stress σ_m this results in $\sigma_m = \frac{1}{3}\sigma_{kk} = \frac{1}{2}(\sigma_I + \sigma_{III}) = \sigma_{II}$. The deviatoric parts are

$$\sigma_I' = \sigma_I - \sigma_m = \frac{1}{2}(\sigma_I - \sigma_{III}), \quad \sigma_{II}' = \sigma_{II} - \sigma_m = 0 \quad \text{and} \quad \sigma_{III}' = \sigma_{III} - \sigma_m = \frac{1}{2}(\sigma_{III} - \sigma_I).$$

Using the von Mises equivalent stress as yield criterion we obtain

$$\sigma_{vM}* = \frac{1}{2}\sqrt{3}(\sigma_I - \sigma_{III})$$

Since for a von Mises material $Y = \sqrt{3}k$ the yield criterion also reads $k = \frac{1}{2}(\sigma_I - \sigma_{III})$. The same expression results for a Tresca material. Hence under plane strain conditions it is immaterial whether we use the von Mises or Tresca criterion if we express it in terms of k.

The second aspect to consider is the stress distribution. Consider at an arbitrary point P the stress matrix for a plane strain situation

$$\sigma_{ij} = \begin{pmatrix} \sigma_{xx} & 0 & \sigma_{xz} \\ 0 & \sigma_{yy} & 0 \\ \sigma_{zx} & 0 & \sigma_{zz} \end{pmatrix}$$

where $\sigma_{yy} = \frac{1}{2}(\sigma_{xx} + \sigma_{zz})$. The mean stress $\sigma_m = \frac{1}{3}\sigma_{kk}$ is thus given by $\sigma_m = \sigma_{yy}$. Diagonalizing the stress matrix leads to the principal stresses

$$\sigma_I = \frac{1}{2}\left[\sigma_{xx} + \sigma_{zz} + \sqrt{(\sigma_{xx} - \sigma_{zz})^2 + 4\sigma_{xz}^2}\right] = \sigma_m + \tau$$

$$\sigma_{II} = \sigma_{yy} = \sigma_m \qquad\qquad\qquad\qquad (14.52)$$

$$\sigma_{III} = \frac{1}{2}\left[\sigma_{xx} + \sigma_{zz} - \sqrt{(\sigma_{xx} - \sigma_{zz})^2 + 4\sigma_{xz}^2}\right] = \sigma_m - \tau$$

where $\tau^2 = \frac{1}{4}(\sigma_{xx} - \sigma_{zz})^2 + \sigma_{xz}^2$. The stress τ is also described by $\tau = \frac{1}{2}(\sigma_I - \sigma_{III})$ and thus can be interpreted as the shear stress. The stress matrix becomes

$$\sigma_{ij} = \begin{pmatrix} \sigma_m & 0 & 0 \\ 0 & \sigma_m & 0 \\ 0 & 0 & \sigma_m \end{pmatrix} + \begin{pmatrix} \tau & 0 & 0 \\ 0 & 0 & 0 \\ 0 & 0 & -\tau \end{pmatrix} = \begin{pmatrix} \sigma_m & 0 & 0 \\ 0 & \sigma_m & 0 \\ 0 & 0 & \sigma_m \end{pmatrix} + \begin{pmatrix} 0 & 0 & \tau \\ 0 & 0 & 0 \\ \tau & 0 & 0 \end{pmatrix}$$

showing that for plane strain the stress can be divided in a pure hydrostatic part $\sigma_m = \frac{1}{3}(\sigma_I + \sigma_{II} + \sigma_{III}) = \frac{1}{2}(\sigma_I + \sigma_{III}) = \sigma_{II}$ and a pure shear part $\tau = \frac{1}{2}(\sigma_I - \sigma_{III})$. The yield criterion becomes $k = \tau$ or, equivalently but easier in use, $k^2 = \tau^2$.

Now we are in a position to introduce slip-line theory. If we take a material, which behaves as rigid-perfectly plastic and consider a loading situation so that plasticity sets in (the material is 'plastified'), the shear stress τ is constant in magnitude and equal to

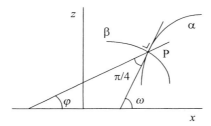

Fig. 14.15: The slip-lines α and β, perpendicular to each other, at an arbitrary point P. The angle ω indicates the orientation of maximum shear stress (α-line) with respect to the x-direction while the angle φ indicates the orientation of the maximum normal stress with respect to the x-direction.

k all through the material and can only vary in direction. The mean stress σ_m can, of course, only vary in magnitude. Using still the same axes convention for plane strain the stress is constant in the y-direction. In the xz-plane the stresses do vary and at every point the principal shear stress directions can be determined. Since in this plane there are two principal shear stresses perpendicular to each other we can draw two sets of trajectories of principal shear stresses, the so-called *slip-lines*. Conventionally they are denoted as the α- and β-lines. It will be clear that the tangents to the trajectories indicate the direction of the principal shear stresses. Since the principal shear stresses at every point are perpendicular but can vary in direction the tangents to the α- and β-lines form an orthogonal curvilinear axes system. The convention is that moving counter clockwise (counted as positive) from a α-line one meets the maximum principal stress σ_1 before encountering the β-line.

From Fig. 14.15 it is clear that

$$\partial z/\partial x = \tan\omega \;(\alpha\text{–line}) \quad \text{and} \quad \partial z/\partial x = -\cot\omega \,(\beta\text{–line})$$

Since the principal normal stress direction φ forms an angle of $\pi/4$ with the principal shear stress direction ω, it holds that $\omega = \varphi + \pi/4$.

The general expressions for the stresses in the xz-plane are given by (either using the Mohr circle representation or Eq. (14.52))

$$\sigma_{xx} = \sigma_m + \tfrac{1}{2}(\sigma_1 - \sigma_{III})\cos 2\varphi$$
$$= \sigma_m + k\sin 2\omega$$

$$\sigma_{zz} = \sigma_m - \tfrac{1}{2}(\sigma_1 - \sigma_{III})\cos 2\varphi$$
$$= \sigma_m - k\sin 2\omega$$

$$\sigma_{xz} = \tfrac{1}{2}(\sigma_1 - \sigma_{III})\sin 2\varphi = -k\cos 2\omega$$

where the last step can be made using the relations $\omega = \varphi + \pi/4$ and $k = \tau$. Inserting these stresses in the (reduced) equilibrium condition

$$\frac{\partial\sigma_{xx}}{\partial x} + \frac{\partial\sigma_{xz}}{\partial z} = \frac{\partial\sigma_{zx}}{\partial x} + \frac{\partial\sigma_{zz}}{\partial z} = 0$$

we obtain

$$\frac{\partial\sigma_m}{\partial x} + 2k\left(\cos 2\omega\,\frac{\partial\omega}{\partial x} + \sin 2\omega\,\frac{\partial\omega}{\partial z}\right) = 0 \tag{14.53}$$

$$\frac{\partial \sigma_m}{\partial z} - 2k \left(\cos 2\omega \frac{\partial \omega}{\partial z} - \sin 2\omega \frac{\partial \omega}{\partial x} \right) = 0 \qquad (14.54)$$

Now since the Cartesian axes system can take any orientation we can orient it in such a way that at a certain position P the x-direction coincides with the α-direction and the z-direction with β-direction. In this case $\omega = 0$. Moreover, since at P the tangents to the α- and β-lines form an orthogonal axes system, $dx = d\alpha$ and $dz = d\beta$. Hence the above equations reduce to

$$\frac{\partial}{\partial \alpha}(\sigma_m + 2k\omega) = 0 \qquad \text{and} \qquad \frac{\partial}{\partial \beta}(\sigma_m - 2k\omega) = 0 \qquad (14.55)$$

Since the point P was chosen arbitrarily these equations hold for all points in the xz-plane. Integration yields

$$\sigma_m + 2k\omega = C_1(\beta) \qquad \text{and} \qquad \sigma_m - 2k\omega = C_2(\alpha) \qquad (14.56)$$

where the constants C_1 and C_2 may depend on the value of β and α in view of the fact that Eqs. (14.55) are partial differential equations. Eqs. (14.56) are known as *Hencky's equations*. They imply that the change in σ_m between two points along an α- (or β-) line is proportional to the change in angle ω with $\pm 2k$ the proportionality constant. Hence, if the value of σ_m is known at a certain point on a slip-line the mean stress can be determined at any point on the slip-line. Typically such a point can be found at the surface where $\sigma_{xz} = 0$. From $\sigma_{xz} = -k \cos 2\omega$ it follows that $\cos 2\omega = 0$ or $\omega = \pm \pi/4$. The slip-line thus intersects the free surface at an angle of 45°. Another typical boundary condition can be found at a surface in contact with another material where no slippage occurs. In this case $|\sigma_{xz}| = k$. Hence from $\sigma_{xz} = -k \cos 2\omega$ it follows that $\cos 2\omega = 1$ or $\omega = \pi/2$ and $\omega = 0$. One set of slip-lines thus intersects the contact surface at an angle of 90° while the other is parallel to that surface. Finally, we note that a similar exercise as above can be made for axi-symmetric configurations.

The above brief discussion only shows only the bare essentials of slip-line theory, as applied to a situation where no displacement boundary conditions are applied. In the latter case a more elaborate set of equations including displacements has to be solved. A now classical review is given by Hill (1950) while Lubliner (1990) presents a review also including modern developments. Finally, it should be stated an extensive bibliography[y] of slip-line field solutions for many situations is available. The following example shows a relatively simple but nevertheless important case.

Example 14.1: Indentation

Consider indentation with a flat frictionless indenter (Fig. 14.16). At the edges of the indenter plastic zones develop almost as soon as the indenter is loaded. The indenter does not sink in, though, since most of the material below the indenter is still rigid. With increasing load the plastic zones will grow until they meet at the centre of the indenter. At that point the indenter starts to sink in. The load at which this occurs can be found using slip-line theory. At the free surface the slip-lines meet at $\omega = \pm \pi/4$. Thus following a α- (or β-) line

---

[y] Johnson, W., Sowerby, R. and Venter, R. (1982), *Plane strain slip-line theory and bibliography*, Pergamon, Oxford.

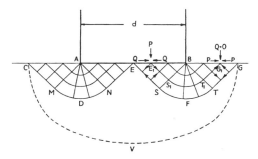

Fig. 14.16: Slip-line field for an indenter.

from A to B ω cannot be less than $\pi/2$. Hence, using the first of Eqs. (14.56), a pressure p under the indenter gives $p - \pi k$ at the free surface. The principal stresses at the free surface are $\sigma_I = 0$, $\sigma_{III} = -2k$ and $\sigma_{II} = \frac{1}{2}(\sigma_I + \sigma_{III}) = -k$. Hence the mean pressure at the free surface is also k. From $p - \pi k = k$ it follows that $p = (1+\pi)k$. The normal stress under the indenter is given by $\sigma_{III} = -p - k = -(2+\pi)k$. This stress is compressive and uniform over the whole face of the indenter. In terms of the uniaxial yield strength Y this results in $-(2+\pi)Y/\sqrt{3} \cong -2.97Y$ for a von Mises material and in $-(2+\pi)Y/2 \cong -2.57Y$ for a Tresca material. The hardness (indentation resistance) as measured with a blunt indenter is thus about three times the uniaxial yield strength due to the elastic constraint. A similar analysis can be made for wedge-shaped indenters and lead to similar results.

Heinrich Hencky (1885-1952)

Hencky received his Diploma in Civil Engineering in the Technical University of Munich in 1908 and a Doctorate from Darmstadt in 1913. He spent a few years with the Alsation railways and then went to Russia just before World War I. He was taken prisoner in Kharkov and interned in the Urals when the war broke out. After the war, he taught at Darmstadt, Dresden and Delft; while at the Technical University in Delft (the Netherlands), he did the work on slip-line theory, plasticity and basic rheology, for which he is best known. In 1930 he went to MIT as an associate professor for one year, and in 1931 he taught what must have been the first regular course entitled "Rheology". After that he returned to Delft and then to Germany and seems to have published less. Interestingly, during World War II, he travelled from Germany to an ASME meeting in Philadelphia in 1941 and gave a paper on plate and shell theory, which was printed in the Journal of Applied Mechanics in 1942. He was described only as 'Mechanical Engineer, Mainz', and was probably engaged in the war effort. After the war he worked in the German industry. He died in a mountain-sport accident.

14.5   Numerical solutions*

In Chapter 8 the finite element method (FEM) was discussed. In fact essentially the same method can also be used for plastic deformations as long as inertia effects can be ignored and the deformations and rotations remain small. It should be remembered that during loading in plastic deformation the hardening modulus h replaces Young's modulus E and that we can take (usually) $v = \frac{1}{2}$. During unloading the material behaves elastically. Let us recall that the incremental equations of plasticity for a normal material obeying the von Mises yield condition read

$$d\boldsymbol{\sigma} = \mathbb{C}^{(e)}{:}(d\boldsymbol{\varepsilon} - d\boldsymbol{\varepsilon}^{(p)}) \quad \text{and} \quad d\boldsymbol{\varepsilon}^{(p)} = \boldsymbol{\sigma}' \, d\lambda \quad \text{with} \quad d\lambda = (3/2)d\varepsilon^*/\sigma^*$$

where $\mathbb{C}^{(e)}$ denotes the tensor of elastic constants. The multiplier $d\lambda$ depends at each position in the structure on the momentary state via the increment in equivalent strain $d\varepsilon^*$ and the (von Mises) equivalent stress $\sigma^*(\varepsilon^{(p)})$, which is dependent on the previously accumulated plastic strain $\varepsilon^{(p)}$ if hardening occurs. For unloading $d\lambda = 0$. These equations can be summarised as

$$d\boldsymbol{\sigma} = \mathbb{C}^{(ep)}{:}d\boldsymbol{\varepsilon} \quad \text{with} \quad d\boldsymbol{\varepsilon} = d\boldsymbol{\varepsilon}^{(e)} + d\boldsymbol{\varepsilon}^{(p)}$$

The formal operator $\mathbb{C}^{(ep)}$ is non-linear since it depends on the direction of $d\boldsymbol{\varepsilon}$ in addition to the current state. For linear hardening the operator is given by $\mathbb{C}^{(ep)} = \{[\mathbb{C}^{(e)}]^{-1} + [\mathbb{C}^{(p)}]^{-1}\}^{-1}$ for loading in the plastic regime, with $\mathbb{C}^{(p)}$ the tensor containing the hardening moduli in $d\boldsymbol{\sigma} = \mathbb{C}^{(p)}{:}d\boldsymbol{\varepsilon}^{(p)}$, and by $\mathbb{C}^{(ep)} = \mathbb{C}^{(e)}$ for loading in the elastic regime and for unloading.

Here we only briefly outline the application of the displacement-based FEM to plasticity problems. We recall from Chapter 10 that the FEM equations[z] are $\boldsymbol{f} = \boldsymbol{p}$, where $\boldsymbol{f}$ and $\boldsymbol{p}$ are columns containing the external and internal nodal loads, respectively. In the elastic case $\boldsymbol{p} = \boldsymbol{Kq}$ where $\boldsymbol{K}$ the (constant) stiffness matrix and $\boldsymbol{q}$ the column containing the nodal displacements. If an initial stress, say σ_0, is present the stress-strain relationship can be written as $\boldsymbol{\sigma} = \boldsymbol{C}^{(e)}\boldsymbol{\varepsilon}^{(e)} + \sigma_0$ where $\boldsymbol{C}^{(e)}$ is the matrix containing the elastic moduli. With the stress σ_0, a traction $\boldsymbol{t}_0 = \sigma_0\boldsymbol{n}$ and an initial strain $\varepsilon_0 = [\boldsymbol{C}^{(x)}]^{-1}\sigma_0$, where $\boldsymbol{C}^{(x)}$ is an appropriate modulus matrix, can be associated. Using the same formalism as before the presence of an initial stress leads to an extra column $\boldsymbol{f}_0$ given by

$$\boldsymbol{f}_0 = -\int \boldsymbol{N}^{\mathrm{T}}\boldsymbol{t}_0 \, dA = -\int \boldsymbol{N}^{\mathrm{T}}\sigma_0\boldsymbol{n} \, dA = -\int \nabla[\boldsymbol{N}^{\mathrm{T}}\sigma_0] dV = -\int \boldsymbol{B}^{\mathrm{T}}\sigma_0 dV$$

so that the FEM equations become $\boldsymbol{f} = \boldsymbol{p}$ with $\boldsymbol{p} = \boldsymbol{Kq} - \boldsymbol{f}_0$. The simplest way to solve a plastic problem is now to use the previously accumulated plastic strain $\boldsymbol{\varepsilon}^{(p)}$ as the initial strain. To that purpose we write the FEM equations in the incremental form

$$\Delta\boldsymbol{f} + \Delta\boldsymbol{f}_0 = \boldsymbol{K}^*\Delta\boldsymbol{q} \quad \text{with} \quad \Delta\boldsymbol{f}_0 = -\int \boldsymbol{B}^{\mathrm{T}}\boldsymbol{C}^{(ep)}\Delta\boldsymbol{\varepsilon}^{(p)} \, dV \quad \text{and} \quad \boldsymbol{K}^* = \int \boldsymbol{B}^{\mathrm{T}}\boldsymbol{C}^{(ep)}\boldsymbol{B} \, dV$$

with $\boldsymbol{C}^{(ep)}$ the matrix corresponding to the tensor $\mathbb{C}^{(ep)}$. Note that the stiffness matrix $\boldsymbol{K}^*$ now contain the matrix $\boldsymbol{C}^{(ep)}$ instead of $\boldsymbol{C}$ and changes during the process being dependent on $\mathbf{q}$. This set of equations can be solved by iteration using a two-step process if the external load is prescribed as follows:

---

[z] Neglecting for simplicity the constraint conditions.

- The first step is the *rate problem*. It is logical to initially assume that $\dot{\lambda}$ has the same sign at the current state as it had at the preceding state. From $\dot{q} = (K^*)^{-1}\dot{f}$ the generalised velocities are calculated and from these the strain rates $\dot{\varepsilon} = B^{\mathrm{T}}\dot{q}$. If the assumptions on the multipliers $\dot{\lambda}$ are satisfied, $\dot{\varepsilon}^{(p)}$ and $\dot{\sigma}$ can be calculated. If not, an implicit iteration scheme[aa] can be applied. Denoting the stiffness matrix based on the original assumption by $(K^*)_{(k)}$, where the subscript (k) indicates an iteration number, the solution of $\dot{q}_{(k+1)} = (K^*)^{-1}_{(k+1)}\dot{f}$ will results in a new set of values for $\dot{\lambda}$ and to a new stiffness matrix. The process is repeated until convergence is arrived.
- The second step is the *load-change problem*. Here it is logical to assume initially that $\Delta\varepsilon^{(p)} = 0$. The linear system $\Delta q_{(i)} = (K^*)^{-1}_{(i)}\Delta f$ is solved and from this the strain increment $\Delta\varepsilon_{(i)} = B^{\mathrm{T}}\Delta q_{(i)}$ is calculated. Next the associated $(\Delta\varepsilon^{(p)})_{(i)}$ and $\Delta\sigma_{(i)}$ using the constitutive equation are determined. In general the increments in the internal load vector $\Delta p_{(i)}$ resulting from $\Delta\sigma_{(i)}$ do not match Δf but if they do, this step is solved. If they do not match, the residual forces which are given by $\Delta r_{(i)} = \Delta p_{(i)} - (\Delta f + \Delta f_0)$ are added to Δf. An iterative scheme $\Delta q_{(i+1)} = \Delta q_{(i)} + (K^*)^{-1}_{(i)}\Delta r_{(i)}$ is now applied until convergence is obtained. Thereafter the whole procedure is repeated for the next load increment.

During plastic deformation, the stiffness matrix has in principle to be modified. In practice this may or may not be done at each step. In the *tangent-stiffness method* the stiffness matrix K^* is recalculated at each iteration providing a fast convergence at the expense of more extensive computations for each cycle. The computations per cycle are reduced in the *modified tangent-stiffness method*, where K^* is only recalculated for each load increase. Finally, in the *initial stiffness method* the elastic stiffness matrix is used throughout at the expense of slower convergence.

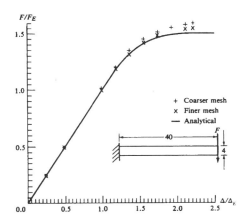

Fig. 14.17: FEM calculation of an end-loaded cantilever in plane stress.

[aa] If a function $dy/dt = A(y,t)$ has to be solved for an arbitrary time t, an explicit iteration scheme, with a subscript (k) indicating the iteration number, is given by $y_{(n+1)} = y_{(n)} + A(y_{(n)},t)\Delta t$ while an implicit scheme is given by $y_{(n+1)} = y_{(n)} + A(y_{(n+1)},t)\Delta t$. While the former equation can be solved directly, the latter must be usually solved by iteration since the function $A(y,t)$ is generally non-linear resulting in a more involved calculation. The advantage is that it is unconditionally stable.

As an example[bb] we show in Fig. 14.17 the bending of an end-loaded cantilever beam using elastic-ideal plastic behaviour in plane stress with $Y/E = 2 \cdot 10^{-3}$. A 40×40 and a 80×80 mesh using four-node quadrilateral elements have been used of which the influence on accuracy can be noticed. This problem still can be solved analytically and this solution is also shown. The overall agreement can be considered as fair.

The above description only barely touches on the possibilities and problems associated with displacement-based FEM. In spite of the intensive research in this area, there remain differences over such issues as the use of many low-order (e.g. constant strain) elements versus fewer higher-order elements and the use of local constitutive equations in rate form versus the use of variational inequalities in the derivations of the discrete equations. Apart from the classical displacement-based formulation also various other approaches have been proposed. Lubliner (1990) and Maugin (1992) have given a compact introduction while the monograph by Zienkiewicz[cc] is considered as a classic in this area.

14.6   Bibliography

Derby, B., Hills, D. and Ruiz, C. (1992), *Materials for engineering*, Longman Scientific & Technical, Harlow, UK.

Fischer-Cripps, A.C. (2000), *Introduction to contact mechanics*, Springer, Berlin.

Hill, R. (1950), *The mathematical theory of plasticity*, Oxford University Press, London.

Lubliner, J. (1990), *Plasticity theory*, McMillan, New York.

Martin, J.B. (1975), *Plasticity*, MIT Press, Cambridge MA.

Maugin, G.A. (1992), *The thermomechanics of plasticity and fracture*, Cambridge University Press, Cambridge.

Timoshenko, S.P. and Gere, J.M. (1973), *Mechanics of materials*, SI ed., Van Nostrand Reinhold Company, New York.

[bb] Simo, J.C. and Taylor, R.L. (1986), Int. J. Num. Methods Eng. **22**, 649.
[cc] Zienkiewicz, O.C. and Taylor, R.L. (1989), *The finite element method*, 4th ed., McGraw-Hill, London.

15

Dislocations

In Chapter 13 it was for the macroscopic flow criteria assumed that flow occurs if a certain critical shear stress is reached. For the Tresca criterion this is the maximum shear stress while for the von Mises criterion a kind of average shear stress, the octahedral shear stress, is used. In this chapter the atomic and molecular background of plasticity in crystalline materials, i.e. the existence and nature of dislocations, is discussed. In the following chapter their co-operative nature and interaction with other microstructural elements in single and polycrystalline materials will be introduced.

15.1 Slip in crystalline materials

A crystal will be deformed by the application of a shear stress. For small strains the behaviour will be elastic and reversible. This implies that upon unloading the atoms move to their original positions. With larger strains the behaviour becomes irreversible. *Slip* occurs which implies that one part of the crystal is sliding across the neighbouring part. The slip takes place along certain crystallographic directions, the *slip direction*, on certain crystallographic planes, the *slip planes*. Generally slip is a inhomogeneous phenomenon. It can take place on certain surfaces while nearby parallel surfaces do not slide, the results being that the surface of a crystal, originally smooth, becomes stepped. This step, the intersection of the slip plane with the surface of the crystal, is called a *slip line* and when clustered together a *slip band*. Fig. 15.1 shows how slip leads to elongation of the crystal and Fig. 15.2 slip bands in a Nb single crystal[a].

Slip is also anisotropic. It occurs more readily along certain directions and planes[b] than along others. In Fig. 15.1 the slip systems, i.e. a combination of slip direction and slip plane, for various lattice types are indicated. The slip direction is nearly always in the direction in which the atoms are most closely packed. Thus in FCC metals along the $\langle 110 \rangle$ directions, in BCC metals along the $\langle 111 \rangle$ directions and HCP crystals along the $\langle 11\bar{2}0 \rangle$ directions. Crystals of the rock salt type slip along $\langle 110 \rangle$. Slip is also often occurring on planes with dense packing. FCC metals slip on the $\{111\}$ planes

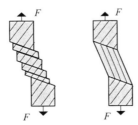

Fig. 15.1: Slip and twinning during tensile loading. The hatching indicates the crystal orientation.

[a] Reid, C.N. (1973), *Deformation geometry for materials scientists*, Pergamon, Oxford.
[b] For reminder: (*hkl*) specific plane, {*hkl*} set of planes, [*hkl*] specific direction, $\langle hkl \rangle$ set of directions.

Table 15.1: Slip systems for various lattices.

Structure	Slip plane	Slip directions	Number of systems	Example
FCC	{111}	⟨110⟩	4·3 = 12	Cu, Ni, Al
BCC	{110}	⟨111⟩	6·2 = 12	α-Fe, Mo, β-brass
	{112}	⟨111⟩	12·1 = 12	
	{123}	⟨111⟩	24·1 = 24	
HCP	{0001}	⟨$2\bar{1}\bar{1}0$⟩	1·3 = 3	Cd, Zn, Mg, Ti,Al$_2$O$_3$
	{$10\bar{1}1$}	⟨$2\bar{1}\bar{1}0$⟩	3·1 = 3	
NaCl	{110}	⟨110⟩	6·1 = 6	NaCl, KCl
	{001}	⟨110⟩	6·1 = 6	
CsCl	{110}	⟨001⟩	6·1 = 6	CsCl

although at high temperatures also slip on the {100} planes occurs. HCP metals slip on the {0001} planes and for some metals also on the {$10\bar{1}1$} and {$10\bar{1}2$} planes at high temperature. Rocksalt crystals slip on the {110} planes. BCC metals have no strongly preferred slip plane but show slip on the {112}, {110} and {123} planes which are all as dense as the FCC ⟨110⟩ planes. In Table 15.1 the number of available slip systems is indicated. For example, for the FCC lattice there are four slip planes in the set {111} and three slip directions ⟨110⟩ on each slip plane. Temperature is important and may change the preferential slip plane. For BCC metals below $T_m/4$, where T_m denotes the melting point, the {112} planes are preferred while between about $T_m/4$ and $T_m/2$ the {110} planes are preferred. At high temperature, say above 0.8 T_m, slip on the {123} planes is favourable. In BCC metals slip can occur at corrugated surfaces, often described as *pencil glide*. Other irregular glide surfaces may be realised by *cross-slip*. This usually occurs when two slip planes, which are parallel but not coincident, extend inwards from opposite sides of the crystal. In the centre, where the planes approach each other, slip bands crossing from one plane to another complete the slip surface (Fig. 15.3). The limited number of slip systems for HCP metals is largely responsible for the orientation dependence of yield and limited ductility.

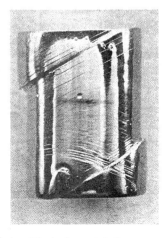

Fig. 15.2: Slip in a Nb single crystal after compressive loading.

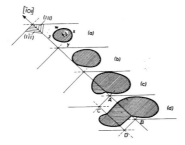

Fig. 15.3: Cross-slip from one 111 plane to another.

For slip to occur in a single crystal tested in tension a certain shear stress on the slip plane is required, usually referred to as the *resolved shear stress*. This stress corresponds to a strain, obviously denoted as the *resolved shear strain*. To estimate the resolved shear stress, consider a single crystal with cross-section A and loaded by a tensile force F (Fig. 15.4). The angle between the slip direction and tensile direction is λ while the angle between the slip plane normal and the tensile direction is ϕ. The angles λ and ϕ are complementary, i.e. $\lambda + \phi = \frac{1}{2}\pi$, only when the slip direction is in the plane defined by the stress axis and the normal to the slip plane. The area of the slip plane is $A/\cos\phi$, while the component of the axial force along the slip plane is $F\cos\lambda$. The *resolved shear stress* τ_R is the ratio of force and area and thus is

$$\tau_R = \frac{F\cos\lambda}{A/\cos\phi} = \frac{F}{A}\cos\lambda\,\cos\phi = \sigma\,\cos\lambda\,\cos\phi \qquad (15.1)$$

where the normal (Cauchy) stress σ is given by $\sigma = F/A$. The factor $m = \cos\lambda\,\cos\phi$ is usually called as *Schmid factor*. This factor is at its maximum at $\lambda = \phi = 45°$ for which $\tau_R = F/2A = \sigma/2$. When the tensile force is perpendicular to the slip plane ($\lambda = 90°$) or parallel to the slip plane ($\phi = 90°$), slip does not occur. In this case single crystals will fracture rather than deform plastically. This is most evident in HCP crystals like Zn where only the basal planes readily show slip. In FCC and BCC crystals slip may be still introduced at larger strains. The minimum resolved shear stress necessary for slip to occur is referred to as *critically resolved shear stress*. In Table 15.2 the value of the critically resolved shear stress for several metals is

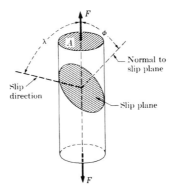

Fig. 15.4: Diagram for the calculation of the 'critically resolved shear stress'.

Table 15.2: The critically resolved shear stress for various metals.

Metal	Crystal type	Purity	Slip plane	Slip direction	τ_R (MPa)
Zn	HCP	99.999	(0001)	$[2\bar{1}\bar{1}0]$	0.18
Mg	HCP	99.996	(0001)	$[2\bar{1}\bar{1}0]$	0.77
Ti	HCP	99.99	(10\bar{1}0)	$[2\bar{1}\bar{1}0]$	14
		99.9	(10\bar{1}0)	$[2\bar{1}\bar{1}0]$	92
Ag	FCC	99.99	(111)	[110]	0.48
		99.97	(111)	[110]	0.73
		99.93	(111)	[110]	1.31
Cu	FCC	99.999	(111)	[110]	0.65
		99.98	(111)	[110]	0.94
Ni	BCC	99.8	(111)	[110]	5.8
Fe	BCC	99.96	(110)	[111]	28
			(112)	[111]	
			(123)	[111]	
Mo	BCC	–	(110)	[111]	50

presented. Obviously large differences in τ_R can occur with slip system and different degrees of purity for the same metal. Not only impurities play a role but also alloying. For example, for pure Cu and Au the critical resolved shear stress $\tau_R \cong 0.6$ MPa and $\tau_R \cong 1.0$ MPa, respectively. For the Cu$_{50}$Au$_{50}$ alloy $\tau_R \cong 5.0$ MPa, in spite of the identical crystal structure and nearly the same atomic radius of Au and Cu.

The calculation of the *resolved (engineering) shear strain* γ_R for a single crystal under tensile loading is more complex and is presented without derivation[c]

$$\gamma_R = \left(\cos\phi_0\right)^{-1}\left\{\left[\left(L/L_0\right)^2 - \sin^2\lambda_0\right]^{1/2} - \cos\lambda_0\right\} \tag{15.2}$$

where ϕ_0 and λ_0 are the value of ϕ and λ at the onset of slip and L_0 and L the length before and after slip, respectively. In deriving this formula only one slip system was assumed to be active. The shear strain thus can be determined from the initial orientation of the slip plane and the elongation of the crystal. If the sample is tested in compression the resolved shear strain γ_R is given by the solution of

$$\left(L/L_0\right)^2 = 1 + 2\gamma_R \cos\varphi_0 \cos\lambda_0 + \gamma_R^2 \cos^2\lambda_0$$

In the stress-strain diagram for a single crystal τ_R and γ_R are used as a measure for the stress and strain, respectively.

In case insufficient slip possibilities are present a metal can also show *twinning*. A deformation twin is a region of a crystalline body, which has undergone a homogeneous shape deformation in such a way that the resulting structure is identical to the original one, apart from a different orientation. The most apparent difference

Table 15.3: Twin planes and directions.

Structure	Example	Plane	Direction
BCC	α-Fe, Ta	(112)	[111]
HCP	Zn, Cd, Mg	(10\bar{1}2)	$[\bar{1}011]$
FCC	Ag, Au, Cu	(111)	[112]

[c] Schmid, E. and Boas, W. (1950), *Plasticity of crystals*, F.A. Hughes & Co., London.

with slip is the change in shape (Fig. 15.1). Twinning also occurs in specific directions on specific planes (Table 15.3). The strains produced by twinning are small so that the overall deformation is limited. The importance is much more that twinning can yield a more favourable orientation for slip. We omit further discussion.

Erich Schmid (1896-1983)

Born in Bruck an der Mur, Austria, he studied mathematics and physics at the University of Vienna receiving his doctorate on the thesis entitled *Über Brown'sche Bewegung in Gasen* (1920). He became an assistant at the Technische Hochschule in Vienna, continuing working on the Brownian motion. In 1922 he was invited to join the Kaiser Wilhelm-Institut für Faserstoffchemie, where he worked with Hermann Mark and Michael Polanyi, publishing together in the same year *Vorgänge bei der Dehnung von Zink-Kristallen*. While Mark and Polanyi rapidly changed from subject, Schmid continued this research increasingly using physical methods. In 1928 he became head of the Institut für Metallforschung where he met Walter Boas with whom he published over 50 papers that made him famous. In 1932 he became a member of the "Vorstand" of the Physikalische Institut of the University of Freiburg. Here he wrote, together with Boas who followed him, the famous book *Kristallplastizität* (1935). It is of interest that an English translation of the book was published in 1950, without the knowledge or approval of the authors, and was reissued without change in 1968. From 1936-1951 he worked as head of the metallurgy laboratory of the Metallgesellschaft AG (Freiburg) on alternative alloys, e.g. Zn and Pb alloys instead of the expensive and rare Sn alloys, for bearings, resulting in the book *Gleitlager* (1953), co-authored by Richard Weber. After the war he moved to Hanau where he reconstructed the fully destroyed research laboratory of the company. In 1951 he returned to Austria where he became full professor and member of the "Vorstand" of the University of Vienna, extending his research on plasticity with various techniques, e.g. with ultrasonic methods and field emission microscopy. He was the one of the first German-speaking scientist studying radiation damage of materials leading to the book *Werkstoffe des Reaktorbaues*, co-authored by Karl Lintner. In 1963 he became President of the Austrian Academy of Sciences and during the next 10 years he was instrumental in the foundation of 12 research institutes.

Problem 15.1

For 4 Mg single crystals of the same diameter but different orientations the yield strength in tension (0.2% strain) is determined. With X-ray diffraction the orientation of the various crystals, as expressed in the angles λ and φ is determined. The data are compiled in Table 15.4.

a) Can the differences in yield strength be attributed to impurities?

b) What is the critically resolved shear stress?

Table 15.4: Yield strength and orientation for 4 Mg single crystals.

Crystal	Y (MPa)	φ (°)	λ (°)
1	20.0	45	54
2	23.0	30	66
3	40.0	60	66
4	100	70	76

15.2   Theoretical shear strength

If we assume that slip occurs via sliding of a crystal plane over a neighbouring plane, a simple estimate, due to Frenkel[d], can be made for the theoretical shear stress necessary for slip. To this purpose we consider two atomic planes with distance a (Fig. 15.5). The distance between the atoms in the plane is b while the shear distance (displacement) is denoted by x. During shear the shear stress τ will change periodically and in first approximation can be described by a sine function with amplitude τ_M and wavelength b

$$\tau = \tau_M \sin\frac{2\pi\,x}{b} \qquad \text{or for } x/b \ll 1, \qquad \tau = \tau_M \frac{2\pi\,x}{b} \tag{15.3}$$

Further for small values of the shear strain $\gamma = x/a$ Hooke's law, $\tau = G\gamma$, holds where G denotes the shear modulus. Combination yields an estimate for the theoretical shear stress τ_M

$$\blacktriangleright \qquad \tau_M = \frac{b}{a}\frac{G}{2\pi} \cong \frac{G}{2\pi} \tag{15.4}$$

since the lattice plane distance a and the atom distance b are approximately equal. The shear modulus for metals is in the order of 50 to 500 GPa and thus a theoretical shear stress of about 10 to 100 GPa is expected. Table 15.2 shows experimental values about 10^2 to 10^4 times smaller. The estimate for τ_M is thus orders of magnitude larger than the experimentally observed values for the critically resolved shear stress and a serious discrepancy is present. More elaborate estimates show that the theoretical shear stress will be somewhat lower as predicted by this simple model[e], say about

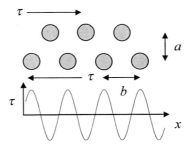

Fig. 15.5: The shear stress between two lattice planes as a function of displacement.

---

[d] Frenkel, J. (1926), Z. Phys. **37**, 572.

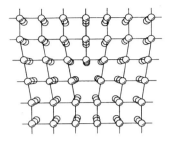

Fig. 15.6: Schematic of an edge dislocation in a simple cubic lattice.

$G/10$ to $G/30$. This improved estimate does not remove the discrepancy. This result is the major indication that some imperfections are present which have a significant influence on slip. These imperfections are the dislocations. Essentially their presence should be able to explain why the experimentally observed yield strength is low as compared to the theoretical yield strength and why slip bands on the surface occur.

15.3   Dislocations

Essential in the estimate of the theoretical shear stress was that the lattice planes moved over each other as a whole. Accepting that slip can occur only partially over a lattice plane leads to two types of imperfections, the edge and screw dislocation.

The idea of an edge dislocation occurring in crystals was at the same time independently proposed by Orowan, Polanyi[f] and Taylor in 1934[g]. A crystal containing an *edge dislocation* can be seen as a crystal cut half open and with an extra half plane inserted in the lattice cut (Fig. 15.6). The lower border of this half plane is the so-called *dislocation line*, at any point characterised by its unit tangent vector **l**. The area around the dislocation line is disturbed. In the area above the dislocation line the lattice is compressed while in the area below the dislocation line the lattice is

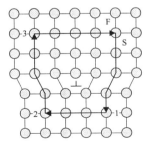
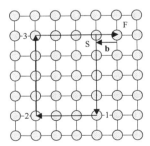

Fig. 15.7: The Burgers circuit in a defect lattice and in a perfect lattice. The dislocation line **l** points into the paper. The start and finish of the loop are indicated by S and F while the vector **b** represents the (true) Burgers vector.

---

[e] For an extensive discussion of the theoretical strength, see Kelly, A. and McMillan, N.H. (1986), *Strong solids*, 3rd ed., Clarendon, Oxford.

[f] Michael Polanyi (1891-1976). A Hungary-born scientist who worked mainly in the UK. He contributed to various topics in physics and chemical kinetics and later turned to philosophy in which he made also important contributions. He emphasised the role of "Personal knowledge" in his book with the same title (1958).

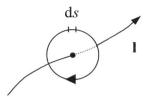

Fig. 15.8: Schematic of the contour integral around the dislocation line l.

under tensile stress. At large distance from the dislocation the lattice is (nearly) undeformed both above and below the dislocation line. An extra lattice plane above the glide plane is often indicated as a *positive dislocation*. Obviously when the extra half plane is below the glide plane one speaks of a *negative dislocation*. An edge dislocation is indicated symbolically by $\perp$ or $\top$ dependent on whether it is positive or negative. The horizontal line in the symbol indicates the slip plane while the vertical line indicates the extra half plane.

A characteristic quantity for the disturbance is the *Burgers vector* **b** (after J.M. Burgers[h]). This vector can be obtained by making a closed, clockwise loop around the dislocation line in the defected crystal (d) looking along its positive direction using an arbitrary number of lattice steps in the x, y and z directions and in the opposite direction, as indicated in Fig. 15.7. If the same loop is made in the perfect crystal (p) one does not return to the original lattice point. The step to close the loop in the perfect lattice is the (true) Burgers vector. This definition is due to Frank[i] and referred to as the (p-)FS/RH (perfect lattice Finish-Start/Right-Hand) convention. It should be noted that various authors use different conventions with respect to **b**. In particular the sign of **b** might be reversed by interchanging one of the members of the pairs (FS, SF), (LH, RH) and (p, d)[j]. The p-FS/RH (or actually the d-SF/RH) convention coincides with the continuum convention of the (local) Burgers vector **b** as given by

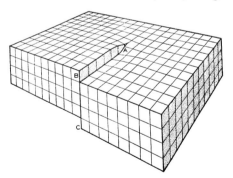

Fig. 15.9: Schematic of a screw dislocation in a simple cubic lattice. The ABC plane is the slip plane. The area ABCD has slipped and AD represents the dislocation line.

[g] Orowan, E. (1934), Z. Phys. **89**, 605; Polanyi, M. (1934), Z. Phys. **89**, 660; Taylor, G.I. (1934), Proc. Roy. Soc. (London) **A145**, 363. See also Hirth and Lothe (1990) and Nabarro (1967).
[h] Burgers, J.M. (1939), Proc. Kon. Ned. Akad. Wetenschap **42**, 293 and **42**, 378.
[i] Frank, F.C. (1951), Phil. Mag. **42**, 1014.
[j] For example, the d-SF/RH convention the sign is the same. However, the circuit is taken in the defected lattice. In this case the shorter the Burgers circuit, the more distorted the lattice in the defected crystal. As a consequence, unless the circuit is taken far away from the position of the dislocation line the *local* Burgers vector as derived from this circuit cannot exactly coincide with the *true* vector in the p-FS/RH convention.

the contour integral (Fig. 15.8)

$$\mathbf{b} = \oint \frac{\partial u}{\partial s} \, \mathrm{d}s$$

where u is the displacement and $\mathrm{d}s$ an element of the contour. For a pure edge dislocation the Burgers vector $\mathbf{b}$ and dislocation line vector $\mathbf{l}$ are perpendicular, e.g. $\mathbf{b} \cdot \mathbf{l} = 0$. The slip plane in this case is spanned by $\mathbf{b}$ and $\mathbf{l}$ and characterised by its normal $\mathbf{g} = \mathbf{b} \times \mathbf{l}$.

Burgers extended the idea of an edge dislocation and introduced the screw dislocation. A crystal containing a (right-handed) *screw dislocation* can be seen as a crystal cut half open and glued together after sliding the cut faces over each other (Fig. 15.9) resulting in a pure shear deformation. In this imperfection the Burgers vector $\mathbf{b}$ is parallel to the dislocation line vector $\mathbf{l}$, e.g. $\mathbf{b} \cdot \mathbf{l} = \|\mathbf{b}\| = b$. The name screw dislocation is due to the fact that the imperfection can be seen as a screw in the lattice (Fig. 15.9). Again it applies that the slip plane contains both the Burgers vector $\mathbf{b}$ and the dislocation line vector $\mathbf{l}$. However, since the Burgers vector and the dislocation line are parallel for a screw dislocation ($\mathbf{g} = \mathbf{b} \times \mathbf{l} = \mathbf{0}$), it is relatively easy for a screw dislocation to move to another slip plane also containing the dislocation line. This is the phenomenon called *cross-slip* and provides a mechanism for screw dislocations to move around an obstacle.

A similar mechanism is not possible for edge dislocations since in that case the dislocation line is perpendicular to the Burgers vector. Therefore edge dislocations change primarily of slip plane by *climb*, a process of diffusion of vacancies (Fig. 15.10) or of interstitials, which is strongly temperature dependent. Since essentially only individual vacancies can move, climb occurs over a short segment of a dislocation line, resulting in the formation of small steps or *jogs*. Climb thus proceeds with formation and motion of jogs and the activation energy is the sum of the formation energy of a jog and the activation energy for self-diffusion. This kind of motion has been called *non-conservative* contrary to glide, which preserves the slip plane. Steps in an edge dislocation line in the glide plane are referred to as *kinks* and they play a role in slip over the glide plane.

For both the edge and screw dislocation it holds that during motion only a few atoms move at the same moment and that the force required doing so is small resulting in a low yield strength as compared to the theoretical yield strength. Moreover, both result in a step at the surface as shown in Fig. 15.11. If sufficient

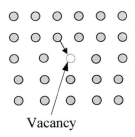

Vacancy

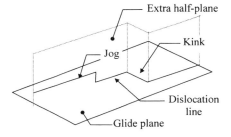

Fig. 15.10: Climb of edge dislocations (left) and jogs and kinks (right). The open circle indicates a vacancy.

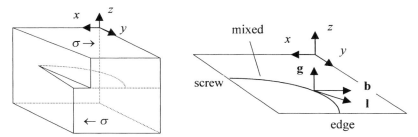

Fig. 15.11: Crystal with a mixed dislocation (left). The direction of the dislocation line vector at a certain position is the tangential vector to the dislocation line (right).

dislocations are present a slip band will arise. It can be expected that the slip band is not a single step but a conglomerate of many small steps since it is unlikely that all dislocations move at the same slip plane. This has been confirmed experimentally.

Edge and screw dislocations are the extremes. Usually a dislocation line is curved and a so-called mixed nature is present, as illustrated in Fig. 15.11. If the crystallographic configuration, apart from a translation, is restored after the dislocation has moved, the dislocation is called a *unit dislocation*. Otherwise one speaks of a *partial dislocation*. If we denote the glide plane, as is usual with a plane, with its normal vector **g** and the dislocation line vector, as before, with **l**, both types of dislocations can be characterised by (**g**)[**l**].

The Burgers vector is the most important invariant characteristic of a dislocation: it is constant along a dislocation line and remains constant if the dislocation moves.

Justification 15.1

Suppose that a dislocation ABCD has a part ABC with the Burgers vector $\mathbf{b}_1$ and another part CDA with the Burgers vector $\mathbf{b}_2$. Then ABCD encloses two regions, which differ in the way they have slipped. By definition they must be separated by a 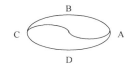 dislocation, e.g. CA with the Burgers vector $\mathbf{b}_3 = \pm(\mathbf{b}_1 - \mathbf{b}_2)$. If ABCD is a single dislocation not joined by others, then $\mathbf{b}_3$ must be zero and $\mathbf{b}_1 = \mathbf{b}_2$. A similar argument shows that the Burgers vector is constant when the dislocation moves.

Because the Burgers vector is constant along a dislocation line, a single

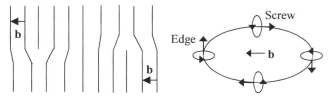

Fig. 15.12: A dislocation loop in a crystal. Schematic of the structure (left) and character as a function of position (right).

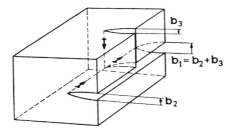

Fig. 15.13: Splitting of a dislocation.

dislocation cannot end in the interior of a crystal. Three possibilities exist:

- The dislocations ends at the surface or at a grain boundary, e.g. as illustrated in Fig. 15.11.
- The dislocation line forms a closed loop. Since the Burgers vector is constant along the loop, the dislocation at two positions must be of a pure edge and screw nature (Fig. 15.12). In two opposing sectors of the loop the dislocation line vector is thus parallel, respectively anti-parallel to the Burgers vector. These parts of the loop differ essentially from each other. Examining the position of the loop at the pure edge dislocation, one observes the half plane in one case above the slip plane and in the other case below the slip plane.
- The dislocation splits in two or more, so-called partial dislocations. If $\mathbf{b}_1$ denotes the Burgers vector of the original dislocation and $\mathbf{b}_2$ and $\mathbf{b}_3$ those of the partial dislocations, conservation of the Burgers vector dictates that $\mathbf{b}_1 = \mathbf{b}_2 + \mathbf{b}_3$. This splitting is energetically advantageous under certain circumstances, i.e. $U(\mathbf{b}_1) > U(\mathbf{b}_2)+U(\mathbf{b}_3)$, where U is the energy of the dislocation, as will be discussed later. A schematic of the realisation of such a splitting is presented in Fig. 15.13.

Apart from the loop just indicated another type of loop is possible. This loop is essentially a part of a crystal plane, either extra or missing (Fig. 15.14). They are referred to as extrinsic and intrinsic *prismatic loops*, respectively. Their character is essentially different from the (regular) loop as sketched in Fig. 15.12: it is always pure edge in character.

Many aspects of dislocations are known and an extensive literature exists. We only mention here the textbooks involved. Well-known classics are Cottrell (1953), Read 1953) and Nabarro (1967). The standard treatise is Hirth and Lothe (1968, 1982). A nice introduction is provided by Weertman and Weertman (1964) while a modern introduction is given by Hull and Bacon (2001). Kovacs and Szoldos (1973) provide a wider overview.

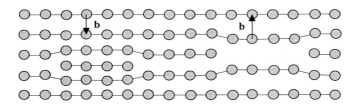

Fig. 15.14: An extrinsic (left) and intrinsic (right) prismatic loop.

Geoffrey Ingram Taylor (1886-1975)

Born in St John's Wood, London, England, he was educated at Trinity College, Cambridge, originally in mathematics but later moved to physics. He won a scholarship to undertake research at Trinity College, where his theoretical study of shock waves won him a Smith's Prize. In 1910 he was elected to a Fellowship at Trinity College and the following year he was appointed to a meteorology post where his work on turbulence in the atmosphere led to his publication *Turbulent motion in fluids* which won the Adams Prize at Cambridge in 1915. At the outbreak of World War I he was sent to Farnborough where he worked on the stress on propeller shafts, which led him to think about the limiting strengths of materials. In 1918 he returned to Cambridge where he worked on the application of turbulent flow to oceanography and on the problem of bodies passing through a rotating fluid. In 1923 he was appointed to a Royal Society research professorship. This enabled him to stop teaching, which he had been doing for the previous four years but he was not much interested in teaching and not a natural lecturer. His investigations in the mechanics of fluids and solids covered an extraordinary wide range, and most of them exhibited the originality and insight. During World War II Taylor again worked on applications of his expertise to military problems such as the propagation of blast waves, studying both waves in the air and underwater explosions. In 1944 he was knighted and appointed to the Order of Merit in 1969. He continued his research after the end of the War and retired in 1952 but he continued his work at Cambridge with little evidence that his status had in any way changed until 1972. In that year he suffered a stroke from which he only partially recovered. He received honorary degrees from more than a dozen universities throughout the world and over 20 Medals for his outstanding contributions to applied mathematics. He published over 250 papers on applied mathematics, mathematical physics and engineering.

Problem 15.2

Sketch the structure of an (100)[010] edge dislocation in NaCl. Why is this dislocation unlikely and is a (110)[001] dislocation much more likely?

Problem 15.3

Check that two dislocations of opposite sign, moving on the same lattice plane, can annihilate each other to result in a perfect lattice. What is the result if these two dislocations move on two different lattice planes?

15.4 Overview of effects

In the previous section dislocations are introduced and their presence explains the basic phenomena associated with slip. However, in plasticity the route from the micro-level (the dislocation) to the meso-level (the yield strength and hardening behaviour) is long and occupied with many problems, solved partially or not at all, making the description of plasticity in terms of atomic phenomena complex. This stands in strong contrast to the elasticity problem where knowledge at the micro-level (second derivative of the potential) leads almost directly to knowledge at the meso-level (the elastic constants). This complexity warrants a survey of the many aspects of dislocations on this route. In this section we do so.

The first aspect to consider is energy. From the presence of the internal strain it is clear that with dislocations excess energy is associated. In principle this energy can be calculated, either quantum-mechanically or by the pair potential method. In both cases, however, a complex calculation results. Therefore, researchers have been using continuum (isotropic) elasticity theory to make a simple model and using this model to estimate the *strain energy* U_{ela} of a dislocation. This estimate results in $U_{ela} \cong \frac{1}{2}\alpha Gb^2 h$ with α a factor of order unity, G the shear modulus, $b = \|\mathbf{b}\|$ the length of the Burgers vector and h the length of the dislocation line (see Section 15.6). For isotropic crystals it holds that for edge dislocations $\alpha = 1/(1-\nu)$, while for screw dislocations $\alpha = 1$. It is obvious that dislocations with a small Burgers vector are relatively easy to produce. Nevertheless the elastic energy of dislocations is rather high. Taking as an example Cu with $G \cong 50$ GPa, $\nu = \frac{1}{3}$ and $b \cong 0.25$ nm results in about 2.3×10^{-9} J/m for an edge dislocation. This corresponds to about 3.7 eV per atom in the dislocation line or 350 kJ/mol, to be compared with the heat of melting which is about 13 kJ/mol. During the formation of point defects the loss in energy is compensated by the gain in configurational entropy, making the free energy change negative upon point defect formation. That is why point defects are a thermodynamic necessity. For dislocations, however, the entropy gain is small and can be neglected as compared with the energy loss and their elastic energy is nearly their free energy (for some further details, see Chapter 8 as well as Cottrell, 1953). Thermodynamically dislocations are thus not stable. Nevertheless, they are nearly always present as grown-in defects.

Second, let us consider the stress field of a dislocation. A screw dislocation only produces a shear stress field with mean stress zero. This implies that screw dislocations with the same sign repel each other while screw dislocations with a different sign attract each other. A positive edge dislocation, on the other hand, produces a compressive stress field above the glide plane and a tensile stress below the glide plane. This makes the interaction between edge dislocations not only dependent on their sign but also on their orientation.

Third, let us now consider some crystallographic aspects of dislocations. Already mentioned is the fact that different crystallographic structures posses different slip systems and that the preferred slip system may change as a function of temperature. Although not discussed, it is also clear that the precise atomic arrangement of a dislocation in different crystal structures is different. Since the energy of a dislocation per unit length is $\frac{1}{2}\alpha Gb^2$ it is favourable to split a (unit) dislocation in two other (partial) dislocations by a so-called *dislocation reaction* if the total elastic energy decreases in this process. For example, the reaction $\mathbf{b}_1 \rightarrow \mathbf{b}_2 + \mathbf{b}_3$ is energetically favourable if $U(\mathbf{b}_1) > U(\mathbf{b}_2) + U(\mathbf{b}_3)$. In addition to the condition $\mathbf{b}_1 = \mathbf{b}_2 + \mathbf{b}_3$ has to be satisfied, of course. As an example consider the FCC crystal with slip system $\{111\}\langle110\rangle$. A possible reaction is

$$a[01\bar{1}]/2 \rightarrow a[\bar{1}2\bar{1}]/6 + a[11\bar{2}]/6$$

The Burgers vector squared of the $[01\bar{1}]$ dislocation is $a^2/2$ while their sum for the $[\bar{1}2\bar{1}]$ and $[11\bar{2}]$ dislocations is $a^2/3$. Energy is thus released making the splitting

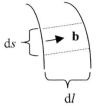

ds

b

d*l*

Fig. 15.15: Force on a dislocation.

favourable. However, by the formation of these partial dislocations also a stacking fault is formed for which energy is required. Equilibrium is obtained at the distance where the two energy contributions cancel. This equilibrium configuration of two partials is referred to as an *extended dislocation*. If a newly formed dislocation does not belong to the primary slip system, the new dislocation is no longer mobile and one speaks of a *sessile* dislocation. On the other hand if the newly formed dislocation does belong to the primary slip system, the dislocation is mobile and called *glissile*. Similar but different reactions are possible in the BCC and HCP lattices.

Since the ultimate goal is to estimate the σ-ε curve, we need to consider the effect of an applied force. If an external shear stress τ is exerted on a crystal, the dislocation experiences a force. The following oversimplified argument provides the magnitude (Fig. 15.15). Consider an element of a dislocation line in a homogeneous shear stress. If a dislocation line element of length ds moves over a distance dl the area swept is dA = dsdl. If A represents the total area of the crystal glide plane, the displacement is du = b dA/A, while the total applied external force $F = \tau A$. The work done dW is given by d$W = F$ d$u = \tau b$ d$A = \tau b$ dsdl or per unit length of dislocation by d$w = \tau b$ dl. Let the normal force on the dislocation line element be f ds, where f denotes the normal force per unit length on the dislocation element due to the externally applied shear stress τ. Then

$$f = \mathrm{d}w/\mathrm{d}l = \tau b$$

This force is perpendicular to the dislocation line and directed towards the unslipped part of the crystal. Since the Burgers vector is the same for all elements of the dislocation, the force has the same value everywhere along the dislocation line.

Due to the fact that dislocations will try to minimise their (free) energy, they experience a *line tension*, comparable to the surface tension of liquids, or, perhaps more to the point, a tension in a rope. For a single dislocation in an unloaded crystal this means that the dislocation line tends to minimise its total length. As an example, pure edge and pure screw dislocations tend to become straight and closed dislocation

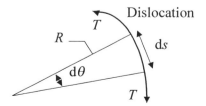

Dislocation

T

R

d*s*

dθ

T

Fig. 15.16: Line tension of a dislocation.

loops tend to form a circle, to contract and disappear (unless other factors prevent this). As a consequence a certain shear stress τ is required to maintain a certain radius of curvature R (Fig. 15.16). A dislocation line element ds can be described by $ds = Rd\theta$ and this element experiences an outward force $\tau b\,ds$. Due to the line tension T an inward force $2T \sin d\theta/2 \cong T\,d\theta$ is also present. In equilibrium the two forces are equal and this results in

$$\tau = \frac{T}{bR}$$

Since the tension is equal to the energy per unit length $T = \partial U_{\mathrm{ela}}/\partial h = \frac{1}{2}\alpha Gb^2$, the shear stress τ necessary to maintain a radius R is given by $\tau = \alpha Gb/2R$.

After having indicated some of the aspects dealing with a single dislocation, we realise that generally many dislocations are present. The amount of dislocations is characterised by the *dislocation density*, i.e. the ratio $\rho = L/V$, where L denotes the total length of dislocation lines in a volume V. It can be shown that ρ can be estimated as $\rho = 2n/A$, twice the number of dislocation lines n that protrude an area A. In good crystals the dislocation density can be as low as $\rho \cong 100$ cm^{-2} while in strongly deformed crystals the dislocation density can rise to $\rho \cong 10^{11}$ to 10^{12} cm^{-2}. By annealing the dislocation density decreases to $\rho \cong 10^6$ to 10^8 cm^{-2}. A typical value is thus[k] $\rho \cong 10^{10}$ cm^{-2}.

If we have a crystal with $\rho \cong 10^{10}$ cm^{-2} and $b \cong 0.3$ nm and if we assume that all dislocations move in the same direction, the maximum permanent strain is approximately $\rho^{1/2}b \cong (10^{10})^{1/2} \times 30 \times 10^{-9} = 3 \times 10^{-3} = 0.3\%$. During deformation, however, much larger values of strain are observed and thus there must be a constant multiplication resulting in new dislocations. Several mechanisms are known of which the *Frank-Read* source, as discussed in Section 15.5, is most frequently operative.

The next aspect to consider is the interaction of dislocations with each other and with other features in the crystal. In fact these interactions are due to the energy minimisation. We first consider the interaction of edge dislocations with each other.

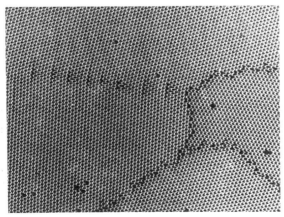

Fig. 15.17: Soap-bubble analogue of microstructural features including vacancies, di-vacancies, low- and high-angle grain boundaries. Note that the low-angle grain boundary (boundary in the upper left half) can be considered as a head-on superposition of edge dislocations.

---

[k] Realise that this means 10^{10} cm/cm^3 = 10^5 km/cm^3!

The interaction energy is due to one dislocation in the stress field of the other. Putting an edge dislocation in the proper neighbourhood of another, one can relieve strain resulting in attraction. A particularly important stable configuration is the head-to-tail configuration, which forms the basis for small-angle grain boundaries (Fig. 15.17[1]). Obviously for other configurations, repulsion might occur as well. Since screw dislocations possess a pure shear stress field, they always attract each other if they have an opposite sign and repel each other if of like sign. All these interactions form the basis of *strain hardening*. During deformation more and more dislocations are formed in the crystal, they entangle and after some relaxation form a *network*. A network is characterised by a more or less ordered configuration of connected dislocations intertwined with relatively dislocation-free areas (see Fig. 15.23). Dislocation motion through this network is increasingly difficult with increasing dislocation density and the yield strength increases. The yield strength is thus dependent on the size of the network mesh and leads to the *Hall-Petch relation*, as discussed in Chapter 16. Grain boundaries provide a similar barrier for dislocations. Surfaces may provide another barrier. Since a dislocation can annihilate at a free surface, in principle a dislocation is attracted towards a free surface because annihilation lowers the energy. However, the surface of many metals is covered with a thin oxide film with a high elastic modulus, which leads to repulsion.

Dislocations also interact with point defects. Since a point defect as a first approximation produces only a hydrostatic stress, there is no (or very limited) interaction between screw dislocations and point defects. On the contrary a hydrostatic stress is associated with the edge dislocation so that point defects do interact with edge dislocations. Obviously, vacancies are attracted to the compressive stress side and substitutional/interstitial atoms are attracted to the tensile stress side.

The same is also true for substitutional atoms. Positioning a point defect with radius smaller than the host lattice in the compressive region may relieve strain. Similarly for a point defect with a larger radius than the host lattice, strain is relieved by a position in the tensile region. This leads to clustering of defects around a dislocation, the so-called *Cottrell cloud*. The result is that around the dislocation line a local energy valley is created. A dislocation provided with a Cottrell cloud requires a higher shear stress to move it, because moving the line outside the cloud entails motion against an energy barrier. Macroscopically this results in higher yield strength. By applying a sufficiently high shear stress, dislocations can breakaway from this cluster of point defects resulting in a lower shear stress to move it further and in this way leading to the yield point phenomenon, as mentioned in Chapter 2. If sufficient point defects are present, dislocation motion can be severely restricted, again leading to hardening. This typically can be achieved by realising a solid solution (remember the Cu-Au alloy) and this hardening mechanism is called *solid-solution hardening*. This mechanism is strongly dependent on the concentration of point defects, interstitial atoms being much more effective then substitutional atoms. Since screw dislocations posses a pure shear stress field, interaction with spherically symmetric point defects (typically substitutional atoms), having only a hydrostatic stress, is absent. If the point defect has asymmetric point symmetry (typically an interstitial atom), interaction is still possible. Some further details are discussed in Chapter 16.

Finally, the presence of precipitates plays a role. Cooling down a saturated solid solution so that an over-saturated solution arises generally results in precipitates.

[1] Smith, C.S. (1950), Metal Progress, October, page 479.

Since precipitates generally have a different elastic modulus and specific volume as compared to the host lattice, dislocations may be hindered or blocked by precipitates. This leads to *precipitation hardening*, which is dependent on the number of precipitates but also strongly on many other factors. We just mention: size, shape, volume fraction, distribution, strength, ductility, strain hardening behaviour of the precipitates, crystallographic misfit and interfacial/surface energy of the precipitates. Maximum hardening occurs for a certain size and number of precipitates. Since precipitates grow in time (Ostwald ripening) and the volume fraction at constant temperature is constant, generally with increasing annealing time first an increase in yield strength is observed later followed by a decrease. In the decreasing regime one speaks of *over-ageing*. These matters are discussed in Chapter 16.

This concludes our overview of dislocation effects. The character of the analysis during this route from micro- to meso-level changes from rather exact to semi-empirical, as will become clear by studying the various topics involved.

Problem 15.4

A piece of Cu is work-hardened. Annealing in a calorimeter shows that 12.6×10^6 J/m^3 was stored in the specimen. Estimate the dislocation line density before annealing given that $G = 48.3$ GPa and $b = 0.255$ nm.

Problem 15.5

Consider an FCC metal containing two dislocations. The first has a Burgers vector $\mathbf{b}_1 = \frac{1}{2}a[101]$ and a dislocation line characterised by $\mathbf{l}_1 = [1\overline{2}\overline{1}]$. For the second dislocation $\mathbf{b}_2 = \frac{1}{2}a[01\overline{1}]$ and $\mathbf{l}_2 = [\overline{2}11]$. Estimate, when these two dislocations meet, whether a new dislocation will be formed and, if yes, whether it is mobile.

Fig. 15.18: Spiral growth on a polyoxymethylene crystal due to two screw dislocations of opposite nature.

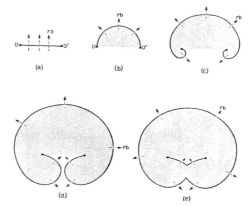

Fig. 15.19: Dislocation multiplication by the Frank-Read mechanism.

15.5 Formation, multiplication and observation of dislocations

Although dislocations are not thermodynamically required, they are always present. This implies that they are present either as a result of the processing procedure or nucleate during thermo-mechanical loading.

Dislocations can nucleate spontaneously by thermal fluctuations in a specimen experiencing a mechanical stress. However, normally the stress-temperature combination is unfavourable to do so, i.e. either a shear approaching the theoretical shear strength or a far too high temperature is required. It is possible though to order dislocations by stress, e.g. during bending thereby producing a regular pattern.

Dislocations often originate during crystal growth. In fact screw dislocations provide a mechanism for crystal growth by providing a step in the crystal surface at which growth is easy. This step rotates during the growth (Fig. 15.18[m]) so that further growth can take place easily.

In the previous section we have noticed that for an average dislocation density a maximum strain of about 0.3% can be reached. In practice much larger values of strain are observed and thus there must be a constant multiplication of new dislocations. The simplest mechanism is the *Frank-Read mechanism*. Fig. 15.19 shows the slip plane for a dislocation line which is assumed to be pinned at the

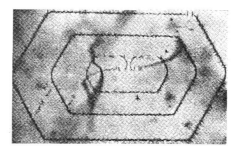

Fig. 15.20: An example of dislocation loops in single crystal Si. Note the hexagonal shape of the loops, due to the anisotropy of the crystal lattice.

---

[m] Reneker, D.H. and Geil, P.H. (1960), J. Appl. Phys. **31**, 1916.

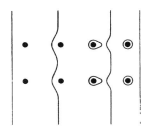

Fig. 15.21: The generation of Orowan loops.

positions D and D'. Under the influence of the applied shear stress the dislocation bows to a loop of which the ends approach each other. At contact the two parts of the loop annihilate each other locally since the character at the ends is opposite (Fig. 15.19). The result is a loop and new line. The energy of the loop becomes minimal if the loop becomes a circle. The energy of the dislocation line is minimised by straightening the line. After that the whole process repeats itself and the two pinning points act as a (double ended) dislocation source. A critical shear stress exists since a loop will only form if the shear stress is sufficiently high to bow the initially straight line to an approximately semi-circular arc. This critical stress is of the order

$$\tau = \alpha Gb/l$$

where l is the distance between D and D'. A similar mechanism can also operate with a single pinning point in which case a spiral lying in the planes of slip results. According to both mechanisms a single dislocation can produce a large amount of slip. Fig. 15.20[n] provides a particularly clear example of a Frank-Read process in single crystal Si.

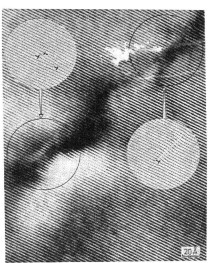

Fig. 15.22: TEM observation of dislocations on (111) in a Ge single crystal. The inserts show a schematic of the dislocated area.

[n] Dash, W.C. (1957), in *Dislocations and mechanical properties of crystals*, J.C. Fisher, W.G. Johnston, R. Thomson and T. Vreeland, eds., Wiley, New York.

Also inclusions can lead to an increase in dislocation density. Fig. 15.21 shows a slip plane with two inclusions. If a dislocation crosses the inclusions, the dislocation is retarded at the inclusion. Similarly as in the Frank-Read mechanism the parts of the dislocation line that bow around the inclusions join to form a loop while the original line is restored. The loops become larger through the action of the applied stress and when they meet a next pair of inclusions, bow out and join again, thereby effectively increasing the dislocation density. These loops are known as *Orowan loops*.

Dislocations can be observed in various ways. A direct observation is possible by transmission electron microscopy (TEM) using either the image or diffraction mode. An example of the micrograph using the image mode is given in the next chapter. In Fig. 15.22[o] a TEM micrograph is given which shows the images of the (111) lattice planes with a spacing of 0.327 nm in a crystal of Ge. Several dislocations can be observed. A more indirect way is by means of the observation of growth on a crystal surface. As indicated before, growth of a crystal takes preferentially place at a dislocation line and results in a growth spiral. This growth spiral can be observed by either electron or optical microscopy. Still another method is by etching. In the neighbourhood of the dislocation line the crystal has a higher energy. At this spot the crystal etches faster than elsewhere and an etch pit arises. The surface of the crystal can be observed by optical microscopy (see Fig. 15.23[p], left) and the dislocation becomes visible through the etch pit. If a certain stress is applied to the crystal for some time and the crystal is etched again, the shifted dislocations show up as new pits while original pits have been etched a bit deeper. From this information the velocity as a function of stress can be determined. Finally, one can take advantage of the fact that impurities segregate along the dislocation line. Using transparent crystals one can observe these dislocations lines directly by optical microscopy. The technique is known as *decoration* and has revealed a great deal of information on dislocation networks (see Fig. 15.23[q], right).

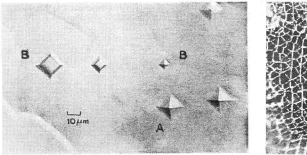 

Fig. 15.23: Left: Etch pits on LiF produced by dislocations. After three times etching dislocation A has not moved while dislocation B has moved from left to right. Right: Decoration of the dislocation network in KCl by Ag particles examined in an optical microscope.

[o] Philips, V.A. and Hugo, J.A. (1976), Acta Metall. **18**, 123.
[p] Gilman, J.J. and Johnston, W.G. (1957), page116 in *Dislocations and mechanical properties of crystals*, J.C. Fisher, W.G. Johnston, R. Thomson and T. Vreeland, eds., Wiley, New York.
[q] Amelinckx, S. (1958), Acta Metall. **6**, 34.

Problem 15.6

Consider the simplest possible nucleation configuration, a dislocation loop with radius r. Use the strain energy and work done to obtain the total energy expression. Show that the required energy for nucleation $U_{cri} = \frac{1}{2}\alpha Gb^2 r_{cri}$ with the critical radius $r_{cri} = \frac{1}{2}\alpha Gb/\tau$ and τ the shear stress. Show that neither thermal nucleation (use the probability of nucleating a loop with Boltzmann's equation) nor mechanical nucleation (compare the stress necessary with the theoretical shear strength) of dislocations is likely.

15.6   Stress and energy

Around a dislocation a stress field is present. From Fig. 15.6 it is clear that above an edge dislocation the lattice is compressed while below an edge dislocation the lattice is tensile loaded. The strain energy can be estimated easily once the stress distribution is known. A simple elastic model for the screw dislocation is given in Fig. 15.24. At some distance from the centre of the dislocation ($r > r_0$) we assume that continuum elasticity theory is valid. In this case we only need the deformation along the dislocation line (z-direction). The shear strain is given by

$$\gamma = 2\varepsilon \cong \tan\gamma = b/2\pi r \tag{15.5}$$

since for small deformations $\gamma \cong \tan\gamma$. As before $b = \|\mathbf{b}\|$ indicates the length of the Burgers vector. Furthermore, we use Hooke's law

$$\tau = G\gamma \tag{15.6}$$

where G denotes the shear modulus and τ the shear stress. From the above equations it follows that

$$\tau = \frac{Gb}{2\pi r} \tag{15.7}$$

The elastic strain energy density is given by

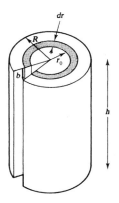

Fig. 15.24: Elastic model of the deformation in a screw dislocation.

$$w = \frac{\mathrm{d}U_{\mathrm{ela}}}{\mathrm{d}V} = \tfrac{1}{2}\tau\gamma = \frac{Gb^2}{2(2\pi r)^2} \tag{15.8}$$

For a cylinder with thickness $\mathrm{d}r$ and length h, hence of volume $\mathrm{d}V = 2\pi h r\,\mathrm{d}r$, the energy becomes

$$\mathrm{d}U_{\mathrm{ela}} = \frac{G}{2}\left(\frac{b}{2\pi r}\right)^2 2\pi r h\,\mathrm{d}r = \frac{Gb^2 h}{4\pi}\frac{\mathrm{d}r}{r} \quad \text{and} \quad U_{\mathrm{ela}} = \int_{r_0}^{R}\mathrm{d}U_{\mathrm{ela}} \tag{15.9}$$

The integral over $\mathrm{d}U_{\mathrm{ela}}$ diverges if $r_0 \to 0$ and if $R \to \infty$. Therefore for this model to be applicable a proper choice has to be made for both r_0 and R.

Let us make first an estimate for lower limit r_0. The domain for which the deformation can be described by continuum elasticity theory starts at about 4 to 5 times the Burgers vector. A logical choice is thus $r_0 = 4b$ or $r_0 = 5b$. The interior of the dislocation, the *core*, cannot be described by continuum elasticity. Using detailed models an estimate of the core energy can be made and this results in 7% to 15% of the elastic energy. A reasonable way to incorporate this contribution is to use a lower limit $r_0 = b$.

For an estimate of the upper limit we need the dislocation density of which a typical value is $\rho \cong 10^{10}$ cm^{-2}. Since mobile dislocations will take positions with net resolved shear stress close to zero and the stress of a dislocation decreases rapidly with increasing distance, the average distance between dislocations $\rho^{-1/2}$ is often used for an estimate of the upper limit R. For a typical Burgers vector of length $b = 0.2$ nm and the typical value $\rho^{-1/2} = 10^{-5}$ cm the upper limit is thus estimated as $R \cong 500b$. The total energy thus becomes

$$U_{\mathrm{screw}} = \int_{r_0}^{R}\mathrm{d}U_{\mathrm{ela}} = \int_{r_0}^{R}\frac{Gb^2 h}{4\pi}\frac{\mathrm{d}r}{r} = \frac{Gb^2 h}{4\pi}\ln r\Big|_{r=r_0=b}^{r=R=500b} \cong \frac{Gb^2 h}{2} \tag{15.10}$$

For an edge dislocation a somewhat more complicated calculation yields

$$U_{\mathrm{edge}} = \frac{Gb^2 h}{4\pi(1-\nu)}\ln\frac{R}{r_0} \tag{15.11}$$

where ν, as usual, denotes Poisson's ratio. For a mixed dislocation, representing a screw and an edge dislocation with the Burgers vector $b_{\mathrm{screw}} = b\cos\gamma$ and $b_{\mathrm{edge}} = b\sin\gamma$ superimposed on each other, the strain energy becomes the sum of two contributions

$$\blacktriangleright \qquad U_{\mathrm{mixed}} = \frac{Gb^2 h}{4\pi}\frac{(1-\nu\cos^2\gamma)}{(1-\nu)}\ln\frac{R}{r_0} \tag{15.12}$$

where γ is the angle between the tangent to the dislocation line and the Burgers vector. The average of the orientation factor $(1-\nu\cos^2\gamma)$ is $\tfrac{1}{2}(1-\tfrac{1}{2}\nu)$ so that on average for $\nu = \tfrac{1}{3}$ the factor $(1-\nu\cos^2\gamma)/(1-\nu) = 0.63$.

The estimate for the elastic energy $\alpha Gb^2/2$ as made in Section 15.4 is now easily understood. Just note that the energy is dependent on the geometry through the factor $\ln(R/r_0)$ and that taking that factor as $\ln(500b/b)$ yields for the total numerical factor $(4\pi)^{-1}\ln(R/r_0) \cong 0.50$. Since the dependence of U on $R = \rho^{-1/2}$ is weak, the energy is only weakly dependent on the presence of other dislocations and usually taken as constant per unit length. Therefore, one often takes the specific energy $u = U/h$ as proportional to $\alpha Gb^2/2$, where α is a constant of the order of one ($\alpha = 1$ for a screw and $\alpha = 1/(1-v)$ for an edge dislocation). Also note that the energy is by no means concentrated at the centre of the dislocation. For $R/r_0 = 500b/b$, half of the energy is in the region between about $22b$ and $500b$.

Finally let us return briefly to the line tension T, given by $T = \partial U_{ela}/\partial h$ where h denotes length of the dislocation line. Obviously in the present model the line tension is dependent on the orientation by its character (screw or edge), on the dislocation density and on the curvature of the dislocations. Accepting the approximation $U_{ela} = \alpha Gb^2 h/2$, the line tension becomes $T = \alpha Gb^2/2 = U_{ela}/h = u$.

Problem 15.7*

The length of the dislocation can also be increased by changing the angle γ with respect to the Burgers vector **b**. Show that for small changes of γ this leads to $T = u + \partial^2 u/\partial \gamma^2$ by expanding the energy in a Taylor series in γ.

15.7 Dislocation motion

As the Frenkel model has shown, in a perfect crystal a high stress is required to produce plastic deformation. This is due to the fact that a whole plane of atoms must move over an energetically unfavourable position. In the presence of dislocations this is no longer the case. In Fig. 15.25 the influence of an edge dislocation is shown. For slip only one row of atoms has to move at the time and only a small local displacement is required. In fact in the presence of kinks only a few atoms have to move at the same moment (see Fig. 15.8). If a kink moves all along the dislocation line, the net effect is that the dislocation has moved from one equilibrium position on the glide plane to the next. In this way the dislocation moves as a whole entity through the crystal under the influence of a shear stress. In the following we discuss subsequently ideal dislocation motion and displacement via kink motion.

In dislocation motion the width of a dislocation plays an important role as can be seen as follows. A dislocation can be interpreted as the boundary on a glide plane, moving in time, between a slipped and an unslipped area and during motion of a

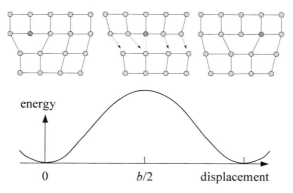

Fig. 15.25: Dislocation motion. Upper part indicating schematically the movement of the atoms (one darker for reference) and the lower part the associated potential energy profile.

dislocation the slipped region grows at the expense of the unslipped region[r]. The larger the width of that boundary, the lower the elastic strain energy because then the atomic spacing in the dislocation is close to the ideal (equilibrium) atomic spacing. On the other hand, with atoms in that boundary a high energy transition energy is associated and, obviously, the smaller the width, the lower the transition energy. The equilibrium width is thus determined by the balance between these two opposing energy changes and can be obtained at the minimum of the sum of these contributions. The result is that the larger the width, the lower the shear stress required to move the dislocation.

The presence of just a few line defects in perfect single crystalline material lowers the critical shear stress considerably. This critical shear stress at low concentration of defects is usually called the *Peierls*[s] *stress*. It can be considered as the shear stress to move a dislocation over a distance b from one equilibrium position to another on the glide plane in an otherwise dislocation free crystal. The Peierls stress τ_P strongly depends on the *width* $w = 2\xi$ of a dislocation on the slip plane (Justification 15.2)

$$\tau_P = \frac{2G}{(1-v)} \exp\left(\frac{-4\pi\xi}{b}\right) = \frac{2G}{(1-v)} \exp\left(\frac{-2\pi a}{b(1-v)}\right) \quad \text{with} \quad \xi = \frac{a}{2(1-v)} \quad (15.13)$$

and a the lattice spacing perpendicular to the glide plane (Fig. 15.26). For a lattice with $a = b$ this results for $v = \frac{1}{3}$ in $\tau_P = 2.4 \times 10^{-4} G$. This number is somewhat larger than the lowest observed yield strength value ($\cong 10^{-5} G$). However, generally for slip on closed-packed planes $a > b$. Taking $a = b\sqrt{2}$ results already in $\tau_P = 4.9 \times 10^{-6} G$. Moreover, the experimental values are strongly dependent on the impurity content so that an intrinsic value has not been reached. Therefore, we expect that τ_P is significantly smaller than the smallest observed value. On the other hand for $a < b$ significantly higher stresses result indicating that slip occurs preferentially on close-packed planes. Hence, although the Peierls model is highly sensitive to the actual a/b ratio and approximate, it clearly shows that slip most easily occurs at widely spaced (a large), densely packed (b small) planes with wide ($a/(1-v)$ large) dislocations. Nowadays more realistic models are available, a few of which of briefly indicated in Justification 15.2.

Justification 15.2: Peierls equation and somewhat beyond*

In this justification block we outline the derivation and some related concepts of Eq. (15.13), following Cottrell (1953) and Hirth and Lothe (1968)[t]. Suppose that we have two crystals, A and B, shifted along the x-axis by $\frac{1}{2}b$ where b is the lattice constant, and join the faces together (Fig. 15.26). The smaller crystal (B) is extended while the larger one (A) is compressed in this operation, thus forming a positive edge dislocation. There are two contributions to the forces acting on the atoms in the surface of crystal A: forces from A itself which try

---

[r] Cottrell, A.H. (1967), *An introduction to metallurgy*, Edward Arnold, London.
[s] Rudoph Peierls (1907-1995). German physicist who did fundamental work during the early years of quantum mechanics. He also studied the physics of lattice vibrations (phonons) and developed the concept or the electron "hole" in solid-state theory. After moving to England, he concluded with Frisch that a ^{235}U atomic bomb would be feasible and worked on the American Manhattan Project.
[t] The original paper is by R.E. Peierls (1940), Proc. Phys. Soc. **52**, 23 which was extended by F.R.N. Nabarro (1947), Proc. Phys. Soc. **59**, 256.

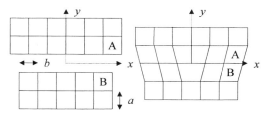

Fig. 15.26: A positive edge dislocation formed by joining two crystals.

to distribute the compression as even as possible, i.e. make the dislocation wider, and forces from crystal B which try to bring the atoms in alignment with those of crystal B, i.e. make the dislocation narrower. To reach equilibrium the atoms have to move slightly from their original position. Vertical movements are ignored and the displacement along the x-axis for the atoms in crystal A is indicated by $u(x)$. The atoms in crystal B are supposed to move by an opposite but equal amount. Two assumptions are made. First, the material above and below the joint can be represented as an elastic continuum with shear modulus G and Poisson's ratio v and its energy contribution is the elastic strain energy associated with this part of the crystal. Second, there is a misfit energy associated only with the distorted bonds across joint plane. A simple sinusoidal force-displacement relation, as used in the Frenkel model for the theoretical shear strength, is used.

By the first assumption this part of the problem has reduced to one of elasticity. Eshelby[u] provided a useful concept to deal with the problem in a sophisticated way. In his view a single dislocation of strength b is to be regarded as a continuous distribution of elastic dislocations of infinitesimal strength b', i.e. $\int_{-\infty}^{+\infty} b'(x)\,dx = b$. Using this concept the elastic strain energy for edge and screw dislocations for a length R appears to be

$$W_{e,e} = \frac{Gb^2}{4\pi(1-v)} \ln\frac{R}{\xi} + \frac{Gb^2}{8\pi(1-v)} \quad \text{and} \quad W_{s,e} = \frac{Gb^2}{4\pi} \ln\frac{R}{\xi} \quad (15.14)$$

respectively, with $\xi = a/[2(1-v)]$. The misfit energy for edge and screw dislocations appears to be

$$W_{e,m} = \frac{Gb^2}{4\pi(1-v)} \quad \text{and} \quad W_{s,m} = \frac{Gb^2}{4\pi} \quad (15.15)$$

The total energy is then given by the sum of Eqs. (15.14) and (15.15). Note the similarity between the elastic energy expression and the expression as derived from a fully elastic calculation

$$W_e = \frac{Gb^2}{4\pi(1-v)} \ln\frac{R}{r_0} \quad \text{and} \quad W_s = \frac{Gb^2}{4\pi} \ln\frac{R}{r_0}$$

These expressions $W_{e,e} + W_{e,m}$ ($W_{s,e} + W_{s,m}$) can be matched to W_e (W_s) by taking

---

[u] Eshelby, J.D. (1949), Phil. Mag. **40**, 903.

$$r_0 = \frac{b}{\gamma} = \frac{2\xi}{e^{3/2}} = \frac{a}{e^{3/2}(1-v)} \qquad \text{and} \qquad r_0 = \frac{b}{\gamma} = \frac{a}{e}$$

for edge and screw dislocations, respectively, and where $e = 2.718\ldots$ denotes the Napier base of logarithms and γ a constant. For example, for an FCC crystal with a $\frac{1}{2}\{110\}$ dislocation on the (111) slip plane we have for a screw dislocation $\gamma = 3.3$ and for an edge dislocation $\gamma = 6.5$ $(v = \frac{1}{3})$. Atomistic calculations indicate $\gamma \cong 2$ for close-packed structures and $\gamma \cong 4$ for covalently bonded materials. Hirth and Lothe (1968) simply recommend to use $\gamma \cong 4$ for non-metallic materials and $\gamma \cong 1$-2 for metals. This partially rationalises the choice $r_0 = b$.

The periodic variation due to the lattice discreteness is lost though in Eq. (15.15) for the misfit energy because it was calculated using a continuous function for the displacement. In fact this energy must be calculated by summing over the atoms in the glide surface. If that is done the result for a displacement $u(x) = \alpha b$, where α denotes a fraction of the Burgers length b, is

$$
\begin{aligned}
W(\alpha) &= \frac{Gb^2}{4\pi(1-v)} + \frac{Gb^2}{2\pi(1-v)}\exp\left(\frac{-4\pi\xi}{b}\right)\cos(4\pi\alpha) \\
&= \frac{Gb^2}{4\pi(1-v)} + \frac{W_P}{2}\cos(4\pi\alpha) \qquad \text{with} \qquad \xi = \frac{a}{2(1-v)}
\end{aligned}
\tag{15.16}
$$

The corresponding stress $\tau_P = -(1/b^2)dW(\alpha)/d\alpha$ is maximum at $\sin 4\pi\alpha = 1$ and is given by

$$\tau_P = \frac{2G}{(1-v)}\exp\left(\frac{-4\pi\xi}{b}\right) = \frac{2G}{(1-v)}\exp\left(\frac{-2\pi a}{b(1-v)}\right)$$

The range of x within which the displacement $u(x)$ is less than one-half of its limiting value is the width w, which appears to be 2ξ. For $v = \frac{1}{3}$, $w = 2\xi = 1.5a$. Consequently the model results in a relatively narrow dislocation. Further analysis showed that this is an effect of the force-displacement relation used. On the basis of other interaction laws[v] it has been concluded that w is given approximately by $w = Gb/2\pi(1-v)\tau_M$, where τ_M is the theoretical shear strength. For $\tau_M = Gb/2\pi a$, as used in the Frenkel model, this expression reduces to the previous one. However, for $\tau_M = G/30$ the width becomes fives times as large. The width is thus extremely sensitive to the assumed interaction law. Finally, it can be shown that for large distances the stress field of a Peierls dislocation reduces to that of a classic elastic dislocation.

Hirth and Lothe (1968) criticise this model because the symmetrical positions of the dislocations during motion are maxima while the position with the largest asymmetry at $\alpha = \frac{1}{4}$ corresponds to a minimum where the opposite is expected. They argue that is due to the physically unrealistic decrease in energy for a displacement larger the $b/2$. Using a modified interrow potential

$$W = \frac{Gb^3}{4\pi a}\left(1-\cos\frac{2\pi\phi}{b}\right) \quad \text{for} \quad \phi \leq \frac{b}{2} \quad \text{and} \quad W = \frac{Gb^3}{4\pi a} \quad \text{for} \quad \phi > \frac{b}{2}$$

[v] Foreman, A.J., Jaswon, M.A. and Wood, J.K. (1951), Proc. Phys. Soc. **64A**, 156.

they showed that the approximate result is

$$\Delta W = -\frac{Gb^3\alpha^2}{\pi^2 a} \sum_{n=1}^{n=\infty} \frac{[(\xi/b)^2 - (n - \tfrac{1}{4})^2](\xi/b)^2}{[(\xi/b)^2 + (n - \tfrac{1}{4})^2]^3}$$

with again the displacement $u(x) = \alpha b$. For wide dislocation $\xi \gg b$ the sum can taken as an integral which results in

$$\Delta W = -\frac{Gb^4\alpha^2}{8\pi\xi a}$$

With $\alpha = \tfrac{1}{4}$ the Peierls barrier becomes

$$W_P \cong Gb^4/400\xi a$$

to be compared with the expression for the original model, Eq. (15.16). For narrow dislocation the symmetrical position is the one with the lowest energy. When $\xi = \tfrac{3}{4}b$, the term with $n = 1$ vanishes and all other terms are negative making the energy positive. The result is

$$W_P \cong Gb^3\alpha^2/300a$$

With $\alpha = \tfrac{1}{4}$ this is much smaller then Eq. (15.16) and opposite in sign in agreement with the qualitative expectation.

Finally one should realise that these considerations all pertain to absolute zero. We mention only two effects. First, there will some degrees of freedom associated with the dislocation implying that some lattice modes will disappear. The differential heat capacity associated with the local dislocation modes will lower the Peierls barrier. Second, due to thermal activation double kinks can be created. This will make dislocation motion easier.

---

*Kink motion**

As indicated in the beginning of this section, dislocation motion can be further eased by the presence of kinks (Fig. 15.27) since less energy (and thus a lower stress) is necessary to move a kink than a complete dislocation. The energy of a kink depends on the value of the Peierls energy U_P and is minimum for a kink length $m = 0$. A balance between a straight dislocation (low dislocation energy U_{ela}) with $m \gg a$ and a highly kinked dislocation (low kink energy U_{kin}) with $m = 0$ is the result. Pre-existing

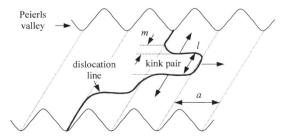

Fig. 15.27: Schematic of a double kink along a dislocation line.

kinks can move laterally at low applied stress to the nodes of the dislocation so that a more or less straight dislocation is the result if the subsequent nodes are lying predominantly in the same Peierls valley. As a result at 0 K at least a stress τ_P is required to move the dislocation. With increasing temperature *double-kink nucleation* may occur due to thermal fluctuations in which a 'positive' and 'negative' kink are formed. Hirth and Lothe (1968) show using elastic deformation modelling that the energy U_{dk} of a double kink is approximately given by

$$U_{dk} = 2U_{kink} + U_{int} = 2\left\{ \frac{Gb^2a}{4\pi(1-v)}\left[\ln\frac{\gamma a}{eb} - (1-v) \right] \right\} + \left\{ -\frac{Gb^2a^2}{8\pi l}\frac{(1+v)}{(1-v)} \right\}$$

where γ is the constant as used in Justification 15.2. The interaction energy between the two kinks is negative, hence two kinks attract each other and a double kink is not stable under low applied stress unless the length l (Fig. 15.27) is large, say $l > 20b$ and $l \gg m$. The activation energy for the thermal activation is the energy necessary for creating a double kink.

If the Peierls stress is exceeded, the dislocations can move. According to Gilman[w], their velocity v is dependent on the temperature and the shear stress τ and for some materials can be described (fully empirically) over a certain velocity range by

$$v = A\tau^n = A_0 \exp\left(-\frac{U_{act}}{kT} \right)\tau^n \tag{15.17}$$

where A_0 and n are material constants. For example, for the metal[x] Fe (3% Si) $n = 35$ for the range 10^{-7} to 10^{-2} cm/s at 293 K and $n = 44$ for the same range at 78 K. For the ionic compound LiF $n = 25$ for the range 10^{-7} to 10^{-1} cm/s at 293 K while for the semi-conductor Si $n = 1.5$ for the range 10^{-4} to 10^{-2} cm/s at 1074 K. The mobility increases with increasing temperature through the Arrhenius factor given by $\exp(-U_{act}/kT)$, where U_{act} is an activation energy and k and T have their usual meaning. The mobility of edge dislocations in the low velocity range is generally larger than the mobility of screw dislocations, typically by a factor of 50. Above ~10 cm/s the velocity increases more slowly since the maximum velocity of a dislocation is the shear wave velocity and a signal cannot move faster than the wave by which it is carried.

If the dislocation velocity v is controlled by kink motion (Hirth and Lothe, 1982) v is proportional to the kink velocity v_k, the distance over which dislocations move, i.e. the Peierls distance a and inversely to the width of the kink pair l. The latter is inversely proportional to the (linear) kink pair density n_k. Hence $v = (a/l)v_k = 2n_kv_ka$. The kink velocity v_k is the product of the driving force F, per unit length h given by $F/h = \tau b$, and the kink mobility D_k/kT where D_k is the kink diffusion coefficient. The kink pair density can be derived from 'mass action' law (neglecting entropy effects) to be $n_k = (2/m)\exp(-U_{kink}/kT)$ with m the kink width. At low stress kinks move and annihilate each other and nucleation replenishes them. The dislocation velocity becomes

$$v = (2\tau bha/m)(D_k/kT)\exp(-U_{kink}/kT) \cong 2\tau b^2(D_k/kT)\exp(-U_{kink}/kT) \tag{15.18}$$

[w] Johnston, W.G. and Gilman, J.H. (1959), J. Appl. Phys. **30**, 129.
[x] Stein, D.F. and Low, J.R. (1960), J. Appl. Phys. **31**, 362.

where the last step is made by approximating m, a and h by b. At larger stress, where the kinks are swept to the end of a segment L of the dislocation before they can be replenished, double kink nucleation becomes controlling. In this case $v = hXJ$, where $X = 2v_k t_k$ is the distance a kink pair separates before it is annihilated by another pair. Here t_k is the mean lifetime of a kink pair and J the nucleation rate of kink pairs given by $J = (\tau bh/m^2 kT)D_k\exp(-U_{dk}/kT)$. In steady state in the time t_k also a new kink pair must be generated and thus $t_k = 1/(\tfrac{1}{2}XJ)$ because $\tfrac{1}{2}X$ is the average length available for nucleation during the lifetime of a kink pair. Therefore $X = 2(v_k/J)^{1/2}$ and by combining with $v = hXJ$, we obtain $v = 2h(v_k J)^{1/2}$ which is approximately the same result as Eq. (15.18). If, however, the dislocation segment has a finite length L the correction factor $L/(L+X)$ should be introduced in the previous equation. When the segment is much shorter than the distance between the kinks, i.e. $L \ll X$, the final result becomes

$$v = (\tau bhaL/m^2)\,(D_k/kT)\,\exp(-U_{dk}/kT) \cong (\tau bL)\,(D_k/kT)\,\exp(-U_{dk}/kT)$$

where the last step is made by the same approximations as before. Note that the exponent is approximately twice as large as in Eq. (15.18).

15.8   Exact solutions*

We have seen that dislocations contain internal stress. It is possible to produce internal stress of single-valued and continuous nature in an elastic body if the body is *multiply connected*. A ring provides a simple example of a doubly connected body. In this case internal stress can be set up by cutting the ring, displacing the cut surfaces relative to each other and joining them again. In principle this operation produces *multiple-valued displacements*. This can be easily seen by considering two points, one on each of the cut faces, which are brought to coincidence when making the joint. Since they did not coincide before joining, the two points must have experienced different displacements. In general the displacement is multi-valued when, by starting from an arbitrary point in the ring and summing all the displacements along any continuous closed path, the sum is not zero. The sum is called the *cyclic constant* of cut. To avoid problems with multi-valued displacements one usually supposes that the displacement changes discontinuously when passing through the surface of the joint by an amount equal and opposite to the cyclic constant. This renders the displacement single-valued.

Volterra already in 1907 and Love in 1927 studied the influence of such 'cuts' in homogeneous isotropic elastic media. They showed that, if the stresses are to be

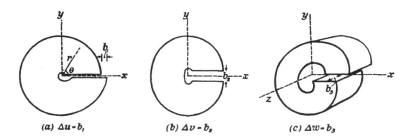

(a) Δu - b_1 (b) Δv - b_2 (c) Δw - b_3

Fig. 15.28: Volterra dislocations.

continuous, the surfaces of the ring must be displaced rigidly with respect to each other, implying certain conditions for the displacement vector **u**. The simplest condition appears to be **u** = constant, as illustrated in Fig. 15.28. These discontinuities precisely represent dislocations. Below we describe in some detail the stress field and strain energy of dislocations. Although real crystals are nearly always anisotropy we limit ourselves to isotropic media and refer to the literature (e.g. Hirth and Lothe, 1968, 1982) for dislocations in anisotropic media.

The edge dislocation

A more precise description of the stress field of an edge dislocation in an isotropic continuum can by obtained by using Airy's stress function Φ since the elastic representation of an edge dislocation exhibits plane deformation. The equation to solve reads

$$(\nabla^2)^2\Phi = 0 \quad \text{where} \quad \nabla^2 a = \frac{\partial^2 a}{\partial x^2} + \frac{\partial^2 a}{\partial y^2} = \frac{\partial^2 a}{\partial r^2} + \frac{1}{r}\frac{\partial a}{\partial r} + \frac{1}{r^2}\frac{\partial^2 a}{\partial \theta^2}$$

The stresses follow from

$$\sigma_{xx} = \frac{\partial^2 \Phi}{\partial y^2} \qquad \sigma_{yy} = \frac{\partial^2 \Phi}{\partial x^2} \qquad \sigma_{xy} = -\frac{\partial^2 \Phi}{\partial x \partial y} \qquad \text{or}$$

$$\sigma_{rr} = \frac{1}{r}\frac{\partial \Phi}{\partial r} + \frac{1}{r^2}\frac{\partial^2 \Phi}{\partial \theta^2} \qquad \sigma_{\theta\theta} = \frac{\partial^2 \Phi}{\partial r^2} \qquad \sigma_{r\theta} = -\frac{\partial}{\partial r}\left(\frac{1}{r}\frac{\partial \Phi}{\partial \theta}\right)$$

Using cylindrical co-ordinates we suppose $\Phi(r,\theta) = R(r)\Theta(\theta)$ where $R(r)$ is a function of r only while $\Theta(\theta)$ is a function of θ only. It appears that for a positive edge dislocation one should take

$$\Phi_0 = -Dr\ln r\,\sin\theta = -Dy\ln\sqrt{x^2+y^2} \qquad \text{with} \qquad D = \frac{Gb}{2\pi(1-v)}$$

resulting in an increase of **b** for **u** in one complete closed circuit. The stresses are

$$\sigma_{xx} = -D\frac{y(3x^2+y^2)}{(x^2+y^2)^2} \qquad \sigma_{yy} = D\frac{y(x^2-y^2)}{(x^2+y^2)^2} \qquad \sigma_{xy} = D\frac{x(x^2-y^2)}{(x^2+y^2)^2} \qquad (15.19)$$

▶ $$\sigma_{rr} = \sigma_{\theta\theta} = -D\frac{\sin\theta}{r} \qquad \sigma_{r\theta} = -D\frac{\cos\theta}{r} \qquad (15.20)$$

The stress $\sigma_{zz} = v(\sigma_{xx}+\sigma_{yy}) = v(\sigma_{rr}+\sigma_{\theta\theta})$ results from the plane strain condition. The largest normal stress is σ_{xx}, which is compressive above and tensile below the slip plane ($y = 0$). The shear is maximal at the slip plane changing sign with x. This solution matches the intuitive expectations based on Fig. 15.6.

Since the stresses decrease with $1/r$, they become only zero at infinity. The solution is thus appropriate for an infinite cylinder and in order to have a finite a stress at the surface of a finite cylinder we have to add another function Φ_1 to compensate for this. It appears to be of the form $\Phi_1 = Ar^3\sin\theta$, where A is determined by the condition $\sigma_{rr} = 0$ at the surface of the finite cylinder, $r = R$. Although the resulting additional stresses compensate for the stress at the surface, their contributions Dr/R^2 are small in the bulk of the cylinder. A compensation for $\sigma_{\theta\theta}$ is not required because they act in the surface of the cylinder.

The stresses diverge for $r = 0$. If we want to realise a stress-free inner cut-off radius we have to add another stress function $\Phi_2 = Br^{-1}\cos\theta$, where B is determined by the condition $\sigma_{rr} = 0$ at the inner surface of the cylinder, $r = r_0$. The associated stresses are of the order of magnitude Dr_0^2/r^3. Similar to the outside correction, the contribution in the bulk of the cylinder is small. Summarizing, the stress field of an edge dislocation in an isotropic medium can be adequately described by Eqs. (15.19) or (15.20).

The screw dislocation

A similar but simpler operation can be done for a screw dislocation. The displacement w is constant in the z-direction and zero for the other directions. Therefore only shear stresses in the x-direction arise are given by

$$\sigma_{\theta z} = \frac{G}{r}\frac{\partial w}{\partial \theta} \qquad \sigma_{rz} = G\frac{\partial w}{\partial r} \tag{15.21}$$

The equilibrium conditions $\sigma_{ij,j} = 0$ lead to

$$\nabla^2 w = 0$$

The solution should give an increase of b in the displacement w for an increase of 2π in θ. The solution, representing a right-handed screw, is simply

$$w = \frac{b\theta}{2\pi} = \frac{b}{2\pi}\tan^{-1}\frac{y}{x}$$

The associated stresses are

$$\blacktriangleright \qquad \sigma_{xz} = -\frac{Gb}{2\pi}\frac{y}{(x^2+y^2)} \qquad \sigma_{yz} = \frac{Gb}{2\pi}\frac{x}{(x^2+y^2)} \qquad \text{or} \qquad \sigma_{\theta z} = \frac{Gb}{2\pi r} \tag{15.22}$$

As expected this stress field is pure shear.

The solution represents again the stress in a cylinder. Because the stresses σ_{rr} and $\sigma_{r\theta}$ are zero at the (inner and outer) surfaces of the cylinder, the solution represents the case of a cylinder of which the surfaces are free. However, $\sigma_{z\theta}$ acts on the end faces (z = constant) producing a couple on them with moment

$$\int_{r_0}^{R}(r\sigma_{z\theta})2\pi r\ dr = \tfrac{1}{2}Gb(R^2 - r_0^2)$$

This couple has to be balanced by a field $\sigma_{z\theta} = -G\tau r$ to give single-valued displacements. Choosing

$$\tau = \frac{b}{\pi(R^2 + r_0^2)}$$

this field gives a couple that is equal and opposite to the one produced by the internal stress. Since these stresses are of the order of magnitude Gbr/R^2, they can be safely neglected with respect to those of Eq. (15.22).

The strain energy

Although the strain energy can be calculated straightforward from the stresses and strains, this is generally a tedious exercise and a simpler method often can be used.

The method uses the geometric representation of a dislocation and considers the amount of work applied when making a cut in the body and joining it back together in a displaced way. This work W enters the body and becomes the strain energy U.

If the position dependent surface forces or traction t to obtain a final displacement u is given, the total work done is

$$W = \tfrac{1}{2} \int_S t \cdot u \, dS$$

The factor $\tfrac{1}{2}$ results because the forces start at zero ending at their final value t during the creation of the displacement u over the surface S. Assuming that the forces at the surfaces $r = R$ and $r = r_0$ are zero, the integral has to be calculated over the cut faces only. For both the edge and screw dislocation only a shear stress exists on this plane, $\tau = \sigma_{\theta r}$ and $\tau = \sigma_{\theta z}$ respectively, acting in the direction in which the plane is moved. Per unit thickness in the z-direction the shear force on an element of the cut faces dr is τdr while each plane moves over a distance $\tfrac{1}{2}b$. The strain energy thus becomes $\tfrac{1}{2}\int b\,\tau dr$. Using Eqs. (15.20), taking into account that $\cos\theta = 1$ along the slip direction, the strain energy U_{edge} for an edge dislocation becomes

$$U_{\text{edge}} = \tfrac{1}{2} \int_{r_0}^{R} \frac{Db}{r} \, dr = \frac{Gb^2}{4\pi(1-v)} \ln\frac{R}{r_0}$$

Taking not only Φ_0 but $\Phi_0 + \Phi_1 + \Phi_2$ one obtains

$$U_{\text{edge}} = \frac{Gb^2}{4\pi(1-v)}\left[\ln\frac{R}{r_0} - \frac{(R^2 - r_0^2)}{(R^2 + r_0^2)} \right] \cong \frac{Gb^2}{4\pi(1-v)}\left[\ln\frac{R}{r_0} - 1 \right]$$

For a screw dislocation one obtains similarly

$$U_{\text{screw}} = \tfrac{1}{2} \int_{r_0}^{R} \frac{Gb}{2\pi r} \, dr = \frac{Gb^2}{4\pi} \ln\frac{R}{r_0}$$

or taking into account the balancing couple

$$U_{\text{screw}} = \frac{Gb^2}{4\pi}\left[\ln\frac{R}{r_0} - \frac{(R^2 - r_0^2)}{(R^2 + r_0^2)} \right] \cong \frac{Gb^2}{4\pi}\left[\ln\frac{R}{r_0} - 1 \right]$$

Also for the energy the contributions resulting from the surface corrections are entirely negligible. As discussed, the factor $\ln(R/r_0)$ is ill determined anyway.

15.9 Interactions of dislocations*

In Section 15.4, we briefly discussed the effect of an external force on a dislocation, their mutual interaction and the interaction with other crystallographic features. Here we reconsider the matter more generally.

Force by an external stress

Suppose we have an unconstrained body with no dislocation in it. A traction t_1 on that body produces a single valued displacement u_1 and results in a strain energy in that body given by

$$U_1 = \tfrac{1}{2} \int_S \mathbf{t}_1 \cdot \mathbf{u}_1 \, dS$$

where the integral is over the external surface S. If we now form a dislocation we must apply a traction $\mathbf{t}_1$ on the cut faces of area A to restore the configuration of the body and add a traction $\mathbf{t}_2$ to produce the rigid displacement $\mathbf{u}_2$ of the cut faces. The strain energy of the dislocation is

$$U_2 = \tfrac{1}{2} \int_A \mathbf{t}_2 \cdot \mathbf{u}_2 \, dA$$

For the total work on the body we must also account for the work done by the forces $\mathbf{t}_1$ when the displacement $\mathbf{u}_2$ occurs. The traction $\mathbf{t}_1$ does work over the external surface S as well as the cut surfaces A. The total strain energy thus becomes

$$U = U_1 + U_2 + \int_S \mathbf{t}_1 \cdot \mathbf{u}_2 \, dS + \int_A \mathbf{t}_1 \cdot \mathbf{u}_2 \, dA \qquad (15.23)$$

where the factor $\tfrac{1}{2}$ in the integrals is absent because the force $\mathbf{t}_1$ remains constant as the displacement $\mathbf{u}_2$ occurs. Introducing the dislocation first and then applying the traction $\mathbf{t}_1$ one can find an alternative expression given by

$$U = U_1 + U_2 + \int_S \mathbf{t}_2 \cdot \mathbf{u}_1 \, dS + \int_A \mathbf{t}_2 \cdot \mathbf{u}_1 \, dA$$

However, in this case the integral over S is zero because $\mathbf{t}_2$ is zero on S. The integral over A is also zero because $\mathbf{t}_2$ is equal and opposite on both sides of the cut while $\mathbf{u}_1$ is single-valued, which means that the work gained on one face is the work lost on the other. The strain energy thus reduces to

$$U = U_1 + U_2 \qquad (15.24)$$

independent of the position of the dislocation. Physically this implies that the elastic properties of the material are unchanged by the presence of the dislocation and unless the dislocation moves, we cannot infer its presence by measuring the materials response. Therefore the force on the dislocation must be entirely due to the external work done when the dislocation moves given by $W = \int \mathbf{t}_1 \cdot \mathbf{u}_2 \, dS$ and this results according to Eqs. (15.23) and (15.24) in

$$W = \left(\int_S \mathbf{t}_1 \cdot \mathbf{u}_2 \, dS \right) = U_{\mathrm{dis}} = \left(- \int_A \mathbf{t}_1 \cdot \mathbf{u}_2 \, dA \right)$$

The specific dislocation energy $u_{\mathrm{dis}} = U_{\mathrm{dis}}/l$, where l is the length of the dislocation line. If the dislocation line vector is in the y-direction, the force per unit length on the dislocation is $f = -\partial u_{\mathrm{dis}}/\partial x = -\partial U_{\mathrm{dis}}/\partial A = \mathbf{t}_1 \cdot \mathbf{u}_2$. For a dislocation with Burgers vector length b and a homogeneously applied shear stress τ to the slip plane, we regain $f = \tau b$.

Forces between dislocations

The force between two dislocations can also be obtained by the previously employed method. We suppose that one dislocation is present in a body, free of tractions at the external surface S. We then introduce another dislocation. The total energy is then sum of the self-energies of dislocations 1 and 2 and the interaction energy. The self-energy is the energy a dislocation would have when it existed alone in the body. The interaction energy is the work done by the traction $\mathbf{t}_1$ produced on the faces A by dislocation 1 acting through the displacement $\mathbf{u}_2$ produced on A during the formation of dislocation 2. A force results if the interaction energy varies with the relative position of the two dislocations.

Therefore consider a body with a positive edge dislocation with self-energy U_1 along the z-axis through the origin. Now we introduce a second dislocation with self-energy U_2 at position x,y, also along the z-direction, by applying a displacement $\mathbf{u}_2$ along a cut parallel to the x-axis. The total strain energy of these two parallel edge dislocations is

$$U = U_1 + U_2 + \int_A \mathbf{t}_1 \cdot \mathbf{u}_2 \, \mathrm{d}A = U_1 + U_2 + U_{\mathrm{int}}$$

Since $\mathbf{u}_2$ has only a component b along the y-direction, the only component of $\mathbf{t}_1$ to consider is σ_{xy}, the shear stress caused by the other dislocation. Both $\mathbf{t}_1$ and $\mathbf{u}_2$ change sign going from one face of the cut to another so that $\mathbf{t}_1 \cdot \mathbf{u}_2$ has to same sign throughout. The interaction energy for the configuration described is thus $U_{\mathrm{int}} = \int \sigma_{xy} b \, \mathrm{d}A$. This term is positive for dislocations of the same sign and negative for dislocations of opposite sign. Similar as before the force per unit length in the x-direction on dislocation 2 at position x,y by dislocation 1 is given by

$$f_x = -\frac{\partial u_{\mathrm{int}}}{\partial x} = \frac{\partial U_{\mathrm{int}}}{\partial A} = \frac{\partial}{\partial A} \int_A \sigma_{xy} b \, \mathrm{d}A = \sigma_{xy} b$$

The force is thus the same as exerted by a uniform shear stress σ_{xy} on the slip plane. A similar operation for f_y leads to $f_y = \sigma_{yy} b$. Introducing the stresses from Eqs. (15.19) and (15.20) leads to

$$f_x = \frac{Gb^2}{2\pi(1-v)} \frac{x(x^2-y^2)}{(x^2+y^2)^2} \qquad f_y = \frac{Gb^2}{2\pi(1-v)} \frac{y(x^2-y^2)}{(x^2+y^2)^2} \qquad \text{or}$$

$$f_r = f_x \cos\theta + f_y \sin\theta = \frac{Gb^2}{2\pi(1-v)} \frac{1}{r} \qquad f_\theta = f_x \cos\theta - f_y \sin\theta = \frac{Gb^2}{2\pi(1-v)} \frac{\sin 2\theta}{r}$$

Using the full stress equations a slightly different form is obtained. The differences are entirely negligible. The resulting forces are *not* central symmetric and the areas of attraction and repulsion are shown in Fig. 15.29, indicating the 'head-to-tail' stability of two positive edge dislocations.

For screw dislocations one realises that the radial symmetry of their stress field

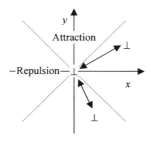

Fig. 15.29: The areas of attraction (upper and lower part) and repulsion (left and right part) between a positive edge dislocation at the origin and one at position x,y.

does lead to central forces between them. Assuming screw dislocation 1 at the origin lying in the z-direction and dislocation 2 at x,y, also lying in the z-direction, a similar exercise using a radial cut for dislocation 2 leads for these two parallel screw dislocations to

$$f_r = -\frac{\mathrm{d}u_{\mathrm{int}}}{\mathrm{d}r} = \frac{\mathrm{d}U_{\mathrm{int}}}{\mathrm{d}A} = \frac{\mathrm{d}}{\mathrm{d}A} \int_A \sigma_{\theta z} b \, \mathrm{d}A = \sigma_{\theta z} b$$

and thus using the stress as given by Eq. (15.22) to

$$f_r = \frac{Gb^2}{2\pi r} \quad (\text{or } f_r = \frac{Gb_1 \cdot b_2}{2\pi r} \text{ if the dislocations differ in Burgers vector})$$

The force is attractive when both dislocations have an opposite sign and repulsive when of the same sign.

Without giving details we note that for two parallel but opposite dislocations (using the full stress solutions) the total energy no longer contains a term dependent on the outer boundary but instead one dependent on their distance. The physical reason is that the stress fields of both dislocations practically cancel at large distances. More general, the aggregation of dislocations decreases the intensity of the long-range stress field, so that the effect of clusters of dislocations, such as grain boundaries, is mainly local. A particular clear example is provided by small-angle grain boundaries, which can be described as a row of regularly spaced edge dislocations (Fig. 15.17, see also Fig. 8.32), not necessarily all in the same orientation. The line density of the dislocations n is given by $n = 1/D$, where D is their spacing, and if θ is the angle between the normal vectors associated with the two grains, we have $\sin(\theta/2) \cong \theta/2 = b/2D$ or $n = 1/D \cong \theta/b$. The interface energy of the grain boundary U_{gb} is approximately the product of the dislocation line density and the specific energy of an individual dislocation u_{edge} and thus

$$U_{\mathrm{gb}} = nu_{\mathrm{edge}} = n\left(\frac{Gb^2}{4\pi(1-v)}\ln\frac{R}{r_0}\right)$$

Approximating the upper limit R by $R \cong D/2$, meanwhile introducing the constant $\theta_0 = b/2r_0$, we find

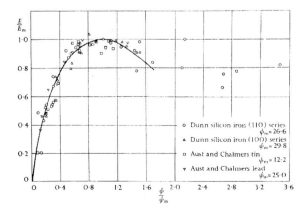

Fig. 15.30: Variation of grain boundary energy with orientation for Fe(Si), Sn and Pb. Note the good overall agreement up to quite high angles.

$$U_{\mathrm{gb}} = \frac{\theta}{b}\left(\frac{Gb^2}{4\pi(1-v)}\ln\frac{D}{2r_0}\right) = \frac{Gb}{4\pi(1-v)}\theta\left(\ln\frac{\theta_0}{\theta}\right) = \frac{Gb}{4\pi(1-v)}\theta(\ln\theta_0 - \ln\theta)$$

A more exact analysis yields the same functional relationship. The form of this relation with an infinite derivative at $\theta = 0$ and a maximum energy U_m at $\theta = \theta_0/e$ agrees fairly well with experimental data up to rather high values of θ (Fig. 15.30), in fact much better and also extending to a much larger tilt angle than can be anticipated from this dislocation model.

More complex situations arise when two dislocations cross. We just show one example. Fig. 15.31 shows a screw dislocation with a vertical dislocation line together with another dislocation (the line EF). If both dislocations cross each other, a so-called *jog* is formed. With jogs also energy is associated and further dislocation motion becomes more difficult.

Interaction with dissolved atoms

An impurity atom dissolved in the crystal will in general have a different radius, giving rise to a stress field around the impurity atom. This results in an interaction with neighbouring dislocations. An atom with a radius that is too large can lower the energy by taking a position in the area of tensile stress of an edge dislocation. Similarly a too small atom can lower the energy by taking a position in the area of compressive stress. Through this mechanism impurities can segregate in the neighbourhood of dislocations. As mentioned before, the interaction between solute atoms and screw dislocations is limited.

To describe this effect we consider a solute atom as an elastic inclusion with radius r' replacing a sphere with original radius r in an isotropic matrix. Using polar angles θ and ϕ, the direction cosines l for an arbitrary orientation are given by

$$l_{rx} = \sin\theta\,\sin\phi \qquad l_{ry} = \cos\theta \qquad l_{rz} = \sin\theta\,\cos\phi$$

The normal stress at the surface of this inclusion in a stress field σ_{ij} is $\sigma_{rr} = l^T \sigma l$. The corresponding force is $\sigma_{rr}\,r^2\,\sin\theta\,d\theta\,d\phi$. If we alter the radius by an amount εr, where $\varepsilon = (r'-r)/r$, the interaction energy U_{int} is equal to minus work done W and is given by

$$U_{\mathrm{int}} = \int \varepsilon r(\sigma_{rr} r^2 \sin\theta)\,d\theta\,d\phi = \frac{4}{3}\pi\varepsilon r^3(\sigma_{xx} + \sigma_{yy} + \sigma_{zz})$$

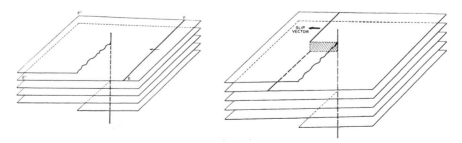

Fig. 15.31: Part of a crystal with a screw dislocation. The crystal is formed by a single plane of atoms in the shape of a spiral. The left shows a dislocation moving over the slip plane as the line EF. The point E moves to E′ while F approaches F′. The right shows the situation after crossing, in which the moving dislocation has obtained a discontinuity, a jog that has crossed the shaded part.

Now we have to insert the stresses due to the edge dislocation, Eq. (15.20), to obtain

$$U_{int} = \frac{4}{3}\frac{(1+v)}{(1-v)}\frac{Gb\varepsilon r^3 \sin\alpha}{R} \qquad (15.25)$$

where R and α are the co-ordinates of the inclusion from the positive edge dislocation. As expected, this energy is positive for the upper side of the dislocation ($\pi > \alpha > 0$) for larger inclusion and negative if smaller. For a smaller inclusion the reverse holds. The resulting force is non-central.

The above expression only includes the interaction energy outside the inclusion. If one also takes into account the interaction energy inside the inclusion, assuming the same elastic moduli, one obtains (see Section 10.3)

$$U_{int} = \frac{4Gb\varepsilon r^3 \sin\alpha}{R}$$

The effect is considerable. Assuming $v = \frac{1}{3}$, U_{int} has increased by a factor 3/2.

Problem 15.8

Sketch the stress field of an edge dislocation and confirm Fig. 15.29.

15.10 Reactions of dislocations*

As mentioned, under certain conditions a dislocation can split into two dislocations or two dislocations can merge to one. The general name for this process is *dislocation reaction*. Since the energy of a dislocation U is proportional to the Burgers vector squared, $U \sim |\mathbf{b}|^2$, a reaction can only occur if

$$|\mathbf{b}_1|^2 + |\mathbf{b}_2|^2 > |\mathbf{b}_3|^2 \qquad (15.26)$$

Of course, Eq. (15.26) is not exact even for isotropic crystals because the character of the dislocation may change, i.e. each term contains a proportionality factor, dependent on the character of the dislocation. However, since the proportionality factor is approximately constant and close to one we neglect this effect here. These processes are strongly dependent on the crystal structure. In a simple cubic lattice after each step of the dislocation the original situation is restored except for the position of the dislocation. In the process of splitting a dislocation in a FCC or BCC lattice, however, the motion of the dislocation can disturb the ideal order of stacking of the lattice planes. We deal briefly with FCC, BCC and HCP elemental crystals only. More detailed information can be found in e.g. Hull (1975) or Hirth and Lothe (1982).

Slip in FCC crystals

The slip system for the FCC lattice is $\{111\}\langle 110\rangle$. Some consideration, however, results in a more complex picture of the slip motion. To that purpose we need the concept of stacking fault. Ideally in an FCC lattice the stacking order of the lattice planes is ..ABCABCA... Two types of *stacking faults* may occur. The first is equivalent to the removal of an atomic plane, e.g. ..ABCBCA.., and denoted as *intrinsic* stacking fault. The second is equivalent to the introduction of an extra atomic plane, e.g. ..ABCACBCA.., and called an *extrinsic* stacking fault.

Now consider the passage of a dislocation between two adjacent planes. The most frequently occurring dislocation in an FCC lattice is the $\frac{1}{2}a\langle 110 \rangle$ dislocation. In fact the extra half plane consists of two different (110) planes, an a- and a b-plane (a simple drawing will reveal this). These planes occur in an abab sequence[y] (see Fig. 15.32). This is the *unit dislocation*. As long as the two lattice planes move together, the unit dislocation is conserved. As soon as these planes start to move independently, partial dislocations are created (see Fig. 15.33).

From an atomic point of view this splitting is expected. In Fig. 15.34 the spheres indicate the a-layer and the other layer rests in the sites marked B. Now consider movement of the layers. If we depict atoms as hard spheres, it is unlikely that an atomic plane will slip over an atom from position B to the next potential energy minimum position B following the vector $\mathbf{b}_1$, but most likely along the 'valleys' between the atoms via position C using vectors $\mathbf{b}_2$ and $\mathbf{b}_3$. The result is a zigzag motion corresponding to the passage of two dislocations immediately after another: the first with the Burgers vector $\mathbf{b}_2$ and the second with the Burgers vector $\mathbf{b}_3$. As always the Burgers vector must be conserved and this leads for $\mathbf{b}_1 = a[110]/2$ to

$$\mathbf{b}_1 \rightarrow \mathbf{b}_2 + \mathbf{b}_3 \qquad \text{or} \qquad a[110]/2 \rightarrow a[211]/6 + a[12\overline{1}]/6 \qquad (15.27)$$

These *Shockley partial dislocations*[z] repel each other since $\mathbf{b}_2$ and $\mathbf{b}_3$ are not perpendicular and the force per unit length is approximately given by

$$f = G\mathbf{b}_2 \cdot \mathbf{b}_3/(2\pi r)$$

Although the passage of the planes is easier in this way, the penalty is that stacking faults arise with associated stacking fault energy γ. Since the lesser stacking fault area is present, the lesser the increase in energy, the effect of the presence of the stacking faults is that the partial dislocations attract each other with a force equal to γ. An equilibrium situation, the *extended dislocation*, with a distance d between the partial dislocations will be reached when the repulsive and attractive forces balance resulting

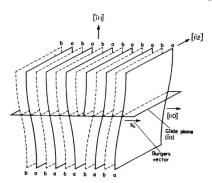

Fig. 15.32: Unit edge $\frac{1}{2}\langle 110 \rangle$ dislocation in the FCC lattice.

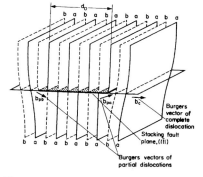

Fig. 15.33: Extended dislocation formed by two partial dislocations.

[y] Seeger, A. (1957), *Glide and work hardening in face-centred cubic materials*, page 243 in *Dislocations and mechanical properties of crystals*, Wiley, New York.
[z] Heidenreich, R.D. and Shockley, W. (1948), Rep. on strength of solids, page 36, Physical Society, London.

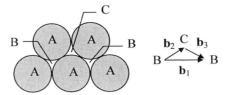

Fig. 15.34: Slip on FCC {111} planes.

in

$$d = \frac{G(\mathbf{b}_2 \cdot \mathbf{b}_3)}{2\pi\gamma} = \frac{Ga^2}{24\pi\gamma} \tag{15.28}$$

Hence the width of the extended dislocation d depends on the stacking fault energy γ. Estimates of γ^{aa} for various metals are given in Table 15.5. A value for γ of about 80 mJ/m^2 as in Cu leads to a distance of about four Burgers vectors between the two partials. Unlike the original dislocation the two partial dislocations define a specific slip plane since the extended dislocation has to move as whole or the partials have to be brought together. Therefore for extended dislocations cross-slip is difficult.

In fact another type of partial dislocation is possible in the FCC lattice (Fig. 15.35), the (negative) *Frank partial dislocation*[bb] characterised by $\mathbf{b} = \frac{1}{3}a[111]$. Since glide is restricted to the (111) plane, this dislocation can only move by climb. This makes this type usually sessile. Generally one supposes that such a dislocation is created via a condensation of vacancies. A positive Frank sessile dislocation can similarly be created by the condensation of interstitial atoms to a lattice plane. For both type of loops also stacking faults are created. Similarly as splitting of a unit dislocation can occur in two Shockley partials, a unit dislocation can also split in a Shockley partial and a Frank partial dislocation, e.g.

Fig. 15.35: The Frank partial dislocation: edge-on view of the (111) planes.

Table 15.5: Stacking fault energy for various metals. Data from various sources

Metal	γ (mJ/m^2)	Metal	γ (mJ/m^2)
Ag	16, 25, 29	Cu	40, 73, 163
Au	10, 30, 55	Al	135, 200, 238
Pd	180	Ni	208, 300, 400

[aa] Since the stacking fault is a coherent interface, their energy is much lower as compared to non-coherent interfaces such surfaces, with a surface energy of say 1.5 J/m^2, and grain boundaries, with an interface energy of say one third of the surface energy or about 0.5 J/m^2.
[bb] Frank, F.C. (1949), Proc. Phys. Soc. **A62**, 202.

$$a[01\bar{1}]/2 \rightarrow a[\bar{2}1\bar{1}]/6 + a[11\bar{1}]/3$$

In this case, however, the elastic energy remains constant but taking into account the elastic anisotropy the reaction may sometimes lead to a decrease.

Other sessile dislocations can be formed via *duplex slip*. In view of the multitude of possible slip systems in FCC crystals, for many orientations a complementary system is present. A complementary or secondary slip system glides on a different $\{111\}$ slip plane in a different $\langle 110 \rangle$ direction then the original $\{111\}\langle 110 \rangle$ dislocation. The corresponding dislocations can react to form a sessile dislocation, called a *Lomer-Cottrell barrier*. For example, consider (Fig. 15.36) the $\frac{1}{2}a[101]$ and $\frac{1}{2}a[\bar{1}10]$ on different $\{111\}$ planes and both parallel to the line of intersection of the $\{111\}$ planes. These dislocations attract each other meanwhile moving towards their line of intersection. Lomer[cc] suggested that they react via

$$\frac{1}{2}a[101] + \frac{1}{2}a[\bar{1}10] \rightarrow \frac{1}{2}a[011]$$

The new dislocation lies parallel to the line of intersection of the two initial $\{111\}$ slip planes in the (100) plane bisecting the slip planes. Its Burgers vector lies in the (100) plane normal to this line of intersection so that it is a pure edge dislocation. Since (100) is not close-packed, the newly formed dislocation will not glide. However, it is still a unit dislocation. Cottrell[dd] further argued that the dislocations on $\{111\}$ planes are normally dissociated in partial dislocations so that the leading partial dislocations can react according to

$$a[\bar{1}2\bar{1}]/6 + a[1\bar{1}2]/6 \rightarrow a[011]/6$$

Also this dislocation lies parallel to the line of intersection of the initial slip planes, it has a pure edge character on the (100) plane and is thus sessile. A triangular shaped barrier is formed with the new dislocation at its apex and two stacking faults bounded by partial dislocations on the original slip planes. These barriers can be overcome only

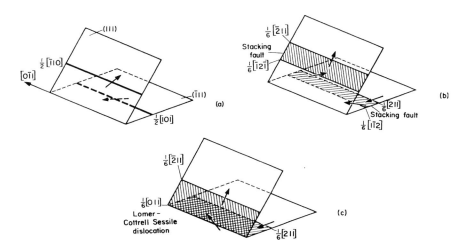

Fig. 15.36: The formation of a Lomer-Cottrell sessile dislocation.

[cc] Lomer, W.M. (1951), Phil. Mag. **42**, 1327.
[dd] Cottrell, A.H. (1952), Phil. Mag. **43**, 645.

at high stress and/or temperature.

Slip in BCC crystals

The slip system for the BCC lattice is $\{110\}\langle 111\rangle$. Recall that BCC crystals do not have a strong preference for a slip plane but only for the slip direction. Extended dislocations are not commonly observed in BCC crystals.

However, Cottrell[ee] suggested a possible mechanism for the formation of a sessile dislocation and the mechanism has been shown to be important in formation of crack nuclei. Consider a BCC crystal in which two dislocations are present. The first is characterised by the dislocation line $\mathbf{l}_1 = [011]$ and the Burgers vector $\mathbf{b}_1 = \frac{1}{2} a[1\bar{1}1]$. Since $\mathbf{b}_1 \cdot \mathbf{l}_1 = 0$, the dislocation is a pure edge dislocation. The accompanying slip plane is given by $\mathbf{g}_1 = \mathbf{b}_1 \times \mathbf{l}_1 = (21\bar{1})$. Assume that the second dislocation line is parallel to the first one with $\mathbf{l}_2 = [011]$, $\mathbf{b}_2 = \frac{1}{2} a[11\bar{1}]$ and $\mathbf{g}_2 = (2\bar{1}1)$. This is also a pure edge dislocation. The dislocations will meet at the junction between $\mathbf{g}_1$ and $\mathbf{g}_2$, i.e. the line [011] as calculated from $\mathbf{g}_1 \times \mathbf{g}_2$. Now a reaction can take place between the two dislocations in which one new dislocation with $\mathbf{b}_3$ is produced but this will only happen if the energy of the new dislocation is lower then the sum of energies for the two originals ones. For our example $\|\mathbf{b}_1\| = \|\mathbf{b}_2\| = \frac{1}{2} a\sqrt{3}$ while the new dislocation has a Burgers vector $\mathbf{b}_3 = \mathbf{b}_1 + \mathbf{b}_2 = \frac{1}{2} a[1\bar{1}1] + \frac{1}{2} a[11\bar{1}] = a[100]$ and thus $\|\mathbf{b}_3\| = a$. This implies that a reaction, in this case association, will take place. Furthermore it holds that $\mathbf{l}_1 = \mathbf{l}_2 = \mathbf{l}_3 = [011]$. Therefore the new dislocation is again a pure edge dislocation with $\mathbf{g}_3 = (01\bar{1})$. The original two dislocations belong to the slip system $\{110\}\langle 111\rangle$, which is a primary slip system in the BCC lattice. The new dislocation belongs to the system $\{011\}\langle 011\rangle$, which is not a primary slip system and is thus a sessile dislocation. The (001) plane is also a preferred cleavage plane in BCC crystals.

Slip in HCP crystals

The slip system for the HCP lattice is $\{0001\}\langle 1\bar{1}20\rangle$. Dislocations with $\langle 1\bar{1}20\rangle$ vectors will follow any line in the crystal but preferably in the basal plane where they can dissociate in two Shockley partial dislocations according to

$$a[1120] \rightarrow a[10\bar{1}0] + a[01\bar{1}0]$$

This extended dislocation is confined to the basal plane. The stacking fault produced by this reaction lies also in the basal plane.

15.11 Bibliography

Cottrell, A.H. (1953), *Dislocations and plastic flow in crystals*, Oxford University Press, Oxford.

Hirth, J.P. and Lothe, J. (1968, 1982), *Theory of dislocations*, 1st and 2nd ed., McGraw-Hill, New York.

Hull, D. and Bacon, D.J. (1975, 2001), *Introduction to dislocations*, 2nd and 4th ed., Pergamon, Oxford.

---

ee Cottrell, A.H. (1958), Trans. Metall. Soc. AIME. **212**, 192.

Kovacs, I. and Szoldos, L. (1973), *Dislocations and plastic deformation,* Pergamon, Oxford.

Nabarro, F.R.N. (1967), *Theory of crystal dislocations*, Oxford University Press, Oxford (see Dover, 1987).

Read, W.T. (1953), *Dislocations in crystals*, McGraw-Hill, New York.

Weertman, J. and Weertman, J.R. (1964), *Elementary dislocation theory*, McMillan, New York.

16

Dislocations and plasticity

In Chapter 15 the origin and properties of dislocations were discussed. In this chapter the influence of dislocations on the plasticity of crystalline materials is discussed. To that purpose, we discuss first hardening in general. Thereafter the stress-strain curves of single crystals and some particular hardening models are briefly discussed. This is followed by a discussion of the influence of the polycrystallinity. Finally the effect of impurities and precipitates is addressed. In both cases we first briefly describe the influencing factors and thereafter deal with theoretical considerations.

16.1 General aspects of hardening

In this section we assess the character of hardening, starting with the relation between dislocation density and strain rate. Thereafter a general description of hardening aspects and the thermal character of plastic deformation is given (Kovacs and Szoldos, 1973).

Strain rate and dislocation density

For macroscopic deformation a large number of dislocations is necessary. A simple expression, usually attributed to Orowan, connects the strain rate $\dot{\gamma}$ to the dislocation density ρ. To that purpose consider Fig. 16.1 in which a crystal with dimensions L_1, L_2 and L_3 is shown. With the displacement of one edge dislocation parallel to the stress τ through the whole crystal, the increase in displacement du is b. For any dislocation i that has moved only partially through the crystal, the contribution to the displacement rate can be estimated as $\dot{u}_i = bv_i/L_2$, where v_i is the velocity of the dislocation. The total displacement rate, due to in total[a] n positive dislocations moving to one side and negative dislocations moving to the other side, is $\dot{\Delta} = \sum \dot{u}_i$. If we use the dislocation density $\rho = 2n/L_1L_2$ and the average dislocation velocity

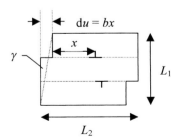

Fig. 16.1: Slip due to the displacement of edge dislocations. The length L_3 is perpendicular to the paper.

---

[a] There is no reason to assume only positive or negative dislocations. On average an equal number of dislocations of opposite nature will be present.

$\bar{v} = (n)^{-1} \sum v_i$, the strain rate $\dot{\gamma} = d\gamma/dt$ is

$$\blacktriangleright \qquad \dot{\gamma} = 2\dot{\varepsilon} = \frac{2\dot{\Delta}}{L_1} = \frac{2b}{L_1 L_2} \sum v_i = b\rho\bar{v} \qquad (16.1)$$

The strain rate is thus proportional to the density ρ of mobile[b] dislocations, to the length of the Burgers vector b and the average velocity of the dislocations $\bar{v}$. The same expression can be derived using dislocations loops. The expression is also used in its integrated form $\gamma = b\rho\bar{x}$, where the density ρ is assumed constant and $\bar{x}$ denotes the average distance a dislocation has moved. Of course, *Orowan's equation*, Eq. (16.1), is derived using particular models. The relation is valid in general if we interpret b as an average Burgers vector length as well as $\bar{v}$ as the average velocity.

Dislocation density and hardening

Before any slip can take place the applied stress must exceed[c] the yield strength Y. The value of Y is determined by many structure dependent factors, e.g. the number of slip systems, grain size, purity, etc. In the early stage of plastic deformation slip is essentially on the primary glide planes. With increasing deformation cross-slip takes place, the dislocations multiply and start to intersect rather soon during plastic deformation. An increase in yield strength arises because the dislocations are hindered in their motion by the increasing number of other dislocations. At the position where the dislocations cross each other, jogs are formed. Jogs can be both sinks and sources of vacancies and interstitials. Since the motion of vacancies and interstitials is a thermally activated process it can be expected that the yield strength is dependent on temperature and strain rate. The dislocations form dense regions or *tangles*, which restructure to a *network* (or *cellular substructure*) characterised by a more or less ordered configuration of tangles intertwined with relatively dislocation-free areas. An example[d] is shown in Fig. 16.2. The network is usually well developed at about 10% strain. At low deformation the size of the cells decreases with increasing strain but at higher deformation the cell size reaches an approximately constant value, indicating

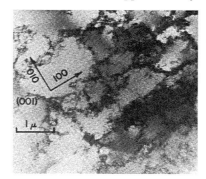

Fig. 16.2: A dislocation network in Ni deformed in tension to 20% at 295 K.

[b] A considerable difference can be present between the total and the mobile dislocation density.
[c] In fact yield is preceded by a small reversible movement of dislocations at a stress below the Frank-Read stress for dislocation multiplication. Since the pinning distance is distributed, the yield point will be fuzzy. Moreover any small irreversible deformation is difficult to detect in the stress-strain curve.
[d] Nolder, R.L. and Thomas, G. (1964), Acta Met. **12**, 227.

that with increasing strain the dislocations sweep through the cells and join the tangle of the cell walls. The majority of dislocations is thus more or less fixed in the network structure and the deformation takes mainly place by a small, roughly constant number of dislocations per cell moving through its 'free interior space'. Also grain and sub-grain boundaries, solute atoms, second-phase particles and surface films play their role as obstacles and the collective effect is known as *hardening*. The nature of the network depends on the material, strain, strain rate and temperature. The network development is less pronounced for low temperature and high strain rate and in materials with low stacking fault energy where cross-slip is more difficult.

A general model for hardening has been given by Taylor[e] and is based on the expression for the stress τ_{dis} due to a dislocation

$$\tau_{dis} = \frac{\alpha G b}{2\pi r} \tag{16.2}$$

with r is the distance to the dislocation line. For isotropic crystals it holds that for edge dislocations $\alpha = 1/(1-v)$, while for screw dislocations $\alpha = 1$. If the dislocations are randomly distributed, in position, character as well as sign, the stress on every dislocation, exerted by all other dislocations, is also given by Eq. (16.2), where now r is interpreted as the average distance between the dislocations. This implies that the stress for all dislocations, except for the nearest ones, must average to zero. Because the average distance r between dislocations is given by the inverse of the square root of the dislocation density ρ, $r \cong \rho^{-1/2}$, the internal stress τ_{int} is

$$\tau_{int} = \frac{\alpha G b \sqrt{\rho}}{2\pi} \tag{16.3}$$

Both the magnitude and sign of the internal stress vary with position in the crystal. The wavelength is of the order of $r \cong \rho^{-1/2}$. Since plastic deformation is the result of the motion of dislocations, a stress at least as large as the one given above is required to displace the dislocations any further. In order to obtain an estimate for the stress-strain curve, Taylor assumed that every newly produced dislocation moves a constant distance l and is then pinned. The distance l is assumed to be constant throughout the deformation process[f]. Furthermore, he assumed all dislocations to be straight and parallel and moving from one side of the crystal to another. In that case the strain is given by the integrated Orowan equation

$$\gamma = \rho b l \tag{16.4}$$

Combining Eq. (16.3) with Eq. (16.4) yields

$$\tau = \frac{\alpha G b}{2\pi} \left(\frac{\gamma}{b l} \right)^{1/2} \tag{16.5}$$

So far it has been assumed that the yield strength[g] τ is only determined by plastic strain. In reality there is also elastic strain and yield will only occurs above the yield strength because the stress has to be larger than the yield strength before the

---

[e] Taylor, G.I. (1934), Proc. Roy. Soc. **A145**, 362.
[f] Although an assumption, for a well-developed network the cell size appears to be experimentally more or less constant for square-root behaviour, for which the model was developed.
[g] In this chapter we denote the yield strength by τ and use subscript as labels for various components.

dislocations can move through the material. Including the initial yield strength τ_0 for dislocation motion, Eq. (16.5) becomes

$$\blacktriangleright \qquad \tau = \tau_0 + \frac{\alpha Gb}{2\pi} \sqrt{\rho} = \tau_0 + \frac{\alpha Gb}{2\pi} \left(\frac{\gamma}{bl} \right)^{1/2} \qquad\qquad (16.6)$$

So, this model predicts square-root hardening[h]. Many cubic (polycrystalline) metals exhibit this square-root behaviour of stress versus strain (see Fig. 16.14) although other values for the exponent other than ½ are frequently observed (Table 2.1). However, it will be clear that the assumptions are oversimplifying the situation considerably and that therefore the coincidence with experimental data is largely fortuitous. In fact careful experiments have shown that generally two types of slip occur leading to rather different hardening behaviour, as discussed in section 16.2.

It is useful to consider another estimate of the τ-γ curve. Consider again the critical bowing stress $\tau = \alpha Gb/l$ between pinning points with distance l. With increasing γ, l will decrease. If we arbitrarily assume $l = \chi/\gamma^n$, where χ and n denote material constants, meanwhile adding the initial yield strength τ_0, we obtain

$$\tau = \tau_0 + \frac{\alpha Gb}{\chi} \gamma^n \qquad\qquad (16.7)$$

As one extreme we take $n = 1$ and then in this model linear hardening is the result. In practice materials with a two-slope stress-strain curve are observed, however, a more general description of the stress-strain curve is frequently necessary in practice and from pragmatic point of view the exponent n is often considered as a parameter leading to Ludwik-type behaviour (see Chapter 2). Of course, the same objections as before can be forwarded and different assumptions lead to different answers.

As will become clear later, theories to derive the stress-strain curve from dislocation models are quite involved. In any case the two simple models presented use a different set of assumptions and yield different answers, illustrating that a certain arbitrariness in the models is hard to avoid. In fact many theories and models have been put forward, though with different degree of success.

The thermal character of plastic deformation

When the stress-strain curve of a metal is measured at two temperatures T_1 and T_2, they usually do not coincide and cannot be made to coincide by dividing the stress by

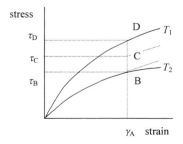

Fig. 16.3: The stress-strain curve at two temperatures.

---

[h] Although, denoting with the τ-γ curve the stress as function of strain, the expression contains clearly a square-root dependence on strain, it is often addressed in the literature as parabolic hardening.

the temperature dependent shear modulus. This is due to two effects[i] (Fig. 16.3). First, at a given strain, the response may differ because of the temperature dependence of the internal structure, where internal structure covers the whole of the dislocation network, tangles, pile-ups and all further elements related to dislocations. Second, if an identical structure is developed when the metal is deformed at another temperature, the stress necessary for further development may still depend on the temperature. To illustrate this, suppose that identical structures are formed at strain γ_A at temperatures T_1 and T_2. In this case any difference in curves must be due to the temperature dependence of the stress required for further deformation. With a sudden change in temperature from T_2 to T_1, the stress will jump from τ_B to τ_D. If, on the other hand, we suppose that the internal structure does vary with deformation temperature but that the temperature has no influence on the flow stress, the crystal deformed to point at B will continue along the dotted line when the temperature is changed to T_1. The slope of this line is different from that at point D because the internal structures differ. If the change is due to both effects, a sudden change in temperature will raise the stress to τ_C. The ratio $(\tau_C/\tau_B)/(\tau_D/\tau_B)$ measures the relative contribution of both effects. The *Cottrell-Stokes law* states that, independent of the actual strain on the material, the ratio

$$\left(\frac{\tau_2}{\tau_3}\right)_a \left(\frac{\tau_3}{\tau_1}\right)_b = \left(\frac{\tau_2}{\tau_1}\right)_c = \text{constant}$$

where $(\tau_2/\tau_3)_a$ is the ratio of flow stresses when the temperature is suddenly changed from T_2 to T_3 at constant strain a. The law was discovered using data on Al for stage II and beyond and may not be valid for BCC metals. It has been found to be valid though for the HCP metals Cd and Mg where it holds in both the easy glide and turbulent flow regions. This suggests that the law should be also valid for stage I in FCC metals. The overall behaviour of the temperature dependence of the flow stress is schematically shown in Fig. 16.4 where we see that a plateau stress, independent of temperature, is reached at sufficiently high temperature, both for pure metals and solid solutions.

Using the Orowan equation

$$\dot{\gamma} = \rho b v$$

the temperature dependence of the yield strength can be explained by a simple model

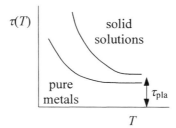

Fig. 16.4: The temperature dependence of the yield strength of pure metals and solid solutions.

[i] Cottrell, A.H. and Stokes, R.J. (1955), Proc. Roy. Soc. A **233**, 17.

due to Seeger[j]. Thereto we consider that in a crystal, barriers hinder the dislocation motion, which makes dislocation motion a thermally activated process with velocity v described by

$$v = v_0 \exp(-U_{act}/RT)$$

with v_0 a velocity pre-factor, U_{act} the activation energy and R and T the gas constant and the temperature, respectively. The activation energy U_{act} is given by

$$U_{act} = U_{bar} - V\tau$$

where V is the 'activation volume' and τ the applied shear stress. The barrier energy U_{bar} consists of two parts. The first contribution is the energy necessary to move the dislocation through the internal stress field of magnitude τ_0 in the lattice, attributed to the long-range interactions between dislocations. The required energy is given by $V\tau_0$, where V is again the activation volume. Since the internal stress fields can be assumed to have a periodicity approximately equal to the dislocation distance, typically say 1 μm, the stress τ_0 is not expected to be assisted significantly by thermal fluctuations and thus to be temperature independent. The second contribution is the energy U_0 necessary to overcome a more local barrier, associated with the actual intersection of dislocations or with dislocation-precipitation interaction. Each dislocation must thus acquire an energy $U_{bar} = U_0 + V\tau_0$ and to a first approximation we can write for the activation energy

$$U_{act} = U_0 - V(\tau - \tau_0)$$

Combining and evaluating for τ, meanwhile using $\dot{\gamma}_0 = \rho b v_0$, results in

$$\blacktriangleright \qquad \tau = \tau_0 + \frac{U_0 - RT \ln(\dot{\gamma}_0 / \dot{\gamma})}{V} = \tau_0 + \tau_T \qquad \text{for} \quad T \le T_0 \qquad (16.8)$$

$$\tau = \tau_0 \quad \text{for} \quad T \ge T_0 \quad \text{with} \quad T_0 = \frac{U_0}{R \ln(\dot{\gamma}_0 / \dot{\gamma})}$$

The necessary shear stress τ to move dislocations, or in other words the yield strength, is thus composed of a temperature independent part τ_0, identified as the plateau stress τ_{pla}, and a temperature-strain rate dependent part τ_T (Fig. 16.5). The yield strength decreases with increasing temperature at constant strain rate. Above T_0, dependent on the strain rate $\dot{\gamma}$, only τ_0 survives. Therefore at sufficiently high temperature the yield

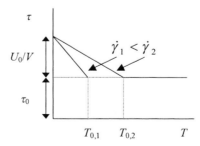

Fig. 16.5: The temperature dependence of the yield strength.

[j] Seeger, A. (1957), page 243 in *Dislocations and mechanical properties of crystals*, J.C. Fisher, W.G. Johnston, R. Thomson and T. Vreeland, eds., Wiley, New York.

strength becomes (approximately) constant, in reasonable agreement with experiments. At constant temperature below T_0 the yield strength increases with increasing strain rate.

In this approach only one thermally activated process is rate controlling. In practice there are more types of barriers, each providing a different kind of resistance to dislocation motion, e.g. *forest dislocations*, i.e. dislocations that intersect the slip plane, dislocation pile-ups, the dislocation network and associated tangles. These barriers precisely provide the basis for strain hardening, as described above.

Egon Orowan (1901-1989)
Born in Budapest, Hungary, he studied from 1920 to 1922 physics, chemistry, mathematics and astronomy in the University of Vienna. In 1922 he began his studies at the Technical University of Berlin, initially mechanical engineering, then electrical engineering and finally physics under the influence of Professor R. Becker. At the end of 1928 he became Becker's assistant and presented his thesis (1932) on the cleavage of mica. Becker asked whether he was interested in checking experimentally a "little theory of plasticity" and this led to the papers *Zur Kristallplastizität* I-V. Orowan worked with the Tungsram Research Laboratory between 1936 and 1937. By 1937 he moved to Birmingham where his main interest was in a theory of fatigue. Then he moved in 1939 to Cambridge until 1950 where he focused on the technology of munitions production. Orowan summarized his work in 1949 in a long, but very condensed, review paper *Fracture and Strength of Solids* (Rep. Progr. Phys. **12**, 185) of which the topics range from Thomas Young's theory of the cohesive strength to currently unsolved problems. In this paper his criticisms of earlier work are expressed forcibly. In 1950 he accepted the invitation to join the Department of Mechanical Engineering at MIT. Here he occupied himself with many of the same problems that he successfully started to investigate while in England. His views on plasticity were expressed in the paper *Creep in Metallic and Non-metallic Materials* (1952). Thereafter, he focused on fracture and the applications of plasticity and fracture in geology. After his retirement he added to these considerations other more philosophical questions of: the stability of the Western industrial economies, aging of societies and problems of higher education.

Problem 16.1

Estimate the number of dislocations on an (111) plane necessary for 5% plastic deformation of an Al bar of length 1 cm. Recall that Al has an FCC lattice with a lattice constant of 0.4 nm.

Problem 16.2

In Cu the dislocation velocity $v = 200$ m/s and the length of the Burgers vector $b = 0.3$ nm. Estimate the dislocation density ρ at a strain rate of $\dot{\gamma} = 5\,\mathrm{s}^{-1}$.

Problem 16.3

Derive Orowan's equation in an alternative way. Assume that in a single crystal block with height h (Fig. 16.6), containing a glide plane PQRS of surface area A_0, a small ring-shaped dislocation of radius R_{loop} is formed with the Burgers vector b. Take the displacement of the upper part Δ proportional to the surface area A swept out by the dislocation ($\Delta = Ab/A_0$) and

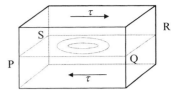

Fig. 16.6: The extension of dislocation loops by shear stress on a glide plane.

calculate the shear strain $\Delta/h = Ab/A_0 h = Ab/V$, where V denotes the volume of the crystal. Denote the length of the dislocation line by L and show that a small displacement of the dislocation dl results in $dA = L\,dl$. Introduce the dislocation density $\rho = L/V$ and arrive at $\dot{\gamma} = \rho b v$ where $v = \dot{l}$.

16.2   Stress-strain curves for single crystals

We now turn to a description of the stress-strain curve for single crystals as a prelude to the stress-strain curve for polycrystalline materials constructed from more detailed dislocation models. We use the shear stress and strain to characterise the deformation behaviour. An early but extensive overview[k] of the deformation behaviour of metallic single crystals is available. The volume edited by Mughrabi (1993) provides a wealth of further information.

HCP metals

A schematic stress-strain curve for HCP single crystals is shown in Fig. 16.7. An elastic region is present where the slope is described by the shear modulus G. For crystals with a lattice constant ratio c/a close[l] to the ideal value of $c/a \cong 1.633$ or a larger value, plastic deformation is primarily by slip on the basal plane {0001} with

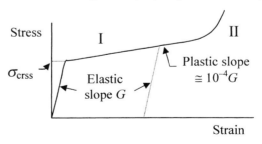

Fig. 16.7: Schematic stress-strain curve for HCP single crystals.

---

[k] Berner, R. and Kronmüller, H. (1965), page 126 in *Moderne Probleme der Metallphysik*, Seeger, A., ed., Springer, Berlin.
[l] For crystals with a lattice constant ratio $c/a < 1.633$ slip generally occurs on the non-basal planes {10$\bar{1}$0}, {10$\bar{1}$1} or {11$\bar{2}$2} with slip direction either ⟨11$\bar{2}$0⟩ or ⟨11$\bar{2}$3⟩.

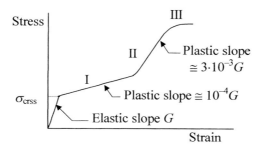

Fig. 16.8: Schematic stress-strain curve for FCC single crystals.

slip direction $\langle 11\bar{2}0 \rangle$ (stage I) and occurs at a stress of about $10^{-5}G$ to $10^{-4}G$. The order of magnitude of the hardening modulus is about $10^{-4}G$. In this region no cross-slip occurs and the deformation is due to slip on parallel planes. Consequently the amount of hardening is small and therefore one also speaks of *easy glide*. It has also been called *laminar flow*. Only after a deformation of 100% to 200% a significant hardening occurs (stage II). This region has been named by Cottrell, using the hydrodynamic terminology further, *turbulent flow*. In this region cross-slip does occur and therefore significant interaction between the dislocations is present.

FCC metals

The schematic stress-strain relation for FCC metals is shown in Fig. 16.8. These metals slip on the {111} planes in each of the $\langle 1\bar{1}0 \rangle$ directions. Analogously to HCP metals an elastic and easy glide region occurs. The hardening modulus for the easy glide region is again about $10^{-4}G$ (stage I). However, the region of easy glide is only about 15% to 20% (although occasionally extending to about 40%) and is thus much smaller than for HCP metals. After that region considerable hardening occurs with a typical slope of about $3 \times 10^{-3}G$ (stage II), which is rather constant for different metals. At still higher strain an approximately parabolic region is present starting at about

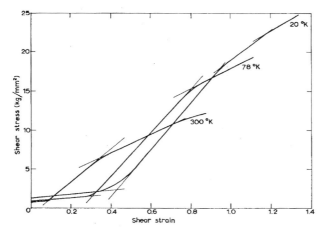

Fig. 16.9: The temperature dependence of the stress-strain curve for Ni. Not shown is the decreasing value for the elastic modulus with temperature.

30% to 50% deformation (stage III). The length of stage I and the onset of stage III are sensitive to temperature, composition and size, whereas the slope of stage II is insensitive to these parameters. The region of easy glide occurs only if the specimen is properly oriented so that the resolved shear stress in one direction on one slip plane is greater than the resolved shear stress in any other slip direction on that plane or in any other slip direction on any other plane. In early experiments this requirement was not always fulfilled leading to the generally accepted parabolic behaviour.

The temperature has a significant influence on the behaviour. Fig. 16.9[m] shows the effect for high purity single crystal Ni. All three stages (easy glide or stage I, turbulent flow or stage II and stage III) can be discerned at all three temperatures. The yield strength as well as the hardening modulus decreases with increasing temperature. Moreover the regions shift with temperature. Not shown in this figure is the decreasing value for the elastic modulus with temperature.

Impurities can influence the extent of stage I. If the impurities are clustered they tend to reduce and finally eliminate the hardening stage I. On the other hand impurities which are dissolved tend to increase stage I hardening. For example, in Ag single crystals an impurity content of 10^{-4}, 3×10^{-4} and 7×10^{-4} leads to stage I with a strain of 1.5%, 2.5% and 5%, respectively.

BCC metals

High purity BCC crystals exhibit a similar behaviour as FCC crystals. These metals slip also in the $\langle 111 \rangle$ directions but on many planes, e.g. the $\{110\}$, $\{112\}$ and $\{123\}$ planes. The slip lines, which can be observed at the surface, are wavy indicating that slip is not confined to unique slip planes. With increasing impurity content the characteristic features of stages I, II and III become increasingly vague. The shape of the stress-strain curve is sensitive to temperature, strain rate and composition.

In Fig. 16.10 a schematic stress-strain curve typical for BCC metals containing a slight amount of impurity atoms is shown. With increasing strain an elastic region and an approximately linear region (stage II) are present. The latter typically has a hardening modulus of about $10^{-4}G$. In between is the characteristic of an upper and lower yield point extending over a strain range of typically 0.05 to 0.1. In this range the deformation proceeds heterogeneously via Lüders bands (Chapter 2) while in later stages it proceeds homogeneously. This region arises from the interaction of the

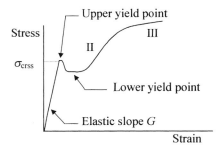

Fig. 16.10: Schematic stress-strain curve for slightly impure BCC single crystals and polycrystals.

---
[m] Haasen, P. (1958), Phil. Mag. **3**, 384.

dislocations with impurities. An example is provided by BCC metals with interstitial impurity atoms, like Fe at low carbon content. The carbon atom is residing in octahedral holes and causes a considerable misfit stress. Therefore the carbon atoms form an impurity cluster, known as *Cottrell cloud*, around the dislocations, thereby diminishing the stress around the impurity atoms. During application of a stress, the dislocations have to unlock and this occurs at the upper yield stress. As soon as the dislocations have moved away from the interstitial atoms, a lower stress, the lower yield stress, is required to move them further. This process also occurs, e.g. in β-brass, Mo and V. Moreover also the HCP metals Cd and Zn show this phenomenon. On the other hand, Fe with a higher C-content (austenitic steel) does not show this behaviour. For this metal the crystal structure is FCC with much larger octahedral interstitial holes, which results in a lower misfit energy. Moreover, diffusion is more difficult in austenitic Fe. Still there are examples of FCC lattices with yield point phenomena, e.g. in some Cu and Al alloys. In that case the impurity atoms are substitutionally dissolved so that they can diffuse much easier through the lattice and pin the dislocations.

16.3   Models for hardening*

The calculation of the stress-strain curve for single crystals as well as polycrystals in terms of dislocations has proven to be a difficult task in which many attempts have been tried. Eminent researchers in the area (e.g. Read, 1953) have stated that the prediction of these curves is loaded with too many assumptions and personal preferences to yield generally acceptable results but others, equally eminent[n], would disagree. Of the many attempts we only discuss two approaches. Modelling of work hardening has been often pursued either via extension of specific structural models focusing on long-range stress fields or an approach dealing with short-range interactions and therefore focusing on the tangles. While both approaches rely on experimental information as obtained by TEM, the first also incorporates data from these observation in the theory. Several overviews are available in the literature, ranging from early[o] to recent[p]. A critical review is given in the paper by Kuhlmann-Willsdorf in which she discusses also some other models. We first discuss a very simple model indicating the reason for heterogeneity, then the long-range stress approach and thereafter the short-range interaction models.

The reason for heterogeneity

Many of the models are complex and loaded with assumptions. From an experimental point of view the square root dependence of the yield strength on dislocation density as well as the formation of cell structure are the most evident facts. Based on only these two elements a simple deformation theory[q] can be build. Assuming a cell structure with a volume fraction f_w of cell walls having a yield

[n] Kuhlmann-Willsdorf, D. (1985), Metall. Trans. **16A**, 2091.
[o] Hirth, J.P. and Weertman, J., eds. (1968), *Work hardening*, Gordon and Breach, New York; Hirsch, P. B. (1975), page 189 in *The physics of metals*, vol. 2: *Defect*s, P.B. Hirsch, ed., Cambridge University Press, Cambridge.
[p] Sevillano, J.G., page 19 in Mughrabi (1993) and Kuhlmann-Willsdorf, D. (1999), Phil. Mag. A79, 955.
[q] Mughrabi, H. (1987), Mater. Sci. Eng. **85**, 15.

strength τ_w and a volume fraction f_c of cell interiors having a yield strength τ_c, we take for the overall yield strength τ

$$\tau = f_w\tau_w + f_c\tau_c$$

Here the yield strengths of the wall and the cell interior are given by

$$\tau_w = \alpha Gb\sqrt{\rho_w} \qquad \text{and} \qquad \tau_c = \alpha Gb\sqrt{\rho_c}$$

where ρ_i indicates the dislocation density. In fact the dislocation densities in the cell and the wall in this model can be seen as internal parameters describing the structure of the dislocation network. For a homogeneous dislocation distribution we would have a yield strength

$$\bar{\tau} = \alpha Gb\sqrt{\bar{\rho}}$$

where $\bar{\rho} = f_w\rho_w + f_c\rho_c$ is the average dislocation density. A simple calculation leads to

$$\tau^2 = \bar{\tau}^2 - (\alpha Gb)^2 f_w f_c\left(\sqrt{\rho_w} - \sqrt{\rho_c}\right)^2$$

This shows that a heterogeneous dislocation distribution results in a lower yield strength. It appears also that the total energy is lower for a heterogeneous structure than for a homogeneous one if the applied load is taken into account. Moreover, it appears that, using this model, the relation between cell size and flow stress can be predicted as well. For this and many other details we refer to the original paper.

Long-range stress models

The long-range stress models are largely due to the research in the 1950s and 1960s of Seeger and his associates. Also Hirsch contributed in this direction. They are in a way an extension of the work-hardening model of Taylor. Basic to this approach is the stress field by a dislocation at a radial distance r, which is given by $\alpha Gb/2\pi r$ where, as before $\alpha = 1$ for screw dislocations and $\alpha = 1/(1-v)$ for edge dislocations (v Poisson' ratio), G the shear modulus and b the length of the Burgers vector. We have noticed that, if the distribution of dislocations is random both with respect to position and sign, the stress exerted on a particular dislocation must be of the same order with r denoting the mean distance between dislocations. Taking $r = \rho^{-1/2}$ with ρ the dislocation density, the internal stress is given by $\alpha Gb\rho^{1/2}/2\pi$. In order to move a dislocation a stress of this magnitude must be applied and this consideration leads to the same expression for the momentary yield strength of a material. Indeed the proportionality between yield strength and square-root dislocation density is one of the best experimentally verified expressions of dislocation theory. However, this does not lead to stage I, II or III behaviour without further considerations. Experiments on single crystals have shown that in stages I and II primarily dislocations on parallel planes of a single slip system are active. Hence in the long-range stress models it is supposed that the long-range stresses dominate while the local stresses due to dislocation crossing are much less important. An extensive overview of the long-range stress approach is given by Kronmüller[r], which is here followed to some extent.

[r] Kronmüller, H. (1965), page 126 in *Moderne Probleme der Metallphysik*, Seeger, A., ed., Springer, Berlin.

For stages I and II it is thus assumed that only a single slip system with a few slip directions is active. The dislocation structure is characterised by its density of sources N, the average path length for dislocations, L_s and L_e for screw and edge dislocations, respectively, and the number of dislocations emitted per source n. Dislocation multiplication is assumed to occur via the Frank-Read mechanism on all spots where the applied stress is not fully screened by the neighbouring dislocations. Since during the emission of a new dislocation the dislocation structure in the neighbourhood of the source does not change, the only change in shear stress τ working on the source is introduced by the stress of the newly emitted dislocations τ_n. To allow for further dislocation emission the relation

$$\tau - \tau_n > \tau_{FR} = \alpha Gb/l$$

must be satisfied, where τ_{FR} is the shear stress to activate the Frank-Read source with a pinning point distance l. The shear stress by the newly formed screw dislocation at a distance L from the source exerted on the source is

$$\Delta \tau_n = \frac{Gb}{2\pi} \frac{1}{L}$$

Hence the increase in applied stress $\Delta \tau$ must either match or exceed this and thus

$$\Delta \tau \geq \frac{Gb}{2\pi} \frac{1}{L} \qquad \text{or} \qquad \frac{\partial \tau}{\partial n} \geq \frac{Gb}{2\pi} \frac{1}{L} \tag{16.9}$$

It is supposed that the emitted dislocation at a distance L from the source encounters the stress field of a single dislocation (stage I) or a pile-up of dislocations on a parallel slip plane (stage II). It can be shown that for many purposes a pile-up of n dislocations with strength b can be considered a single dislocation with strength nb. The stress needed to overcome the barrier provided by the single dislocation or the pile-up on neighbouring slip planes results in hardening. For a single dislocation moving on a slip plane in the middle between two other slip planes with a distance d containing either a single dislocation or a pile-up, the stress to move the dislocation is given by

$$\tau_{max} = \frac{nbG}{2\pi d} \tag{16.10}$$

where $n = 1$ is used for a single neighbouring dislocation (stage I) and $n = m$ for a pile-up of m dislocations (stage II). For the next step to relate d to L we need the average distance R between groups of screw dislocations with the same sign given by

$$R = (NL_s)^{-1/2}$$

with N the source density and L_s the average loop length. From probabilistic arguments one can show that, if interaction between the new dislocation and the dislocation(s) on the neighbouring slip plane is to take place, we must have approximately

$$d = \frac{2aR}{nL} \tag{16.11}$$

with $2a = \frac{1}{2}L_e$ for a single dislocation (stage I) or $2a = nL$ for a pile-up (stage II).

After these preliminaries, we now can estimate the work-hardening coefficient for easy glide or stage I. Experimentally it has been shown that in this stage N, L_s and L_e

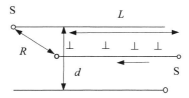

Fig. 16.11: The long-range stress model after Seeger. Shown is a cross-section with the slipped area caused by dislocation loops originating from the sources S.

are approximately constant and slip always occurs with a single slip system. During easy glide, dislocations can move over relatively large distances without meeting an obstacle and they do not pile-up at the slip plane under consideration. Most of the dislocations escape at the crystal surface. Assuming n square loops for each of the N sources and referring to Fig. 16.11, we have for each loop a contribution to the dislocation density given by $L_e/V = L_eN$. The same loop contributes a term L_s to the integrated Orowan equation. The total strain becomes

$$\gamma = \eta^2 nbL_sL_eN \qquad (16.12)$$

The factor η takes into account that for a pile-up of m dislocations with strength b on the neighbouring slip plane, considered as a single dislocation of strength mb, the effective path length is 3/4 of the real path length. Furthermore, stereology tells us that the quantity L_sL_eN can be estimated from the experimentally accessible perpendicular distance x of the slip lines at the surface as $L_sL_eN = x^{-1}$. We therefore obtain

$$\frac{d\gamma}{dn} = \frac{9}{16}\frac{b}{x} \qquad (16.13)$$

Using Eqs. (16.10) and (16.11) with $2a = \frac{1}{2}L_e$, the required stress becomes

$$\tau_{max} = \tau - \tau_0 = \frac{bG}{4\pi d} = \frac{bG}{4\pi}\frac{nL}{2Ra} = \frac{bG}{4\pi}\frac{2nL}{RL_e} = \frac{nbGL}{2\pi RL_e}$$

Solving for L, substituting in Eq. (16.9) and integrating one obtains

$$\tau - \tau_0 = \frac{bG}{2\pi}\left(\frac{1}{RL_e}\right)^{1/2} n$$

Finally for $\theta = \partial\tau/\partial\gamma = (\partial\tau/\partial n)(\partial n/\partial\gamma)$ we obtain, inserting Eq. (16.13) and using $R = (NL_s)^{-1/2}$ and $N = (xL_sL_e)^{-1}$, the result

$$\theta_I = \frac{d\tau}{d\gamma} = \frac{d\tau}{dn}\left(\frac{d\gamma}{dn}\right)^{-1} = \frac{8G}{9\pi}\left(\frac{x}{L_e}\right)^{3/4}$$

For example, for Cu, with a shear modulus of 40 GPa, the experimentally observed average slip distance is about 30 nm while the dislocation loop size is about 600 μm. This leads to an estimated hardening modulus $d\tau/d\gamma \cong 6.7$ MPa, which compares favourably with the experimental value of 7 MPa. More general, plotting the experimental data of θ/G versus $(x/L_e)^{3/4}$ for several metals indeed the slope appears to be close to $8/9\pi$. However, the temperature dependence as can be derived from this expression does not give unequivocal agreement with experiment.

Turning now to stage II, secondary slip systems start to be active. Hence a shorter path length of the dislocations results, since Lomer-Cottrell locks can be formed. It has been experimentally observed that

$$L_i = \frac{\Omega_i}{\gamma - \gamma_{II}} \tag{16.14}$$

with a value for the constant Ω_i of about 4 μm (i indicating either s or e) and where γ_{II} denotes the strain at the onset of region II. Furthermore, it is assumed that new sources become active. Again denoting the source density by N, using Eq. (16.12) with $\eta = 1$ or $d\gamma = nbL_sL_edN$, where n is the number of dislocations, inserting Eq. (16.14) and integrating we easily obtain

$$(\gamma - \gamma_{II})^3 /3 = nb\Omega_s\Omega_e N \tag{16.15}$$

Using Eq. (16.11) with $2a = nL$ we have for stage II $d = R$. Using Eq. (16.10) and $R = (NL_s)^{-1/2}$ again, we have

$$\tau_{max} = \frac{nbG}{2\pi d} = \frac{nbG}{2\pi R} = \frac{nbG}{2\pi}(NL_s)^{1/2}$$

Making use of Eq. (16.15) and differentiating to obtain $\theta = \partial\tau/\partial\gamma$ the result is

$$\theta_{II} = \frac{d\tau}{d\gamma} = \frac{G}{2\pi}\left(\frac{nb}{3\Omega_e}\right)^{1/2} \tag{16.16}$$

which indicates work-hardening with a coefficient θ proportional to G. Agreement with experiment is observed for $n \cong 20$ to $n \cong 30$. Since it is experimentally observed that the distance of the slip lines in region II is of the order of $20b$ to $30b$, the agreement is considered to be satisfactory.

According to the long-range stress approach stage III starts when other slip systems become active and cross-slip occurs. This happens only at relatively high stress. Since cross-slip is aided by thermal fluctuations the stress to initiate stage III is temperature dependent and is expected to decrease with increasing temperature. This is indeed observed experimentally. Through cross-slip the hardening rate becomes smaller since the movement of the screw dislocation onto cross-slip planes relaxes the high stress fields produced by the dislocation pile-ups. The quantitative theory is complex.

Although the line of reasoning of long-range interactions has influenced modelling of dislocation behaviour considerably, a number of serious critiques have been uttered. The first criticism is that pile-ups as suggested by the long-range stress approach are not stable. They dissociate in dislocation dipoles (closely spaced dislocations of different sign on two parallel slip planes) of which the long-range stress is considerably less than for the pile-up[s]. The second set of criticism[t] indicates that the estimates of stress and the concept of back stress used are not altogether clear. Nabarro et al.[u] have discussed certain other difficulties in connection with this theory.

[s] Kuhlmann-Wilsdorf, D., van der Merwe, J.H. and Wilsdorf, H.G.F. (1952), Phil. Mag. **43**, 632.
[t] Hirsch, P. B. (1975), page 189 in *The physics of metals*, vol. 2: *Defects*, P.B. Hirsch, ed., Cambridge University Press, Cambridge.
[u] Nabarro, F.R.N., Basinski, Z.S. and Holt, D.B. (1964), Adv. Phys. **13**, 190.

Short-range interaction approach

In short-range interaction models the hardening is attributed to much more local forces than the long-range stress field. According to Kuhlmann-Wilsdorf stage I occurs via one of two mechanisms. The first mechanism is a gradual filling of a crystal with dislocations. In this stage the dislocation density is shown experimentally to be very non-uniform, implying that some regions are dislocation free and others have a high density. Stage I ends when the dislocation density has become quasi-uniform. In stage II the density remains then quasi-uniform but it increases with increasing strain. The second mechanism for stage I is the formation of irregular patterns of tangles and stage I ends when their formation is more or less complete. For stage II the tangle structure remains but their spacing decreases in stage II. The latter description is the preferred view since using electron microscopy dislocation tangles[v] are frequently observed in cold-worked metals. In stage II hardening then occurs because the segments of dislocation lines that act as Frank-Read sources become increasingly smaller. Stage III is explained using the same model but with constant cell size.

The expression for the momentary yield strength is again considered a basic equation since it has an overwhelming experimental support and reads

$$\tau = \tau_0 + \alpha Gb/l \tag{16.17}$$

where τ_0 is the friction stress and $\alpha Gb/l$ the contribution necessary for the bowing out of dislocations. In all cases $\alpha \cong \frac{1}{2}$ within a factor of two or so. As before, l is the distance between the pinning points or link distance. It is also used to derive an expression for stage II hardening using the 'principle of similitude', which says that once Nature has chosen a particular geometry, increasing stress will cause the structure to remain essentially similar to itself apart from a shrinkage such that $l \sim \tau^{-1}$. This principle can be rationalised by considering that for an applied stress at any position of a dislocation line, the sum of all stresses due to the other dislocations and the friction stress (which can be usually neglected) must be equal to the applies stress. Once an optimal structure has been reached, the same structure will do at any other stress since the dislocation stresses scale inversely with the distance to their axis. Now it is clear that independent of the structure of the network, the average radius to which a loop can expand before merging with the network is proportional to the momentary link distance l. Therefore, we may write

$$l = m/\sqrt{\rho} \tag{16.18}$$

with m a numerical parameter. The increment in shear strain $d\gamma$ due to the emission of dn new dislocations is

$$d\gamma = qbl^2 dn \tag{16.19}$$

where q is another parameter. The remainder of the dislocation loops increases the dislocation density with

$$d\rho = p/dn = (p/qbl)d\gamma = (p\sqrt{\rho}/qmb)d\gamma \tag{16.20}$$

with p still another parameter. From Eqs. (16.17) and (16.20) we obtain

---

[v] The existence of dislocations tangles can also be rationalised by long-range stress argument, see e.g. de Wit, R. (1963), Trans. Met. Soc. AIME **227**, 1443 or Weertman, J. (1963), Trans. Met. Soc. AIME **227**, 1439.

$$\frac{\partial \tau}{\partial \gamma} = \frac{\partial \tau}{\partial \rho} \frac{\partial \rho}{\partial \gamma} = \frac{\alpha p G}{2qm} \qquad \text{or} \qquad \frac{\partial \tau}{\partial \gamma} = cG = \theta \qquad (16.21)$$

i.e. the familiar linear work-hardening law with a coefficient θ proportional to the shear modulus G. According to the model this type of hardening must be expected under a wide range of conditions provided the dislocations primarily interact with themselves and the principle of similitude is valid so that the parameters p, q and m remain constant, in agreement with the wide-spread occurrence of stage II behaviour with an approximate slope of $G/300$ under many different conditions. By means of stereology the parameters as occurring in the theory were quantitatively determined[w] for 99.99% pure Fe for a strain range of 22% to 740% true strain. The theoretically computed work hardening curve reproduced the shape of the experimental work hardening curve within about 2% using only the friction stress τ_0 as a parameter since it cannot be determined from the micrographs of the dislocation structure. It appeared that the geometrical parameters appear only in one grouping and the resulting value for this parameter was 1/1780 while from experimental work hardening curve 1/2240 was obtained. This agreement was held within the limits of the experiment, also given that the influence of the friction stress was neglected.

Stage II continues as long as the principle of similitude is obeyed but this will break down at some moment. It has been proposed that stage II terminates because the cells do not shrink any further at a certain moment. If this is the case[x] the average cell diameter L_{cell} replaces $q^{1/2}l$ in Eqs. (16.19) and (16.20) so that we have for stage III

$$\frac{\partial \tau}{\partial \gamma} = \frac{\partial \tau}{\partial \rho} \frac{\partial \rho}{\partial \gamma} = \frac{\alpha G p}{2(q\rho)^{1/2} L_{cell}} = \frac{\alpha^2 G^2 bp}{2q^{1/2} L_{cell}(\tau - \tau_0)} \qquad (16.22)$$

where the last step is made using Eq. (16.17). By integration we obtain

$$\tau - \tau_{III} = \alpha G \sqrt{bp/L_{cell}\sqrt{q}} \sqrt{\gamma - \gamma_{III}} \qquad (16.23)$$

which is the well-known parabolic law starting at τ_{III} and γ_{III}. In fact, if in the derivation for stage II and stage III the more accurate expression for the Frank-Read stress $\tau = \tau_0 + (\alpha Gb/l)\ln(l/b)$ is used, a small curvature for stage II is introduced while stage III becomes not strictly parabolic. Although the deviations are small, they have been observed experimentally. We refer to the original paper[y] for details.

In recent work Kuhlmann-Willsdorf emphasizes[z] that the dislocation network is formed almost instantaneously at small strain and that at low or moderate temperature once straining stops the dislocation network remains essentially intact. The dislocation network is thus formed by the highest stress applied in the plastic regime. From this she concludes that the dislocations are located in wells of relative minimum energy, the position of which is only marginally dependent on stress changes between the previously applied highest stress and the same stress but with inverse sign. Therefore, during plastic deformation the dislocations are in fact always close to equilibrium but

---

[w] Kuhlmann-Willdsorf, D. (1970), Metall. Trans. **1**, 3173.
[x] It is appears to be the most logical and likely assumption, which has been tested experimentally only limitedly.
[y] Kuhlmann-Willdsorf, D. (1968), page 97 in *Work hardening*, Hirth, J.P. and Weertman, J., eds., Gordon and Breach, New York.
[z] Kuhlmann-Wilsdorf, D. (2001), Mater. Sci. Eng. **A315**, 211. See also Kuhlmann-Wilsdorf, D. (1999), Phil. Mag. **A79**, 955.

upon unloading the structures are frozen-in so that a structure relatively far from equilibrium remains. On these premises a variety of phenomena can be explained. The approach as advocated by her was originally addressed for obvious reasons as *mesh-length theory* but nowadays, due to the different emphasis, generally is referred to as *Low Energy Dislocation Structure* (LEDS) *theory*.

Apart from this line of reasoning, several other short-range interaction approaches have been forwarded. We mention briefly four. Gilman's theory[aa] considers hardening to be due to the increasing amount of 'debris' that is left when a screw dislocation moves through the crystal. This 'debris' consists of dislocation dipoles (closely spaced dislocations of different sign on two parallel slip planes), which interact with the dislocations that follow. In Mott's and Hirsch's theory[bb] the major part of hardening in stage II is due to the presence of jogs in dislocation lines, formed when a dislocation cuts through the forest dislocations. It has also been considered that the interaction with the forest dislocations is the main cause[cc] for hardening. An extensive statistical-thermodynamical theory[dd] has been developed by Kocks et al.. More recent general reviews on this intricate topic are presented by Sevillano[ee] and Argon[ff]. These reviews emphasize the thermal activated nature of the formation of the dislocation network. This brings us to another essential difference between the long-range and short-range interaction models. In fact the image as provided with the LEDS theory indicates that the formation of the dislocation structure can be considered to be due more or less directly as a result of the applied stress. This is in contrast to the long-range stress models where thermal activation of each part of the dislocation network plays an important role. Finally, we note that, in spite of the common opinion that work hardening is poorly understood, the LEDS theory is capable of explaining many of the phenomena encountered.

Simulations

Nowadays simulation of dislocation structures can be done using, e.g. the elastic interaction energy for each dislocation line segment with all other segments. After adding the applied stress and introducing thermal motion the total energy can be minimised. In short, two approaches, both using the stress field of dislocation segments as obtained from elasticity theory, are possible. First, a continuum approach in which the position of the dislocation in space is completely free and, second, a lattice approach in which the dislocations are supposed to move on a discrete lattice. The former approach usually yields more accurate results in cases where the local stresses at certain positions are important but at the expense of elaborate calculations. The latter approach results for situations where the overall configuration is determining in sufficiently reliable answers but requires less computing power. Another approach[gg] is a hybrid method in which a representative unit of a discrete dislocation configuration is embedded in a continuum matrix. Within the representative unit the dislocations dynamics are solved, at a time scale of say 10^{-9} s,

[aa] Gilman, J.J. (1962), J. Appl. Phys. **33**, 2703.
[bb] Mott, N.F. (1960), Trans. Met. Soc. AIME **218**, 962 and Hirsch, P.B. (1962), Phil. Mag. 7, 67.
[cc] Basinski, S.J. and Basinski, Z.S., page 261 in Nabarro (1979).
[dd] Kocks, U.F., Argon, A.S. and Ashby, M.F. (1975), Progr. Mater. Sci. **19**, 1.
[ee] Sevillano, J.G., page 19 in Mughrabi (1993).
[ff] Argon, A.S. (1996), Page 1877 in *Physical metallurgy*, vol. 3, Cahn, R.W. and Haasen, P., eds., North-Holland, Amsterdam.
[gg] Devincre, B., Kubin, L.P., Lemarchand, C. and Madec, R. (2001), Mater. Sci. Eng. **A309-310**, 211.

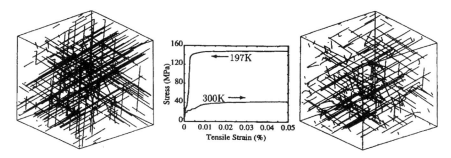

Fig. 16.12: Simulation of the dislocation structure in BCC Ta single crystal with initial dislocation density of 10^{11} cm^{-2} and the associated stress-strain diagram at 197 and 300 K.

for a given state of stress yielding the resulting plastic strain increment. At the continuum level the equations of motion are solved using FEM on a time scale of say 10^{-8} s, given the input of the strain increment of the representative unit and resulting in a new overall stress distribution. This process is then iterated. Thus hybrid approach, although fully numerical, seems to be a most promising one.

An example of a simulation is shown in Fig. 16.12[hh], which shows the dislocation structure for Ta metal and the resulting stress-strain curve at two temperatures. At low temperature the structure contains mostly elongated screw dislocations while at higher temperature the fraction short edge dislocations increases. Kubin[ii] has given an extensive review of dislocation patterning.

Finally, it should be noted that these simulations are based on models based on thermally activated processes, precisely the type of processes to which Kuhlman-Wilsdorf objects for work hardening. Therefore, not every researcher in the field will subscribe the results of these simulations.

16.4 Plastic deformation in polycrystals

For polycrystals various other aspects are important in relation to the understanding and prediction of the plastic behaviour. First, the effect of the polycrystallinity itself on the calculation of the stress-strain curve of a polycrystal from that of a single crystal but also the influence of the interaction between dislocations and grain boundaries, impurities and precipitates. These aspects will be discussed in the next paragraphs (Cotttrell, 1953; Kovacs and Szoldos, 1973).

In polycrystals the number of slip system available is important because a particular slip system in neighbouring grains has a different orientation with respect to the tensile direction. Von Mises[jj] has shown that for a polycrystalline material minimally 5 independent slip systems are required for plastic deformation, continuous over the polycrystal. A slip system is independent when it causes a change in shape that cannot be reached via (a combination of) other slip systems. The number of 5 independent slip systems arises because for an arbitrary deformation 6 independent strain components are present. However, generally during plastic deformation the volume is conserved so that only 5 independent strain components are required. FCC metals have 12 slip systems but of these 12 there are only 5 independent ones. Table

[hh] Tang, M., Kubin, L.P. and Canova, G.R. (1998), Acta Mat. **46**, 3221.

[ii] Kubin, L.P., page 131 in Mughrabi (1993).

[jj] Von Mises, R. (1928), Z. Angew. Math. Mech. **8**, 161.

15.1 shows the slip systems for various structures. From that table it is clear that for inorganic materials generally secondary slip systems are necessary for plastic deformation to occur. These slip systems only become active at elevated temperature. On the other hand in many metals sufficient slip systems are available so that extensive plastic deformation at room temperature is possible.

The requirement of simultaneous availability of slip systems has been named *slip flexibility* by Kelly[kk]. MgO provides a good example of the many requirements that has to be satisfied before plasticity in polycrystalline material can occur. Below 350 °C $\{110\}\langle1\bar{1}0\rangle$ slip occurs. This yields only two independent slip systems and the polycrystal is brittle. Above 350 °C slip occurs on the $\{001\}$ planes and this does yield 5 independent slip systems. However, the critically resolved shear stress for slip on $\{001\}$ is much larger than for slip on $\{110\}$ so that still fracture occurs relatively easy. At 1500 °C the shear stresses become approximately equal so that the polycrystal should be ductile, however, cross-slip cannot occur. Only at 1700 °C this becomes possible resulting in plastically deformable polycrystalline MgO.

The stress-strain curve for a polycrystal can be calculated from the one of a properly oriented single crystal. To that purpose consider first plastic deformation only. The resolved shear stress τ for the single crystal, as discussed in Chapter 15, is given by

$$\tau = \sigma \cos\lambda \cos\phi = \sigma\, m = \frac{\sigma}{M} \tag{16.24}$$

where m denotes the *Schmid factor* and its reciprocal M the *orientation factor* and σ the applied normal stress. For the polycrystal a proper average orientation factor $\overline{M}$ has to be estimated. Averaging over a uniform distribution[ll] yields $\overline{M} \cong 2.2$, which does not corresponds with experiment. For FCC metals the best estimate $\overline{M} \cong 3.1$ is based on the von Mises condition of 5 independent slip systems and the assumption of uniform strain[mm]. The same factor is present between the increment in plastic shear strain $d\gamma_{\text{pla}}$ and the increment in plastic normal strain $d\varepsilon_p$ since the strain is assumed to be uniform and continuous. For the polycrystal it thus holds that

$$\frac{\sigma}{\tau} = \overline{M} = \frac{d\gamma_{\text{pla}}}{d\varepsilon_{\text{pla}}} \tag{16.25}$$

If the distribution of orientation factors is not changing during deformation, the expression $d\gamma_{\text{pla}} = \overline{M}d\varepsilon_{\text{pla}}$ can be integrated to $\gamma_{\text{pla}} = \overline{M}\varepsilon_{\text{pla}}$. The plastic part of the stress-strain curve in tension for the polycrystal is then given by

$$\sigma = \overline{M}\tau(\gamma_{\text{pla}}) = \overline{M}\tau(\overline{M}\varepsilon_{\text{pla}}) \tag{16.26}$$

where $\tau = \tau(\gamma)$ is the stress-strain relation for the corresponding single crystal. The hardening modulus for the FCC polycrystal is thus about 9.5 times the one for the single crystal because

$$\blacktriangleright \qquad \frac{d\sigma}{d\varepsilon} = \overline{M}^2 \frac{d\tau}{d\gamma} \tag{16.27}$$

[kk] Kelly, A. and McMillan, N.H. (1966), *Strong solids*, 3rd ed., Clarendon, Oxford.
[ll] Sachs, G. (1928), Z. Ver. Deut. Ing. **72**, 734.
[mm] Taylor, G.I. (1938), J. Inst. Met. **62**, 307.

The influence of the microstructure on plasticity is thus considerable. In general elastic effects have to be incorporated, as the above description is valid only after plastic deformation sets in. The most appropriate single crystal stress-strain curve for calculating the polycrystal stress-strain curve are those on which several slip systems are active right from the beginning of the deformation[nn]. In particular the [001] and [$\bar{1}$11] orientations give a much steeper stress-strain curve than other orientations. In this way reasonable agreement with experimentally determined curves is possible.

Problem 16.4

For a single crystal the stress-strain curve in terms of the shear stress τ and shear strain γ is given in Fig. 16.13. The single crystal satisfies initially $\tau = G\gamma$ with a Poisson's ratio $\nu = 0.25$ and starts to yield at τ_0 on the relevant crystallographic plane. Thereafter hardening occurs with a hardening modulus $g = d\tau/d\gamma$. A fully dense, isotropic homogeneous polycrystal of the same composition is made. This polycrystal is tested in uniaxial tension.

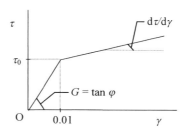

Fig. 16.13: Schematic τ-γ curve (not to scale) of a shear test on a single crystal.

Calculate for that polycrystal Young's modulus E, the yield strength Y and the hardening modulus $h = d\sigma/d\varepsilon$.

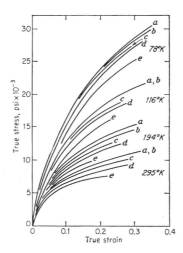

Fig. 16.14: The influence of the grain size on stress-strain curve for Al. The mean grain size is: $a = 0.11$, $b = 0.13$, $c = 0.21$, $d = 0.33$ and $e = 0.53$ mm.

[nn] Kocks, U.F. (1958), Acta Met. **6**, 85.

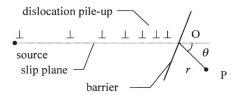

Fig. 16.15: Schematic view of dislocation pile-up at a barrier.

16.5   The influence of boundaries

Generally the mechanical behaviour is strongly influenced by the presence of grains. In the previous section we estimated the influence on the hardening modulus. At intermediate temperature polycrystalline metals generally are stronger and harder than the corresponding single crystal metals. For example, the critical shear stress for Cu single crystal is ~1 MPa whereas for polycrystalline Cu it is ~16 MPa. The presence of the grain boundaries hinders the motion of the dislocations. Inter alia, at elevated temperature the presence of grain boundaries becomes a disadvantage since the creep rate is generally inversely proportional to the grain size. Fig. 16.14[oo] shows the large influence of both the grain size and temperature for pure Al. Small grain size and low temperature favour hardening, the effect of temperature being the most pronounced. Obviously the presence of grain boundaries as obstacles for the motion of dislocations plays an important role. We will see that other obstacles do play a significant role as well and we discuss these effects in the next paragraphs. Stroh[pp] has provided an early but still very readable review of dislocations in relation to barriers.

A grain boundary is an obstacle for dislocations since dislocations move over specific slip planes. Because the slip planes do not continue into the neighbouring grain, a running dislocation is stopped at the boundary (Fig. 16.15). The walls of the cells in a dislocation network provide a similar barrier and we will in this section refer to both as boundaries. Dislocations of the same sign do exert a repulsive force upon each other and hence the dislocations do pile-up at the boundary. This process continues until the externally applied stress equals the back stress field of the n dislocations in the pile-up. The number of dislocations that can occupy a distance L under an effective resolved shear stress τ_{eff} along a slip plane between the source and an obstacle is given[qq] by $n = \alpha\pi\tau_{\text{eff}}L/2Gb$ with as before $\alpha = 1$ for a screw dislocation and $\alpha = 1-\nu$ for an edge dislocation. A factor ½ is introduced if the back stress on the source arises from dislocations piled up on both sides of the source. We have, so to speak, obtained a giant dislocation with Burgers vector of length nb. It appears that the centre of gravity of this pile-up is located at $3L/4$ and the total slip therefore can be thought to be due a single dislocation of strength nb moving over a distance $3L/4$. With reference to Fig. 16.15 the tensile stress σ normal to the plane OP (which is perpendicular to the plane of the paper) at a distance r is given by

$$\sigma = \frac{3}{2}\left(\frac{L}{r}\right)^{1/2}\tau_{\text{eff}}\sin\theta\cos\frac{\theta}{2}$$

[oo] Dorn, J.E., Pietrokowsky, P. and Tietz, T.E. (1950), Trans. AIME **188**, 933.
[pp] Stroh, A.N. (1958), Adv. Phys. **6**, 418.
[qq] Eshelby, J.D., Frank, F.C. and Nabarro, F.R.N. (1951), Phil. Mag. **42**, 351.

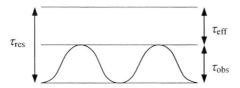

Fig. 16.16: The effective stress τ_{eff} as a result of the resolved stress τ_{res} and the obstacle stress τ_{obs}.

The maximum of σ occurs at $\cos\theta = \frac{1}{3}$ or $\theta = 70.5°$ for which $\sigma = 2\tau_{eff}(L/r)^{1/2}/\sqrt{3}$. Similarly[π] the shear stress τ on the plane OP at a distance r is $\tau = \beta\tau_{eff}(L/r)^{1/2}$ with β an orientation dependent factor close to unity.

If the stress concentration ahead of such a giant dislocation is large enough, a dislocation source in the neighbouring cell or grain can be activated, which at his turn will emit dislocations. This occurs when the total shear stress τ_{tot} generated by the pile-up equals the nucleation stress τ_{nuc}, i.e. $\tau_{tot} = \tau_{nuc}$. The effective resolved stress, i.e. the resolved stress τ_{res} minus the stress necessary for passing obstacles on the glide plane τ_{obs} (Fig. 16.16), is $\tau_{eff} = \tau_{res} - \tau_{obs}$. The total stress on the new glide plane is given by $\tau_{tot} = \tau_{obs} + \tau$ and thus we obtain finally

$$\tau_{obs} + \beta(\tau_{res} - \tau_{obs})\left(\frac{L}{r}\right)^{1/2} = \tau_{nuc} \tag{16.28}$$

For L we now take the grain diameter $2R$ while for r we take the distance from the head of the pile-up to the nearest source in the neighbouring grain ($r \ll R$). Solving for τ_{res} we obtain

$$\blacktriangleright \qquad \tau_{res} = \tau_{obs} + \frac{(\tau_{nuc} - \tau_{obs})r^{1/2}}{\beta\sqrt{2}}R^{-1/2} = \tau_{obs} + kR^{-1/2} \tag{16.29}$$

Assuming that on average the shear stress τ equals half the tensile stress σ, i.e. $\tau = \sigma/2$, a similar relation results for the tensile stress. From TEM experiments the nucleation of the new dislocations appears to occur at the grain boundary in the new grain, consistent with that the factor k experimentally appears to be independent of temperature. The obstacle stress τ_{obs} depends strongly on temperature, strain and impurity content.

Eq. (16.29) is the *Hall-Petch relation*, first obtained experimentally[ss], which tells us that the yield strength is inversely proportional to the square root of the cell or grain size. In the initial phase of the deformation process the relevant size is the grain size since at that stage the dislocation network has not yet formed and describes the influence on the initial yield strength. The effect is illustrated in Fig. 16.17[tt] for various types of steel. In heavily worked metals the cell size of the network replaces the grain size. It should be noted that the Hall-Petch expression must be used with care. If the dislocation pile-up model is valid, sufficient dislocations should be able to pile-up and this is only possible for sufficiently large grain size.

π Stroh, A.N. (1954), Proc. Roy. Soc. London **223**, 404.
ss Hall, E.O. (1951), Proc. Phys. Soc. London **643**, 717 and Petch, N.J. (1953), J. Iron Steel Inst. London **173**, 25.
tt Cracknell, A. and Petch, N.J. (1955), Acta Met. **3**, 186.

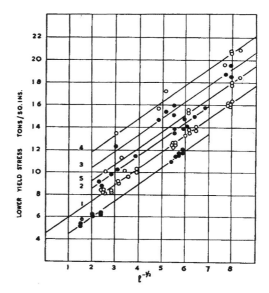

Fig. 16.17: The Hall-Petch relation showing the (lower) yield strength as a function of the mean grain size *l* for several steels.

Another explanation for the Hall-Petch expression concentrates on the influence of grain size on the dislocation density ρ. The yield strength (see Section 16.1) in terms of ρ is $Y = Y_{obs} + \alpha G b \rho^{1/2}/2\pi$, where Y_{obs} is a frictional stress due to obstacles and $\alpha = 1-\nu$ for edge and $\alpha = 1$ for screw dislocations. Experimentally $\frac{1}{2}\alpha$ is indeed a constant that appears to have a value of 0.3 to 0.6. Further it is found that the dislocation density is reciprocally dependent on the grain size D, i.e. $\rho = 1/D$. This leads to $Y = Y_{obs} + \frac{1}{2}\alpha G b D^{-1/2}$ without any reference to dislocation pile-up.

16.6   Yield point phenomena*

In Section 16.2, we have seen that slightly impure BCC metals show an upper and lower yield point in the stress-strain curve. In fact, this phenomenon may also be present in other impure metals. The explanation for BCC metals, in particular Fe, is in terms of the impurities concentrating around a dislocation, as proposed by Cottrell (1953).

However, for other materials another explanation may be more relevant and this one was originated by Johnston and Gilman[uu] and further elaborated by Hahn[vv]. It started with the observation that for LiF dislocations, already present at the beginning of a test, were strongly pinned so that the applied stress never made them move. Instead new dislocations were created during deformation. If a material is tested at constant strain rate $\dot{\gamma}$, the strain rate consists of an elastic component $\dot{\gamma}_{ela} = \dot{\tau}/G$ with shear modulus G and rate of change of stress $\dot{\tau}$ and a plastic component $\dot{\gamma}_{pla} \cong b\rho v$ with b the magnitude of the Burgers vector, ρ the dislocation density and v their velocity. In Chapter 15 we have seen that v is a strong function of the applied stress τ,

[uu] Johnston, W.G. and Gilman, J.J. (1959), J. Appl. Phys. **30**, 129 and (1962), J. Appl. Phys. **33**, 2716.
[vv] Hahn, G.T. (1962), Acta. Met. **10**, 727.

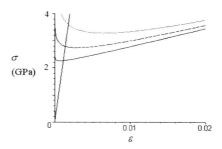

Fig. 16.18: The stress-strain curve illustrating the upper and lower yield point as calculated from the Johnston-Gilman-Hahn theory for $n = 5$, 10 and 35. For $n = 35$ a clear yield point is observed.

which can be described by $v = v_0(\tau/\tau_0)^n$. The value of n is typically high, say 35. It has been experimentally observed that $\rho = C\gamma_{\text{pla}}^a + \lambda$, where C and $a \cong 0.7\text{-}1.5$ are two parameters and λ the density of the unpinned dislocations initially present. Assuming linear strain hardening, there is also a back stress acting on the dislocations, described by $q\gamma_{\text{pla}}$ where q is a constant. The strain rate thus becomes

$$\dot{\gamma} = \frac{\dot{\tau}}{G} + bv_0\left(C\gamma_{\text{pla}}^a + \lambda\right)\left(\tau - q\gamma_{\text{pla}}\right)^n$$

Because of the high value of n, the expression is dominated[ww] by either the first term or the second term on the left-hand side. Hence the stress is either

$$\tau = G\gamma \quad \text{or} \quad \tau = q\gamma_{\text{pla}} + \left[\frac{\dot{\gamma}}{bv_0(C\gamma_{\text{pla}}^a + \lambda)}\right]^{1/n}$$

The behaviour is illustrated in Fig. 16.18. The total stress-strain curve is a combination of these two curves. The lower yield point occurs because the plastic contribution initially decreases with increasing strain. For smaller values of λ, a larger initial decrease occurs. It should be noted that there is also a large influence of the parameter n and the dislocation density ρ. In general the occurrence of a yield drop requires a low initial dislocation density, a dislocation velocity not too rapidly increasing with applied stress and a high dislocation multiplication rate. Using this line of thought Gilman and Johnston were able to explain the yield point behaviour and hardening of LiF. Hahn extended this approach also to BCC metals (Fe-3.25% Si, mild steel and W at 250 °C), where impurities readily can lock dislocations. In fact he concluded that the importance of unlocking dislocations in BCC metals is possibly overstated. The yield drop effect can also be realised in FCC metals via alloying or irradiation.

16.7   Solid solutions and dislocations

Often metals are not used in a pure state but on purpose as alloys. Since a solute atom introduces a deviation from the lattice regularity one can expect that in solid solutions dislocations interact with the solute atoms. Therefore the stress to move the dislocations, i.e. the yield strength, increases. The temperature dependence of the yield

---

[ww] With a change of a factor two in $\tau - q\gamma_{\text{pla}}$, the last factor changes by about 10^{10}.

strength is already briefly indicated in Section 16.1 and schematically given in Fig. 16.4. Generally it is more pronounced for solid solutions than for pure metals. The temperature dependence of the yield strength $\tau(T)$ can be described by

$$\tau(T) = \tau_T(T) + \tau_{pla}$$

where τ_{pla} is the value of the plateau. Most of the earlier theories deal only with the plateau stress τ_{pla}. From a phenomenological point of view it is important to distinguish between (*spherically*) *symmetrical* and *non-symmetrical* defects. We recall from Chapter 8 that an atom can dissolve as a *substitional* atom or as an *interstitial* atom. Generally symmetrical defects are either substitutional atoms or vacancies while interstitial atoms form non-symmetric defects. The effect on the hardening behaviour is quite different. Symmetric defects typically show a value for $d\tau/dc$ (with c the solute mole fraction) of $G/10$ to $G/20$ while for non-symmetric defects this values ranges from $2G$ to $10G$, where G is the shear modulus. The reason behind this is that symmetric stress fields interact with the edge dislocations only since their stress field contains a dilatational component. Non-symmetric defects interact also with screw dislocations because in this case the stress field of the dislocation is purely shear.

Various factors determine the magnitude of the hardening effect with increasing solute mole fraction c. They are:

- The *size* (or *paraelastic*) *effect*, which is due to the (usually) different size of a solute atom as compared with the atoms of the matrix. This introduces an internal elastic deformation, characterised by $\delta = a^{-1}\partial a/\partial c$ with a the lattice constant.
- The *modulus* (or *diaelastic*) *effect*, due to the fact that a solute material has a different shear modulus as the matrix, characterised by $\eta = G^{-1}\partial G/\partial c$ with G the shear modulus. This contribution is of second order but since η is usually much larger then δ both the size and modulus effect make a significant contribution.
- An *electrical effect* can be present if the solute atom introduces an excess or deficit on electrons. For example, if the solute atom provides an extra electron, the charge will be smeared out leaving the solute atom with a slight positive charge, which on its turn interacts with the dislocation.
- A *chemical effect* (or *Suzuki effect*), largely due to the fact that the solubility for a solute atom can be different in a stacking fault as compared with the matrix. Since partial dislocations locally introduce stacking faults, this leads to extra interaction.
- Finally, a *configuration* (or *Fisher*) *effect*, due to the fact that solutes atoms usually are not really randomly distributed in the matrix. They can either form clusters or produce locally short-range order. If a dislocation moves through e.g. a short-range ordered volume, a change in like-unlike bonds occurs which lead to an energy change since like-unlike bonds are preferred. A similar argument can be given for like-like bonds in a cluster of solute atoms.

We will discuss in the next section briefly the various theories that exist for solid solution hardening. In many cases only the size and modulus effect are explicitly taken into account. Nabarro[xx] has given an early review of solid solution hardening. Haasen[yy] has given an extensive overview of solid solution hardening in FCC metals,

[xx] Nabarro, F.R.N. (1975), page 152 in *The physics of metals*, vol. 2 *Defects*, Hirsch, P.B. ed., Cambridge University Press, London.
[yy] Haasen, P., page 155 in in Nabarro (1979). See also (1993), *Physical metallurgy*, 3rd ed., Cambridge University Press, Cambridge.

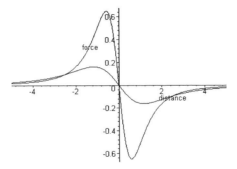

Fig. 16.19: The interaction of an edge dislocation and a solute atom for $z = 1$ and $z = 2$ due to the size effect.

which we follow here to some extent, while Suzuki[zz] has done the same for BCC metals. Neuhauser and Schwink[aaa] provide a more recent and extensive overview.

Relevant interactions*

In this section we discuss briefly the interaction between dislocations and solute atoms using elastic modelling. We deal with the size and modulus effect only.

The size effect introduces an internal elastic deformation, which is characterised by $\delta = a^{-1}\partial a/\partial c$ with a the lattice constant. In Section 15.9 the interaction energy for a solute atom with a straight edge dislocation using elastic modelling was found to be

$$U_{\text{size}} = p\Delta V\xi$$

with the pressure

$$p = \left(\sigma_{xx} + \sigma_{yy} + \sigma_{zz}\right)/3 = -\frac{Gb}{3\pi}\frac{1+v}{1-v}\frac{\sin\alpha}{r}$$

where $\alpha = \sin^{-1}(z/r)$ is the angle between the glide plane and the vector **r** with length $r = (x^2+y^2+z^2)^{1/2}$ representing the distance between the dislocation and solute atom. The volume difference ΔV is given by

$$\Delta V = 4\pi\varepsilon r_s^3/3 \quad \text{with} \quad \varepsilon = (r_s'-r_s)/r_s \quad \text{or} \quad \Delta V = 3\Omega\delta$$

where r_s' is the radius of a solute and r_s is the radius of a matrix atom, respectively and $\Omega = 4\pi r_s^3/3$ the original volume of the matrix atom at the position of the solute atom. The factor

$$\xi = 3(1-v)/(1+v)$$

takes into account the energy stored in the volume of the defect. The force on the dislocation in the slip direction x due to a solute atom at (x,z) therefore becomes

$$f_{\text{size}} = -\frac{\partial U_{\text{size}}}{\partial x} = -\frac{Gb\Delta V}{\pi z^2}\frac{2(x/z)}{[1+(x/z)^2]^2} \equiv -\frac{Gb\Delta V}{\pi z^2}\varphi(x/z) \qquad (16.30)$$

This expression is plotted in Fig. 16.19 for two values of constant z. To obtain the maximum force we take $z = b/\sqrt{6}$, corresponding to one half separation of the {111}

[zz] Suzuki, H., page 191 in Nabarro (1979).
[aaa] Neuhauser, H. and Schwink, C., page 191 in Mughrabi (1993).

planes in the FCC lattice and obtain after some calculation $f_{size}^{max} \cong Gb^2\delta$. From the values of δ for several metal atoms in Cu as given in Table 16.1 it appears that typically $\delta \cong 0.1$.

Since a screw dislocation does not possess a dilatational stress field, the interaction with a symmetrical defect is zero. However, for a non-symmetrical defect, e.g. with a tetragonal deformation such as carbon in α-Fe, there is a finite interaction. In this case the strain is given by ε_{11} and $\varepsilon_{22} = \varepsilon_{33}$ with all other components zero. The interaction energy is given by

$$U = \varepsilon_{ij}\sigma_{ij}\Omega$$

with the stress field for a screw dislocation given by Eq. (15.21). This leads to a similar expression as Eq. (16.30) but with $\delta = (\varepsilon_{11} - \varepsilon_{22})/3$. In the specific case of carbon in α-Fe the strains are given by $\varepsilon_{11} = 0.38$ and $\varepsilon_{22} = -0.03$ and this misfit provides a strong obstacle for dislocation motion in Fe.

The modulus effect is characterised by $\eta = G^{-1}\partial G/\partial c$ with G the shear modulus. It can be shown that the deformational interaction energy due to the shear interaction of a dislocation with a solute atom is given by

$$U_{mod} = \eta e\Omega = Gb^2\eta\Omega/8\pi^2(1-v)^2r^2$$

where e is the energy density, so that the force becomes

$$f_{mod} = -\frac{\partial U_{mod}}{\partial x} = -\frac{Gb^2\eta\Omega}{8\pi^2(1-v)^2z^3}\varphi(x/z)$$

In principle this contribution is of second order since U_{mod} decreases as r^{-2}. However, since η is usually much larger than δ (see Table 16.1), both the size and modulus effect do contribute significantly. The size and modulus interactions have a different sign for $z < 0$ and the same sign for $z > 0$. They cannot therefore be simply added.

Modelling of solid solution hardening*

Although the detailed theories about solid solution hardening are diverse, it has become clear more or less right from the beginning[bbb] that one has to distinguish between three different regimes. Referring with $\lambda = a/c^{1/2}$ (where c is the solute mole fraction and a^2 the area per atom on the slip plane) to the mean distance of the solute atoms and with $l = \alpha Gb/\tau$ to the pinning distance of a dislocation bow-out under a

Table 16.1: Interaction parameters for Cu at room temperature.

	δ (%)	c_{lim} (at%)	η_p	η_s
Al	6.7	10	−0.58	−0.68
Ge	9.2	10	−1.2	−2.0
Mn	10.5	15	−0.55	−0.65
Ni	−2.9	30	0.60	0.63
Si	1.8	10	−0.70	−1.2
Zn	6.0	15	−0.48	−0.51

Data from Neuhäuser and Schwink (1993). The parameter c_{lim} denotes the maximum solubility. The labels p and s denote the values for a polycrystal and single crystal, respectively.

[bbb] See e.g. Cottrell (1953).

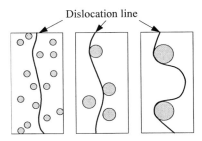

Fig. 16.20: The three different regimes for the interactions between dislocations and solute atoms.

stress τ, we have essentially $\lambda \cong l$, $\lambda \ll l$ and $\lambda \gg l$. Labusch[ccc] introduced a dimensionless number to characterise these situations, also taking into account the forces involved. If the range of the interaction is given by z, the maximum interaction force by f_{max} and the line tension by $T \cong \frac{1}{2}\alpha Gb^2$, he defined the interaction parameter β by

$$\beta = \frac{z}{a}\left(\frac{2cT}{f_{max}}\right)^{1/2}$$

For $\beta \ll 1$, hardening is by dilute, strong obstacles (Fleischer-Friedel regime) while for $\beta \gg 1$ implies concentrated, weak obstacles (Mott-Labusch regime). Fig. 16.20 provides a schematic view of these three regimes.

For the dilute regime the mean spacing between two solute atoms touched by the dislocation under a stress τ can be estimated as follows[ddd]. The area of the grey segment in Fig. 16.21 is given by

$$A = R^2\theta - R^2\cos\theta\sin\theta \cong \frac{2R^2\theta^3}{3} \cong \frac{\lambda^3}{12R} = \frac{\lambda^3\tau b}{12T}$$

where[eee] the line tension $T \cong \frac{1}{2}\alpha Gb^2$ is introduced. One now assumes that the dislocation touches another atom when the area A equals half of the area available per solute atom. The latter is given by $1/(2c_A)$ where $c_A = c/a^2$ is the areal density of the solute atoms. Solving the above equation for λ leads to

$$\lambda = \left(\frac{6T}{b\tau c_A}\right)^{1/3}$$

The force balance is taken at break-through and reads

$$\tau b\lambda(\tau) = f_{max} \tag{16.31}$$

and therefore we have for the Fleischer-Friedel yield strength τ_F for dilute solutions

▶ $$\tau_F b = f_{max}^{3/2} c_A^{1/2} / \sqrt{6T} \tag{16.32}$$

[ccc] Labusch, R. (1970), Phys. Stat. Sol. **41**, 659 and (1972), Acta Metall. **20**,917.
[ddd] Friedel, J. (1964), *Dislocations*, Pergamon, Oxford.
[eee] The first step is the difference between the sector area $\frac{1}{2}R^2 2\theta$ and the triangle area $\frac{1}{2}R^2 \sin2\theta = \frac{1}{2}R^2 2\sin\theta\cos\theta$. In the second step the Taylor series $\cos\theta = 1-\theta^2/2+\cdots$ and $\sin\theta = \theta-\theta^3/6+\cdots$ are used. In the third step the contour length $s = 2\theta R$ is approximated by λ since bow-out is small. In the fourth step with $\tau = \alpha Gb/\lambda$ the pinning distance λ is approximated by $2R$, again because bow-out is small.

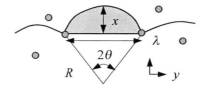

Fig. 16.21: The bowing-out of a dislocation under an applied stress τ to a radius R.

Experimental results show various exponents for c_A, but is must be stated that the measuring τ as a function of c_A is usually loaded with many difficulties (inhomogeneous distribution of solutes, phase separation, influence of solutes on the dislocation structure, inhomogeneous deformation). Computer simulations[fff] support this model though for moderate mole fractions.

From the overall force balance equation (16.31) it is clear that in the dilute regime model the dislocation either experiences no force or the full interaction. For the concentrated regime this cannot be maintained and a distribution function $\rho_f(f)df$ is introduced for the number of atoms touched by a dislocation of unit length moving in the x-direction and having an interaction force with the solute atom at $(x,y) = (0,0)$ between f and $f + df$. When the force-distance profile $f(x)$ of the obstacles is known, the force distribution $\rho_f(f)$ can be transformed to a distance distribution $\rho(x)$. The force balance is then given by

$$\tau b = \int \rho_f(f) f \, df = \int \rho(x) f(x) \, dx \qquad (16.33)$$

When the dislocation meets the solute atom at $y = 0$, a segment of the dislocation close to $x(0)$ lags behind to the average position X of the dislocation line (Fig. 16.22) according to

$$dX - dx(0) = F(0)(df/dx)dx(0)$$

The function $F(0)$ describes the shape of the dislocation under a unit applied stress at $y = 0$ in the $-x$-direction. It appears to be given by

$$F(0) = \frac{1}{2\sqrt{T\kappa}} \quad \text{with} \quad \kappa = \int \rho(x) \frac{df}{dx} \, dx \qquad (16.34)$$

representing the average curvature of the obstacle potential. It is plausible that the maximum interaction takes place where the curvature of the interaction potential is

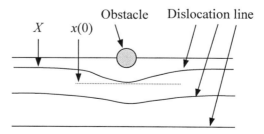

Fig. 16.22: Schematic of the lagging behind of a dislocation line before an obstacle.

[fff] Foreman, A.J.E. and Makin, M.J. (1966), Phil. Mag. **14**, 131; Kocks, U.F. (1966), Phil. Mag. **13**, 541.

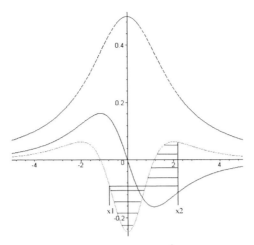

Fig. 16.23: The force f (——), its derive f' (....) and integral $\int f \, dx$ (- - -). The parameters x_1 and x_2 are chosen in such a way that the integral of $\rho = c_A\{1+f'/F(0)\}$ remains constant.

maximal. In a stationary state of overcoming obstacles at the critical stress τ the ratio of the density $\rho(0)$ of dislocation segments waiting in front of the solute atom to the mean density $\bar{\rho} = c_A$ is equal to the inverse of the ratio of the velocities at these points. Therefore, $\rho(0)/\bar{\rho} = \rho(0)/c_A = \bar{v}/v(0) = (dX/dt)/[dx(0)/dt]$. The distribution function $\rho(x)$ can then be calculated as

$$\rho = c_A\left(1+\frac{1}{2\sqrt{\kappa T}}\frac{df}{dx}\right) \quad \text{for } x < x_1 \text{ and } x > x_2$$

$$\rho = 0 \quad \text{for } x_1 < x < x_2 \tag{16.35}$$

Since the distribution function cannot be negative, it is put zero between x_1 and x_2 (Fig. 16.23). The values x_1 and x_2 are determined in such a way that the integral $\int \rho(x) dx$ remains unchanged. Neglecting the term 1 in Eq. (16.35), Eq. (16.33) can be integrated using Eq. (16.34) with as result the Labusch yield strength τ_L for concentrated solutions given by

$$\blacktriangleright \quad \tau_L b = f_{max}^{4/3} c_A^{2/3} z^{1/3} / 2(4IT)^{1/3} \quad \text{and where } I = \int_0^1 \frac{\partial (f/f_{max})}{\partial (x/z)} \, d(f/f_{max}) \tag{16.36}$$

represents a pure number of order one. Experiments do indeed show a preference for the exponent ⅔ instead of ½ as for the Fleischer regime but the remark about experimental difficulties must be repeated. A more complete analysis shows that the neglect of the term 1 can be justified for not too high mole fractions of obstacles and not too weak obstacles. Computer simulations[ggg] show that there is a continuous transition from the Fleischer to the Labusch regime. These simulations therefore allow for an interpolation formula between the two regimes, which reads

$$\tau = \tau_F\left(1+2.5\beta\right)^{1/3}$$

and has, as expected, the property $\tau \to \tau_F$ for $\beta \to 0$ and $\tau \to \tau_L$ for $\beta \gg 1$.

[ggg] Labusch, R. and Schwarz, R.B. (1976), Nucl. Metall. **20**,650.

The model is in reasonable good agreement with some experimental data. For example, for Cu, Ag and Au alloy crystals the $c_A^{2/3}$-dependency is well obeyed and the slopes of the experimental curves are explained satisfactory by the solute parameters δ and η. This agreement is not always obtained though. Further it appears that if τ and T are scaled properly, the expression for τ becomes a unique function, independent of β, as long as $\beta > 1$. This so-called stress-equivalence is also observed experimentally.

An enormous advantage of this theory is that it allows one to superimpose the effects of various obstacles by adding the distribution ρ for each of them. If the range of interaction z is the same a straightforward calculation for two interactions with maximum force $f_{max,1}$ and $f_{max,2}$ and mole fraction c_1 and c_2, respectively, leads to

$$\tau b = \frac{z^{1/3}}{2(4IT)^{1/3} a^{1/3}} \left(c_1 f_{max,1}^2 + c_2 f_{max,2}^2 \right)^{2/3}$$

For the special case of size and modulus effect of one type of solute characterised by

$$f_{size}^{max} \cong Gb^2 |\delta| \quad \text{and} \quad f_{mod}^{max} \cong Gb^2 |\eta| / \zeta$$

respectively, we have as maximum force above the slip plane $f_{size}^{max} + f_{mod}^{max}$ while for the atom below the slip plane it is $f_{size}^{max} - f_{mod}^{max}$. The total is an effective interaction force given by

$$f_{eff}^{max} \cong Gb^2 \left(\delta^2 + \alpha^2 \eta^2 \right)^{1/2} \quad \tau b = \frac{z^{1/3} c_A^{2/3}}{2(4IT)^{1/3}} \mu^{4/3} b^{8/3} \left(\delta^2 + \eta^2 / \zeta^2 \right)^{2/3}$$

with $\zeta = (1-\nu^2)24\pi(z/b) \cong 20$ for a foreign atom in a FCC lattice next to the slip plane. This quadratic addition has been also verified experimentally. Since δ and η/ζ are of the same order of magnitude it is imperative to take both effect into account.

Finally also the temperature dependence of the yield strength can be estimated using this theory. Describing the dislocation velocity v by $v = v_0 \exp(-E/kT)$ it appears that in the Fleischer approach the temperature dependent part of the yield strength is given approximately by

$$\tau(T,v) = \tau_0 \left[1 - \left(\frac{T}{T_0} \right)^{2/3} \right]^{3/2} \quad \text{with} \quad kT_0 = \frac{z f_{max}}{\ln(v_0/v)}$$

According to this model $\tau(T,v)$ should rapidly decrease with temperature and become zero at T_0. In the Labusch approach with $\beta > 1$ the equation of motion has to be solved numerically. It appears that a non-zero plateau becomes visible at $\tau/\tau_0 \cong 0.2$. For details we refer to the references quoted.

16.8   Particles and dislocations

Similar to dissolved atoms, particles dispersed in a matrix can interact with dislocations and thus contribute to an increased yield strength and hardening. Dislocations may be either retarded but move through the particles, occurring in particular if the particle is soft and coherent, or the dislocation may bow out and move around the particle, occurring in particular if the particle is brittle and non-coherent. There are two principally different routes to realise the introduction of particles in a matrix. The first route is via dispersion of small particles in the matrix during the

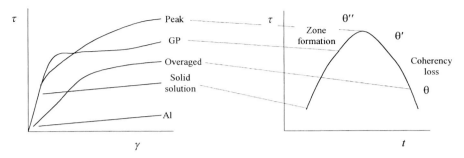

Fig. 16.24: Stress-strain curves and yield strength versus aging time (particle size) for precipitation-hardened alloys.

fabrication process. This leads to *dispersion hardening*. Normally particles of a limited solubility are used and the dispersion leads generally to non-coherent interfaces (see Chapter 1, Fig. 1.5). In principle many options for the particle material exist, e.g. carbides, oxides, nitrides and borides. The second route is via precipitation of dissolved atoms to particles in the matrix via controlled cooling leading to *precipitation* (or *age*) *hardening*. Here a decreasing solubility with decreasing temperature is required which puts limitations on the choice of useful systems. Quenching a super-saturated solution, followed by annealing, creates precipitates and these precipitates interact with dislocations. The precipitates often have coherent interfaces and are mainly expected if the precipitates have the same crystal structure as the matrix and are small. In practice precipitation hardening is applied quite often. The strongest type of metal is obtained by a combination of dispersion hardening and work hardening, e.g. cold drawn steel (piano) wires. Both Nabarro[hhh] and Gerold[iii] have given an early review of precipitation hardening, while Reppich[iij] provides a more recent and extensive overview of particle strengthening.

A typical, though somewhat complex, example[kkk] of precipitation hardening is provided by the Al-Cu system also known as Duralumin®. In Al up to 5 wt% Cu can be dissolved at 550 °C, decreasing to almost zero at room temperature. For the solid solution stage the yield strength is higher as for the pure Al. Slip lines are widely spaced and the behaviour is characteristic for easy glide. In the early stage of aging clusters of Cu atoms are formed, also known as Guinier-Preston zones (GP1). Slip lines become more narrowly spaced. The increase in yield strength and the yield drop suggests that in this stage dislocations move through the particles. The local strains that arise make the alloy harder. During further aging the clusters order on the {100} planes and the result is known as GP2 (or θ''). The hardening rate is increased and the short slip lines suggest that in this stage the dislocations move around the particles. Still further aging lead to coherent $CuAl_2$ particles (or θ') on the {100} planes and the yield strength becomes lower but the hardening rate increases. Finally, relatively coarse incoherent $CuAl_2$ (or θ) particles are formed from the lattice of θ' and this results in a lower strain and therefore lower hardness. In this stage tangles are formed

---

[hhh] Nabarro, F.R.N. (1975), page 152 in *The physics of metals*, vol. 2 *Defects*, Hirsch, P.B. ed., Cambridge University Press, London.
[iii] Gerold, V., page 219 in Nabarro (1979).
[iij] Reppich, B., page 311 in Mughrabi (1993).
[kkk] For a more extended description, see Haasen, P. (1993), *Physical metallurgy*, 3rd ed., Cambridge University Press, Cambridge.

and secondary slip occurs leading to work hardening of the matrix. In this stage the particles experience a large elastic strain due to the plastic deformation of the matrix and the strength is determined by either the yield or the fracture of the particles. The whole process is illustrated in Fig. 16.24.

The amount of hardening due to particles is dependent on many factors, e.g. the distributions of particles, their volume fraction ϕ, their average radius r, their shape and the mean interparticle spacing λ. Similarly as for solute atoms various mechanisms play a role. They are:

- The *size effect*, similar as for solute atoms. The maximum interaction force is approximately $f_{max} \cong Gbr/|\delta|$.
- The *modulus effect*, also similar as for solute atoms. The maximum interaction force is approximately $f_{max} \cong Gb^2|\eta|$.
- For coherent particles the dislocation can pass through the particle. Since the particle is sheared by a distance b, a new interface between particle and matrix with specific energy γ_{pm} is formed leading to a force $f_{max} \cong \phi b \gamma_{pm}$ (Kelly-Nicholson).
- When a dislocation passes through an ordered particle, an antiphase boundary is formed with specific energy γ_{pp}. This leads to a force $f_{max} \cong \phi A \gamma_{pp}$ with A the area involved (Williams).
- When a dislocation passes through a particle with a stacking fault with specific energy γ_p different from the one for the matrix γ_m, a force $f_{max} \cong \phi A(\gamma_m - \gamma_p)$ arises with A again the relevant area (Gleiter-Hornbogen).

In the next section we deal briefly with a few aspects of particle hardening.

Modelling of particle hardening*

A simple model illustrates the dependence of the yield strength on the particle radius r and volume fraction ϕ. It contains two parts, both dependent on r and ϕ. For small particle size, since for particle hardening the volume fraction is usually small, the Fleischer equation as derived for dilute solid solutions

$$\tau_F = f_{max}^{3/2} c_A^{1/2} / b\sqrt{6T}$$

can be used. If the forces mentioned in the previous section are introduced for mechanisms 1, 4 and 5 we obtain

$$\tau_F = \gamma^{3/2} c_A^{1/2} r^{1/2} / b\sqrt{6T} \tag{16.37}$$

where γ represents $Gb|\delta|$ (mechanism 1: size effect), γ_{pp} (mechanism 4: antiphase boundary effect) or $\gamma_m - \gamma_p$ (mechanism 5: stacking fault effect). For mechanism 2 (modulus effect) for some alloys the expression quoted is not in agreement with experiment. However, the situation is complex and will not be discussed here. We see that in this regime the yield strength increases with particle size and concentration.

For large particle size the Orowan mechanism sets in. Assume that a moving edge dislocation meets two obstacles separated by a distance l in the slip plane. The shear stress necessary to make the dislocation pass, meanwhile creating an Orowan loop, is $\tau = \alpha Gb/l$. The number of precipitates per unit volume n is given by

$$n = \frac{\phi}{4\pi r^3/3} \tag{16.38}$$

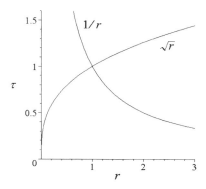

Fig. 16.25: Schematic of particle hardening as the result of the Fleischer and Orowan mechanisms in dimensionless units so that the maximum occurs at (1,1).

while their average distance a is

$$a = n^{-1/3} = (4\pi/3\phi)^{1/3} r \qquad (16.39)$$

and we obtain for the maximum shear stress τ_{max}

$$\tau_{max} = \alpha G b \phi^{1/3} \frac{1}{r} \qquad (16.40)$$

Equating this stress with the yield strength we see that the yield strength in this regime increases with an increasing volume fraction and decreasing radius of the precipitates. The particle-particle distance is overestimated in this way though and seems better to approximate l by $r/c_A^{1/2}$. The qualitative conclusion does not change though. An improved estimate takes into account that the particles have a finite size and that their distance is better characterised by the mean free path $\lambda = (1-\phi)L/\phi$, as defined in Section 8.13. Here $L = Cl$ is the mean intercept for lumps of particles with individual mean intercept l (not to be confused with the pinning distance). The parameter C is the contiguity, where for separated particles it holds that $C = 1$. For spherical particles $l = 4r/3$. Using $\tau = \alpha G b/\lambda$ one obtains

$$\tau_{max} = \alpha G b \frac{3\phi}{4(1-\phi)} \frac{1}{r} \qquad (16.41)$$

A schematic of this model is shown in Fig. 16.25. It explains the initial increase and subsequent decrease of the yield strength with particle radius r and volume fraction ϕ.

Combining the expression for the creation of Orowan loops $\tau = \alpha G b/l$ with the Fleischer expression (16.37) allows one to estimate the optimum hardening state. If l is approximated by $l = r/c_A^{1/2}$ and the line tension $T \cong \frac{1}{2}\alpha G b^2$ is used, the result is

$$\frac{\alpha G b}{l} = \left\{ \frac{\alpha G b c_A^{1/2}}{r} \right\} = \frac{\gamma^{3/2} c_A^{1/2} r^{1/2}}{b\sqrt{6T}} = \left\{ \frac{\gamma^{3/2} c_A^{1/2} r^{1/2}}{b\sqrt{3\alpha G b^2}} \right\} \quad \text{or} \quad r_{cri} \cong \frac{\left(\sqrt{3\alpha}\right)^{2/3} G b^2}{\gamma}$$

in approximate agreement with experimental data. In this approximation r_{cri} is independent of the volume fraction ϕ. In practice a slight dependence on ϕ is observed. Using another approximation for the pinning distance l, e.g. $l = \lambda$, introduces this dependence on ϕ.

Cyril Stanley Smith (1903-1992)

Born in Birmingham, England, he received the B.Sc. degree in metallurgy from the University of Birmingham in 1924 and the Sc.D. degree from MIT in 1926. From 1927 until 1942 he was a research metallurgist at the American Brass Company, where he received some 20 patents and contributed numerous papers on the electrical and thermal conductivity, and the mechanical and magnetic properties of copper alloys. After brief service with the War Metallurgy Committee in Washington, he joined the Manhattan Project where he directed the preparation of the fissionable metal for the atomic bomb and other materials for nuclear experiments for which he received the Presidential Medal for Merit in 1946. He joined the University of Chicago in 1946 where he became the founder and first director of the Institute for the Study of Metals, the first interdisciplinary research organization dealing with materials in the United States. According to him it was "a natural outgrowth of the close association of metallurgists with chemists and physicists on the Manhattan Project." He taught from 1961 until his retirement in 1969 at MIT. He was renowned for his research in physical metallurgy, particularly in areas such as the role of interface energy and topology in the structure of polycrystalline materials and the application of metallography to the study of artifacts. His important contributions to the nature of structure in inorganic matter began with the application of simple topology to the shapes of metal grains and then, by extension, to all levels of the structural hierarchy. Eventually his work included exploration of the structures, on different scales, underlying patterns in both art and science. He was a pioneer in the application of materials science and engineering to the study of archaeological artifacts but was also recognized as an authority on the historical relationships between people from the beginning of human history and the materials they came to understand and use. His books include *A History of Metallography: The Development of Ideas on the Structure of Metals to 1890*, *Sources for the History of the Science of Steel*, and *A Search for Structure*.

Problem 16.5

For an alloy annealing results in 2 volume percent precipitates. The corresponding critical shear stress is 80 MPa. Show that the critical shear stress for the same alloy but with a different annealing procedure resulting in 8 volume percent precipitates is 127 MPa.

16.9   Final remarks

In the previous sections a compact survey has been given of the most important aspects of dislocations with respect to plastic deformation. We have seen that the large discrepancy between the theoretical and experimental shear strength can be explained by the presence of dislocations. The plastic deformation of single crystals can be understood in terms of dislocations. The strong influence of the grain size for polycrystals on the yield strength has been explained. Moreover some typical polycrystalline aspects, such as increased hardening modulus and yield strength, can be rationalised. We dealt briefly with the theories of work hardening, solid solution hardening and particle hardening. On the other hand, we explicitly paid no attention to fibre reinforcement and martensite hardening, which is very important for steels. For these topics we refer to the literature[III]. As a summary we can say that the yield strength is given by the additional effect of the Peierls (lattice) resistance, the Hall-Petch grain size effect, the dislocation density effect and the solid solution effect, i.e.

$$\tau = \tau_0 + k / \sqrt{d} + \alpha Gb\sqrt{\rho} + \beta\sqrt{c}$$

where all symbols have the same meaning as before. Finally we remark that, although the existence of dislocations is essential for the understanding of plasticity in metallic and inorganic materials, a generally accepted quantitative theory for plastic deformation on the basis of dislocations does not (yet) exist, although some authors would claim differently. This is largely due to the many, complex factors to be taken into account, not in the least the geometry of the dislocation configuration. The volume by Nabarro (1967) provides further information while several other aspects are discussed by Meyers et al. (1999).

16.10 Bibliography

Cottrell, A.H. (1953), *Dislocations and plastic flow in crystals*, Oxford University Press, Oxford.

Kovacs, I. and Szoldos, L. (1973), *Dislocations and plastic deformation,* Pergamon Press, Oxford.

Meyers, M.A., Armstrong, R.W. and Kirchner, H.O.K. (1999), *Mechanics and materials*, Wiley, New York.

Mughrabi, H. (1993) volume ed., *Plastic deformation and fracture of materials*, vol. 6 in Materials science and technology, Cahn, R.W., Haasen, P. and Kramer, E.J., eds., VCH, Weinheim.

Nabarro, F.R.N. ed. (1979), *Dislocation in solids,* vol. 4, North-Holland, Amsterdam.

Read, W.T. (1953), *Dislocations in crystals*, McGraw-Hill, New York.

[III] Fibre reinforcement: Kelly, A. and McMillan, N.H. (1986), *Strong solids*, 3rd ed., Clarendon, Oxford; Metallurgy of steels: Hornbogen, E. (1983), page 1075 in *Physical metallurgy*, 3rd ed., Cahn, R.W. and Haasen, P., eds., North-Holland, Amsterdam.

17

Mechanisms in polymers

In Chapters 15 and 16 the nature and applications of dislocations were discussed providing a micro- and a mesoscopic picture of the deformation in crystalline materials. Although the macroscopic aspects of amorphous materials and polymers do not differ widely from those of crystalline materials, the micro- and mesoscopic differences are considerable. It must be stated clearly right from the beginning that plastic and visco-elastic effects for polymers are not easy to separate. In this chapter first some experimental data on amorphous materials and polymers are discussed and thereafter the molecular background for the plastic deformation of these materials is addressed. We first discuss the motion of vacancies or holes in amorphous materials. After that the entanglement for polymers and some of the consequences are briefly dealt with. Finally the influence of the semi-crystallinity is addressed.

17.1   A brief review of data

Amorphous materials, as briefly discussed in Chapter 1, do not possess long-range order. In fact an amorphous material can be considered as a 'frozen' liquid. Therefore defects like dislocations cannot occur in amorphous materials. Consequently, the yield strength is often quite high and a clear competition between plastic deformation and fracture is present. With this competition we deal later. Originally it was thought that plastic deformation in polymers was mainly due to the increase in temperature in a neck during deformation. Since the thermal conductivity of polymers is relatively low, the heat generated was supposed to change the material behaviour due to temperature increase in the neck only. The process is referred to as *adiabatic heating*. Although an increase in temperature is certainly present for sufficiently fast deformations, it has become clear that also for quasi-static deformation real plastic effects are present. Moreover the yield point phenomenon, as indicated in Chapter 1, occurs for several polymers. This makes a differentiation between yield strength (upper yield point) and flow or drawing stress (lower yield point) necessary. We follow the line of discussion as given by Ward (1983). See also Gedde (1995) and Strobl (1997).

A main eye-opener was the experimental work on polyethylene terephthalate[a] (PET) where the yield strength as well as drawing stress was measured as a function of strain rate (Fig. 17.1). With increasing strain rate both the yield strength and drawing stress increase but at a certain strain rate the drawing stress starts to decrease while the yield strength continues to increase. It was argued that with increasing strain rate the drawing process becomes more and more adiabatic, thereby increasing the effective temperature of the drawing zone. This increase in temperature will cause the lowering of the drawing stress. From the known decrease in yield strength with temperature and assuming a similar decrease for the drawing stress an estimate can be made for the temperature increase during drawing. This led, together with the results from some other experiments, to the conclusion that above a strain rate of 0.1 min^{-1} the adiabatic heating effect becomes important.

---

[a] Allison, S.W. and Ward, I.M. (1967), Brit. J. Appl. Phys. **18**, 1151.

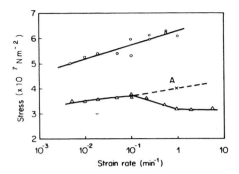

Fig. 17.1: Yield strength (o) and drawing stress (Δ) for PET as a function of strain rate.

Although adiabatic heating plays a significant role under sufficiently fast deformation conditions, plastic deformation also occurs under quasi-static conditions under which adiabatic heating plays no role. The two explanations advanced are geometric softening and an intrinsic yield drop. *Geometric softening* during tension is due to the fact that the decrease in cross-section is not sufficiently compensated by material hardening and therefore attributed to the decrease of slope of the stress-strain curve with increasing strain. It was, however, shown with compression tests[b] that there is a true decrease in yield strength, i.e. an *intrinsic yield drop*. In these studies a.o. polystyrene (PS), polymethyl methacrylate (PMMA) and polycarbonate (PC) were examined using various tests which can, together, indicate the shape of the yield locus in the σ_1-σ_2 plane. It was concluded that neither the Tresca, nor the von Mises criterion was adequate to describe the behaviour but that the yield strength[c] τ could be described by the Coulomb criterion $\tau = \tau_0 - \mu \sigma_N$, where σ_N is the normal stress.

For example, using the plane strain compression test (Fig. 17.2), the yield strength in compression σ_1 was measured as a function of the tensile stress σ_2 and for PMMA resulted[d] in $\sigma_1 = -11.1 + 1.365\sigma_2$ or $\tau = 4.74 - 0.158\sigma_N$ using the true stress in units of MPa. Since for the Tresca criterion it holds that $\sigma_1 - \sigma_2 =$ constant, this criterion obviously does not fit. Similar the von Mises criterion does not fit, neither for an incompressible material ($\nu = \frac{1}{2}$), which reads also $\sigma_1 - \sigma_2 =$ constant, nor in the general case. Although the expression for the yield strength is well obeyed by the Coulomb

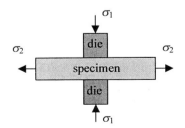

Fig. 17.2: Plane strain compression test.

---

[b] Whitney, W. and Andrews, R.D. (1967), J. Polymer Sci. **C16**, 2981; Brown, N. and Ward, I.M. (1968), J. Polymer Sci. **A2**, 607.

[c] We use here τ for the yield strength in order to avoid confusion with Boltmann's constant k.

[d] Bowden, P.B. and Jukes, J.A. (1968), J. Mater. Sci. **3**, 183.

criterion, the prediction of the direction of the deformation as given by $\theta = \pi/4 - \phi/2$ (see Section 13.4) is not very well obeyed. This was attributed amongst other things to the influence of the hydrostatic pressure.

Studies on the tensile behaviour of PMMA, PS, Nylon 6, PE and a few other materials indicated that with increasing pressure the yield strength increases. For isotropic PS a plot of both the yield and fracture strength as a function of pressure[e] indicates the transition from fracture to yield above about 300 MPa. Moreover the yield strength increases linearly with increasing pressure. Similar results have been obtained for PMMA. The behaviour can be represented as a series of Mohr circles, like in Fig. 13.8, or more directly as the yield strength τ as a function of pressure p via $\tau = \tau_0 + \alpha p$, with τ_0 the yield strength at zero pressure and α a parameter. The value of α is temperature dependent and increases rapidly near a visco-elastic transition. It will be clear that a conventional von Mises or Tresca criterion is incapable of describing the experiments sufficiently and that a pressure dependent term is required.

For anisotropic polymers regularly, when the tensile direction is not parallel to the initial draw direction, the deformation concentrates in a narrow band. These bands are of two types: one relatively narrow, approximately parallel to the initial draw direction, which looks similar to a slip band in a metal, and one of a more diffuse nature making a larger angle with the initial draw direction and which is similar to a kink band in metals. These bands, first observed[f] in nylon 66 and nylon 610, were explained by slip on (010) planes. For highly oriented and about 85% crystalline PE at small angles between the initial draw direction and the test direction, the band appeared to be in or close to the [001] direction. It was suggested that the deformation was due to a combination of (001) slip and twinning. For larger angles a kink band appears in which gross reorientation occurs. For this type of PE the yield strength was measured[g] as a function of the angle θ between the tensile testing direction and the initial drawing angle. It was concluded that the data could be described by the Coulomb criterion, which in this case reads $\tau = \sigma(\sin\theta \cos\theta + c\sin^2\theta)$, where c is a parameter. The minimum yield strength was obtained at about 60°, to be compared with the value of 45° for isotropic materials. The angle of the plane on which yield occurs is not very well predicted.

For an about 30% crystalline PET it was found that the direction of the deformation band differed significantly from the drawing direction for most test directions. In fact the yield strength continuously decreases from a high value of about 350 MPa at $\theta = 0°$ to a quite low value of about 50 MPa at $\theta = 90°$. The simplest description would be in terms of Schmid's law $\tau = \sigma \sin\theta \cos\theta$, but since this expression is symmetric around 45°, the data for PET cannot be fitted by this expression, in particular not for values larger than 45°. It has been shown, though that the data can be fitted well with the Hill criterion, as described in Section 13.2. May be more convincing, the prediction of the orientation of the deformation band direction appeared to be in good agreement with the experimental values for the whole range of θ. Simple shear tests were done as well for PET as a function of the angle ϕ with the initial draw direction. Resolving the shear in a compressive and a tensile component, we have the initial draw direction parallel to the compressive stress at 45° while it is

[e] Biglione, G., Baer, E. and Radcliffe, S.V. (1969), page 520 in *Fracture* 1969, Pratt, P.L., ed., Chapman and Hall, London.

[f] Zaukelies, D.A. (1961), J. Appl. Phys. **33**, 2797.

[g] Keller, A. and Rider, J.G. (1966), J. Mater. Sci. **1**, 389.

parallel to the tensile stress at 135°. Plotting the yield strength as a function of the angle ϕ one obtains two maxima, one at 45° and one at 135°. According to the Hill criterion these two maxima should have an equal value. Experimentally it appears that at 45° a substantial lower yield strength is present, as compared with the Hill theory, which is not the case at 135°. There is thus a significant Bauschinger effect. Modifying the term σ_{xx} to $\sigma_{xx} - \sigma_{Bau}$, where σ_{Bau} represents the difference in yield strength in compression and tension makes a good fit with the experimental data possible. It is argued that compressive deformations of the chains is relatively easy while tensile deformation is relatively difficult since it might involve e.g. bond breaking and that this can explain the behaviour observed.

As indicated before, the temperature and strain rate do have a significant influence. For PMMA plotting the yield strength and drawing stress as a function of the temperature for various strain rates one obtains approximately straight lines with increasing slope for increasing strain rate. These lines seem to converge and extrapolate to zero yield strength at about the glass transition temperature.

Hermann Staudinger (1881-1965)
Born in Worms, Germany, and receiving his Ph.D. from the University of Halle in 1903, Hermann Staudinger had mainly been interested in ketene chemistry, only beginning his investigations of natural rubber in 1910. This new line of research led him to propose that polymers were made of macromolecules in 1917. The idea was controversial for reasons both scientific and political. World War I was raging and, living in neutral Switzerland at the time, he was a vocal critic of German use of chemical weapons on the battlefield. This made him very unpopular in his native country. Meanwhile, some scientists felt that there was an upper limit to the number of atoms that could be joined together in a single molecule. With dogged persistence he championed his theory, and in the 1930s the work of scientists like Herman Mark and Wallace Carothers would confirm Staudinger's theory. Further validation came in 1953 when he was awarded the Nobel Prize in Chemistry for his discoveries in the field of macromolecular chemistry.

Problem 17.1

In early ideas a temperature rise to above T_g, where T_g is the glass transition temperature, was held responsible for yielding. Show that for PC with a T_g of approximately 140 °C during a tensile test up to the fracture strain of 65% less than 1/5 of the energy needed to increase the temperature from 25 °C to 150 °C is delivered. Assume that the viscosity at the $T_g + 10$ °C is 10^{11} Pa and that the $C_p = 0.35$ cal/g °C.

17.2   Yield strength

To describe the temperature and strain rate dependence of the yield strength for glassy polymers, two basic approaches have been forwarded. The Eyring activated complex model and the Robertson liquid-like structure model. We describe them both in the next sections.

Activated complex theory

For the modelling of the deformation behaviour of amorphous materials the Eyring or activated complex theory is frequently used. For a general discussion of Eyring's theory for deformation processes, see Chapter 7 or Krausz and Eyring (1975). In this theory plastic deformation is considered to be a thermally activated process where a unit (an atom, molecule or part of a molecule) jumps in a *hole* in the structure (Fig. 17.3). The hole in the structure is part of the free volume in the amorphous material. The free volume increases with increasing temperature. The frequency v for these jumps is given by

$$v = A\exp\left(\frac{-\Delta H}{RT}\right) \tag{17.1}$$

where A is a constant with dimension s^{-1} and ΔH the activation energy. For mechanical deformation we have also to consider the externally applied stress. If we consider only plastic deformation (i.e. no elastic deformation), per unit volume an amount of energy w is dissipated given by

$$w = \tfrac{1}{2}\sigma:\varepsilon = \tfrac{1}{2}\sigma_{ij}\varepsilon_{ij} \tag{17.2}$$

There is no information on a microscopic scale on the values for the stress and strain tensor. Consequently all terms are undetermined. In order to proceed, one usually assumes a macroscopic (uniaxial) yield strength Y and a so-called *activation volume* V^* in such a way that the total work W is given by

$$W = wV_{mol} = YV^* \tag{17.3}$$

where V_{mol} refers to the molar volume. The externally applied stress is now supposed to produce a shift in the original symmetric potential energy barrier as indicated in Fig. 17.3. The strain rate $\dot{\varepsilon}$ will be proportional to the difference in jump frequency in the forward direction v_{for} and reverse direction v_{rev}

$$\dot{\varepsilon} \sim v_{for} - v_{rev} = A\exp\left[\frac{-(\Delta H - YV^*)}{RT}\right] - A\exp\left[\frac{-(\Delta H + YV^*)}{RT}\right]$$

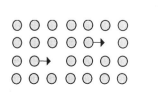
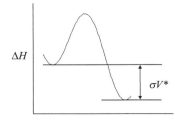

Fig. 17.3: Migration process of units via a jump in holes in the structure and the accompanying potential energy curve under an applied stress.

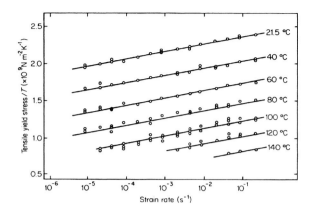

Fig. 17.4: The yield strength as a function of the strain rate for polycarbonate.

$$= 2A \exp\left(\frac{-\Delta H}{RT}\right) \sinh\left(\frac{YV^*}{RT}\right) \tag{17.4}$$

For sufficiently large argument x, $\sinh x = \frac{1}{2}\exp x$ and Eq. (17.4) can be approximated to

$$\dot{\varepsilon} = A \exp\left[\frac{-(\Delta H - YV^*)}{RT}\right] \quad \text{leading to} \quad \frac{Y}{T} = \frac{\Delta H}{V^*T} + \frac{R}{V^*}\ln\frac{\dot{\varepsilon}}{A} \tag{17.5}$$

From an Arrhenius plot of Y/T versus $\ln\dot{\varepsilon}$ the quantities ΔH and V^* can be determined. An example, showing data for PC, is provided in Fig. 17.4[h]. Eq. (17.5) represents the behaviour as observed experimentally quite well.

The approach as presented provides a quasi-atomistic explanation for the yield in amorphous materials. Interpreting the holes as vacancies, the model is also used for creep in metals. The basic entity that jumps is different for different materials. For example, for amorphous metals the entity is the metal atom itself that jumps, for silica glasses it is the SiO_4 (tetrahedral) unit, while for amorphous polymers it is a certain part of the chain. It is tempting to interpret the activation volume in molecular terms. Although this is possible in a number of cases, in general one should be careful in doing so. Values obtained from a plot like Fig. 17.4 typically provide estimates for V^* by a factor of 10 or so larger than single molecular units. This indicates that the deformation process is not the result of the motion of a single unit but more likely to be due to the collective motion of several units.

The effect of pressure on the yield strength of polymers can be rationalised in this approach as follows. The stress term $W = YV^*$ can be split in a part dependent on the shear strength τ with an associated activation volume V_s^* plus a contribution of the hydrostatic pressure p resulting in change in hole size V_h^*, i.e. $W = \tau V_s^* - pV_h^*$. For the shear strength τ we can take the octahedral shear stress. Moreover we may take for the strain rate $\dot{\varepsilon}$ the octahedral strain rate $\dot{\gamma}$ if the constant A is modified properly. Inserting this in Eq. (17.4) and, again using $\sinh x = \frac{1}{2}\exp x$, leads to

---

h Bauwens-Crowet, C., Bauwens, J.C. and Homès, G.A. (1969), J. Polymer Sci. A **2**, 735.

$$\dot{\gamma} = A \exp\left(-\frac{\Delta H - \tau V_s{}^* + p V_h{}^*}{RT}\right)$$

At constant strain rate we obtain

$$\tau = \frac{\Delta H + RT \ln\dfrac{\dot{\gamma}}{A}}{V_s{}^*} + \frac{V_h{}^*}{V_s{}^*}p \qquad \text{or} \qquad \tau = \tau_0 + \alpha p$$

which is the empirical description we encountered in Section 17.1.

In principle more than one thermally activated mechanism may be present. This leads a curved representation in the yield strength versus ln strain rate plot. A sum over expressions like Eq. (17.4) can describe this behaviour. For some materials, e,g, PMMA, PVC and PC, the behaviour can be described pretty well by just two terms. At high temperature and low strain rate one of the two terms can be approximated as before by the exponential expression. The other term becomes important at low(er) temperature and high strain rate.

The liquid-like structure model

The liquid-like structure model is also derived using the expectation that the stress interacts with the polymer chains. If a shear stress is applied the molecules seek a new arrangement and it is supposed that the structure that develops becomes more liquid-like. While in the Eyring theory the final state after each jump has the same energy as the initial state, in the liquid-like structure theory it is assumed that the final state after a jump has a slightly higher energy than the initial state (at zero stress). Experimentally a time lag is often observed between the application of the stress and the onset of yielding. This behaviour is interpreted as the time required for rearranging towards a higher energy, more liquid-like structure.

A simple model has been provided[i] which considers the polymer to have identical backbone bonds of equal length, a constant bond angle and only *cis* and *trans* conformational states for each bond (see Chapter 7). The molecules are thus planar and the trans state is assumed to have a lower energy ΔE, which is not only due to the preferred intramolecular state but also the intermolecular interactions dependent on the molecular packing. Above T_g the molecules are in equilibrium with respect to rotational conformations and the distribution between *cis* and *trans* states is given by the Boltzmann distribution. The fraction *cis* states at temperature $T > T_g$ is

$$\varphi_{cis} = \sigma(T)/[1 + \sigma(T)] \qquad \text{with} \qquad \sigma(T) = \exp(-\Delta E / kT)$$

where k is the Boltzmann constant. For $T < T_g$ the configuration is thought to be frozen in at T_g so that the fraction φ_{cis} is given by the same expression but with T replaced by T_g. For a chain to keep the same overall direction two *trans-cis* rotations are required so that the energy involved for such a transition without applied stress is $2\Delta E$. If we apply now a shear stress τ the ratio of *cis* and *trans* states is altered because the energy difference between them is now described by $\Delta E \pm v\tau \cos\theta$, where v is the activation volume (associated with the volume swept out during the *cis-trans* transition, i.e. the average volume of a chain segment containing two single non-collinear bonds) and θ the angle between applied shear stress and overall chain

[i] Robertson, R.E. (1966), J. Chem. Phys. **44**, 3950; (1968), Appl. Polym. Symp. **7**, 201.

direction. The sign depends on whether the stress eases the transition (from *trans* to *cis*) or not (from *cis* to *trans*). Thus the conformational energy depends on the orientation of the molecule with respect to the shear stress. For a collection of randomly oriented molecules the energy for half of them is lowered and the other half is increased and the fraction of *cis* states changes accordingly. Obviously the fraction of flexed bonds increases for orientations with

$$(\Delta E - \tau v \cos\theta)/kT \le \Delta E/kT_g$$

For one part of the *cis-trans* distribution the applied stress assists the approach to the equilibrium distribution of *cis-trans* states and increases the number of flexed bonds, in effect implying an increase in temperature. For the other part, the applied stress effectively decreases the temperature. Because of the strong dependence of the rate of conformational change with temperature, Robertson argued that only changes towards equilibrium need to be taken into account for the calculation of the maximum fraction of flexed bonds under applied stress. The maximum fraction φ_{max} corresponds to arise in effective temperature to Θ. This stress-dependent structural temperature Θ, characterising the rotational conformations, can be calculated from the value of φ_{max} assuming a zero-stress equilibrium of *cis*-states at Θ and therefore from

$$\varphi_{max} = \sigma(\Theta)/[1+\sigma(\Theta)] \qquad \text{with} \quad \sigma(\Theta) = \exp(-\Delta E/k\Theta)$$

Now Robertson further assumed that the viscosity of that state is given by

$$\eta(\Theta,T) = \eta_g \left\{ 2.303 \left[\left(\frac{c_1^g c_2^g}{\Theta - T_g + c_2^g} \right) \left(\frac{\Theta}{T} \right) - c_1^g \right] \right\}$$

which is basically the WLF equation (see Chapter 20) modified for the structure temperature Θ which is obviously different from the thermal bath temperature T characterising the vibrations. In this expression η_g is viscosity at T_g, which can be taken[j] as 10^{13} Pa, and $c_1^g = 17.44$ and $c_2^g = 51.6\,°C$ are the standard WLF coefficients. The yield strength Y as a function of the strain rate $\dot{\varepsilon}$ then becomes

$$Y = \eta(\Theta,T)\dot{\varepsilon}$$

The energy required to reach the structural temperature θ is given by $\rho(\Theta - T_g)\Delta c_p$ with ρ the density and Δc_p the difference between the heat capacity in the liquid and the glass state at T_g. This c_p-difference probes the structural change since the vibrational component is contained in both the liquid state value and the glass state value. Since five of the parameters required, namely v, T_g, c_1^g, c_2^g and η_g are well defined, we need only an estimate for ΔE and its value is set at $3.81\,T_g$ cal/mol, a value that follows from the glass-transition model of Gibbs and Di Marzio[k]. Using these values a reasonable agreement was obtained for the yield strength as a function the strain rate for PMMA, PS and PC. However, the experimental data points lie systematically below the calculated theoretical line. A better agreement can be obtained when the pressure dependence is taken explicitly into account by adding a term[l] pV to $\Delta E \pm v\tau$. Since the liquid-like model deals mainly with intramolecular

[j] Miller, A. A. (1964), J. Polym. Sci.**A2**, 1095.
[k] Gibbs, J.H. and Di Marzio, E.A. (1958), J. Chem. Phys. **28**, 373.
[l] Duckett, R.A., Rabinowitz, S. and Ward, I.M. (1970), J. Mater. Sci. **5**, 909.

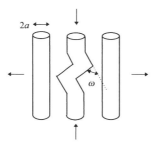

Fig. 17.5: Schematic view of the kinking model due to Argon.

interactions, it might be supposed to apply at $T \le T_g$. However, for $T \ll T_g$ the intermolecular interactions are expected to dominate so that in this regime the activated complex theory is arguably more appropriate.

Finally, it should be said that there are other structural theories for the plastic deformation of polymers that do incorporate a kind of defect, akin to the dislocation. In this way it is possible to calculate the temperature and strain rate dependence in terms of molecular parameters that can be independently determined. One example is the theory by Argon[m], which states that yield is the result of kinking of the polymer chain (Fig. 17.5). The kink angle ω and chain radius a are essential molecular ingredients. In his model the (stress dependent) activation enthalpy volume ΔH^* for the formation of pair of kinks in a polymeric glass with shear modulus G and Poisson's ratio v under the influence of a shear stress τ is given by

$$\Delta H = \frac{3\pi G \omega^2 a^3}{16(1-v)} \left[1 - 6.75(1-v)^{5/6} \left(\frac{\tau}{G} \right)^{5/6} \right] \tag{17.6}$$

The shear strain rate then becomes

$$\dot{\gamma} = \dot{\gamma}_0 \exp(-\Delta H / kT) \tag{17.7}$$

Substitution of Eq. (17.6) leads to

$$\tau = \frac{0.102G}{(1-v)} \left[1 - \frac{16(1-v)}{3\pi G \omega^2 a^3} kT \ln \frac{\dot{\gamma}_0}{\dot{\gamma}} \right]^{6/5} \tag{17.8}$$

Independent, sensible estimates for the various parameters involved show that there exists good agreement between theory and the extensive experimental data for PET. However, note that this expression resembles the Eyring expression and becomes of the same form if the exponent 6/5 is replaced by 1 (as in the Eyring theory) so that a curve-fit procedure cannot distinguish between them.

Semi-crystalline polymers

In amorphous polymers plasticity is restricted to $T < T_g$, but in semi-crystalline polymers the regime $T_g < T < T_m$ is also relevant. Semi-crystalline polymers crystallised from the melt have a rather complex microstructure or *morphology* containing crystallised parts with folded chains and amorphous parts with random

[m] Argon, A.S. (1973), Phil. Mag. **28**, 839.

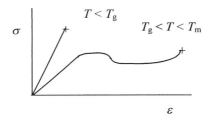

Fig. 17.6: Stress-strain curve for semi-crystalline polymers.

chains, organised in a super structure, the so-called *spherulites*. Consequently many variables influence the deformation behaviour, e.g. the temperature, the strain rate, the degree of crystallinity, the crystal size, the spherulite size and the entanglement density. The most important one is probably the temperature and thus also the strain rate. The behaviour is important since several commodity polymers, e.g. PE and isotactic PP, are semi-crystalline and often used at $T > T_g$.

In a polymer crystal the same dislocation mechanism can operate as in metal and inorganic materials with one important restriction: the chain cannot break. Therefore, the required five independent slip systems are generally not available. Consequently for a semi-crystalline polymer below the glass transition temperature both the crystalline part and amorphous part cannot flow resulting in a high yield strength and limited plastic flow. At a temperature above T_g, the yield strength of the amorphous part is virtually zero and the deformation behaviour is dominated by the crystalline regions. Moreover the crystalline parts then can orient themselves and deform without fulfilling the requirement of the five slip systems since their environment has become rubber-like. As a result the macroscopic yield strength decreases significantly while the strain to failure increases. A schematic stress-strain curve for a semi-crystalline polymer below and above T_g is given in Fig. 17.6.

The need to avoid molecular fracture of the chains restricts slip in polymeric crystals to those planes, which contain the chain direction, i.e. (hk0) planes if [001] is the chain direction. Normally slip on those planes is considered only in parallel or perpendicular directions to the chain direction, denoted as chain direction slip and transverse slip, respectively. After crystallisation the molecules are approximately normal to the lamellar surface but during deformation molecules slip along each other and become tilted in the crystallite. This slip process is associated with dislocations. As discussed in Chapter 15, the energy U of a dislocation of length l is given by

$$U = \frac{\alpha G b^2 l}{4\pi} \ln\left(\frac{R}{r_0}\right)$$

with G the shear modulus, b the length of the Burgers vector, R the size of the crystal, r_0 the core radius, often taken as $r_0 \cong b$ and $\alpha = 1$ for screw and $\alpha = 1/(1-v)$ for edge dislocations. The energy of a dislocation is proportional to its length l and the shorter l, the lower U. Since a dislocation cannot end within a crystal a dislocation line perpendicular to the lamellar surface and therefore parallel to the chain direction has the shortest length. Moreover the need to avoid excessive deformation favours that b is also parallel to the chains. Consequently screw dislocations are strongly favoured. These dislocations are present in solution grown crystals (see e.g. Fig. 15.18). Another possibility is the creation of a dislocation during deformation. The Helmholtz energy

F for a mechanically loaded crystallite then becomes $F = U + \tau b Rl$. The activation energy for nucleating a dislocation U^* is calculated from $\partial F/\partial R = 0$ at R^* and leads to

$$R^* = \frac{Gb}{4\pi\tau^*} \qquad \text{and} \qquad U^* = \frac{Gb^2 l}{4\pi}\left[\ln\left(\frac{R^*}{r_0}\right) - 1\right]$$

Assuming thermal nucleation a barrier of about $50kT$ yields a shear stress τ^* approximately equal to the experimental yield strength in shear. The temperature dependence of the yield strength as predicted is, however, much smaller as compared with the experimental dependence. As an alternative, more complex helical motion, possible associated with α-relaxations (See Chapter 20), has been proposed[n].

Finally, we note that also twinning and martensitic-like transformations can take place in polymer crystals. For these effects we refer to the literature, e.g. Young (1981).

Problem 17.2

Show that for PC, using the same data as in Problem 17.1, the structural temperature $\theta = 180\ °C$ and that this requires only 2.5 cal/g, well within the 8 cal/g of mechanical energy available during drawing.

Problem 17.3

Show, using the data of Fig. 17.4, that the yield strength of PC can be described by an activation energy $\Delta H = 309$ kcal/mol and an activation volume $V^* = 3.9\times10^{-3}$ m^3/mol. Also show that at a strain rate of 10^{-1} s^{-1} and 10^{-5} s^{-1} the yield strength is $\tau = 70$ MPa and 60 MPa, respectively.

Wallace Carothers (1896-1937)

Born near Burlington, Iowa, in 1896 and grown up in Des Moines, Carothers studied chemistry first at Tarkio College in Missouri and later earned his masters and doctoral degrees at the University of Illinois. After teaching for three semesters at Harvard, Carothers left to work at DuPont in 1928. Given nearly free reign to investigate whatever interested him, Carothers set about to answer a question central to the controversy over Staudinger's theory. Since some felt macromolecules could not exist, Carothers set out to test this notion by attempting to create synthetic macromolecules. He succeeded in 1930, and his success silenced one of the main criticisms of Staudinger's theory. It was merely coincidence that the synthetic macromolecules Carothers created happened to behave like natural silk. The next five years were spent looking for a synthetic polymer that would be a practical silk substitute. Success came in 1935 in the form of a polymer that would become known as nylon. While nylon was an instant hit when it first went on sale in the form of women's hosiery in 1940, Carothers did not live to see this success. A lifelong sufferer of clinical depression, he took his own life in

---

[n] Galeski, A. (2003), Prog. Polym. Sci. **28**, 1643.

1937. Best known as the inventor of nylon, Wallace Carothers was much more than the creator of one useful polymer. The research that led to the invention of nylon also demonstrated the existence of macromolecules, greatly bolstering the macromolecular theory of Staudinger.

17.3   Flow behaviour

Having discussed the yield strength of amorphous and semi-crystalline polymers, we now turn to the flow behaviour of these materials. In order to do so we need to discuss first some structural considerations. These considerations are necessarily brief and in Chapter 20 dealing with visco-elasticity a somewhat more elaborate discussion is presented. There after we consider the flow behaviour itself.

Entanglements and plateau modulus

In crystalline materials the yield strength can be low as a consequence of the presence of dislocations. In amorphous materials the yield strength generally is high and fracture occurs before yielding. As a consequence the stress-strain curve is linear up to the point of fracture. In the low-molecular weight, crystalline materials generally only one type of bonding, typically covalent or ionic, is important. In high-molecular weight, polymeric materials, however, two types of bonding are important: the strong, primary intramolecular covalent bonds and the weak, secondary intermolecular van der Waals bonds. During yielding the secondary van der Waals bonds are broken but the primary covalent bonds mainly stay intact. In order to assess the behaviour properly, we will first discuss the behaviour of long-chain polymers in the melt where all secondary bonds are broken.

In a polymeric melt all secondary bonds are broken and the flow behaviour is highly determined by the chain-like shape of the polymer. These chains have formed coils. The overall picture of a polymer melt is a bowl of wriggling spaghetti with a length-diameter ratio of 10^4 or more. The effect of the shape shows in various experiments. In rheological measurements one measures the intrinsic viscosity η_0, i.e. the viscosity measured when the shear stress approaches zero. For short chain polymers this parameter increases linearly with the length of the chain and thus with the molecular weight M. Above a certain critical molecular weight M_{cri}, however, the dependence is much stronger and can be described by $\eta_0 \sim M^\delta$, where the exponent δ ranges from 3.2 to 3.6. The effect occurs for all polymers whereby M_{cri} varies between 300 to 700 main chain atoms. Above M_{cri} a strongly enhanced steric hindrance is present. The chains cannot any longer move freely and get entangled, as schematically illustrated in Fig. 17.7. This behaviour also occurs for amorphous polymers.

If for a monodisperse amorphous polymer of varying molecular weight the shear modulus is measured as a function of the frequency ω, a result as shown schematically in Fig. 17.8 is obtained. At high frequencies, the material behaves glass-like, independent of the molecular weight. At lower frequency there are significant differences. Low-molecular weight materials melt directly while for higher molecular weight first a plateau arises. The length of this plateau increases with increasing molecular weight. Here also the steric hindrance is becoming more and more important and limits the deformation at lower frequencies.

The modelling of the steric hindrance is a central problem in the rheology of polymers. One way to do this is to assume that the entangled chains in an amorphous polymer can be compared with a chemically cross-linked rubber. In a rubber the individual chains hinder each other via chemically bonded nodes. This leads to a

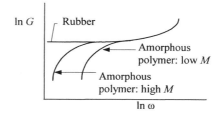

Fig. 17.7: Schematic view of an entanglement.

Fig. 17.8: Shear modulus for rubbers and amorphous polymers of low and high molecular weight M.

plateau in the elastic modulus as a function of frequency, also shown in Fig. 17.8. The plateau continues to low frequency since yield in a chemically cross-linked rubber is impossible. In amorphous polymers this plateau is also present, but a drop occurs at low frequency since at low frequency enough time is available for the entanglements to rearrange. The height of this *rubber plateau*, the rubber modulus G_0, is a function of the molecular weight of the sub-chains, i.e. the chains between the cross-links M_{sub}. In Chapter 11 it was shown that for rubbers the shear modulus G_0 is given by $G_0 = \rho RT/M_{sub}$, where ρ, R and T have their usual meaning. Now it is possible to model the hindrance in the polymeric melt, which in reality is diffuse, as a set of discrete physical cross-links that cannot disentangle on the time scale of the experiment. These discrete cross-links (Fig. 17.7) represent some average hindrance and do not exist in reality. With this concept and using the plateau modulus of rubber elasticity, an average molecular weight M_{ent} between the physical cross-links or entanglements can be defined. According to this model it holds that

$$M_{ent} = \rho RT/G_0 \qquad (17.9)$$

where the molecular weight M_{ent} is, like the critical molecular weight M_{cri}, a measure for the steric hindrance in the melt. Approximately it holds that $M_{cri} = 2M_{ent}$. Table 17.1 shows the typical values for M_{ent} as obtained from shear modulus measurements for several polymers. From this table one can conclude that PS is a loosely entangled polymer whereas PE and PC are more severely entangled. Note that for a fair comparison on the basis of the number of monomer units, the difference in molecular mass should be taken in to account.

Table 17.1: Typical molecular weight M_{ent} between entanglements for several polymers.

Polymer	M_{ent} (g/mol)
PC	1800
PE	2000
PS	20000

Influence of the entanglements on the flow behaviour

In the usual range of temperature and strain rate, polymers have a relatively high yield strength, roughly approximating the theoretical yield strength, which is of the order of $G/10$ to $G/30$. Absolutely these values are not that high in view of the fact that the elastic parameters for polymers are typically a few GPa's. We recall that the value of the initial yield strength is highly dependent on the temperature and strain rate. As indicated, with decreasing strain rate and/or increasing temperature the yield strength decreases. At the glass transition temperature the material becomes a viscous liquid with zero yield strength.

For relatively long-chain polymers entanglements are the determining factor for the structure. Entanglements as have arisen in the melt largely survive the solidification process. A main indication is given by the following behaviour. If for a plastically deformed bar, the temperature is increased to slightly above the glass transition temperature, the secondary bonds break and the chains recover their most probable state. This results in recovery of the original dimensions of the bar. This implies that normally only the shape of the polymer coils change while their centre of gravity remains fixed. This behaviour is in contrast with low molecular weight materials where the relative position of the units does change with respect to each other and, generally, a thermal treatment only recovers the original flow behaviour but not the shape.

Assuming that the entanglements remain present in a solidified polymer, a simple estimate can be made for the maximum stretch (or *natural draw ratio*) λ_{max} that can be reached. In a first approximation the maximum stretch is equal to the stretch for the part of the chain between the entanglements. The latter is equal to the projected length of the chain L_{pro} divided by the coil dimension $\langle R_0^2 \rangle^{1/2}$. If there are on average N_{ent} bonds with length l and enclosed angle α present between two entanglements, λ_{max} is given by

$$\lambda_{max} = \frac{L_{pro}}{\sqrt{\langle R_0^2 \rangle}} = \frac{N_{ent} l \cos\alpha}{\sqrt{CN_{ent} l^2}} \sim \sqrt{N_{ent}} \sim \sqrt{M_{ent}} \qquad (17.10)$$

This proportionality with the chain length between the nodes N_{ent} is experimentally observed. The maximum stretch is also proportional to the molecular weight M.

So high values of M_{ent} leads to a high value of the maximum stretch. For UHMW-PE ($M_w > 3 \times 10^6$ D) a value of $\lambda_{max} \sim 4$-5 is predicted, in reasonable agreement with experiment. For PS we expect on this basis a high toughness. However, it is well known that PS is brittle and fractures upon deformation, usually far below 10% strain. The reason is that, although PS is locally ductile (fibrils in a craze can be elongated up to 400%), the material does not deform homogeneously. This is due to the presence of significant strain softening and the absence of sufficient strain hardening in which case any initial local deformation will become unstable (Chapter 2). It has been shown[o] that it is possible to make PS samples tough by cold rolling with a thickness reduction of 32%. The rolling also leads to a length increase of 36%, which implies that molecular orientation has been induced. This rolling leads to a much lower (upper) yield strength (removal of strain softening), as shown by compression tests, which on its turn leads to macroscopic toughness. In tension these specimens can be elongated up to 20%, to be compared with 2% for the virgin material. The orientation will also lead to increased strain hardening, but the effect is here predominantly related to the removal of strain softening, as supported by tensile tests on mechanically preconditioned PS samples at 10, 20 and 30 min after rolling. After 20 min the yield point has returned and the material becomes brittle again after 30 min. Since the orientation cannot be expected to change in this time at room temperature,

[o] Govaert, L.E., van Melick, H.G.H. and Meijer, H.E.H. (2001), Polymer **42**, 1271.

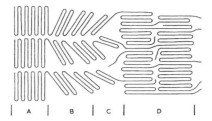

Fig. 17.9: Schematic representation of the deformation at molecular level in a semi-crystalline polymer changing from the ideal structure (A) via crystallites deforming by slip, twinning and/or transformation (B) to break-up and molecular pull-out (C) to a new fibrillar structure (D).

the toughening must be attributed to the reduction of strain softening. Unfortunately, the recovery of the upper yield point is relatively fast.

Semi-crystalline polymers

While for amorphous polymers the flow (and hardening) behaviour is largely dictated by the entanglements, for semi-crystalline materials the deformation and degradation of the crystallites is determining. The slip systems that can be activated depend on the local orientation of the crystallite with respect to applied stress. During deformation these orientations change and a major contribution to work hardening is due to the rotation of these crystallites in the rubbery matrix. During this process the shear stress required for slip decreases. In both stages the sperulitic structure is lost resulting in complete break-up of the microstructure and leading to a new, fibre-like structure. This results ultimately in a fibrous texture with the crystallites having the molecular axis parallel to the stretch. A possible mechanism is illustrated in Fig. 17.9.

It appears that the deformation of the spherulites in a semi-crystalline polymer is usually considerably less than the overall deformation. For example, for nylon 66 it was observed[p] that for a specimen elongated by 400% the spherulites with a size of about 50 μm elongated only 180% with a lateral contraction of about 40%. The non-crystalline layers within the spherulites also seem to flow by interlamellar slip. According to electron microscopy observations[q] during creep experiments on PE interlamellar slip is less free than interspherulitic slip at −10 °C and below but at high temperature probably occurs approximately to the same extent, while in thin films[r] at 95 °C a large amount of interlamellar slip is observed.

During deformation the crystallites degrade and a more fibrillic structure arises. This whole process occurs in the range up to a stretch of 3 to 5. At higher stretches two factors are of importance: the strength of the secondary bonds and the entanglement density. If the secondary bonds are too strong, like in nylon where hydrogen bonds are present, the deformation stops. If the secondary bonds are not that strong, the entanglement network inherited from the melt is the only limiting factor and high maximum stretches can be obtained. The example of polyethylene is shown in Fig. 17.10. At low-molecular weight the entanglement network is mainly determined by the crystallisation conditions. During slow cooling the chains are able to fold thereby destroying the entanglement network largely. In this case large

---

[p] Crystal, R.G. and Hansen, D. (1968), J. Polym. Sci, **6**, 981.
[q] Nakafuku, C., Minato, K. and Takemura, T. (1968), Trans. Soc. Rheol. **5**, 261.
[r] Kobayashi, K. and Nagasawa, T. (1966), J. Polym. Sci. **C15**, 163.

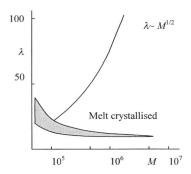

Fig. 17.10: Maximum attainable stretch as function of molecular weight for polyethylene.

maximum stretches, in the order of 20 or so, can be reached. During fast cooling the entanglement network remains largely intact and accordingly much smaller values of λ_{max}, namely a value corresponding to the original entanglement network of the order of 5, are obtained. At high molecular weight the crystallisation rate has barely any influence on the entanglement density and the corresponding λ_{max} is again about 5.

However, recently these concepts have been applied[s] for PE using extreme slow cooling so that crystalline regions virtually consisting of a single molecule are created. By careful and controlled melting such polymers, a heterogeneous melt with more entangled regions (due to the amorphous parts of the solid) and less entangled regions (due to the crystalline part of the solid) can be created. It is found that chain reptation, necessary for the homogenisation of the entanglement distribution, is considerably hindered. The long-lived heterogeneous melt shows a decreased viscosity and provides enhanced drawability on crystallisation.

Apart from slow cooling, another way to modify the entanglement network is by crystallisation of the polymer from a diluted solution. This is the basis of the *gel-spin process*. From the solution the polymer is crystallised so that the low but concentration dependent entanglement network is fixed by evaporation or extraction of the solvent. The maximum stretch λ_{max} is now only determined[t] by the projected chain length L_{pro} or equivalently by the molecular weight M. Extremely high values of λ_{max}, of the order of 200 at laboratory scale, can be obtained in this way (Fig. 17.10). Since the entanglement density is again determined by the molecular weight M, the same consideration as before lead to a $M^{1/2}$-dependence.

17.4   Bibliography

Gedde, U.W. (1995), *Polymer physics*, Chapman and Hall, London.

Krausz, A.S. and Eyring, H. (1975), *Deformation kinetics*, Wiley, New York.

Strobl, G. (1997), *The physics of polymers*, 2nd ed., Springer, Berlin.

Ward, I.M. (1983), *Mechanical properties of polymers*, 2nd ed., Wiley, Chichester.

Young, R.J. (1981), *Introduction to polymers*, Chapman and Hall, London.

---

[s] Rastogi, S., Lippits, D.R., Peters, G.M.W., Graf, R., Yao, Y. and Spiess, H.W (2005), Nature Materials **2**, 635.

[t] Lemstra, P.J., Kirschbaum, R., Ohta, T. and Yasuda, H. (1987), page 39 in Ward, I.M., *Developments in oriented polymers 2*, Elsevier Applied Science.

Continuum visco-elasticity

In the previous chapters, we have discussed reversible deformation (elasticity) and irreversible deformation (plasticity). Both types of phenomena are essentially time-independent and time played only the role of ordering parameter. In Chapters 1 and 2, however, we briefly encountered that there are also time-dependent irreversible phenomena, usually addressed as visco-elasticity and visco-plasticity for which the continuum aspects are treated in this chapter. First, we provide some general considerations illustrating creep and relaxation and then deal with a simple, essentially one-dimensional formulation of both phenomena. Thereafter, a systematic generalisation is given and we discuss the extension to three dimensions, including an alternative integral formulation. Finally, the matter is treated from the thermodynamic point of view.

18.1   General considerations

If we put a stress on an elasto-plastic material it will initially deform instantaneous elastically and above the yield strength plastically at a rate controlled by the deformation process. The behaviour so far is considered to be essentially time-independent. While this is essentially true for elasticity, in fact for plasticity the processes inside the material are considered to be so fast that they react instantaneously to the deformation. However, instantaneous response is an idealisation and in general there exists a certain (relative) temperature where the mobility of structural elements (such as point defects, dislocations, entanglements, etc.) within the material becomes high enough so that the time to adapt the structure of the material becomes comparable with the time of observation. Significantly below this temperature the structure does not change while significantly above that temperature the adaptation is instantaneous. In the intermediate regime the deformation behaviour of materials becomes time-dependent where the rate of deformation depends on the applied stress and temperature. Similarly, above a certain deformation rate the material cannot follow the deformation more or less instantaneously and the behaviour becomes more elastic. On the other hand, at sufficiently low deformation rate the relaxation becomes instantaneous. So, temperature and deformation rate are exchangeable to a certain extent. This behaviour is addressed as *visco-elasticity* if no stress threshold is present and as *visco-plasticity* in the presence of a stress threshold (yield strength). As indicated in Chapter 1, for different type of materials this behaviour occurs at a rather different temperature, given a certain deformation rate. For example, at normal deformation rate (say 1 to 100 s^{-1}) for polymers often room temperature is sufficiently high while for inorganics and metals at least a few hundred degrees is required. Two loading situations are commonly encountered. The first is the sudden application of a stress at time $t = 0$ and keeping the stress constant at later times. This process results in an increasing strain with increasing time and is called *creep*. The second is the sudden application of a strain at time $t = 0$ and keeping this strain constant afterwards. In this case the resulting stress decays from an initial value and the process is defined as *relaxation*. In the following, we briefly describe the material response for these cases but limit ourselves to linear response meaning that the response is proportional to the driving force. It also implies that solutions of the governing differential equations can be added, a fact usually referred to as the *Boltzmann superposition principle*.

The time-dependent response of a material can be depicted in Fig. 18.1. The stress-strain curve, as given on the left, indicates that in general with increasing strain rate the resulting stress is higher at the same strain, thus illustrating the time-dependent response. The middle part shows the creep behaviour. After loading a material to a certain stress, elastic strain is instantaneously present. Keeping the stress level constant leads to creep and the creep rate increases with increasing stress. After unloading recovery takes place. Elastic recovery takes place more or less instantaneously while delayed recovery occurs thereafter. Delayed recovery may lead to zero or non-zero final strain, in which cases we speak of full or partial recovery. The right part shows the relaxation behaviour. After deforming up to a certain strain, a stress results and this stress increases with increasing strain. Keeping the strain constant, the stress relaxes with a rate of relaxation dependent on the strain level.

It has been mentioned already that the material response is strongly dependent on temperature and therefore the parameters involved in the description of these processes are, like other material properties, functions of temperature[a]. Recall the example of an inorganic glass. While at low temperature the material reacts as an elastic solid, at high temperature it behaves as a viscous liquid. At intermediate temperature the behaviour is visco-elastic. It thus becomes clear that the description of the material response is strongly dependent, not only on temperature, but also on the time of observation. Comparing the observation time τ_{obs} with the relaxation time τ_{rel} of the material, Reiner in 1964 introduced the ratio $D \equiv \tau_{rel}/\tau_{obs}$ named *Deborah's number*[b]. A material seems to behave

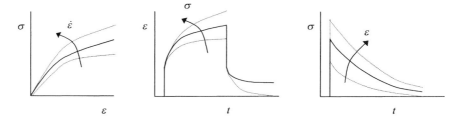

Fig. 18.1: Possible time-dependent response of a material. Left: the stress-strain curve, showing increasing stress with increasing strain rate at a certain strain. Middle: creep behaviour, showing increasing creep rate with increasing applied stress level and partial or full recovery. Right: relaxation behaviour, showing increasing stress with increasing applied strain level and decay to zero stress.

---

[a] Additionally one should be aware that due to the mobility within the material the microstructure or morphology of the material could change considerably so that during the process the material properties itself change.

[b] After the prophetess Deborah who prophesied that "the mountains would flow before the Lord" (Judges 5.5, AV). This expresses vividly the fact that objects, which are seemingly unchangeable against time scale of the human life span, deform when measured on a geological time scale.

elastic if $D \gg 1$ and   viscous if $D \ll 1$

As indicated, we deal in this chapter with the region where $D \cong 1$, i.e. with visco-elasticity.

18.2   Analogous models

To describe the time-dependent behaviour of materials frequently analogous models are used. These models are essentially 1D and are constructed as an assembly of elements such as springs, dashpots and skid blocks. The elastic component of the deformation is described by a spring with the spring constant k, for which, when a strain $\varepsilon^{(e)}$ is applied, the stress $\sigma^{(e)}$ is given by $\sigma^{(e)} = k\varepsilon^{(e)}$. The viscous component of the deformation is described by a dashpot with the damper constant (viscosity) η, for which, when a strain rate $\dot{\varepsilon}^{(v)}$ is applied, the stress $\sigma^{(v)}$ is given by $\sigma^{(v)} = \eta\dot{\varepsilon}^{(v)}$. Threshold behaviour is represented by a skid block which is an element defined by $\varepsilon = 0$ if $\sigma < Y$ and possibly $\varepsilon \neq 0$ if $\sigma = Y$. The latter element is used to describe plasticity. We will limit ourselves here to the spring and dashpot. Combination of these elements leads to a model for visco-elastic behaviour. In fact we met the simplest combinations, a series connection or a parallel connection of a single spring and a single dashpot already mentioned in Chapter 6. These models are labelled as *analogous models* in view of the analogy of the deformation behaviour of materials with the behaviour of these constructions. In this section we use stress σ, strain ε, elastic constant k and viscosity η as generic terms. As an example for shear one should read the shear stress τ, the shear strain γ, the shear modulus G and shear viscosity η while for uniaxial extension one should read the uniaxial stress σ, uniaxial strain ε, elastic modulus E and extensional viscosity η_{ext}. The use of a 1D description, as used in this section will be extended to 3D in Section 18.4.

The *Maxwell model* is represented by a series connection of a single spring and a single dashpot (Fig. 18.2). For this element the stress σ is the same in both components and the total strain is $\varepsilon = \varepsilon^{(e)} + \varepsilon^{(v)}$. Since only the total strain ε is observable, $\varepsilon^{(v)}$ is usually considered as an internal variable. The time dependence of this model can be found by differentiating $\sigma^{(e)} = k\varepsilon^{(e)}$ with respect to time so that adding $\dot{\varepsilon}^{(v)}$ and $\dot{\varepsilon}^{(e)}$ yields the total strain rate

$$\dot{\varepsilon} = \dot{\varepsilon}^{(e)} + \dot{\varepsilon}^{(v)} = \frac{\dot{\sigma}}{k} + \frac{\sigma}{\eta} \quad \text{or} \quad \sigma + \tau\dot{\sigma} = \eta\dot{\varepsilon} \tag{18.1}$$

where the *relaxation time* τ is given by $\tau = \eta/k$. This differential equation can be solved given the boundary conditions. We distinguish between relaxation and creep.

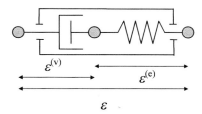

Fig. 18.2: The Maxwell model.

To describe relaxation we suddenly apply a strain ε_0 at time $t = 0$ (alternatively we write $\varepsilon = \varepsilon_0 H(t)$ where $H(t)$ denotes the Heaviside step function, see Section 3.15). Therefore, $\dot{\varepsilon} = 0$ except at $t = 0$ (alternatively, $\dot{\varepsilon} = \varepsilon_0 \delta(t)$ where $\delta(t)$ denotes the Dirac delta function, see Section 3.15) and Eq. (18.1) becomes

$$0 = \frac{\dot{\sigma}}{k} + \frac{\sigma}{\eta}$$

Solving we obtain

▶ $$\sigma(t) = \sigma_0 \exp(-t/\tau) = k\varepsilon_0 \exp(-t/\tau) \qquad (18.2)$$

Hence for relaxation as described by the Maxwell model the stress at time $t = 0$ is given by $\sigma_0 = k\varepsilon_0$ and the stress decays to zero exponentially with time.

To describe creep we suddenly apply a stress σ_0 at time $t = 0$ (or $\sigma = \sigma_0 H(t)$). Therefore, $\dot{\sigma} = 0$ except at $t = 0$ (or $\dot{\sigma} = \sigma_0 \delta(t)$) and Eq. (18.1) becomes

$$\dot{\varepsilon} = \sigma/\eta$$

Solving we obtain

▶ $$\varepsilon(t) = \frac{\sigma_0}{k} + \frac{\sigma_0}{\eta} t \qquad (18.3)$$

Hence for creep as described by the Maxwell model the strain at time $t = 0$ is given by σ_0/k and the strain increases linearly with time. The behaviour of this model is thus liquid-like.

James Clerk Maxwell (1831-1879)
Educated at the University of Edinburgh and Saint Peter's College in Cambridge. In 1854 he won the Smith's Prize examination in which one of the questions was the proof of Stokes theorem, which appeared in print on this exam for the first time. Maxwell returned to Scotland in 1856 in the chair of natural philosophy at Marischal College in Aberdeen before becoming professor of physics and astronomy at King's College in London from 1860 to 1865. The next years were spent at his family home in Glenlair and in these years he wrote his famous *Treatise on electricity and magnetism*, which was published in 1873. In 1870 the Cavendish Laboratory was founded through a gift from the seventh Duke of Devonshire and a descendant of chemist and natural philosopher Henry Cavendish. Maxwell became the first director of the laboratory, being the third choice after Thomson and Helmholtz each declined, and professor of experimental physics at Cambridge University. In that period he spent a great deal of attention to editing the electrical manuscripts of Cavendish and setting up the Cavendish Laboratory as a laboratory of precision electrical measurements.

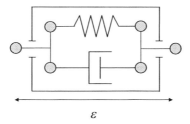

Fig. 18.3: The Kelvin model.

The *Kelvin model* (also called as the *Voigt* or *Kelvin-Voigt* model) is represented by a parallel connection of a single spring and a single dashpot (Fig. 18.3). For this element the strain is the same in both components and the total stress σ is given by $\sigma = \sigma^{(e)} + \sigma^{(v)}$. The total strain ε equals the elastic strain $\varepsilon^{(e)}$ and this model contains no internal variable. The time dependence of this model can be found from

$$\sigma = \sigma^{(e)} + \sigma^{(v)} = k\varepsilon + \eta\dot\varepsilon \qquad \text{or} \qquad \frac{\sigma}{k} = \varepsilon + \tau\dot\varepsilon \qquad (18.4)$$

where $\tau = \eta/k$ is the *retardation time*. For a solution we have to set the boundary conditions and again we distinguish between creep and relaxation.

For creep, we apply a stress σ_0 at time $t = 0$ and keep it constant afterwards (or $\sigma = \sigma_0 H(t)$. Solving Eq. (18.4) then leads, for $t > 0$, using again $\tau = \eta/k$, to

$$\blacktriangleright \qquad \varepsilon(t) = \frac{\sigma_0}{k}\left[1 - \exp(-t/\tau)\right] \qquad (18.5)$$

The strain increases with decreasing rate and approaches asymptotically the value σ_0/k for $t \to \infty$. Creep in this model is therefore restricted and the behaviour is solid-like. If the stress σ_0 is removed at $t = t_1$ (or equivalently $-\sigma_0$ is added) the strain resulting from this stress is

$$\varepsilon(t) = -\frac{\sigma_0}{k}\left\{1 - \exp[-(t - t_1)/\tau]\right\}$$

and the superposition principle yields for the total strain at time $t > t_1$.

$$\varepsilon(t) = \frac{\sigma_0}{k}\exp(-t/\tau)\left[\exp(t/\tau) - 1\right]$$

The strain for this model thus tends to zero for $t \to \infty$ and shows therefore *full recovery*. In reality some materials show full recovery while others only partially recover.

To describe relaxation we have $\varepsilon = \varepsilon_0 H(t)$ and $\dot\varepsilon = \varepsilon_0\delta(t)$ so that

$$\blacktriangleright \qquad \sigma(t) = k\varepsilon_0 H(t) + \eta\varepsilon_0\delta(t) \qquad (18.6)$$

Only at $t = 0$ there a viscous contribution to the stress $\sigma(t)$ but for $t > 0$ the stress has a constant value $k\varepsilon_0$. The Kelvin model shows no time-dependent relaxation.

 **Lord Kelvin, William Thomson (1824-1907)**
Scottish mathematician and physicist who contributed to many branches of physics. He was known for his self-confidence, and as an undergraduate at Cambridge he thought himself the sure "Senior Wrangler" (the name given to the student who scored highest on the Cambridge mathematical Tripos exam). After taking the exam he asked his servant, "Oh, just run down to the Senate House, will you, and see who is Second Wrangler." The servant returned and informed him, "You, sir!". Another example of his hubris is provided by his 1895 statement "heavier-than-air flying machines are impossible", followed by his 1896 statement, "I have not the smallest molecule of faith in aerial navigation other than ballooning...I would not care to be a member of the Aeronautical Society." Kelvin is also known for an address to an assemblage of physicists at the British Association for the advancement of science in 1900 in which he stated, "There is nothing new to be discovered in physics now. All that remains is more and more precise measurement." Kelvin argued that the key issue in the interpretation of the second law of thermodynamics was the explanation of irreversible processes. He noted that if entropy always increased, the universe would eventually reach a state of uniform temperature and maximum entropy from which it would not be possible to extract any work. He called this the Heat Death of the Universe. With Rankine he proposed a thermodynamical theory based on the primacy of the energy concept, on which he believed all physics should be based. He said the two laws of thermodynamics expressed the indestructibility and dissipation of energy. He also tried to demonstrate that the equipartition theorem was invalid. Thomson also calculated the age of the earth from its cooling rate and concluded that it was too short to fit with Lyell's theory of gradual geological change or Charles Darwin's theory of the evolution of animals though natural selection. He used the field concept to explain electromagnetic interactions. He speculated that electromagnetic forces were propagated as linear and rotational strains in an elastic solid, producing "vortex atoms" which generated the field. He proposed that these atoms consisted of tiny knotted strings, and the type of knot determined the type of atom. Kelvin's theory said ether behaved like an elastic solid when light waves propagated through it. He equated ether with the cellular structure of minute gyrostats. With Tait, Kelvin published *Treatise on Natural Philosophy* (1867), which was important for establishing energy within the structure of the theory of mechanics.

Neither the Maxwell nor the Kelvin element accurately describes the behaviour of most materials. The Kelvin element does not show instantaneous strain on loading or unloading (*impact response*), nor does it describe a permanent strain after unloading. The Maxwell model shows no time-dependent recovery and does not show a decreasing strain rate at constant stress.

A model showing all the qualitative features mentioned above is the *Burgers model*, which consists of a Maxwell model in series with a Kelvin model (Fig. 18.4). The total strain at time t is the sum of the strain ε in each of the three elements, where the spring and the dashpot in the Maxwell element are considered as two elements. We thus have

$$\varepsilon = \varepsilon_D + \varepsilon_S + \varepsilon_K$$

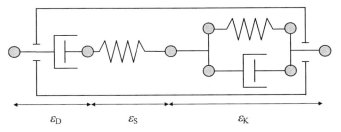

Fig. 18.4: The Burgers model.

where $\dot{\varepsilon}_D = \sigma/\eta_M$, $\varepsilon_S = \sigma/k_M$ and ε_K can be derived from the differential equation for the Kelvin element $\dot{\varepsilon}_K + (k_K/\eta_K)\varepsilon_K = (1/\eta_K)\sigma$. Eliminating ε_D, ε_S and ε_K from these four equations yields the constitutive equation for the Burgers model

$$\sigma + \left(\frac{\eta_M}{k_M} + \frac{\eta_M}{k_K} + \frac{\eta_K}{k_K}\right)\dot{\sigma} + \frac{\eta_M\eta_K}{k_Mk_K}\ddot{\sigma} = \eta_M\dot{\varepsilon} + \frac{\eta_M\eta_K}{k_K}\ddot{\varepsilon} \qquad (18.7)$$

Solving this equation by direct integration is cumbersome. For creep under a constant stress σ_0 the solution is

$$\varepsilon(t) = \frac{\sigma_0}{k_M} + \frac{\sigma_0}{\eta_M}t + \frac{\sigma_0}{k_K}\left[1 - \exp(-k_Kt/\eta_K)\right]$$

and is thus the sum of the Maxwell and Kelvin creep behaviour. The first two terms represent the instantaneous elastic response and the viscous flow while the last term represents the delayed elasticity. However, solving differential equations via the Laplace transform is advantageous since this transform changes a differential equation of the original variable into an algebraic equation of the transformed variable (see Section 3.15). Therefore, we demonstrate the solution as obtained via the Laplace transform in Justification 18.1.

Justification 18.1*

Indicating the Laplace transform $L[f(t)]$ transform by $\hat{f}(s)$ we obtain for the four equations given above

$$\hat{\varepsilon} = \hat{\varepsilon}_D + \hat{\varepsilon}_S + \hat{\varepsilon}_K , \quad s\hat{\varepsilon}_D = \sigma/\eta_M , \quad \hat{\varepsilon}_S = \hat{\sigma}/k_M , \quad (s + k_K/\eta_K)\hat{\varepsilon}_K = \hat{\sigma}/\eta_K$$

Solving for $\hat{\varepsilon}$ yields

$$\hat{\varepsilon} = \frac{\hat{\sigma}}{k_M} + \frac{\hat{\sigma}}{\eta_M s} + \frac{\hat{\sigma}}{\eta_K(s + k_K/\eta_K)}$$

which is, if rewritten as,

$$\hat{\sigma} + \left(\frac{\eta_M}{k_M} + \frac{\eta_M}{k_K} + \frac{\eta_K}{k_K}\right)s\hat{\sigma} + \frac{\eta_M\eta_K}{k_Mk_K}s^2\hat{\sigma} = \eta_M s\hat{\varepsilon} + \frac{\eta_M\eta_K}{k_K}s^2\hat{\varepsilon}$$

the transforms of Eq. (18.7). We further use the abbreviations

$$p_1 = \frac{\eta_M}{k_M} + \frac{\eta_M}{k_K} + \frac{\eta_K}{k_K} \qquad p_2 = \frac{\eta_M \eta_K}{k_M k_K} \qquad q_1 = \eta_M \quad \text{and} \quad q_2 = \frac{\eta_M \eta_K}{k_K}$$

The creep behaviour can be obtained by applying a constant stress σ_0 and taking the initial conditions as $\varepsilon = \varepsilon_S = \sigma_0/k_M$, and $\varepsilon_D = \varepsilon_K = 0$ for $t = 0$. The solution is

$$\varepsilon(t) = \frac{\sigma_0}{k_M} + \frac{\sigma_0}{\eta_M} t + \frac{\sigma_0}{k_K}\left[1 - \exp(-k_K t/\eta_K)\right]$$

and is the sum of the Maxwell and Kelvin creep behaviour. The relaxation behaviour can be obtained by applying a constant strain ε_0 at time $t = 0$. The final solution is

$$\sigma(t) = \frac{\varepsilon_0}{A}\left[(q_1 - q_2 r_1)\exp(-r_1 t) - (q_1 - q_2 r_2)\exp(-r_2 t)\right]$$

where $A = (p_1^2 - 4p_2)^{1/2}$, $r_1 = (p_1 - A)/2p_2$ and $r_2 = (p_1 + A)/2p_2$.

There are various other possibilities to combine four elements in one model and in fact one may show that the behaviour of several of these models is identical. In other words, these models are *degenerate*. Quite generally the behaviour of all possible models can be described by the differential equation

▶ $\qquad p_0\sigma + p_1\dot\sigma + p_2\ddot\sigma + \cdots = q_0\varepsilon + q_1\dot\varepsilon + q_2\ddot\varepsilon + \cdots \qquad$ or $\qquad P\sigma = Q\varepsilon \qquad$ (18.8)

where P and Q are linear differential operators with respect to time t and are given by

$$P = \sum_{r=0}^{a} p_r \frac{\partial^r}{\partial t^r} \qquad \text{and} \qquad Q = \sum_{r=0}^{b} q_r \frac{\partial^r}{\partial t^r}$$

The coefficient p_0 can be taken without loss of generality as $p_0 = 1$.

Using this description the Maxwell model is given by $p_0 = 1$, $p_1 = \eta/k$ and $q_1 = \eta$, while the Kelvin model is represented by $p_0 = 1$, $q_0 = k$ and $q_1 = \eta$. In both cases all other coefficients are equal to zero. As indicated above, neither the Maxwell model nor the Kelvin model is adequate to describe realistic material behaviour. By retaining more terms a better description might be obtained but only at the cost of introducing more parameters. For further details, see Findlay et al. (1976) and Tschoegl (1989).

Problem 18.1

Derive the creep solution for the Kelvin model.

Problem 18.2*

Derive the relaxation equation for the Burgers model.

18.3   Generalisation*

Obviously, the problems of the individual Maxwell and Kelvin models can be overcome by the combination of them as in the Burgers model. This model shows all the qualitative features generally encountered. However, generalisation is done in an ad-hoc way and we need a systematic way of doing this. Let us start with a more formal statement of the superposition principle (Ward, 1983; Christensen, 1971).

The Boltzmann superposition principle

The Boltzmann superposition principle can be formalised as follows. Consider a visco-elastic body on which the stress is changed in several steps $\Delta\sigma_i$ at various times θ_i. Defining the *creep function* $k\varepsilon(t)/\Delta\sigma = c(t;\boldsymbol{p})$, where k and $\boldsymbol{p}$ denote the elasticity modulus and material parameters, respectively, we have

$$k\varepsilon(t) = \Delta\sigma_0 c(t) + \Delta\sigma_1 c(t-\theta_1) + \Delta\sigma_2 c(t-\theta_2) + \cdots = \sum_{i=0}^{n} \Delta\sigma_i c(t-\theta_i)$$

For infinitesimal stress steps we have

$$\Delta\sigma \to d\sigma = \frac{\partial\sigma}{\partial t}\,dt$$

yielding the *hereditary integral*

$$k\varepsilon(t) = \int_{-\infty}^{t} c(t-\theta)\,d\sigma = \int_{-\infty}^{t} \frac{\partial\sigma(\theta)}{\partial\theta} c(t-\theta)\,d\theta \tag{18.9}$$

Integration by parts and using $\sigma(-\infty) = 0$ we obtain the creep response

$$\blacktriangleright \qquad k\varepsilon(t) = c(0)\sigma(t) - \int_{-\infty}^{t}\sigma\,dc = c(0)\sigma(t) - \int_{-\infty}^{t}\sigma(t)\frac{\partial c(t-\theta)}{\partial t}\,d\theta \tag{18.10}$$

In a similar way, considering the same body whose strain is now changed in steps $\Delta\varepsilon$ and using the *relaxation function* $\sigma(t)/k\Delta\varepsilon = r(t;\boldsymbol{p})$, where $\boldsymbol{p}$ again denotes the material parameters, one obtains the hereditary integral

$$\frac{\sigma(t)}{k} = \int_{-\infty}^{t} r(t-\theta)\,d\varepsilon = \int_{-\infty}^{t} \frac{\partial\varepsilon(\theta)}{\partial t} r(t-\theta)\,d\theta \tag{18.11}$$

and the stress response

$$\blacktriangleright \qquad \frac{\sigma(t)}{k} = r(0)\varepsilon(t) - \int_{-\infty}^{t}\sigma\,dc = r(0)\varepsilon(t) - \int_{-\infty}^{t}\varepsilon(t)\frac{\partial r(t-\theta)}{\partial t}\,d\theta \tag{18.12}$$

If we assume that the first change in stress occurs at $t = 0$, the lower limit of the integrals in Eqs. (18.9) and (18.11) may be taken as $t = 0$ instead of $t = \infty$.

Since both the creep function $c(t)$ and the relaxation function $r(t)$ describe the response of visco-elastic material, there must be a relation between them. To clarify this we use the Laplace transforms

$$L[c(t)] = \hat{c}(s) = \int_{0}^{\infty} c(t)\exp(-st)\,dt \quad \text{and} \quad L[r(t)] = \hat{r}(s) = \int_{0}^{\infty} r(t)\exp(-st)\,dt$$

and the convolution theorem (Section 3.15)

$$L[f(t)]L[g(t)] = L\left[\int_{0}^{t} f(t-\lambda)g(\lambda)\,d\lambda\right] = L\left[\int_{0}^{t} f(\lambda)g(t-\lambda)\,d\lambda\right]$$

Applying this to Eqs. (18.9) and (18.11) we obtain

$$k\hat{\varepsilon}(t) = [s\hat{\sigma}(s) - \sigma(0)]\hat{c}(s) = s\hat{\sigma}(s)\hat{c}(s) \qquad \text{and}$$

$$\frac{\hat{\sigma}(t)}{k} = [s\hat{\varepsilon}(s) - \varepsilon(0)]\hat{r}(s) = s\hat{\varepsilon}(s)\hat{r}(s)$$

where the latter step can be made since $\sigma(0) = \varepsilon(0) = 0$. Combining yields

$$s^2\hat{c}(t)\hat{r}(t) = 1 \tag{18.13}$$

Transforming back results in

$$\int_0^t c(t-\theta)r(\theta)\,d\theta = \int_0^t c(\theta)r(t-\theta)\,d\theta = t \tag{18.14}$$

Using Eqs. (18.13) and (18.14) $c(t)$ can be determined from $r(t)$ and vice versa. Table 18.1 shows the various steps for the Maxwell and Kelvin elements. The advantages of using the Laplace transform for these problems should be clear by now.

Finally we note that from the relaxation function easily the zero-shear rate viscosity η_0 can be calculated. To that purpose we assume a constant shear rate $\dot{\gamma} = C$ and identify the elastic constant in Eq. (18.11) with the zero-time modulus G_0. Hence we have $\tau = \eta_0\dot{\gamma}$ or $\tau = \eta_0 C$. Inserting $\dot{\gamma} = C$ in Eq. (18.11) right away delivers

$$\eta_0 = \int_{-\infty}^t G(t)\,dt \tag{18.15}$$

The generalised Kelvin model

For the Kelvin model we obtained $k\varepsilon + \eta\dot{\varepsilon} = \sigma$. A general solution can be found by dividing by η and introducing an integrating factor $\exp(t/\tau)$, where $\tau = \eta/k$ is addressed as the *retardation time*. The result is

$$\dot{\varepsilon}\exp(t/\tau) + \frac{k}{\eta}\varepsilon\exp(t/\tau) = \left\{\frac{d}{dt}[\varepsilon\exp(t/\tau)]\right\} = \frac{1}{\eta}\sigma(t)\exp(t/\tau)$$

By integration and rearranging we obtain the general solution

$$\varepsilon = \frac{1}{\eta}\int_{-\infty}^t \sigma(\theta)\exp[-(t-\theta)/\tau]\,d\theta$$

Using for the load the step function $\sigma = \sigma_0 H(t)$ we regain

$$c(t) \equiv \frac{k\varepsilon(t)}{\sigma_0} = 1 - \exp(-t/\tau) \tag{18.16}$$

A simple but straightforward generalisation is a model with n Kelvin elements and a

Table 18.1: Creep and relaxation function for the Maxwell and Kelvin elements.

Maxwell element	Kelvin element
$r(t) = \exp(-Gt/\eta)$	$c(t) = 1 - \exp(-Gt/\eta)$
$L[r(t)] = [s + (G/\eta)]^{-1}$	$L[c(t)] = \{s[1 + (\eta s/G)]\}^{-1}$
$L[c(t)] = s^{-1} + (G/\eta s^2)$	$L[r(t)] = s^{-1} + (\eta/G)$
$c(t) = 1 + (Gt/\eta)$	$r(t) = 1 + (\eta\delta(t)/G)$

spring in series[c]. If $n = 1$ we have the so-called *standard model* (sometimes addressed as the *Zener model*), characterised by k_0, k_1 and η_1. The *creep compliance* for this model is (Betten, 2002)

$$\frac{\varepsilon(t)}{\sigma_0} \equiv S(t) = S_0 + S_1[1 - \exp(-t/\tau)]$$

where $S_0 = 1/k_0$, $S_1 = 1/k_1$ and $\tau_1 = \eta_1/k_1$ and is the sum of the responses of the spring and of the Kelvin element. For $n > 1$, we similarly obtain the creep compliance as the sum of the responses of the individual elements

$$S(t) = S_0 + \sum_{m=1}^{m=n} S_n[1 - \exp(-t/\tau_m)]$$

with $S_m = 1/k_m$ and $\tau_m = \eta_m/k_m$. Such a model has a discrete spectrum of retardation times τ_k. In the limit of $n \to \infty$ the discrete spectrum becomes a continuous spectrum $f(\tau)$ and the creep compliance becomes

$$S(t) = S_0 + \alpha \int_0^\infty f(\tau)[1 - \exp(-t/\tau)]\,\mathrm{d}\tau$$

The factor α can be obtained from $S(\infty) = S_\infty$ if the spectrum is normalised, i.e. if the relation

$$\int_0^\infty f(\tau)\,\mathrm{d}\tau = 1$$

holds, resulting in $\alpha = S_\infty - S_0$, so that the (reduced) creep function becomes

$$C(t) \equiv \frac{S(t) - S_0}{S_\infty - S_0} = 1 - \int_0^\infty f(\tau)\exp(-t/\tau)\,\mathrm{d}\tau \tag{18.17}$$

Using once more the Laplace transform, in this case in combination with the substitution $\tau = 1/\xi$ so that $\mathrm{d}\tau = -\xi^{-2}\mathrm{d}\xi$ and interchange of the limits to remove a negative sign, the final result is

$$\blacktriangleright \quad C(t) = 1 - \int_0^\infty f(\xi^{-1})\exp(-t\xi)\xi^{-2}\,\mathrm{d}\xi = 1 - L[f^\#(\xi)] \tag{18.18}$$

$$\text{with } f^\#(\xi) = \xi^{-2}f(\xi^{-1})$$

From this expression the creep compliance can be calculated given the retardation spectrum.

Example 18.1

As an example we mention the following results for the (normalised) Maxwell distribution of retardation times:

---

[c] In general materials do show immediate response upon mechanical loading, the so-called *impact response*, while the Kelvin model does not. By introducing the extra spring this flaw is remedied.

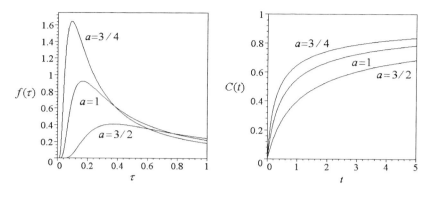

$$f(\tau) = \frac{1}{2\sqrt{\pi}} \frac{a\exp[-a^2/4\tau]}{\tau^{3/2}} \qquad \Leftrightarrow \qquad C(t) = 1 - \frac{1}{2}a\left(t - \frac{a}{4}\right)^{-1/2}$$

The figure shows the results for $a = \frac{3}{4}$, 1 and 3/2.

Instead of calculating the creep function from the retardation spectrum, one can also calculate the retardation spectrum from the (experimental) creep function. A frequently encountered creep curve is

$$\varepsilon(t) = \alpha + \beta[1 - \exp(-\gamma\sqrt{t})]$$

so that $L[f^{\#}(\xi)] = \exp(-c\sqrt{t})$. The inverse Laplace transform yields

$$f^{\#}(\xi) = \frac{1}{2\sqrt{\pi}} \frac{\gamma\exp(-\gamma^2/4\xi)}{\xi^{3/2}}$$

from which we obtain the retardation spectrum

$$f(\tau) = \frac{1}{2\sqrt{\pi}} \frac{\gamma\exp(-\gamma^2\tau/4)}{\tau^{-1/2}}$$

For an interpretation we recall that for the linear (Newton-like) damper we have

$$\varepsilon = \frac{\sigma}{\eta}t \qquad \text{or} \qquad \sigma = \eta\dot{\varepsilon}$$

We replace the linear damper by a non-linear damper depending on $\sqrt{t}$ instead of t. Using a slightly different notation for later convenience, we write

$$\varepsilon = \gamma\frac{\sigma}{k}\sqrt{t} \qquad \text{and} \qquad \sigma = \frac{2}{\gamma}k\sqrt{t}\dot{\varepsilon}$$

so that for the differential equation of this non-linear Kelvin element

$$\varepsilon + \frac{2}{\gamma}\sqrt{t}\dot{\varepsilon} = \frac{\sigma}{k}$$

results. Introducing an integrating factor $\exp(\gamma\sqrt{t})$ we obtain in a similar way as before

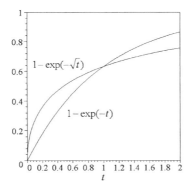

Fig. 18.5: The creep response of the Kelvin and non-linear Kelvin model.

$$c(t) = \frac{k\varepsilon(t)}{\sigma} = (1 - \exp(-\gamma\sqrt{t}))$$

which is the creep function for this non-linear damper. Adding a spring in series to realise impact response the final result becomes

$$\varepsilon(t) = \alpha + \beta(1 - \exp(-\gamma\sqrt{t}))$$

This is to be compared with the result for the linear Kelvin model given by $\varepsilon(t) = \alpha + \beta(1-\exp(-\gamma t))$ (Fig. 18.5).

The generalised Maxwell model

For the Maxwell model we obtained $\dot{\varepsilon} = \dot{\sigma}/k + \sigma/\eta$. Similar as for the Kelvin model we can obtain a general solution by introducing the integrating factor $\exp(t/\tau)$, where $\tau = \eta/k$ is called the *relaxation time*. This results in

$$\sigma(t) = k \int_{-\infty}^{t} \exp[-(t-\theta)/\tau]\dot{\varepsilon}(\theta)\,\mathrm{d}\theta$$

so that the relaxation function becomes, taking into account $\dot{\varepsilon} = k\varepsilon_0\delta(t)$,

$$r(t) \equiv \frac{\sigma(t)}{k\varepsilon_0} = \exp(-t/\tau) \tag{18.19}$$

In this case generalisation is done by taking n Maxwell elements with a spring in parallel. The model with $n = 1$ is the *Poynting-Thomson model*, characterised by k_∞, k_1 and η_1. The *relaxation modulus* becomes (Betten, 2002)

$$\frac{\sigma(t)}{\varepsilon_0} \equiv E(t) = E_\infty + E_1 \exp(-t/\tau_1)$$

where $E_\infty = k_\infty$, $E_1 = k_1$ and $\tau_1 = \eta_1/k_1$. Obviously for n Maxwell elements the relaxation function reads

$$E(t) = E_\infty + \sum_{j=1}^{n} E_j \exp(-t/\tau_j)$$

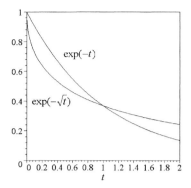

Fig. 18.6: The relaxation response of the Maxwell and non-linear Maxwell model.

with $E_j = k_j$ and $\tau_j = \eta_j / k_j$. Taking the limit $n \to \infty$, we obtain in a similar way as for the generalised Kelvin model the relaxation modulus

$$E(t) = E_\infty + (E_0 - E_\infty) \int_0^\infty g(\tau) \exp(-t/\tau) \, d\tau \tag{18.20}$$

Since the range of relaxation times covers usually several decades, a logarithmic time-scale is often used which is defined by

$$g(\tau) \, d\tau = \tau g(\tau) \, d\ln\tau \equiv g^\&(\tau) \, d\ln\tau$$

so that we can also write

$$E(t) = E_\infty + (E_0 - E_\infty) \int_0^\infty g^\&(\tau) \exp(-t/\tau) \, d\ln\tau$$

Using the Laplace transform again on Eq. (18.20) with the substitution $\tau = 1/\xi$ and therefore $d\tau = -\xi^{-2} \, d\xi$ we, similarly as before, obtain the reduced relaxation modulus

$$\blacktriangleright \qquad R(t) \equiv \frac{E(t) - E_\infty}{E_0 - E_\infty} = L[g^\#(\xi)] \qquad \text{with} \quad g^\#(\xi) = \xi^{-2} g(\xi^{-1}) \tag{18.21}$$

Also as before t may be replaced by $\sqrt{t}$ yielding a relaxation function given by

$$r(t) = \exp(-c\sqrt{t})$$

which, similar as in the case of the Kelvin model, is to be compared with the result for the linear Maxwell model $\exp(-ct)$ (Fig. 18.6). The generalisation of this expression is

$$r(t) = \exp[-(t/\tau)^b]$$

with b a parameter and which is often called the *Kohlrausch function* or the *stretched exponential*. For stabilised glasses experimentally the value of $b \cong 0.5$ is often obtained.

Dynamic response

So far only the quasi-static response of materials during creep and relaxation have been discussed. However, frequently a sinusoidal load with certain frequency v is applied. If the strain ε varies as (Findlay et al., 1976)

$$\varepsilon = \varepsilon_0 \sin\omega t \tag{18.22}$$

where $\omega = 2\pi\nu$ is the *angular frequency*, the stress σ generally varies as

$$\sigma = \sigma_0 \sin(\omega t + \delta) = \sigma_0 \left[\sin\omega t \cos\delta + \cos\omega t \sin\delta \right] \tag{18.23}$$

where δ is the *phase angle*. The stress, therefore, can be considered as being resolved in two parts. The first part with amplitude $\sigma_0 \cos(\delta)$ in phase with the strain and the second part with amplitude $\sigma_0 \sin(\delta)$, which is $\pi/2$ out of phase with the strain.

One can also write

$$\sigma = E'\varepsilon_0 \sin\omega t + E''\varepsilon_0 \cos\omega t$$

where $E' = (\sigma_0/\varepsilon_0) \cos \delta$ and $E'' = (\sigma_0/\varepsilon_0) \sin \delta$

It appears to be expedient to use a complex representation. In complex notation the expression for the strain and stress become

$$\varepsilon = \varepsilon_0 \exp(i\omega t) \quad \text{and} \quad \sigma = \sigma_0 \exp[i(\omega t + \delta)] \tag{18.24}$$

with $i = \sqrt{-1}$. The complex modulus E^* is given by

$$E^* = \frac{\sigma}{\varepsilon} = \frac{\sigma_0}{\varepsilon_0} \exp i\delta = \frac{\sigma_0}{\varepsilon_0} \left[\cos\delta + i\sin\delta \right]$$

Hence we have $E^* = E' + iE''$, where E' and E'' are the *real* and *imaginary* part of the modulus, respectively. The real part E' is often called the *storage modulus* while the imaginary part E'' is referred to as *loss modulus*. This terminology stems from the energy dissipated per cycle given by

$$\Delta U = \int \sigma \, d\varepsilon = \int_0^{2\pi/\omega} \sigma\dot\varepsilon \, dt$$

Inserting Eqs. (18.22) and (18.23) we easily obtain

$$\Delta U = \int_0^{2\pi/\omega} \omega\varepsilon_0 \cos\omega t \, \sigma_0 \left[\sin\omega t \cos\delta + \cos\omega t \sin\delta \right] dt$$

$$= \sigma_0\varepsilon_0 \sin\delta \int_0^{2\pi/\omega} \omega \cos^2\omega t \, dt = \pi\varepsilon_0^2 E''(t)$$

where

$$\int_0^{2\pi/\omega} \omega \cos^2\omega t \, dt = \pi \quad \text{and} \quad \sigma_0\sin(t) = \varepsilon_0 E''(t)$$

are used. Usually it holds that $E'' \ll E'$.

A similar analysis for the compliance $S^* = 1/E^*$ yields[d] $S^* = S' - iS''$. It holds that

$$E^* = E' + iE'' = \frac{1}{S^*} = \frac{1}{S' - iS''} = \frac{S' + iS''}{S'^2 + S''^2} \tag{18.25}$$

The energy dissipated is also obtained similarly yielding

---

[d] The negative sign for the imaginary part is a convention, leading to positive values for both the imaginary components of S^* and E^*.

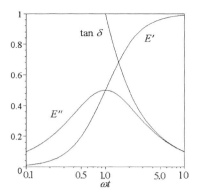

Fig. 18.7: The (reduced) dynamic response of the Maxwell model as a function of $\tau\omega$.

$$\Delta U = \pi\sigma_0^2 S''(t) \tag{18.26}$$

Another way to characterise visco-elastic behaviour is by using the complex viscosity η^* defined by

$$\sigma = \eta^* \dot{\varepsilon} \qquad \text{where} \quad \eta^* = \eta' - i\eta''$$

Using Eqs. (18.24) we have $\dot{\varepsilon} = i\omega\varepsilon$ and therefore $\eta^* = E^*/i\omega$. Hence $\eta' = E''/\omega$ and $\eta'' = E'/\omega$.

Let us consider the dynamic response of some of the models considered before. We start with the Maxwell model and deal thereafter with the Kelvin model.

We recall that for the Maxwell model the behaviour is described by $\dot{\varepsilon} = \dot{\sigma}/k + \sigma/\eta$ with characteristic relaxation time $\tau = \eta/k$ so that we can also write

$$k\tau\dot{\varepsilon} = \tau\dot{\sigma} + \sigma$$

Using

$$\varepsilon = \varepsilon_0 \exp(i\omega t) \text{ and } \sigma = \sigma_0 \exp[i(\omega t + \delta)]$$

the result is

$$k\tau i\omega\varepsilon_0 \exp(i\omega t) = \tau\sigma_0 i\omega \exp[i(\omega t + \delta)] + \sigma_0 \exp[i(\omega t + \delta)] \quad \text{or} \quad k\tau i\omega\varepsilon = \sigma(i\omega\tau + 1)$$

Solving for $E^* = \sigma/\varepsilon$ we obtain

$$E^* = E' + iE'' = \frac{\sigma}{\varepsilon} = \frac{ik\tau\omega}{1 + i\tau\omega} = k\frac{\tau^2\omega^2}{1 + \tau^2\omega^2} + ik\frac{\tau\omega}{1 + \tau^2\omega^2}$$

and therefore to

$$\tan\delta = \frac{E''}{E'} = \frac{1}{\tau\omega}$$

The response is shown in Fig. 18.7.

Generalising the dynamic relaxation response in a similar way as for the static relaxation response as described in Section 18.3, we obtain

$$\blacktriangleright \qquad E'(\omega) = k_\infty + (E_0 - E_\infty)\int g^\&(\tau)\frac{\tau^2\omega^2}{1 + \tau^2\omega^2}\, d\ln\tau \qquad \text{and} \tag{18.27}$$

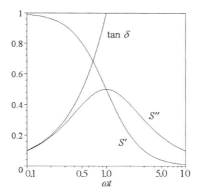

Fig. 18.8: The (reduced) dynamic response of the Kelvin model as a function of $\tau\omega$.

$$E''(\omega) = E_\infty + (E_0 - E_\infty) \int g^\&(\tau) \frac{\tau\omega}{1+\tau^2\omega^2} \, d\ln\tau \qquad (18.28)$$

Both the real and imaginary part of the relaxation can be determined from the relaxation time distribution, i.e. either $g(\tau)$ or $g^\&(\tau)$.

For the Kelvin model we obtained

$$\sigma = k\varepsilon + \eta\dot{\varepsilon}$$

Inserting similarly as before

$$\varepsilon = \varepsilon_0 \exp(i\omega t) \text{ and } \sigma = \sigma_0 \exp[i(\omega t + \delta)]$$

the result becomes

$$\sigma = k\varepsilon + i\eta\omega\varepsilon = \varepsilon(k + i\eta\omega)$$

so that the retardation response becomes

$$S(t) = S' - iS'' = \frac{\varepsilon}{\sigma} = \frac{1}{k+i\eta\omega} = \frac{k-i\eta\omega}{k^2+\eta^2\omega^2} = \frac{1}{k}\frac{1-i\tau\omega}{1+\tau^2\omega^2} \qquad (18.29)$$

and therefore

$$\tan\delta = \frac{S''}{S'} = \tau\omega$$

The results are displayed in Fig. 18.8.

Generalising as before the dynamic creep response becomes

$$S'(\omega) = S_\infty + (S_0 - S_\infty) \int f^\&(\tau) \frac{1}{1+\tau^2\omega^2} \, d\ln\tau \quad \text{and} \qquad (18.30)$$

$$S''(\omega) = S_\infty + (S_0 - S_\infty) \int f^\&(\tau) \frac{\tau\omega}{1+\tau^2\omega^2} \, d\ln\tau \qquad (18.31)$$

Both the real and imaginary part of the creep function can be calculated from the retardation time distribution.

Problem 18.3

Show that the retardation spectrum for the creep function $\varepsilon = \sigma(k+mt^n)$ is given by $g(\tau) = nm\tau^{n-1}/\Gamma(1-n)$, where Γ is the gamma function $\Gamma(1-n) = \int_0^\infty x^{-n}\exp(-x)\,dx$.

18.4 A thermodynamic extension to 3D and thermal effects*

So far the description of visco-elasticity has been 1D and was based on the use of the analogous models. However, these models oversimplify reality since in real materials the separation between various contributions is usually not clear-cut at all. Moreover they are 1D. In this section we extend the description to 3D, still restricting ourselves to linear visco-elastic materials, but we will also include thermal effects (Ziegler, 1983).

Let us start with isothermal conditions. For an isotropic elastic material, using the Lamé constants λ and μ and the density ρ, the specific Helmholtz function f as a function of the strain ε_{ij} is given by (see Chapter 6)

$$\rho f = \tfrac{1}{2}\lambda\varepsilon_{ii}\varepsilon_{jj} + \mu\varepsilon_{ij}\varepsilon_{ij}$$

so that the quasi-conservative stress $\sigma_{ij}^{(q)}$ reads

$$\sigma_{ij}^{(q)} = \rho\frac{\partial f}{\partial\varepsilon_{ij}} = \lambda\varepsilon_{kk}\delta_{ij} + 2\mu\varepsilon_{ij}$$

For a Newtonian fluid, on the other hand, there is no quasi-conservative stress but only a dissipative stress, derived from a specific dissipation function φ as a quadratic function of strain rate $\dot{\varepsilon}_{ij}$ (actually rate of deformation d_{ij})

$$\rho\varphi = \lambda'\dot{\varepsilon}_{ii}\dot{\varepsilon}_{jj} + 2\mu'\dot{\varepsilon}_{ij}\dot{\varepsilon}_{ij}$$

where λ' and μ' are the viscosity constants. Hence the dissipative stress $\sigma_{ij}^{(d)}$ is given by

$$\sigma_{ij}^{(d)} = \tfrac{1}{2}\rho\frac{\partial\varphi}{\partial\dot{\varepsilon}_{ij}} = \lambda'\dot{\varepsilon}_{kk}\delta_{ij} + 2\mu'\dot{\varepsilon}_{ij}$$

The simplest visco-elastic body, without any internal variables, is constructed by analogy of the Kelvin model by adding $\sigma_{ij}^{(q)}$ and $\sigma_{ij}^{(d)}$. In this case the total stress is

$$\sigma_{ij} = \sigma_{ij}^{(q)} + \sigma_{ij}^{(d)} \tag{18.32}$$

In general we refer to this type of material, where one adds the quasi-conservative and dissipative stress, as a *Kelvin material*.

If we, on the other hand, assume, inspired by the Maxwell model, that the total strain is the addition of an elastic strain $\varepsilon_{ij}^{(e)}$ related to the quasi-conservative stress $\sigma_{ij}^{(q)}$ and a viscous strain $\varepsilon_{ij}^{(v)}$ due to the dissipative stress $\sigma_{ij}^{(d)}$ we have

$$\varepsilon_{ij} = \varepsilon_{ij}^{(e)} + \varepsilon_{ij}^{(v)}$$

In this case the total stress becomes

$$\sigma_{ij} = \lambda \varepsilon_{kk}^{(e)} \delta_{ij} + 2\mu \varepsilon_{ij}^{(e)} = \lambda' \dot{\varepsilon}_{kk}^{(v)} \delta_{ij} + 2\mu' \dot{\varepsilon}_{ij}^{(v)} \tag{18.33}$$

To show that this model contains internal variables we rewrite the model as follows. We consider the total strain as external variable and the viscous strain $\varepsilon_{ij}^{(v)}$ as internal variable α_{ij} so that the Helmholtz function becomes

$$\rho f = \tfrac{1}{2}\lambda(\varepsilon_{ii} - \alpha_{ii})(\varepsilon_{jj} - \alpha_{jj}) + \mu(\varepsilon_{ij} - \alpha_{ij})(\varepsilon_{ij} - \alpha_{ij})$$

while the dissipation function reads

$$\rho\varphi = \lambda' \dot{\alpha}_{ii} \dot{\alpha}_{jj} + 2\mu' \dot{\alpha}_{ij} \dot{\alpha}_{ij}$$

Using the analogous expressions for the internal forces

$$\beta_{ij}^{(q)} = \rho \frac{\partial f}{\partial \alpha_{ij}} \quad \text{and} \quad \beta_{ij}^{(d)} = \tfrac{1}{2}\rho \frac{\partial \varphi}{\partial \dot{\alpha}_{ij}}$$

we obtain

$$\begin{aligned}
\sigma_{ij}^{(q)} &= -\beta_{ij}^{(q)} = \lambda(\varepsilon_{kk} - \alpha_{kk})\delta_{ij} + 2\mu(\varepsilon_{ij} - \alpha_{ij}) \\
\sigma_{ij}^{(d)} &= 0 \quad \text{and} \quad \beta_{ij}^{(d)} = \lambda' \dot{\alpha}_{kk} \delta_{ij} + 2\mu' \dot{\alpha}_{ij}
\end{aligned} \tag{18.34}$$

Using $\sigma_{ij} = \sigma_{ij}^{(q)} + \sigma_{ij}^{(d)}$ and $\beta_{ij}^{(d)} = -\beta_{ij}^{(q)}$ we regain Eq. (18.33). In general we refer to this type of material, where one adds the elastic and viscous strain, as a *Maxwell material*.

In order to describe thermal effects, we use for the elastic part the same procedure as described in Chapter 9, where we obtained for isotropic materials

$$\begin{aligned}
\rho f = \rho f_0 - \rho s_0 (T - T_0) + \tfrac{1}{2}\lambda \varepsilon_{ii} \varepsilon_{jj} + \mu \varepsilon_{ij} \varepsilon_{ij} \\
- (3\lambda + 2\mu)\alpha \varepsilon_{kk}(T - T_0) - \frac{\rho c}{2T_0}(T - T_0)^2
\end{aligned} \tag{18.35}$$

where f_0 is the reference Helmholtz energy, s_0 is the entropy in the reference state, α is the (linear) thermal expansion coefficient and c is the heat capacity at the reference temperature T_0. This resulted for the quasi-conservative stress in

$$\begin{aligned}
\sigma_{ij} &= \rho \frac{\partial f}{\partial \varepsilon_{ij}} = \lambda \varepsilon_{kk} \delta_{ij} + 2\mu \varepsilon_{ij} - (3\lambda + 2\mu)\alpha \delta_{ij}(T - T_0) \\
&= [\lambda \varepsilon_{kk} - (3\lambda + 2\mu)\alpha(T - T_0)]\delta_{ij} + 2\mu \varepsilon_{ij}
\end{aligned} \tag{18.36}$$

The dissipation function has to be extended by the heat flow and we take for the dissipation function

$$\rho\varphi = \lambda' \dot{\varepsilon}_{ii} \dot{\varepsilon}_{jj} + 2\mu' \dot{\varepsilon}_{ij} \dot{\varepsilon}_{ij} + \frac{1}{\kappa T} q_k q_k$$

where q_k denotes the heat flow and κ the thermal conductivity. The force associated with the heat flow is $-T_{,k}/T$ so that

$$-\frac{T_{,k}}{T} = \rho \frac{\partial \varphi}{\partial q_k} = \frac{1}{\kappa T} q_k \quad \text{or} \quad q_k = -\kappa T_{,k} \tag{18.37}$$

which represents *Fourier's law*.

Applying the same procedure as for the isothermal case we can obtain the expressions for the stresses for a Kelvin and Maxwell material. They read for the Kelvin material

$$\sigma_{ij} = [\lambda \varepsilon_{kk} - (3\lambda + 2\mu)\alpha(T - T_0) + \lambda'\dot{\varepsilon}_{kk}]\delta_{ij} + 2\mu\varepsilon_{ij} + 2\mu'\dot{\varepsilon}_{ij} \tag{18.38}$$

and for the Maxwell material the result is

$$\begin{aligned}\sigma_{ij} &= [\lambda(\varepsilon_{kk} - \alpha_{kk}) - (3\lambda + 2\mu)\alpha(T - T_0)]\delta_{ij} + 2\mu(\varepsilon_{ij} - \alpha_{ij}) \\ &= \lambda\dot{\alpha}_{kk}\delta_{ij} + 2\mu\dot{\alpha}_{ij}\end{aligned} \tag{18.39}$$

where, as before, the external variable is ε_{ij} and the internal variable is α_{ij}. The detailed derivation we leave as an exercise for the reader.

A further generalisation of the Maxwell material can be obtained with admitting more than one internal variable tensor. Assume that we have n such tensors $\alpha_{ij}^{(r)}$, where we indicate them with a superscript label $r = 1, ..., n$. To insure impact response we added in Section 18.3 an extra elastic element in parallel. For the same reason we also add here an extra elastic contribution with label (0), so that $\alpha_{ij}^{(0)} = 0$. In that case the dissipation function reads

$$\rho\varphi = \sum_{r=1}^{r=n} \lambda^{(r)}{}'\dot{\alpha}_{ii}^{(r)}\dot{\alpha}_{jj}^{(r)} + 2\mu^{(r)}{}'\dot{\alpha}_{ii}^{(r)}\dot{\alpha}_{ii}^{(r)} + \frac{1}{\kappa T} q_k q_k$$

while the Helmholtz function is given by

$$\begin{aligned}\rho f &= \rho f_0 - \rho s_0 (T - T_0) \\ &+ \sum_{r=0}^{r=n} \left[\tfrac{1}{2}\lambda^{(r)}(\varepsilon_{ii} - \alpha_{ii}^{(r)})(\varepsilon_{jj} - \alpha_{jj}^{(r)}) + \mu^{(r)}(\varepsilon_{ij} - \alpha_{ij}^{(r)})(\varepsilon_{ij} - \alpha_{ij}^{(r)}) \right] \\ &- \left[\sum_{r=0}^{r=n} (3\lambda^{(r)} + 2\mu^{(r)})\alpha^{(r)}(\varepsilon_{kk} - \alpha_{kk}^{(r)}) \right](T - T_0) - \frac{\rho c}{2T_0}(T - T_0)^2\end{aligned}$$

The response can be obtained in the way as described before. Elimination of the internal variables occurs similarly as in the 1D case yielding a 3D generalisation of Eq. (18.8). The first step is to separate the stress and strain in an isotropic and deviatoric part. This leads for the stress to

$$\sigma_{kk} = 3\sum_{1}^{n+1} K^{(r)} \left[\varepsilon_{kk} - \alpha_{kk}^{(r)} - 3\alpha^{(r)}(T - T_0) \right] \quad \text{and} \quad \sigma_{ij}' = 2\sum_{1}^{n+1} \mu^{(r)}(\varepsilon_{ij}' - \alpha_{kk}^{(r)'})$$

where for both equations we define $\alpha_{ij}^{(n+1)} = 0$ and the bulk modulus $K^{(r)} = \lambda^{(r)} + \tfrac{2}{3}\mu^{(r)}$. For the strain similarly we have

$$K^{(r)}{}'\dot{\alpha}_{kk}^{(r)} = K^{(r)}{}'\left[\varepsilon_{kk} - \alpha_{kk}^{(r)} - 3\alpha^{(r)}(T - T_0) \right] \quad \text{and} \quad \mu^{(r)}{}'\dot{\alpha}_{ij}^{(r)}{}' = \mu^{(r)}(\varepsilon_{ij}' - \alpha_{ij}^{(r)}{}')$$

where $r = 1, ..., n$ and the bulk viscosity $K^{(r)'} = \lambda^{(r)}{}' + \tfrac{2}{3}\mu^{(r)}{}'$. Differentiating the isotropic stress n times and making use of the expression for the isotropic internal strain in each step, we obtain $\dot{\sigma}_{kk}$ as a linear function of $\dot{\varepsilon}_{kk} - 3\alpha^{(r)}\dot{T}$ and equivalent expressions for the higher derivatives for $r = 1, ..., n+1$ as well as $\varepsilon_{kk} - \alpha_{kk}^{(r)} - 3\alpha^{(r)}(T - T_0)$ for $r = 1, ..., n$. Eliminating the latter functions leads to

$$\sigma_{kk} + p^{(1)}\dot{\sigma}_{kk} + \cdots + p^{(n)}\sigma_{kk}^{(n)} = q^{(0)}\varepsilon_{kk} + q^{(1)}\dot{\varepsilon}_{kk} + \cdots + q^{(n)}\varepsilon_{kk}^{(n)}$$
$$+ r^{(0)}(T - T_0) + r^{(1)}\dot{T} + \cdots + r^{(n)}T^{(n)} \tag{18.40}$$

Similarly differentiating the deviatoric stress n times and making use of the expression for the deviatoric internal strain in each step, we obtain $\dot{\sigma}_{ij}'$ as a linear function of $\dot{\varepsilon}_{ij}'$ and $\varepsilon_{ij}' - \alpha_{ij}^{(r)\prime}$. Eliminating the latter functions leads to

$$\sigma_{ij}' + p^{(1)\prime}\dot{\sigma}_{ij}' + \cdots + p^{(n)\prime}\sigma_{ij}^{(n)\prime} = q^{(0)\prime}\varepsilon_{ij}' + q^{(1)\prime}\dot{\varepsilon}_{ij}' + \cdots + q^{(n)\prime}\varepsilon_{ij}^{(n)\prime} \tag{18.41}$$

The actual procedure is quite cumbersome though. The fact that the behaviour is controlled by two independent expressions, Eqs. (18.40) and (18.41), is due to the assumed isotropy. For the general anisotropic case we obtain a single expression. In any case, all the coefficients in these expressions are determined by the Helmholtz and dissipation function.

Problem 18.4

Derive Eqs. (18.38) and (18.39).

18.5 The heriditary integral formulation*

In Section 18.3 we have seen that a 1D integral representation based on the Boltzmann superposition integral has led us to the creep description (Findlay et al. 1976)

$$\varepsilon(t) = \int_{-\infty}^{t} c(t-\theta)\,d\sigma = \int_{-\infty}^{t} \frac{\partial\sigma(\theta)}{\partial\theta} c(t-\theta)\,d\theta \tag{18.42}$$

where the elastic constant k has been incorporated in the function c and which generalises for a Kelvin system to the expression

$$\varepsilon(t) = \sigma_0 S(t) \quad \text{with} \quad S(t) = S_0 + (S_\infty - S_0)\int_0^\infty f(\tau)[1 - \exp(-t/\tau)]\,d\tau \tag{18.43}$$

Similarly a relaxation description

$$\sigma(t) = \int_{-\infty}^{t} r(t-\theta)\,d\varepsilon = \int_{-\infty}^{t} \frac{\partial\varepsilon(\theta)}{\partial t} r(t-\theta)\,d\theta \tag{18.44}$$

is obtained which generalises for a Maxwell system to

$$\sigma(t) = \varepsilon_0 E(t) \quad \text{with} \quad E(t) = E_\infty + (E_0 - E_\infty)\int_0^\infty g(\tau)\exp(-t/\tau)\,d\tau \tag{18.45}$$

A similar reasoning for anisotropic materials leads to

$$\sigma_{ij} = \int_0^t C_{ijkl}(t-\tau)\frac{\partial\varepsilon_{kl}}{\partial\tau}\,d\tau \tag{18.46}$$

where the fourth-order tensor functions C_{ijkl} are called the *relaxation moduli*. The symmetry of stress and strain implies that

$$C_{ijkl}(t) = C_{jikl}(t) = C_{jilk}(t) \tag{18.47}$$

while thermodynamics requires that

$$C_{ijkl}(t) = C_{klij}(t) \tag{18.48}$$

so that these function have the same symmetry as the elastic stiffness constants. The time dependence of the relaxation moduli is generally given by

$$C_{ijkl}(t) = \sum_p C_{ijkl}^{(p)} \exp(-t/\rho_p) + C_{ijkl}$$

where the tensors $C_{ijkl}^{(p)}$ and the elastic stiffness constants C_{ijkl} are material constants. They are positive semi-definite, i.e.

$$C_{ijkl}^{(p)} \varepsilon_{ij} \varepsilon_{kl} \geq 0 \qquad \text{and} \qquad C_{ijkl} \varepsilon_{ij} \varepsilon_{kl} \geq 0 \qquad \text{if} \qquad \varepsilon_{ij} \varepsilon_{kl} \geq 0$$

but their sum is positive definite, i.e.

$$\left(C_{ijkl}^{(p)} + C_{ijkl} \right) \varepsilon_{ij} \varepsilon_{kl} > 0 \qquad \text{if} \qquad \varepsilon_{ij} \varepsilon_{kl} \geq 0$$

The parameters ρ_p are the *relaxation times* and determine together with the constants $C_{ijkl}^{(p)}$ the time and rate-dependence of the material. Since often there are a large number of relaxation times which are closely spaced, the summation is frequently replaced by an integration with respect to the relaxation times.

The inverse of Eq. (18.46) reads

$$\varepsilon_{ij} = \int_0^t S_{ijkl}(t-\tau) \frac{\partial \sigma_{kl}}{\partial \tau} \, d\tau \tag{18.49}$$

where the fourth-order tensor functions $S_{ijkl}(t)$ are the *creep compliances*. The symmetry of the $S_{ijkl}(t)$ is the same as for the $C_{ijkl}(t)$ and its time dependence is given by

$$S_{ijkl}(t) = \sum_q S_{ijkl}^{(q)} \left[1 - \exp(-t/\tau_q) \right] + S_{ijkl} + S_{ijkl}' t$$

where the parameters τ_q are the *retardation times*. The fourth-order tensor constants $S_{ijkl}^{(q)}$, S_{ijkl} and S_{ijkl}' obey the same rules as given by Eqs. (18.47) and (18.48).

The mechanical response of a material is, in general, significantly affected by temperature, which modifies the density through thermal expansion and the relaxation (or retardation) times. The latter ones are particularly sensitive. If all ρ_p's and τ_p's are influenced equally by temperature, a material is called *thermorheologically simple*. In this case the temperature dependence can be introduced relatively easily.

The thermal extension of Eq. (18.46) reads

$$\sigma_{ij} = \int_0^t C_{ijkl}(\xi - \xi') \frac{\partial \varepsilon_{kl}}{\partial \xi'} \, d\xi' - \int_0^t \beta_{ij}(\xi - \xi') \frac{\partial \Delta T}{\partial \xi'} \, d\xi' \tag{18.50}$$

in which ξ is the reduced time defined by $d\xi \equiv dt/a_T$, so that we have

$$\xi \equiv \int_0^t \frac{dt}{a_T} \qquad \text{and} \qquad \xi' \equiv \int_0^\tau \frac{dt}{a_T}$$

where $a_T = a_T(T)$ is a temperature dependent material function (e.g. as occurring in the WLF theory, Chapter 20). The temperature relaxation functions $\beta_{ij}(\xi)$ have also the exponential form

$$\beta_{ij}(\xi) = \sum_p \beta_{ij}^{(p)} \exp(-\xi/\rho_p) + \beta_{ij}$$

where the constants $\beta_{ij}^{(p)}$ and β_{ij} define the thermal stress behaviour of the material. These coefficients are not necessarily positive definite.

The inverse of Eq. (18.50) is

$$\varepsilon_{ij} = \int_0^t S_{ijkl}(\xi - \xi') \frac{\partial \sigma_{kl}}{\partial \xi'} \, d\xi' - \int_0^t \alpha_{ij}(\xi - \xi') \frac{\partial \Delta T}{\partial \xi'} \, d\xi' \qquad (18.51)$$

where the functions $\alpha_{ij}(\xi)$ are given by

$$\alpha_{ij}(\xi) = \sum_q \alpha_{ij}^{(q)} \left[1 - \exp(-\xi/\tau_q)\right] + \alpha_{ij}$$

All these relations reduce to the ones for isotropic materials making the proper substitutions.

Finally, it should be remarked that, although thermorheological simplicity is common among many materials, is certainly does not always hold. For example, for composites where both phases have a different thermal response, it does not hold unless one phase is essentially only elastic.

This concludes the continuum discussion about visco-elasticity. The bibliography lists a number of books in which aspects for various materials are treated in different ways. For polymers Ferry (1970) and Strobl (1997) are a classic and a good introduction, respectively, while for inorganic glasses Scherer (1986) can be recommended. Visco-elasticity of metals, usually called creep, is treated by Betten (2002).

18.6 Bibliography

Betten, J. (2002), *Creep mechanics*, Springer, Berlin.

Christensen, R.M. (1971), *Theory of viscoelasticity*, Academic Press, New York.

Ferry, J.D. (1970), *Viscoelastic properties of polymers*, 2nd ed., Wiley, New York.

Findlay, W.N., Lai, J.S. and Onaran, K. (1976), *Creep and relaxation of nonlinear viscoelastic materials*, North-Holland, Amsterdam.

Scherer, G. W. (1986), *Relaxation in glass and composites*, Wiley, New York.

Strobl, G. (1997), *The physics of polymers*, 2nd ed., Springer, Berlin.

Tschoegl, N.W. (1989), *The phenomenological theory of linear viscoelastic behavior*, Springer, Berlin.

Ward, I.M. (1983), *Mechanical properties of polymers*, 2nd ed., Wiley, Chichester.

Ziegler, H. (1983), *An introduction to thermomechanics*, 2nd ed., North-Holland, Amsterdam.

19

Applications of visco-elasticity theory

In the following chapter some applications of visco-elastic theory are discussed. We start with the correspondence principle that allows the prediction of visco-elastic behaviour when the solution for the corresponding elastic problem is known. This principle is thereafter illustrated by the example of a pressurised thick-walled tube. An important application of visco-elasticity is in the testing of the creep of materials. After a brief general consideration the phenomenological modelling of creep deformation is treated, followed by some remarks on creep failure. Since indentation has become an important method for materials testing, the last section deals with indentation creep.

19.1 The correspondence principle*

For the treatment of the correspondence principle, we first briefly iterate the relevant equations. Then we discuss their Laplace transformed counterparts. Considering the boundary conditions we then arrive at the correspondence principle, which implies that under certain conditions elastic analysis methods can be used to derive transformed visco-elastic solutions.

Visco-elasticity theory applications

Visco-elastic processes are abundantly present in industry. Almost all shaping of polymers is done via visco-elastic processes. For many polymer items injection moulding is used and the design of such a product benefits greatly from visco-elastic modelling. Blowing of glass products is another example where visco-elasticity plays an important role. May be one of the most demanding applications is the prediction of the deformation (creep) and the longevity of (thermo)-mechanically loaded structures. This not only holds for products constructed from metals but also those made from inorganic materials or polymers.

Relaxation and creep

We have seen that within the framework of the small displacement gradient formulation of continuum mechanics the general relation between a stress σ_{ij} and strain ε_{ij} for anisotropic materials can be written by the Boltzmann superposition integral

$$\sigma_{ij} = \int_0^t C_{ijkl}(t-\tau)\frac{\partial \varepsilon_{kl}}{\partial \tau}\,d\tau \tag{19.1}$$

where the fourth-order tensor functions C_{ijkl} are called the *relaxation moduli*. The time dependence of the relaxation moduli is generally given by

$$C_{ijkl}(t) = \sum_p C_{ijkl}^{(p)}\exp(-t/\rho_p) + C_{ijkl} \tag{19.2}$$

where the tensor $C_{ijkl}^{(p)}$ and the elastic stiffness constants C_{ijkl} are material constants. The parameters ρ_p are the *relaxation times* and determine together with the constants $C_{ijkl}^{(p)}$ the time and rate-dependence of the material. As before, since often there are a large number of relaxation times, which are closely spaced, the summation is frequently replaced by integration with respect to the relaxation times.

The inverse of Eq. (19.1) reads

$$\varepsilon_{ij} = \int_0^t S_{ijkl}(t-\tau)\frac{\partial\sigma_{kl}}{\partial\tau}\,\mathrm{d}\tau \tag{19.3}$$

where the fourth-order tensor functions $S_{ijkl}(t)$ are the *creep compliances*. Its time dependence is given by

$$S_{ijkl}(t) = \sum_q S_{ijkl}^{(q)}\left[1-\exp(-t/\tau_q)\right]+S_{ijkl}+S_{ijkl}'t \tag{19.4}$$

where the tensor $S_{ijkl}^{(p)}$, the elastic compliance constants S_{ijkl} and tensor S_{ijkl}' are material constants and where the parameters τ_q are the *retardation times*.

We write the Laplace transform of a function $f(t)$ as

$$\hat{f} = \hat{f}(s) = \int_0^s \exp(-st)f(t)\,\mathrm{d}s$$

and involving the convolution theorem, we can write Eqs. (19.1) and (19.3) as

$$\hat{\sigma}_{ij} = \widetilde{C}_{ijkl}\hat{\varepsilon}_{kl} \qquad\text{and}\qquad \hat{\varepsilon}_{ij} = \widetilde{S}_{ijkl}\hat{\sigma}_{kl} \tag{19.5}$$

where the so-called operational moduli $\widetilde{C}_{ijkl} = s\hat{C}_{ijkl}$ and $\widetilde{S}_{ijkl} = s\hat{S}_{ijkl}$ are the s-multiplied transforms of $C_{ijkl}(t)$ and $S_{ijkl}(t)$, respectively. They obey $\widetilde{C}_{ijkl} = (\widetilde{S}_{ijkl})^{-1}$. The transformed stress-strain relationship given by Eqs. (19.5) are identical to those for an elastic body with moduli $\widetilde{C}_{ijkl}$ and compliances $\widetilde{S}_{ijkl}$. For positive s, $\widetilde{C}_{ijkl}$ and $\widetilde{S}_{ijkl}$ are positive definite, as can be seen from the explicit representation

$$\widetilde{C}_{ijkl} = \sum \frac{sC_{ijkl}^{(s)}}{s+1/\rho_s}+C_{ijkl} \qquad\text{and}\qquad \widetilde{S}_{ijkl} = \sum \frac{sS_{ijkl}^{(s)}}{s+1/\tau_s}+S_{ijkl}+\frac{S_{ijkl}'}{s} \tag{19.6}$$

Including the temperature dependence for thermorheologically simple materials explicitly leads to

$$\sigma_{ij} = \int_0^t C_{ijkl}(\xi-\xi')\frac{\partial\varepsilon_{kl}}{\partial\xi'}\,\mathrm{d}\xi' - \int_0^t \beta_{ij}(\xi-\xi')\frac{\partial\Delta T}{\partial\xi'}\,\mathrm{d}\xi' \tag{19.7}$$

where ξ is the reduced time defined by $\mathrm{d}\xi \equiv \mathrm{d}t/a_T$, so that we have

$$\xi \equiv \int_0^t \frac{\mathrm{d}t}{a_T} \qquad\text{and}\qquad \xi' \equiv \int_0^\tau \frac{\mathrm{d}t}{a_T}$$

where $a_T = a_T(T)$ is a temperature dependent material function. With a constant but different from zero temperature appears as a parameter and we obtain

$$\hat{\sigma}_{ij} = \widetilde{C}_{ijkl}(T,s)\hat{\varepsilon}_{kl} - \widetilde{\beta}_{ij}(T,s)\frac{\Delta T}{s} \tag{19.8}$$

$$\tilde{C}_{ijkl}(T,s) = s\hat{C}_{ijkl}(T) \qquad \text{and} \qquad \tilde{\beta}_{ij}(T,s) = s\hat{\beta}_{ij}(T)$$

For thermorheologicaly simple materials Eq. (19.6) is still valid but with ρ_p replaced by $\rho_p a_T$. The same must be done in the expression for $\tilde{\beta}_{ij}(T,s)$. For transient temperature the Laplace transform is used with respect to the reduced time ξ. In this case the transform of Eq. (19.7) becomes

$$\hat{\sigma}_{ij} = \tilde{C}_{ijkl}(T,s)\hat{\varepsilon}_{kl} - \tilde{\beta}_{ij}(T,s)\Delta\hat{T} , \qquad (19.9)$$

with $\tilde{C}_{ijkl}$ given by Eq. (19.6) and

$$\tilde{\beta}_{ij} = s\hat{\beta}_{ij} = \sum \frac{s\beta_{ij}^{(s)}}{s+1/\rho_s} + \beta_{ij} \qquad (19.10)$$

This transformed stress-strain relationship (19.9) is again identical to the one for an elastic body with moduli $\tilde{C}_{ijkl}$ and $\tilde{\beta}_{ij}$.

Basic equations and the correspondence principle

The basic equations in continuum mechanics are the three equilibrium equations and the displacements relations given by, respectively,

$$\frac{\partial\sigma_{ij}}{\partial x_j} + b_i = 0 \qquad \text{and} \qquad \varepsilon_{ij} = \left(\frac{\partial u_i}{\partial x_j} + \frac{\partial u_j}{\partial x_i}\right) \qquad (19.11)$$

which are completed by the constitutive equations that are given for visco-elasticity by Eq. (19.1) or, using the reduced time ξ defined by $d\xi \equiv dt/a_T$, by Eq. (19.7).

To arrive at the correspondence principle we first rewrite the strain-displacement and equilibrium relations in terms of x_i and ξ and for that purpose we represent either σ_{ij} or u_i by f. Transforming x_i and t to the new variables we obtain

$$x_i' = x_i \qquad \text{and} \qquad \xi = \xi(x_i,t) = \int_0^t \frac{dt'}{a_T[T(x_i,t')]}$$

We derive

$$\frac{\partial f}{\partial x_i} = \frac{\partial f}{\partial x_i'}\frac{\partial x_i'}{\partial x_i} + \frac{\partial f}{\partial\xi}\frac{\partial\xi}{\partial x_i} = \frac{\partial f}{\partial x_i} + \frac{\partial f}{\partial\xi}\frac{\partial\xi}{\partial x_i}$$

Therefore when we combine Eqs. (19.11) with Eq. (19.1), all spatial derivatives should be changed according to

$$\frac{\partial}{\partial x_i} \rightarrow \frac{\partial}{\partial x_i} + \frac{\partial\xi}{\partial x_i}\frac{\partial}{\partial\xi}$$

The formulation of a (visco-elastic) problem is completed by the boundary conditions, given by

$$u_i = \bar{u}_i \qquad \text{on } \Gamma_{\mathbf{u}} \qquad \text{and} \qquad \sigma_{ij}n_j = t_i \qquad \text{on } \Gamma_{\mathbf{t}} \qquad (19.12)$$

with $\bar{u}_i$ and t_j the prescribed displacement on the surface part $\Gamma_{\mathbf{u}}$ and traction on the surface part $\Gamma_{\mathbf{t}}$ while n_j denotes the (outer) normal on the surface.

The correspondence principle, as first presented by Lee[a] for isotropic media at constant temperature, subsequently elaborated by Morland and Lee[b] for problems with temperature variations and by Biot[c] for anisotropic materials implies that elastic methods can be used to obtain transformed visco-elastic solutions. Transforming Eqs. (19.11) and (19.12) results in

$$\frac{\partial \tilde{\sigma}_{ij}}{\partial x_j} + \tilde{b}_i = 0 \qquad \tilde{\varepsilon}_{ij} = \left(\frac{\partial \tilde{u}_i}{\partial x_j} + \frac{\partial \tilde{u}_j}{\partial x_i} \right) \qquad \tilde{u}_i = \overline{u}_i \ \text{ on } \Gamma_\mathbf{u} \text{ and } \qquad \tilde{\sigma}_{ij} n_j = \tilde{t}_i \ \text{ on } \Gamma_\mathbf{t}$$

Here it is assumed that the division of the surface Γ in a displacement-controlled part $\Gamma_\mathbf{u}$ and a traction-controlled part $\Gamma_\mathbf{t}$ does not change in time. This set of equations is formally equivalent with an elastic problem and therefore the solutions of elastic problems can be transformed to yield the corresponding solutions for the visco-elastic problem. A correspondence principle does not exist when the boundaries $\Gamma_\mathbf{u}$ and $\Gamma_\mathbf{t}$ do change in time. The in materials science important problem of indentation belongs to this class of problems. In these cases different methods have to be employed.

19.2   Pressurised thick-walled tube*

Consider a thick-walled tube with the inner radius a and the outer radius b with an internal pressure p. This problem shows rotation symmetry and is independent of the axial co-ordinate z (except that the displacement is linear in z). We consider the following two situations:

- The tube is axially clamped, i.e. $u_z = 0$. The tube is in plane strain. In this case $\sigma_{xx} \neq 0$ but it always holds that $\sigma_{rz} = \sigma_{\theta z} = 0$.
- The tube is axially free, i.e. $\sigma_{zz} = \sigma_{rz} = \sigma_{\theta z} = 0$. The tube is in plane stress and in this case $u_z \neq 0$ and given by $u_z = w = Wz$.

For comparison we will deal with the elastic, plastic and visco-elastic solution in some detail. The problem is two-dimensional and the essential unknown variables are $u_r = u(r)$, $\varepsilon_{rr} = du/dr$, $\varepsilon_{\theta\theta} = u/r$, $\sigma_{rr}(r)$ and $\sigma_{\theta\theta}(r)$. For visco-elastic behaviour these variables depend on time t.

The following equations will be taken to always hold for the domain $a \leq r \leq b$:

- The material is incompressible, i.e. $\varepsilon_{kk} = \varepsilon_{rr} + \varepsilon_{\theta\theta} + \varepsilon_{zz} = 0$ or for plane strain $\varepsilon_{rr} + \varepsilon_{\theta\theta} = 0$ and therefore

$$\frac{1}{r}\frac{\mathrm{d}}{\mathrm{d}r}\big[ru(r)\big] = 0 \qquad \text{or} \qquad u(r) = \frac{U}{r} \tag{19.13}$$

For plane stress $\varepsilon_{rr} + \varepsilon_{\theta\theta} + \varepsilon_{zz} = 0$ and using $u_z = Wz$ we obtain

$$\frac{1}{r}\frac{\mathrm{d}}{\mathrm{d}r}\big[ru(r)\big] + W = 0 \qquad \text{or} \qquad u(r) = \frac{U}{r} - \tfrac{1}{2}Wr \tag{19.14}$$

- The equilibrium conditions given by

$$\frac{\mathrm{d}\sigma_{rr}}{\mathrm{d}r} + \frac{1}{r}\big(\sigma_{rr} - \sigma_{\theta\theta}\big) = 0 \qquad \text{or} \qquad \frac{\mathrm{d}}{\mathrm{d}r}(r\sigma_{rr}) - \sigma_{\theta\theta} = 0 \tag{19.15}$$

---

[a] Lee, E.H. (1955), Quart. Appl. Mech. **13**, 183.
[b] Morland, L.W. and Lee, E.H. (1960), Trans. Soc. Rheol., 223.
[c] Biot, M.A. (1958), Proc. 3rd Natl. Congr. Appl. Mech. ASME, 1.

- The boundary conditions given by

$$\text{at } r = a, \ \sigma_{rr} = -p \quad \text{and} \quad \text{at } r = b, \ \sigma_{rr} = 0 \tag{19.16}$$

Elastic solutions

For an elastic incompressible material it holds that Poisson's ratio $v = \frac{1}{2}$ and that the shear modulus G is given by $G = \frac{1}{3}E$, where E is Young's modulus. Thus we have

$$\sigma_{rr} = -p + \frac{E}{3}\varepsilon_{rr} \qquad \sigma_{\theta\theta} = -p + \frac{E}{3}\varepsilon_{\theta\theta} \qquad \sigma_{zz} - p + \frac{E}{3}\varepsilon_{zz} \tag{19.17}$$

The solutions for both plane strain and plane stress are

$$\sigma_{rr} = \frac{pa^2}{\left(b^2 - a^2\right)}\left(1 - \frac{b^2}{r^2}\right) \qquad \sigma_{\theta\theta} = \frac{pa^2}{\left(b^2 - a^2\right)}\left(1 + \frac{b^2}{r^2}\right) \tag{19.18}$$

The displacements u and stress σ_{zz} for plane strain become

$$u = \frac{3pa^2b^2}{2E\left(b^2 - a^2\right)r}\frac{1}{r} \qquad \sigma_{zz} = \frac{pa^2}{\left(b^2 - a^2\right)} \tag{19.19}$$

while for plane stress the corresponding expressions become

$$u = \frac{pa^2}{2E\left(b^2 - a^2\right)}\left(r + \frac{3b^2}{r}\right) \qquad w = -\frac{pa^2}{E\left(b^2 - a^2\right)}z \tag{19.20}$$

Elasto-plastic solutions

In this case we have to distinguish between the plastic area $a < r < c$ and the elastic area $c < r < b$, where c is a still unknown parameter. If we use the von Mises yield criterion, where for the the present conditions we insert $\sigma_{rz} = \sigma_{r\theta} = \sigma_{\theta z} = 0$, we have

$$\left(\sigma_{rr} - \sigma_{\theta\theta}\right)^2 + \left(\sigma_{\theta\theta} - \sigma_{zz}\right)^2 + \left(\sigma_{zz} - \sigma_{rr}\right)^2 \le 6k^2 \tag{19.21}$$

where k is yield strength in shear. For both plane strain and plane stress in the region for $c < r < b$ it holds that

$$\sigma_{rr} = S\left(1 - \frac{b^2}{r^2}\right) \qquad \sigma_{\theta\theta} = S\left(1 + \frac{b^2}{r^2}\right) \tag{19.22}$$

as well as that

$$\sigma_{zz} = S \quad \text{with} \quad S = \frac{kc^2}{b^2} \quad \text{for plane strain and} \tag{19.23}$$

$$\sigma_{zz} = 0 \quad \text{with} \quad S = \sqrt{\frac{3}{3b^2 + c^4}}kc^2 \quad \text{for plane stress} \tag{19.24}$$

In the case of plane stress for $a < r < c$, the von Mises yield condition results in, using $\sigma_{zz} = \left(\sigma_{rr} + \sigma_{\theta\theta}\right)/2$ and $\sigma_{\theta\theta} > \sigma_{rr}$,

$$\sigma_{\theta\theta} - \sigma_{rr} = 2k \tag{19.25}$$

In combination with the equilibrium conditions this leads to

$$\frac{\mathrm{d}\sigma_{rr}}{\mathrm{d}r} = \frac{2k}{r} \tag{19.26}$$

The solution for this equation, using the boundary condition $\sigma_{rr} = -p$, is

$$\sigma_{rr} = -p + 2k \ln\left(\frac{r}{a}\right)$$

$$\sigma_{\theta\theta} = -p + 2k\left[1 + \ln\left(\frac{r}{a}\right)\right]$$

$$\sigma_{zz} = -p + k\left[1 + 2\ln\left(\frac{r}{a}\right)\right] \tag{19.27}$$

The value for c follows from the continuity of σ_{rr} and the condition that at at $r = c$ the material starts to flow resulting

$$\left(\frac{c}{b}\right)^2 - 2c\ln\left(\frac{c}{b}\right) = 1 + 2\ln\left(\frac{b}{a}\right) - \frac{p}{k} \tag{19.28}$$

The value of p at the beginning of flow is

$$p = p^* = k\left(1 - \frac{a^2}{b^2}\right) \tag{19.29}$$

while for complete yielding it is

$$p = p^{**} = 2k \ln\left(\frac{b}{a}\right) \tag{19.30}$$

In the case of plane stress for $a < r < c$, using the von Mises yield condition, together with $\sigma_{zz} = 0$, results in

$$\sigma_{rr}^2 - \sigma_{\theta\theta}\sigma_{rr} + \sigma_{\theta\theta}^2 = 6k^2 \tag{19.31}$$

This leads in combination with the equilibrium conditions to

$$\frac{\mathrm{d}}{\mathrm{d}r}(r\sigma_{rr}) = \frac{1}{2}\left[1 + \sqrt{12k^2 - 3\sigma_{rr}^2}\right] \tag{19.32}$$

with as an implicit solution

$$\ln\left(\frac{b}{a}\right) = \frac{1}{2}\sqrt{3}\left[\sin^{-1}\left(\frac{\sigma_{rr}}{2k}\right) + \sin^{-1}\left(\frac{p}{2k}\right) - \frac{1}{2}\ln\left(\frac{\sqrt{12k^2 - 3\sigma_{rr}^2} - \sigma_{rr}}{\sqrt{12k^2 - 3p^2} - p}\right)\right] \tag{19.33}$$

The remainder of the solution is much more complicated than for the plane strain case. If required, using the Prandtl-Reuss equation, the displacements can be calculated relatively easily for plane strain. For plane stress the calculation is much more complex though.

Visco-elastic solutions

For visco-elastic behaviour the pressure p becomes a function time t and if we choose the Heaviside function $H(t)$ to describe the switching on of the pressure, we have $p(t) = pH(t)$ and therefore as variables become a function of radius as well as time, e.g. $u = u(r,t)$. Independent of the constitutive behaviour chosen we have

$$\sigma_{rr} = \frac{pa^2}{(b^2 - a^2)}\left(1 - \frac{b}{r}\right)H(t) \quad \text{and} \quad \sigma_{\theta\theta} = \frac{pa^2}{(b^2 - a^2)}\left(1 + \frac{b}{r}\right)H(t) \quad (19.34)$$

For the Maxwell model we have

$$\dot{\varepsilon}_{ij} = \frac{1}{2G}\dot{\sigma}_{ij}{}' + \frac{1}{2\eta}\sigma_{ij}{}' \tag{19.35}$$

where, as before, the deviatoric stresses are given by $\sigma_{ij}{}' = \sigma_{kk} - \tfrac{1}{3}\sigma_{kk}\delta_{ij}$. For plane strain we have $\varepsilon_{zz} = 0$ and therefore $\sigma_{ij}{}' = 0$ and $\sigma_{zz} = (\sigma_{zz} - \sigma_{\theta\theta})/2$. As a consequence

$$\sigma_{rr} = -\sigma_{\theta\theta} = -\frac{1}{2}(\sigma_{rr} - \sigma_{\theta\theta}) = -\frac{r}{2}\frac{\partial\sigma_{rr}}{\partial r} \tag{19.36}$$

Using $u(r,t) = U(t)/r$ we have

$$\dot{\varepsilon}_{\theta\theta} = \frac{\dot{U}}{r^2} = \frac{1}{2G}\dot{\sigma}_{\theta\theta}{}' + \frac{1}{2\eta}\sigma_{\theta\theta}{}' \tag{19.37}$$

and therefore

$$u(r,t) = \frac{pa^2 b^2}{(b^2 - a^2)}\left(\frac{1}{2G} + \frac{t}{2\eta}\right) \tag{19.38}$$

For plane stress we have $\sigma_{zz} = 0$ and therefore to $\varepsilon_{rr} = (2\sigma_{rr} - \sigma_{\theta\theta})/3$ and thus using Eq. (19.14) to

$$\dot{\varepsilon}_{rr} = -\frac{1}{2}\dot{W} - \frac{\dot{U}}{r^2} = \frac{pa^2}{3(b^2 - a^2)}\left(1 - 3\frac{b^2}{r^2}\right)\left[\frac{\delta(t)}{2G} + \frac{H(t)}{2\eta}\right] \tag{19.39}$$

which leads to

$$u(r,t) = \frac{pa^2}{3(b^2 - a^2)}\left(\frac{1}{2G} + \frac{t}{2\eta}\right)\left(r + 3\frac{b^2}{r}\right)H(t) \tag{19.40}$$

Problem 19.1

Show that for a Kelvin-Voigt model the displacement behaviour is given
a) for plane strain by

$$u(r,t) = \frac{pa^2 b^2}{2\eta(b^2 - a^2)}[1 - \exp(-\lambda t)]\frac{H(t)}{r}$$

b) for plane stress by

$$u(r,t) = \frac{pa^2 b^2}{6\eta\lambda(b^2 - a^2)}\left(r + 3\frac{b^2}{r}\right)[1 - \exp(-\lambda t)]H(t) \quad \text{and}$$

$$w(r,t) = -\frac{pa^2b^2}{3\eta\lambda(b^2 - a^2)}[1 - \exp(-\lambda t)]zH(t)$$

19.3   The creep curve

Given the proper conditions visco-elasticity can occur in all type of materials. As indicated in Chapter 18 either the deformation at a given load, i.e. *creep*, is studied or the change in load given a certain deformation, i.e. *relaxation*. Often the term creep is loosely used to describe slowly time-dependent behaviour. A creep test is in principle done by loading a specimen at constant stress, meanwhile recording the strain. The resulting strain versus time curve is referred to as the *creep curve*. However, in practice usually the load is taken constant and one has to take extra measures to obtain a constant stress level. For somewhat larger deformations this is even true for tensile specimens.

Most of the phenomenology on creep is from experiments on metals but remains valid for ceramics and polymers. The deformation $\varepsilon(t,T)$ as a function of time t and temperature T of a material is considered to be the sum of an elastic, a plastic and a creep part and can be written as

$$\varepsilon(t,T) = \varepsilon_{ela} + \varepsilon_{pla} + \varepsilon_{cre} = \frac{\sigma}{E(T)} + \varepsilon_{pla}[Y(T),h(T)] + \varepsilon_{cre}[t;p(T)]$$

where $E(T)$ denotes the elastic modulus, $Y(T)$ the yield strength, $h(T)$ the hardening modulus and $p(T)$ the material parameters describing creep. All the material parameters are temperature dependent but in this approach only ε_{cre} is time-dependent. In 1910 Andrade introduced three different stages in a creep curve, whose creep rate can be represented by

$$\dot{\varepsilon}_{cre} = \dot{\varepsilon} - \dot{\varepsilon}_{ela} - \dot{\varepsilon}_{pla} = \dot{\varepsilon} \sim t^{-\kappa}$$

For $0 < \kappa \le 1$ the behaviour is called *primary* (or *transient*) *creep*. For $\kappa = 0$, we have *secondary* (or *steady-state*) *creep*, while for $\kappa < 0$ the behaviour is referred to as *tertiary* (or *accelerating*) *creep* (Fig. 19.1). In primary creep the creep rate $\dot{\varepsilon}_{pri}$ decreases with time. The creep resistance of the material increases by virtue of its own deformation. During secondary creep a nearly constant creep rate $\dot{\varepsilon}_{sec}$ is present due to a balance between the competing processes of hardening and recovery. During tertiary

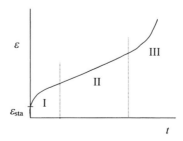

Fig. 19.1: Schematic of a typical creep curve. The primary, secondary and tertiary regimes are indicated by I, II and III, respectively. Creep starts at the static strain $\varepsilon_{sta} = \varepsilon_{ela} + \varepsilon_{pla}$.

creep, which occurs mainly in constant load tests, an effective reduction in cross section arises because of void formation (or necking). This reduction increases the tertiary creep rate $\dot{\varepsilon}_{\text{ter}}$. Findlay et al. (1976) have provided an extensive treatment of visco-elasticity theory, mainly illustrated by metal behaviour. Other general presentations are Christensen (1971) and Tschoegl (1989). Ferry (1970) is the classic reference for polymers while Ward (1983) and Strobl (1997) provide more recent introductions. For crystals and inorganics we refer to Poirier (1985). Finally for inorganic glasses Scherer (1986) provides an extensive review.

19.4   Creep deformation*

In this section we discuss the behaviour of creep of materials, as divided in three regimes and referred to as primary, secondary and tertiary creep, from a phenomenological point of view. We discuss each of the stage in turn, strongly inspired by the treatment of Betten (2002).

Primary creep

To characterise *primary creep* often a power law is used so that for a 1D description the creep strain ε_{cre} is given by

$$\varepsilon_{\text{cre}} = A\sigma^n t^m \tag{19.41}$$

resulting in a creep rate

$$\dot{\varepsilon}_{\text{cre}} = Am\sigma^n t^{m-1} \tag{19.42}$$

For metals typical values are $n \sim 5$ and $m \sim 0.2$. This behaviour is called *time-hardening* since it depends on time (and stress). Inserting Eq. (19.41) in Eq. (19.42) yields

$$\dot{\varepsilon}_{\text{cre}} = mA^{1/m}\sigma^{n/m}\varepsilon_{\text{cre}}^{(m-1)/m}$$

which is addressed as *strain-hardening* since this expression depends on strain (and stress). For a generalisation to a 3D description one conventionally takes into account that for many metals plastic deformation can be considered as incompressible. Therefore this is also assumed for primary creep. A straightforward generalisation of Eq. (19.42), as given by Odquist and Hult[d], yields therefore

$$d_{ij} = \frac{3}{2}K(J_2')^{(n-1)/2}\sigma_{ij}'t^{m-1}$$

where d_{ij} represents the rate of deformation tensor and J_2' is the second invariant of the stress deviator σ_{ij}'.

Secondary creep

It is often assumed that *secondary creep* depends on a creep potential $F = F(\sigma_{ij})$, very similar to the plastic potential. Limiting ourselves to isotropic materials, we recall that for the isotropic behaviour the relation F should obey

---

[d] Odquist, F. and Hult, J. (1962), *Kriechfestigkeit metallischer Werkstoffe*, Springer, Berlin.

$$F(a_{ip}a_{jq}\sigma_{pq}) = F(\sigma_{ij})$$

where a_{ip} are the direction cosines for a change in axis system. Equivalently F should only depend on the invariants J_I of the stress tensor

$$F = F[J_1(\sigma_{ij}), J_2(\sigma_{ij}), J_3(\sigma_{ij})]$$

where, as before, the invariants are given by

$$J_1(\sigma) = \delta_{ij}\sigma_{ji} \qquad J_2(\sigma) = (\sigma_{ij}\sigma_{ji} - \sigma_{ii}\sigma_{jj})/2$$
$$J_3(\sigma) = (2\sigma_{ij}\sigma_{jk}\sigma_{ki} - 3\sigma_{ij}\sigma_{ji}\sigma_{kk} + \sigma_{ii}\sigma_{jj}\sigma_{kk})/6$$

If we also assume incompressibility the above expression reduces further to

$$F = F[J_2'(\sigma_{ij}), J_3'(\sigma_{ij})]$$

where, as usual, the prime indicates that the deviatoric stresses are used so that

$$J_2(\sigma) = \sigma_{ij}'\sigma_{ji}'/2 \qquad J_3(\sigma) = 2\sigma_{ij}'\sigma_{jk}'\sigma_{ki}'/3$$

For anisotropic materials a linear transformation is used reading

$$\tau_{ij} = \beta_{ijkl}\sigma_{kl} \tag{19.43}$$

where τ_{ij} is called the *mapped stress tensor*. This transformation maps the actual stress state σ_{kl} to a fictitious isotropic stress state τ_{ij}. The invariants of the mapped tensor are given by

$$J_1(\tau) = A_{pq}\sigma_{pq} \qquad J_2(\tau) = A_{pqrs}\sigma_{pq}\sigma_{rs}/2 \qquad J_3(\tau) = A_{pqrstu}\sigma_{pq}\sigma_{rs}\sigma_{tu}/3$$

where the following definitions are used

$$A_{pq} = \beta_{iipq} \qquad A_{pqrs} = \beta_{iipq}\beta_{jirs} - \beta_{iipq}\beta_{jjrs}$$
$$A_{pqrstu} = \beta_{iipq}\beta_{jkrs}\beta_{kitu} - 3\beta_{iipq}\beta_{jirs}\beta_{kktu}/2 + \beta_{iipq}\beta_{jjrs}\beta_{kktu}/2$$

In this approximation the creep potential is assumed to read $F = F(\sigma_{ij}, \beta_{ijkl})$ and the invariants of the stress tensor are replaced by the invariants of the mapped stress tensor. Hence the actual creep state is mapped on a fictitious isotropic state with equivalent rate of deformation $d = \dot{\gamma}$ by the transformation $\tau_{ij} = \beta_{ijkl}\sigma_{kl}$. The limiting creep stress states $\sigma_x^{(c)}$ $\sigma_y^{(c)}$ and $\sigma_z^{(c)}$ are defined by the intersection of the flow surface with the coordinates axes (Fig. 19.2) and are taken at a certain, arbitrary creep strain, e.g. 1% strain in 10^5 h.

The theory of the creep potential is very akin to the theory of the plastic potential and based on maximum dissipation rate. Here the description

$$d_{ij} = \lambda\frac{\partial F(\sigma)}{\partial\sigma_{ij}} \qquad \text{or} \qquad d_{ij} = \lambda\frac{\partial F(\tau)}{\partial\tau_{pq}}\frac{\partial\tau_{pq}}{\partial\sigma_{ij}} = \dot{\gamma}_{pq}\frac{\partial\tau_{pq}}{\partial\sigma_{ij}}$$

is used where $\dot{\lambda}$ is a Lagrange multiplier. From the usual assumption of the convexity of the potential surface in both the σ_{ij}-space and τ_{pq}-space, the flow rule is obtained as

$$d_{ij} = \beta_{pqij}\frac{\partial F(\tau)}{\partial\tau_{pq}}\lambda \equiv \beta_{pqij}\dot{\gamma}_{pq} \tag{19.44}$$

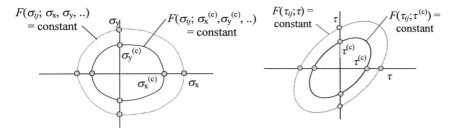

Fig. 19.2: The actual stress space σ_{ij} and mapped stress space τ_{ij}. The limiting stresses (cut-off values with the axes) are only indicated for the first quadrant. While for the actual stress space the limiting stresses are all different, for the actual stress space the limiting stresses are all equal to $\tau^{(c)}$.

To determine the factor $\dot{\lambda}$ the power law as introduced by Norton[e] and Bailey[f] is used, which reads

$$d = K\sigma^n \equiv d^{(c)}\left(\frac{\sigma}{\sigma^{(c)}}\right)^n$$

For the fictitious state defined by $\dot{\gamma}^{(c)} = d^{(c)}$ we have analogously

$$\dot{\gamma} = L\tau^m \equiv \dot{\gamma}^{(c)}\left(\frac{\tau}{\tau^{(c)}}\right)^m = d^{(c)}\left(\frac{\tau}{\tau^{(c)}}\right)^m = d \qquad (19.45)$$

so that, since $d = d_{11}$, we have

$$\dot{\lambda} = \frac{d}{(\partial F / \partial \tau_{ij})_{i=j=1}} \qquad \text{with} \quad \tau_{11} \equiv \tau \qquad (19.46)$$

In the last equation the fictitious isotropic creep stress τ occurs, which can be determined by the hypothesis of equivalent dissipation rate

$$\tau d = \sigma_{ij}d_{ij} = \dot{D} \qquad (19.47)$$

so that using Eqs. (19.44), (19.46) and $\sigma_{ij} = \beta_{ijkl}^{-1}\tau_{kl}$ we obtain

$$\tau\left(\frac{\partial F}{\partial \tau_{ij}}\right)_{i=j=1} = \tau_{ij}\frac{\partial F}{\partial \tau_{ij}}$$

The rate of dissipation $\dot{D}$ is obtained by inserting Eqs. (19.43), (19.44) and (19.46) resulting in

$$\dot{D} = \dot{\lambda}\tau_{pq}\frac{\partial F(\tau)}{\partial \tau_{pq}}$$

As elaborated in Chapter 6, for homogeneous creep potentials of degree r, we obtain from Euler's theorem

[e] Norton, F. (1929), *Creep at high temperatures*, McGraw-Hill, New York.
[f] Bailey, R. (1935), Proc. Inst. Mech. Eng. **131**, 131.

$$F(S\tau_{ij}) = S^r F(\tau_{ij}) \qquad \text{and} \qquad \tau_{ij} \frac{\partial F(\tau)}{\partial \tau_{ij}} = rF(\tau_{ij}) \tag{19.48}$$

and for the dissipation rate

$$\dot{D} = \varphi r \tau^r \qquad \text{assuming} \qquad F(\tau_{ij}) = \varphi \tau^r$$

As an example, assume $F = \tau_{ij}\tau_{ij}/2 = \tau^2/2$ with degree 2 so that $\varphi = \frac{1}{2}$ and $\dot{D} = \dot{\lambda}\tau^2$. Using Eq. (19.46) yields $\dot{\lambda} = d/\tau$ in consonance with Eq. (19.47).

We now apply this formalism to the simplest 3D creep potential, namely a von Mises-type expression for the mapped stress tensor so that we deal with an incompressible material,

$$F = J_2' = \frac{1}{2}\tau_{ij}'\tau_{ij}'$$

where the deviator $\tau_{ij}' = \beta_{\{ij\}pq}'\sigma_{pq}'$ with $\beta_{\{ij\}pq}' \equiv \beta_{ijpq} - \beta_{kkpq}\delta_{ij}/3$ deviatoric in the pair of indices $\{ij\}$. If the invariant J_2' is evaluated we obtain

$$J_2' = \frac{1}{2}\tau_{ij}'\tau_{ij}' = \frac{1}{2}\beta_{\{ij\}pq}'\beta_{\{ji\}rs}'\sigma_{pq}'\sigma_{rs}'$$

The flow rule (19.44) in combination with Eqs. (19.45), (19.46) and (19.48) yields

$$d_{ij} = \Phi\beta_{pq\{ij\}}'\frac{\partial F}{\partial J_2}\tau_{pq}'$$

where the function Φ is defined by

$$\Phi = \frac{1}{2}L\left(\frac{3}{(\partial F/\partial J_2)_{i=j=1}}\right)^{(m+1)/2}\left(\frac{\partial F}{\partial J_2}J_2(\tau')\right)^{(m-1)/2}$$

Note that here $\beta_{pq\{ij\}} \equiv \beta_{pqij} - \beta_{pqkk}\delta_{ij}/3$ is used. The dissipation rate becomes

$$\dot{D} = 2\frac{\partial F}{\partial J_2}J_2(\tau')\Phi = L\left[\frac{3(\partial F/\partial J_2)J_2(\tau')}{(\partial F/\partial J_2)_{i=j=1}}\right]^{(m+1)/2}$$

The parameters L and m have to be determined from experiments. For demonstration we consider orthotropic materials, which provide an important example. Writing

$$d = UX^a \qquad d = VY^b \qquad \text{and} \qquad d = WZ^c$$

for the three orthogonal directions x, y and z we obtain the limiting creep stress state

$$d^{(c)} = (UVW)^{1/3}[(\sigma_x^{(c)})^a(\sigma_y^{(c)})^b(\sigma_y^{(c)})^c]^{1/3}$$

By the concept of the mapping stress the limiting creep stress can be expressed by the fictitious isotropic limiting stress $\tau^{(c)}$

$$\tau^{(c)} = \tau_{xx} = \beta_{xxxx}\sigma_x^{(c)} \equiv l_x\sigma_x^{(c)}$$

with similar expressions for y and z. Comparing with $d^{(c)} = L\tau^m$, we obtain

$$L = [(U/l_x^a)(V/l_y^b)(W/l_z^c)]^{1/3} \qquad \text{and} \qquad m = (a + b + c)/3$$

Tertiary creep

   In *tertiary creep* cavitation is an important process and microscopic cracks and voids appear in the material. Therefore for a 1D description the additional variable ω is introduced representing the *area fraction of damaged material*. Alternatively $\psi = 1 - \omega$ is used. The rate of deformation is then described by $d = d(\sigma, \omega)$. Also a *net stress* $\tilde{\sigma}$ is used defined by $\tilde{\sigma} = \sigma / \psi$. When $\psi = 1$ ($\omega = 0$) the material is in its virgin state while for $\psi = 0$ ($\omega = 1$) the material cannot bear a load any more. Finally, it is assumed that the rate of change of ω is controlled by the present stress and damage state, i.e. $\dot{\omega} = g(\sigma, \omega)$ or equivalently $\dot{\psi} = -g(\sigma, \psi)$.

   For both the functions *f* and *g* power laws are frequently used, i.e. one states that

$$\frac{d}{d_0} = \frac{(\sigma/\sigma_0)^n}{(1-\omega)^m} \qquad \text{and} \qquad \frac{\dot{\omega}}{\dot{\omega}_0} = \frac{(\sigma/\sigma_0)^v}{(1-\omega)^\mu} \tag{19.49}$$

where $n \geq v$, m, μ, d_0, σ_0 and $\dot{\omega}_0$ are material constants. For an undamaged material ($\omega = 0$) the power law of Norton and Bailey is recovered. If we integrate the kinetic equation for $\dot{\omega}$, taking into account the initial condition $\omega(t = 0) = 0$, and insert the result in the expression for *d*, we obtain

$$\frac{d}{d_0} = \left(\frac{\sigma}{\sigma_0}\right)^m \left[1 - k\left(\frac{\sigma}{\sigma_0}\right)^v \dot{\omega}_0 t\right]^{-m/k} \qquad \text{with} \quad k = 1 + \mu$$

Integrating once more, taking into account that $\varepsilon_{\text{ter}}(t = 0) = 0$, leads to

$$\varepsilon_{\text{ter}} = \frac{a}{b(1-c)}\left[1 - (1 - bt)^{1-c}\right]$$

where the abbreviations

$$a = d_0\left(\frac{\sigma}{\sigma_0}\right)^n \qquad b = k\left(\frac{\sigma}{\sigma_0}\right)^v \dot{\omega}_0 \qquad \text{and} \qquad c = \frac{m}{k}$$

are used. Creep rupture occurs when $\omega = 1$ or $d \to \infty$ and therefore the time to rupture is given by

$$t_{\text{rup}} = \left[k\left(\frac{\sigma}{\sigma_0}\right)^v \dot{\omega}_0\right]^{-1}$$

For convenience the parameters *m* and μ are often taken to be equal to the parameters *n* and *v* and this simplifies the expressions to

$$d = K\tilde{\sigma}^n \qquad \text{and} \qquad \dot{\omega} = L\tilde{\sigma}^v$$

where the net stress $\hat{\sigma} = \sigma /(1 - \omega)$ is used. Comparing with Eq. (19.49) we find that

$$K = d_0 / \tilde{\sigma}_0^n \qquad \text{and} \qquad L = \dot{\omega}_0 / \tilde{\sigma}_0^v$$

The first of these relations reduces to the Norton-Bailey expression if we replace the net stress $\tilde{\sigma}$ with the nominal stress σ. This formulation allows an easy generalisation to 3D, similar to that as described before.

Due to the relations $m = n$ and $\mu = v$ the creep rupture time becomes

$$t_{\text{rup}} = \left[(1+v)L\sigma^v\right]^{-1}$$

where the nominal stress is taken as the actual stress at the beginning of the tertiary creep stage. Taking into account the Norton-Bailey law $d = K\sigma^n$, the result is

$$d_{\text{sec}}^{v/n} t_{\text{rup}} = K^{v/n} / L(1+v)$$

where d_{sec} indicates the minimum or secondary creep rate. Finally, also assuming that $v = n$ one obtains

$$d_{\text{sec}} t_{\text{rup}} = K / L(1+n) = \text{constant}$$

a relation also obtained empirically by Monkman and Grant[g]. The above analysis and associated simplifications thus provide a rationalisation on the basis of the net stress concept.

19.5   Creep failure

The creep failure of materials is quite important and therefore has been extensively studied. In Fig. 19.3 experimental results for a typical alloy are shown. The prediction of failure from these data is somewhat involved and therefore in the description of failure by creep several parameters have become in use. Two of them have gained a more widespread use. Both are based on the strain rate expression (Mangonon, 1999)

$$\dot{\varepsilon} = A \exp(-Q/RT)$$

where A is a parameter, Q the apparent activation energy and R and T have their usual meaning (see Chapter 20). Writing this expression as a difference equation, we have

$$d\varepsilon = A(\sigma) \exp(-Q/RT) \, dt$$

Discarding the integration constant, upon integration one obtains

$$\varepsilon = A(\sigma)\left[t \exp(-Q/RT)\right] = A\theta$$

where $A(\sigma)$ is a constant if the applied stress σ is constant and θ is the *temperature-*

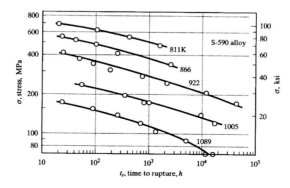

Fig. 19.3: Creep failure data for alloy S-590 at various temperatures and stress.

[g] Monkman, F. and Grant, N. (1959), Proc. ASTM **56**, 593.

compensated time, given by

$$\ln \theta_{rup} = \ln t_{rup} - Q/RT \qquad (19.50)$$

Experimentally it is found that the rupture strain $\varepsilon_{rup} = f(\theta_{rup})$, which states that the rupture time ε_{rup} is constant for given rupture time and the temperature-compensated rupture time θ_{rup} can be used as a design parameter.

The *Sherby-Dorn approach* assumes that θ_{rup} is a function of stress only and that Q is constant, i.e. $\theta_{rup} = \theta_{rup}(\sigma)$. The parameter $P_{SD} = \ln\theta_{rup}$ is then a constant for constant stress. Consequently,

$$\ln t_{rup} = P_{SD} + Q/RT \qquad (19.51)$$

and if we make a plot of $\ln t_{rup}$ versus $1/T$ for various stress levels, we should find straight lines with slope Q and intercept P_{SD}. The slope for the various tresses should be constant and represents the activation energy Q, which is typically about 380 kJ/mol for steels. If this is the case, a plot of the various P_{SD} values versus stress can be constructed (Fig. 19.4). If the data fall on a single line, such a plot can be used to predict the lifetime for particular levels of stress and temperature. It is not necessary to limit oneself to rupture. One may also use an appropriate strain, say 1 or 2%, with the Sherby-Dorn parameter derived for these conditions.

The *Larson-Miller approach* on the other hand assumes that θ_{rup} is constant and that the activation energy is a function of stress, i.e. $Q = Q(\sigma)$. Defining the parameter $P_{LM} = Q$ we have in this case

$$P_{LM} = T(\ln t_{rup} - \ln \theta_{rup}) = T(nt_{rup} - C) \quad \text{or} \quad \ln t_{rup} = P_{LM}/T - C \qquad (19.52)$$

which should lead, if we plot $\ln t_{rup}$ versus $1/T$ at various stress levels to straight lines with slope P_{LM} and intercept C. The intercept of the various lines should be constant and represents the constant C, which has typically a value of about 20 for steels (if t is given in hours, T in Kelvin, Q in kcal/mol and base 10 logarithms are used). If this is the case, a plot of the various PLM values versus stress can be made (Fig. 19.5). Again if all the data fall on a single line, the plot can be used to predict the lifetime for particular stress and temperature levels.

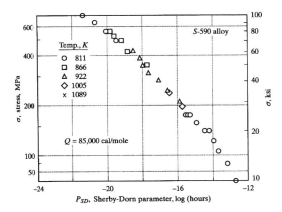

Fig. 19.4: The Sherby-Dorn plot for the data as given in Fig. 19.3.

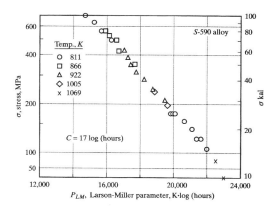

Fig. 19.5: The Larson-Miller plot for the data as given in Fig. 19.3.

Example 19.1

In Fig. 19.3 the experimental data of lifetime versus stress for alloy S-590 are shown. It is required that a component has a minimum lifetime of one year (or 365×24) = 8760 h) at 500 °C (= 773 K). The question is what is the maximum allowable stress.

Using the Sherby-Dorn approach, we calculate from Fig. 19.3 the activation energyh Q as 85000 kcal/mol. Given that $R = 1.98$ cal/K mol we obtain therefore $P_{SD} = \log(8760)–(1/\ln 10)×85000/(773×1.98) = –20.2$. From Fig. 19.4, we see that the maximum allowable stress is about 500 MPa.

Using the Larson-Miller approach we calculate from Fig. 19.3 the constant C as 17 log(h). Therefore, we calculate $P_{LM} = T(\log t_{rup} + C) = 773(\log 8760 + 17) = 16.2×103$. From Fig. 19.5, we see that the maximum allowable stress is about 500 MPa.

In this case there is no significant difference between the two approaches. In practice the maximum allowable stress is taken lower because a safety factor will be applied.

19.6   Indentation creep

Because creep and relaxation measurements on macroscopic specimens are generally relatively expensive, since long people have tried to assess the inelastic behaviour of materials by indentation. The first successful results were for plasticity but also for the field of visco-elasticity (creep and/or relaxation) several attempts to use indentation have been made. In fact, the field is under active investigation due to the extended experimental possibilities related with indentation. In this section we will mention a few results.

---

h Since in the literature t is often given in hours, T in Kelvin, Q in kcal/mol and base 10 logarithms are used, we do here likewise.

The first useful attempts were possibly due to the research group of Tabor in the 1960s, who investigated the creep of the soft metals Pb and In via indentation[i] on single crystals from liquid air temperature to 50 °C below the melting point using a spherical indenter and loading times from 10^{-4} to 10^3 s. Generally the yield pressure p decreased with loading time t. They obtained above $\cong 0.6T_m$ and for times exceeding a few seconds a linear relationship between the $\log p$ and $\log t$. The activation energies obtained were close to those of self-diffusion. The deformation around the indentations corresponds to slip suggesting that the rate-limiting process was climb. For shorter loading times, i.e. below 1 s or so, the yield pressure markedly increased and below about 10^{-3} s an upper limit p_{max} was reached, regardless of the testing temperature. For In also some twinning was observed. For measurements at liquid air temperature, p was almost independent of the loading time and equal to p_{max}. From these observations the authors concluded that the creep was dominated by a self-diffusion mechanism. This investigation was extended to other materials[j] including Sn and Pb single crystals, Al polycrystals, MgO single crystals and WC. The yield pressure appeared to be independent of the shape of the indenter so that the strain rate was associated with the rate at which the elastic-plastic boundary diffuses into the undeformed material. The kinematics of the process was analysed by transient creep model derived by Mott for constant stress conditions, assuming that this model can even be applied when the stress changes. This model results in a power-law for creep with exponent m. The authors obtained a good agreement with the theory and again the activation energies were close to those for self-diffusion. They concluded that it was possible to describe the indentation process in terms of transient creep.

In the 1990s Sakai[k] started to describe the indentation process of inorganics in terms of the energy required to deform the materials. He defined a true hardness as the energy required for to create a unit volume of indentation in ideally plastic materials. These attempts have been extended alter to deal with the complete load P versus depth h curve for indentations[l] on SiC, Si_3N_4, Al_2O_3 and soda-lime glass. Polycrystalline Cu was used as a reference material. The loading P-h curves could be described by the quadratic equation $P = k_l h^2$ and the unloading curve with $P = k_u(h-h_r)^2$, where h_r is the residual indentation depth, within experimental error for Vickers, Berkovich and Knoop indenters. Sakai[m] also extended his work to visco-elasticity. A recent review is given by Li and Bushan[n].

At present there is active research for visco-elastic indentation deformation, also in relation to adhesion. A major problem in a visco-elastic contact problem is that the contact area during loading and unloading is not the same at the same load. The boundary conditions are thus not constant in time so that the correspondence principle cannot be applied. This is often ignored and the analysis for visco-elastic behaviour is usually done in terms of the corresponding elastic-plastic analysis. There are several attempts to remedy this situation but the matter seems not to be resolved. From a recent attempt[o], it appears, however, that at least the (zero time) elastic modulus E can be estimated properly from the jumps in load P and displacement h at the unloading

[i] Mulhearn, T.O and Tabor, D. (1960), J. Inst. Metals **61**, 7.
[j] Atkins, A.G., Silvério, A. and Tabor, D. (1966), J. Inst. Metals **94**, 369.
[k] Sakai, M. (1993), Acta. Mettall. Mater. **41**, 1751.
[l] Sakai, M. Shimizu, S. and Ishikawa, T. (1999), J. Mater. Res.. **14**, 1471.
[m] Sakai, M. (2002), Phil. Mag. **82**, 1841.
[n] Li, X. and Bhushan, B. (2002), Mater. Charac. **48**, 11.
[o] P.G.Th. van der Varst, A.A.F. van de Ven and de With, G. (2005), submitted to J. Mater. Res.

point, similarly to the elastic relation $E \sim \mathrm{d}P/\mathrm{d}h$, irrespective of the material response. The attention for visco-elastic adhesion increased the interest in the Johnson, Kendall and Roberts (or JKR) test in which typically the interaction is studied between a sphere and a plate. After the initial elastic solution of this problem by Hertz (see Appendix E) and some early plastic and visco-elastic attempts, modelling for this configuration is now actively pursued. Shull[p] has given a recent review, while Barthel and Haiat[q] provide, what they call, 'a hand-waving introduction'. Johnson[r] has also presented a short introduction to the topic.

In conclusion, this is an active field where the answers, although not final yet, are highly relevant for the testing of materials at a millimetre scale (JKR-test), the micrometre scale (indentation test) and nanometre scale (AFM test).

19.7   Bibliography

Betten, J. (2002), *Creep mechanics*, Springer, Berlin.

Christensen, R.M. (1971), *Theory of viscoelasticity*, Academic Press, New York.

Ferry, J.D. (1970), *Viscoelastic properties of polymers*, 2nd ed., Wiley, New York.

Findlay, W.N., Lai, J.S. and Onaran, K. (1976), *Creep and relaxation of nonlinear viscoelastic materials*, North-Holland, Amsterdam.

Mangonon, P.L. (1999), *The principles of materials selection for engineering design*, Prentice-Hall, Upper Saddle River, NJ.

Poirier, J.-P. (1985), *Creep of crystals*, Cambridge University Press, Cambridge.

Scherer, G. W. (1986), *Relaxation in glass and composites*, Wiley, New York.

Strobl, G. (1997), *The physics of polymers*, 2nd ed., Springer, Berlin.

Tschoegl, N.W. (1989), *The phenomenological theory of linear viscoelastic behavior*, Springer, Berlin.

Ward, I.M. (1983), *Mechanical properties of polymers*, 2nd ed., Wiley, Chichester.

---

[p] Shull, K.R. (2001), Mater. Sci. Eng. **R36**, 1.
[q] Barthel, E. and Haiat, G. (2004), J. Adhesion **80**, 1.
[r] Johnson, K.L. (1996), Langmuir **12**, 4510.

20

Structural aspects of visco-elasticity

In the previous chapters we have discussed the basics of continuum visco-elasticity and its application to structures and processes. In the present chapter the structural background of visco-elasticity is treated. We start with the creep of inorganics and metals. Thereafter the creep and relaxation of polymers is treated.

20.1 Creep of inorganics and metals

We recall (see Chapter 19) that the deformation $\varepsilon(t,T)$ as a function of time t and temperature T of a material is considered to be the sum of an elastic, a plastic and a creep part and can be written as (Findlay et al., 1976)

$$\varepsilon(t,T) = \varepsilon_{ela} + \varepsilon_{pla} + \varepsilon_{cre} = \frac{\sigma}{E(T)} + \varepsilon_{pla}[Y(T),h(T)] + \varepsilon_{cre}[t;\boldsymbol{p}(T)]$$

where $E(T)$ denotes the elastic modulus, $Y(T)$ the yield strength, $h(T)$ the hardening modulus and $\boldsymbol{p}(T)$ the material parameters describing creep. The creep rate can be represented by

$$\dot{\varepsilon}_{cre} = \dot{\varepsilon} - \dot{\varepsilon}_{ela} - \dot{\varepsilon}_{pla} = \dot{\varepsilon} \sim t^{-\kappa}$$

For $0 < \kappa \leq 1$ we have *primary creep* while for $\kappa = 0$, we have *secondary* (or *steady-state*) *creep*. Finally for $\kappa < 0$ the behaviour is referred to as *tertiary creep*. In primary creep the creep rate $\dot{\varepsilon}_{pri}$ decreases with time. The creep resistance of the material increases by virtue of its own deformation. During secondary creep a nearly constant creep rate $\dot{\varepsilon}_{sec}$ is present due to a balance between the competing processes of hardening and recovery. During tertiary creep, which occurs mainly in constant load tests, an effective reduction in cross-section arises because of void formation (or necking). This reduction increases the tertiary creep rate $\dot{\varepsilon}_{ter}$. Findlay et al. (1976) have provided an extensive treatment of visco-elasticity theory, mainly illustrated by metal behaviour. Ferry (1980) is the classic reference for polymers while for crystals and inorganics we refer to Poirier (1985). Finally for inorganic glasses Scherer (1986) provides an extensive review. Inorganics and metals primarily deform elastically and plastically. However, at high temperature creep also starts to play a role. In a general framework this is a part of visco-elastic behaviour.

Conventional creep modelling

The simplest way to describe any phenomenon is empirically. This has been attempted for creep as well and we follow here to some extent the description by Cottrell (1953). As stated, the creep strain rate $\dot{\varepsilon}$ can be described reasonable accurately by

$$\dot{\varepsilon} = At^{-\kappa}$$

where A and κ are empirical parameters. For low temperature and stress, the extreme case $\kappa = 1$ is applicable resulting in the *logarithmic creep law*

$$\varepsilon_{cre} = A \ln(t/t_0)$$

Early experiments[a] on rubber, glass and various metals indeed show this behaviour. For higher creep rate and larger strain frequently $\kappa < 1$. For a wide range of conditions and materials the value $\kappa = 2/3$ is preferred, which lead to *Andrade creep*[b]. This description, resulting in

$$\varepsilon_{cre} = 3A t^{1/3}$$

appears to be valid for primary creep for a wide range of conditions and materials. Andrade himself showed this for many polycrystalline metals, extended to some 20% in about 30 min. Several other materials follow a similar behaviour.

Another description for primary creep which sometimes yields a better description (Garofalo, 1965) is given by

$$\varepsilon_{cre} = A[1 - \exp(-t/\tau)]$$

where A and τ are empirical constants.

The expressions given so far relate to *transient creep* since the creep rate decays with time. To include the essential linear part of a creep curve a term with $\kappa = 0$ is added. This choice results in *steady state* of *quasi-viscous creep* with

$$\varepsilon_{cre} = At$$

Therefore, generally at low temperature logarithmic creep $\varepsilon_{cre} = \alpha \ln(t/t_0)$ is observed. At intermediate temperature the combination $\varepsilon_{cre} = \alpha \ln(t/t_0) + \beta t^{1/3}$ with parameters α and β, is more suitable while for still higher temperature $\varepsilon_{cre} = \beta t^{1/3} + \gamma t$ with parameters β and γ fits better.

Experimentally it appears that the parameter A increases with temperature T proportional to $\exp(-1/T)$ at constant stress and with stress σ proportional to $\exp(\sigma)$ or to σ^n at constant temperature. Assuming a kinetic mechanism for the strain rate $\dot{\varepsilon}$ this suggests that the activation energy $U(\sigma)$ for sufficiently small stress range can be linearized to $U(\sigma) = U_0 - v\sigma$, where v is denoted as the activation volume, so that

$$\dot{\varepsilon} = A_0 \exp\left[-\left(U_0 - v\sigma\right)/RT\right]$$

where A_0 and v are parameters and R is the gas constant. To illustrate this we quote the results for Zn crystals at a shear stress of about 350 kPa, which read $U_0 \cong 125$ kJ/mol and $v\sigma \cong 42$ kJ/mol[c]. As an example Fig. 20.1 shows the primary and secondary creep for Al.

The most extensive theoretical description is conventionally for secondary creep. Here a multitude of mechanisms for crystalline materials have been proposed and analysed. These mechanisms are based on dislocation and point defect motion as well as the viscous deformation of glassy (secondary) phases at the grain boundaries. In

[a] Phillips, F. (1905), Phil. Mag. **9**, 513; Chevenard, P. (11934), Rev. Mét. **31**, 473; Laurent, P. and Eudier, M. (1950), Rev. Mét. **47**, 39; Davis, M. and Thompson, N. (1950), Proc. Roy. Soc. **B63**, 847.
[b] Andrade, E.N. de C. (1910), Proc. Roy. Soc. **A84**, 1 and (1914), Proc. Roy. Soc. **A90**, 329.
[c] Cottrell, A.H. and Aytekin, V. (1950), J. Inst. Met. **77**, 389.

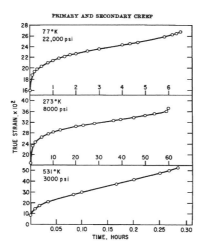

Fig. 20.1: Primary and secondary creep for aluminium at various temperatures (Garofalo, 1965).

general emphasis has been given to the dependence of the secondary creep rate $\dot{\varepsilon}_{sec}$ on stress σ, grain size l and temperature T using a power law

$$\dot{\varepsilon}_{sec} = A\frac{DGb}{RT}\left(\frac{\sigma}{G}\right)^{n}\left(\frac{b}{l}\right)^{m} \tag{20.1}$$

where A is a dimensionless constant and m and n are characteristic constants and $D = D_0\exp(-Q/RT)$ is the diffusion coefficient and Q is the activation energy. The symbols G and b denote the shear modulus and the Burgers vector, respectively. For various mechanisms the values of m and n are given in Table 20.1. It seems that the mechanisms for creep and the related deformation laws are in principle well understood but that in practice numerous problems arise. Simultaneously operating mechanisms makes interpretation of the data often difficult. Moreover, the interpretation of multiaxial stress states using simple, usually bend or tensile data, is an unsolved problem: the matter of which equivalent stress to take has not been settled in general.

Table 20.1: Creep constants and diffusion paths for various mechanisms.

Mechanism	m	n	Diffusion path
Lattice mechanisms			
Dislocation movement, climb controlled	0	4.5	Lattice
Dislocation movement, glide controlled	0	3	Lattice
Dissolution of dislocation loops	0	4	Lattice
Dislocation climb without glide	0	3	Lattice
Dislocation climb by pipe diffusion	0	5	Dislocation core
Boundary mechanisms			
Volume vacancy flow (Nabarro-Herring)	2	1	Lattice
Grain boundary vacancy flow (Coble)	3	1	Grain boundary
Grain boundary sliding with liquid phase	1	1	Second phase
Grain boundary sliding without liquid phase	1	2	Lattice, grain boundary

20.2   Models for primary and secondary creep

In order to be able to compare various materials properly it is advantageous to use a reduced temperature T/T_m, where T_m is the melting point, and shear stress σ/G, where G is the shear modulus. It appears that the different types of behaviour fall roughly in the same areas in a graph of σ/G versus T/T_m. This kind of plot is usually called a deformation map (or creep diagram, see next section). A large amount of plastic deformation occurs only if the critical resolved shear stress is exceeded. In that case there is extensive dislocation multiplication and motion. Creep below the critical resolved shear stress occurs but is due to other mechanisms than dislocation mobility. This may include e.g. the ordering of interstitial atoms due to the applied stress resulting in a preferred occupation of sites. Upon release of the stress the preferred ordering relaxes accompanied with a relaxation of stress. However, above the critical resolved shear stress for metals dislocations are the main entities that cause creep. For inorganics also vacancies are highly relevant for creep due to the relatively low mobility of dislocations. Moreover, also the contribution of viscous deformation of secondary phases at the grain boundaries is important. The effect of grain size and alloying is illustrated in Fig. 20.2 and Fig. 20.3, respectively.

At low temperature dislocations have no difficulty in moving parallel to their slip planes while motion in other directions is difficult for any dislocation, either edge or partly edge in character. Vacancies or interstitial atoms must diffuse towards or away from a dislocation with a (partly) edge character when the motion is perpendicular to the slip plane, the so-called *dislocation climb*. Since diffusion is slow, the motion of dislocation in directions other than parallel to the slip plane is slow as well. At high temperature diffusion becomes fast and the motion of dislocations perpendicular to the slip becomes fast accordingly. In the words of Weertman (1983) one can say that they obtain an extra 'degree of freedom'.

Therefore at relatively low temperature (say $< \frac{1}{2}T_m$, where T_m is the melting point) and low stress the motion of dislocations other than parallel to the slip plane is difficult. Upon arresting the stress, however, not all dislocations stop to move. Thermal fluctuations will produce some additional motion, which by increasing deformation, increases the hardening of the material. The work hardening mechanisms make dislocation motion increasingly difficult and the rate of deformation must decrease with increasing time. In this way we obtain transient creep, in particular *logarithmic creep*.

To rationalise this behaviour a non-specific dislocation model is used as described by Weertman (1983). Imagine a crystal divided into sub-volumes of length L, which is the dimension of the smallest element that can be permanently deformed by the passage of dislocations through it independently of the deformation of neighbouring elements. So, stresses within neighbouring elements cannot reverse the deformation of the volume element once it has occurred. The energy W necessary to move a dislocation through the element is given by $W = \sigma V \, d\varepsilon$. The strain $d\varepsilon$ produced per unit volume when one dislocation segment of length L moves a distance L is approximately bL^2. With σ the applied stress and σ^* the back stress due to the long range contribution of the hardened remainder of the material, the energy W becomes

$$W = (\sigma - \sigma^*)L^3(bL^2/L^3) + U = bL^2(\sigma - \sigma^*) + U \tag{20.2}$$

where b is the length of the Burgers vector. The energy U is the energy to overcome the resistance offered by the short range hardening mechanisms, e.g. the energy

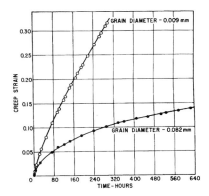

Fig. 20.2: Grain size effect on the creep curve of an Fe austenitic alloy at 705 °C and 9550 psi (Garofalo, 1965).

needed for a dislocation to cut through a forest dislocation, for non-conservative motion of a jog or due to an appreciable contribution of the Peierls stress. Expanding the long range hardening stress σ^* by a Taylor expansion to first order we obtain

$$\sigma^* = \sigma_0^* + h(\varepsilon - \varepsilon_{ela} + \varepsilon_{pla}) \tag{20.3}$$

where h is the hardening modulus. The creep rate is proportional to the frequency with which thermal fluctuations cause deformation of a volume element with volume L^3. Transition state theory states that the probability that such a volume element will be deformed in unit time is equal to $v_0 \exp(-W/kT)$, where v_0 is the vibrational frequency of the entity causing the deformation, in this case the dislocation[d]. This frequency is associated with the strain $d\varepsilon$. Therefore, the creep rate is

$$\dot{\varepsilon} = v_0 \frac{bL^2}{L^3} \exp\left(\frac{-W}{kT}\right) = v_0 \frac{b}{L} \exp\left(\frac{-U_0}{kT}\right) \exp\left[\frac{bL^2}{kT}(\sigma - h\varepsilon)\right] \tag{20.4}$$

with

$$U_0 = U + bL^2\left(\sigma_0^* - h\varepsilon_{ela} - h\varepsilon_{pla}\right) \tag{20.5}$$

Integrating this equation with respect to time, meanwhile assuming $\sigma = \sigma_0^*$, we obtain

$$\varepsilon = \varepsilon_0 \ln(1 + vt) \quad \text{with} \quad \varepsilon_0 = \frac{kT}{hbL^2} \quad \text{and} \quad v = v_0 \frac{hb^2L}{kT} \exp\left(\frac{-U}{kT}\right) \tag{20.6}$$

which is the experimentally observed behaviour[e]. If we make a more specific dislocation model, the results can be made more specific. Experimental results, e.g. for Cu[f] at 90 and 170 K and a stress level ranging from 93 to 173 g/mm^2, indeed show that U_0 is indeed independent of the applied stress.

A similar reasoning can be applied for secondary creep. Here the pertinent facts for metals are the Arrhenius-like temperature behaviour with an activation energy

---

[d] This frequency is probably somewhat lower than the frequency of the atoms in the lattice, say about 10^{10} to 10^{11} s^{-1}.
[e] The difference between $\ln(1+vt)$ and $\ln(vt)$ is negligible since $vt \gg 1$.
[f] Conrad, H. (1958), Acta Met. **6**, 339.

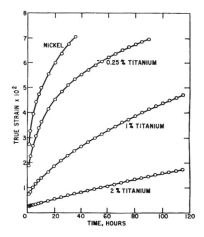

Fig. 20.3: Creep curve for Ni and Ni-Ti alloys tested at 700 °C at a stress level of 5750 psi (Garofalo, 1965).

within experimental error equal to the self-diffusion constant of the metal and a stress power law dependence. We deal with alloys first and thereafter with pure metals.

For solid solutions there are several phenomena that hinder the dislocation motion. Amongst them there are impurity clouds (Cottrell-Jaswon effect), stacking faults (Suzuki effect), stress-induced order (Schoeck effect) and the dislocation induced disorder (Fisher effect). For all these phenomena the dislocation velocity v is given by

$$v = \sigma b / A \qquad (20.7)$$

where A is a temperature-dependent constant whose value depends on the controlling mechanism. The steady-state creep rate can be expressed as

$$\dot{\varepsilon}_{sec} = \rho v b \qquad (20.8)$$

where ρ is the dislocation density and b the length of the Burgers vector. In Chapter 15 we estimated that the stress σ to move a dislocation in a crystal with dislocation density ρ was given by $\sigma \cong G b \rho^{1/2} / 2\pi$ with G the shear modulus. Therefore, we have

$$\rho \cong \left(\frac{2\pi\sigma}{Gb} \right)^2$$

Combining we obtain for the creep rate

$$\dot{\varepsilon}_{sec} \cong \frac{4\pi^2 \sigma^3}{AG^2} \qquad (20.9)$$

The creep rate is thus proportional to the third power of the applied stress, which is approximately the experimentally observed behaviour.

For pure metals the correlation between the activation energy for creep and self-diffusion strongly suggest that self-diffusion is involved in the creep process. Self-diffusion usually takes places via a vacancy mechanism and it is known that vacancies play a role in climb of dislocations. Therefore, it is logical to assume that the rate-controlling process in the high temperature creep of pure metals is the climb of dislocations. We assume that a steady-state dislocation configuration is developed in which vacancies are continuously being created and being destroyed. They are destroyed when dislocations of opposite sign, which were originally on different slip planes climb towards each other and annihilate upon meeting. They are created by climb of dislocations and then diffuse to other dislocations, which destroy them by climb. On average the number of dislocations that destroy vacancies is equal to the number that creates vacancies and the creep rate is controlled by the vacancy diffusion between the dislocations.

Similarly as for primary creep it is possible to derive a model without referring to a specific dislocation model. Assume that L represents the average distance, which a dislocation moves in a direction parallel to a slip plane between the creation and annihilation of a vacancy and d the average distance a dislocation climbs in a direction normal to its slip plane. The amount of work done by the applied stress σ on a unit length of dislocation line from the time it is created until it is destroyed is σbL. Each edge dislocation creates or destroys approximately d/b^2 vacancies per unit length of dislocation line between its creation and annihilation time. We recall that for the creation of a stress-free vacancy we need an energy ΔH_{cre} and that the number density of vacancies is given by $n_0 = n_0^* \exp(-\Delta H_{cre}/kT)$, where n_0^* is the number of lattice sites per unit volume. In this case, however, there is a stress, which does an amount of work equal to σbL during the period a dislocation climbs the distance d. Hence on average an extra energy $\sigma b^3 L/d$ is involved and the average energy to create a vacancy in the presence of a dislocation which creates vacancies is lowered from ΔH_{cre} to $\Delta H_{cre} - (\sigma b^3 L/d)$. In the same way we obtain $\Delta H_{cre} + (\sigma b^3 L/d)$ for the creation of a vacancy at a dislocation which climbs through the annihilation of vacancies.

For the creation of vacancy at a certain dislocation line and the annihilation of another vacancy at a second dislocation an amount of energy equal to $2\sigma b^3 L/d$ is used. The average vacancy difference Δn between vacancy creating dislocations and vacancy annihilating dislocations is thus

$$\Delta n = n_0^* \exp\left(\frac{-\Delta H_{cre}}{kT}\right)\left[\exp\left(\frac{\sigma b^3 L}{dkT}\right) - \exp\left(\frac{-\sigma b^3 L}{dkT}\right)\right] = 2n_0 \sinh\left(\frac{\sigma b^3 L}{dkT}\right)$$

This difference reduces at low stress to

$$\Delta n = 2n_0 \frac{\sigma b^3 L}{dkT}$$

The rate of vacancy flow depends on the actual dislocation distribution. We use a simple model (Fig. 20.4) of a core of radius R_0 containing N straight dislocations of the same sign and climbing in one direction. The enhanced (reduced) concentration

$$n = n_0^* \exp(-\Delta H_{cre}/kT)(1 \pm \sigma b^3 L/dkT)$$

acts to a distance R while for a distance larger than R the concentration becomes again the equilibrium concentration

$$n_0^* \exp(-\Delta H_{cre}/kT)$$

The number of vacancies that flow to or away from the centre area per unit time can be calculated from diffusion theory and is equal to

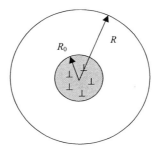

Fig. 20.4: Dislocation distribution model.

$$\frac{2\pi\Delta n D}{\ln(R_0/R)} = 2\pi D n_0 \frac{\sigma b^3 L/dkT}{\ln(R_0/R)}$$

where D is the diffusion coefficient. The average climb velocity v_{cli} for each dislocation in the group is thus

$$v_{\mathrm{cli}} = 2\pi D n_0^* \exp\!\left(\frac{-\Delta H_{\mathrm{cre}}}{kT}\right) \frac{\sigma b^5 L/dkT}{N\ln(R_0/R)} \cong n_0 \frac{D\sigma b^5 L}{NdkT}$$

The average velocity v in the direction of the slip plane is equal to

$$v = v_{\mathrm{cli}} L/d$$

since, if a dislocation moves by L over the slip plane, it must climb a distance d perpendicular to it. Substituting v in Eqs. (20.8) and (20.9) we obtain

$$\dot{\varepsilon}_{\mathrm{sec}} = \frac{4\pi^2 \sigma^3 b^4 L^2 D n_0 \exp(-\Delta H_{\mathrm{cre}}/kT)}{G^2 kT d^2 N}$$

This explains an exponent $n = 3$ and $m = 0$ for dislocation climb. The effect of stress level on the secondary creep rate of Al is illustrated in Fig. 20.1. In a similar way other mechanisms can be dealt with, which leads to the set of exponents given in Table 20.1. For details we refer to the literature.

Alternative creep modelling

It should be remarked that all workers do not subscribe the conventional division of the creep curve in three parts. Notably Evans and Willshire (1985, 1992) propose a division in two parts: a (primary) part decaying with time and a (tertiary) part accelerating with time. This in a natural way results in a typical creep curve. Quantitatively the creep strain ε is related to time t by

$$\varepsilon = \theta_1\big[1 - \exp(-\theta_2 t)\big] + \theta_3\big[\exp(-\theta_4 t) - 1\big]$$

The parameters θ_1 and θ_3 are scaling parameters, which quantify the extent of the primary and tertiary stages, while θ_2 and θ_4 define the curvature of the primary and tertiary stages. All θ parameters are approximately linear dependent on stress. For multiaxial stress states the choice of equivalent stress is again relevant but, as said, no definitive answer has been given.

The parameters θ_1 and θ_3 are nearly temperature independent while θ_2 and θ_4 probably can be described through an activation energy expression. Up to now several metals and alloys, but only a limited number of ceramics has been analysed in this way. An example of the latter is nominally pure and fully dense polycrystalline MgO. Although powerful in terms of extrapolation and predictive capability, at least for metals, the interpretation in terms of microstructure and atomic mechanisms is not at all clear. Nevertheless, this so-called *theta approach* offers a promising engineering alternative to the conventional one.

Deformation mechanism maps

For a number of metals and ceramics, where sufficient information is available, the dominant mechanism for creep in particular regimes of stress, temperature and grain size can be mapped in so-called deformation mechanism maps[g]. In the following we will demonstrate the various aspects of deformation mechanism maps using Al_2O_3 as an example. From the conventional constitutive equation for secondary creep and the exponents as given in Table 20.1, the strain rate $\dot{\varepsilon}$ as a function of stress σ, temperature T and grain size l can be calculated. Conventionally the strain rate at constant grain size as a function of relative temperature is plotted. The disadvantage of plotting for a certain grain size can be circumvented by plotting two of the variables σ, T and l against each other at a constant value for the third variable. In this way three different types of deformation maps are possible as indicated in Table 20.2. Stress σ, temperature T and grain size l are normalized with the shear modulus G, melting point T_{mel} and the Burgers vector b, respectively. For convenience of mapping frequently the inverse of the relative temperature T/T_{mel} is used.

Table 20.2: Types of deformation mechanism maps.

Type	Variables	Constant
I	Grain size l/b and temperature T/T_{mel}	Stress σ
II	Stress σ/G and temperature T/T_{mel}	Grain size l
III	Grain size l/b and stress σ/G	Temperature T

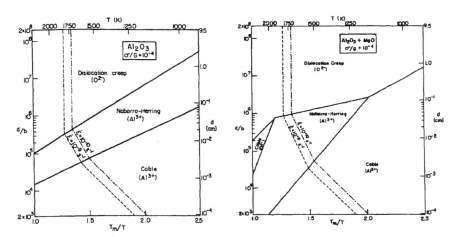

Fig. 20.5: Deformation mechanism map type I. Left: Al_2O_3, right: $Al_2O_3 + MgO$.

[g] Frost, H.J. and Ashby, M.F. (1982), *Deformation mechanism maps*, Pergamon, Oxford.

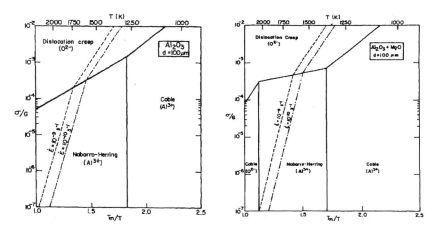

Fig. 20.6: Deformation mechanism map type II. Left: Al_2O_3, right: Al_2O_3 + MgO.

An example of a type I map for Al_2O_3 is shown in Fig. 20.5[h]. Diffusion creep is independent of the stress level but dislocation creep is not. Therefore a certain stress level has to be chosen. Diffusion creep and dislocation creep are independent processes, consequently the strain rates are additive and the fastest mechanism determines the rate-controlling mechanism. In Fig. 20.5 the corresponding map for Al_2O_3 is shown for $\sigma/G = 10^{-4}$. At large stress dislocation creep takes over from diffusion creep.

A type II map for Al_2O_3 with a grain size of 100 µm is shown in Fig. 20.6[h]. From this map it becomes clear that diffusion creep prevails for relatively low stress and dislocation creep for relatively high stress.

These types of representation are simple and convenient means of representing creep data and therefore obviously have high educational and theoretical values. The value for practical materials is, however, limited because the exact location of the various regions is highly dependent on the exact microstructural characteristics and dopant levels. Moreover, in many creep processes microcracks or voids are generated but the interrelation between fracture and creep is not clear at present. Finally, in many ceramics a second phase is present which generally is less refractory than the matrix material. This renders the deformation behaviour of the grains less important.

20.3   Creep and relaxation of polymers

In inorganics and metals the entities responsible for creep and relaxation are atoms, in contrast to polymers where the motion of a part of the molecule or the complete molecule has to be considered. While the relaxation of a small part of the molecule may be fast, conformational relaxation that involves a large part of the molecule may be quite slow. So we have a wide spectrum of response times. Usually in a relaxation time spectrum several peaks can be discerned. They are indicated by α, β, γ, etc. where the slowest rate, appearing at the highest temperature, is indicated by α. This is usually at the glass transition temperature. The next slowest, requiring a lower temperature to be activated, is indicated by β and so on (Ward, 1983).

---

[h] Langdon, T.G. (1980), Ceram. Int. **6**, 11.

It appears, in general, impossible to measure the response of materials completely within a limited time at a single temperature. Therefore often measurements are done at various temperatures, which employing the *time-temperature equivalence principle*, are used to construct a master curve. We first discuss this equivalence and in the next section the various relaxation mechanisms active in polymers.

The time-temperature equivalence

By using creep and relaxation measurements at various temperatures for a limited period of time more or less the complete behaviour can be determined. This is done by shifting the results horizontally for any temperature with a factor a_T to the response at the arbitrarily chosen reference temperature T_0. Moreover, a vertical shift with a factor $b_T = \rho T / \rho_0 T_0$ is applied to correct for density effects. At the reference temperature no shift is applied. For the dynamic frequency-dependent response we have

$$E*(T,\omega) = E*(T_0,\omega_0) = b_T E*(T_0, a_T \omega) \qquad \text{or}$$
$$E*(T,\ln\omega) = b_T E*(T_0, \ln\omega + \ln a_T)$$

while for the regular time-dependent response we similarly write

$$E(T,t) = E(T_0,t_0) = b_T E(T_0, t/a_T) \qquad \text{or} \qquad E(T,\ln t) = b_T E*(T_0, \ln t - \ln a_T)$$

The shift factor b_T is often close to one, therefore neglected and we do so here. The shift factor $\ln a_T$ is often described by the *Williams-Landel-Ferry* (WLF) *equation*

$$\log a_T = \frac{-c_1^0(T-T_0)}{c_2^0 + (T-T_0)} \qquad c_1^0 \text{ and } c_2^0 \text{ constants}$$

Frequently for the reference temperature the glass transition temperature T_g is chosen. For this choice of reference temperature, $c_1{}^g$ and $c_2{}^g$ are in the range as given by

$$c_1^g = [14-18]\,\text{K} \qquad \text{and} \qquad c_2^g = [30-70]\,\text{K}$$

An approximately universal relation is obtained by choosing the constants as $c_1^g = 17.44\,\text{K}$ and $c_2^g = 51.6\,\text{K}$. The constants $c_1{}^0$ and $c_2{}^0$ can be calculated from

$$c_1^0 = \frac{c_1^g c_2^g}{c_2^g} \qquad \text{and} \qquad c_2^0 = c_2^g + T_0 - T_g$$

It appears that the WLF equation is also valid for the viscosity η using for the shift factor $a_T = \eta_T / \eta_{Tg}$. The WLF equation is valid in the approximate temperature range $T_g < T < T_g+100$ while at higher temperature the shift factor behaves Arrhenius-like.

The free volume and other approaches

Originally the WLF equation was an empirical equation but it can be given a background on the basis of free volume theory. The free volume v_f is defined by $v_f = v-v_0$, where v is the total macroscopic volume and v_0 the actual volume of the molecules. A schematic division of the total volume as a function of temperature is shown in Fig. 1.6. It can be argued that the occupied volume increases (to first order) linearly with temperature. At the glass transition temperature the total volume, however, shows an extra expansion corresponding to the free volume suggesting that the visco-elastic processes start at about this temperature. Recall that the glass transition temperature is slightly dependent on the cooling rate. If we neglect this dependence here, the free volume increases also linearly (to first order) with temperature. Consequently, we have for the fractional free volume $f = v_f/v$

$$f = f_g + \alpha_f(T - T_g) \tag{20.10}$$

where f_g is the fractional free volume at T_g and α_f is the (volumetric) thermal expansion coefficient of the free volume. In all visco-elastic processes the relaxation time in the form of the ratio $\tau = \eta/E$ plays an important role. Since the temperature dependence of the viscosity η is large as compared with that of Young's modulus E, we ignore the latter. In this case we may take for the shift factor a_T from temperature T to temperature T_g

$$a_T = \eta_T/\eta_{T_g} \tag{20.11}$$

The next argument is the use of the empirical *Doolittle equation*, which states that the viscosity is related to the free volume as

$$\eta = a\exp(bv/v_f) = a\exp(b/f) \tag{20.12}$$

where a and b are constants with a typical value of $b \cong 1$. From these equations we easily obtain

$$\ln a_T = b\left(\frac{1}{f} - \frac{1}{f_g}\right) \tag{20.13}$$

Substituting Eq. (20.10) we finally have (meanwhile switching from ln to log)

$$\log a_T = -\frac{(b/2.303 f_g)(T - T_g)}{(f_g/\alpha_f) + T - T_g} \tag{20.14}$$

which has the form of the WLF equation. For many amorphous polymers $f_g \cong 0.025$ while a reasonable value for $\alpha_f \cong 4.8 \times 10^{-4}$ K^{-1}. If we substitute these values we obtain

$$\ln\eta_T = \ln\eta_{T_g} - \frac{c_1^g(T - T_g)}{c_2^g + T - T_g} \tag{20.15}$$

with $c_1^g = b/2.302f \cong 17.4$ and $c_2^g = f_g/\alpha_f \cong 52$, providing an interpretation for the constants c_1^g and c_2^g. This leads more or less automatically to the view that at the molecular level the glass transition should be related to $T_g - c_2^g$. There are two approaches along this line. In the first it is proposed[i] that the free volume corresponds to that part of the excess volume, which can be redistributed without energy change, and that this only occurs above a critical temperature T_2, which is to be identified with $T_g - c_2^g$. The second approach is to consider the glass transition as a real thermodynamic transition[j] occurring at a temperature T_2. Using transition state theory the frequency of molecular jumps v is given by

$$v = A\exp(-n\Delta G/kT) \cong \frac{kT}{h}\exp(-n\Delta G/kT) \tag{20.16}$$

[i] Cohen, M.H. and Turnbull, D. (1959), J. Chem. Phys. **31**, 1164.
[j] Gibbs, J.H. and Di Marzio, E.A. (1958), J. Chem. Phys. **28**, 373 and 807; Adam G. and Gibbs, J.H. (1965), J. Chem. Phys. **43**, 139.

where ΔG is the Gibbs energy of a single segment of the n segments rearranging. The other symbols have their usual meaning. If s_n is the entropy of a unit of n segments, assumed to be independent of the temperature, the total entropy S for a mole of segments, using N_A for Avogadro's constant, becomes

$$S = N_A s_n/n$$

Solving for n and substituting in Eq. (20.16) the result is

$$v = A \exp(-N_A s_n \Delta G/SkT) \tag{20.17}$$

The vital assumption is that the entropy becomes $S = 0$ at the thermodynamic transition temperature T_2, implying that n becomes infinite and that there are no configurations left for further rearrangement. The entropy S can now be calculated directly from difference in heat capacity Δc_p between the glass state and the supercooled liquid state at T_g, i.e.

$$S(T) = \Delta c_p \ln\frac{T}{T_2} = \Delta c_p \ln\left(1+\frac{T-T_2}{T_2}\right) \cong \Delta c_p \frac{T-T_2}{T_2}$$

where it is assumed that Δc_p is constant over the temperature range considered. Substituting in Eq. (20.17) results in a relaxation time τ given by

$$\tau = \frac{1}{v} = \tau_0 \exp[bT_2/T(T-T_2)] \cong \tau_0 \exp[b/(T-T_2)]$$

with $\tau_0 = 1/A$ and $b = N_A s_n \Delta G/k\Delta c_p$. This is the WLF equation if we identify T_2 with $T_g - c_2^g$.

The free volume explanations have met serious objections. From experimental data on β-relaxations it is inferred[k] that that the free volume is not constant for constant total volume meanwhile varying temperature or pressure. At the least this implies a significant reinterpretation of the free volume, one that is likely not to be so clearly linked with the geometrical starting point of this theories. A more complete theory of the dynamics of disentangled and entangled polymers is given by Rouse and Doi and Edwards, respectively. These theories are treated in Section 20.5. In the next section we first briefly review the experimental data.

20.4 A brief review of experimental data for polymers

In this section the various relaxation mechanisms active in polymers are briefly reviewed. We distinguish between local and co-operative mechanisms in amorphous polymers, thereafter qualitatively discuss chain motion and conclude with some remarks on partially crystalline materials.

Local and co-operative processes

With local processes those mechanisms are indicated that involve only a small part of the chain. In Fig. 20.7 the loss tangent of polycyclohexylmethacrylate (PCHMA) in the glassy state[l] is shown. The loss angle can be observed to peak in a relatively

[k] Hoffman, J.D., Williams, G. and Passaglia, E. (1966), J. Polym. Sci. C, **14**, 173.
[l] Heijboer, J. (1978), Page 75 in *Molecular basis of transitions and relaxations*, Midland Macromolecular Monographs, vol. 4, Gordon and Breach, London.

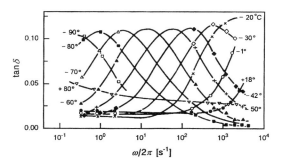

Fig. 20.7: The frequency dependence of the loss tangent of PCHMA at various temperatures.

narrow range, which shifts to higher frequencies with increasing temperature without a significant change in shape. This relaxation process is attributed to the interchange of boat and chair conformations of the cyclohexyl side group in PCHMA. From this graph the temperature dependence of the relaxation rate can be derived and is given in an Arrhenius plot in Fig. 20.8. In this case the complete behaviour more or less can be assessed from one experiment.

With co-operative processes those mechanisms are indicated that involve a large part of the chain. In Fig. 20.9 the creep compliance of polystyrene at various temperatures[m] is shown. Measurement at different temperatures results in probing different parts of the compliance curve and using the time-temperature equivalence principle a master curve can constructed. This curve shows a few characteristics. At short times the material behaves like a glass. With increasing time a transition zone occurs, followed by a plateau with rubbery behaviour. According to time-temperature equivalence principle the location of the transition zone within the time spectrum is strongly determined by the temperature. The behaviour at the plateau is called rubbery since it resembles the behaviour of cross-linked rubber. In this case, however, the entanglements act as cross-links so that given sufficient time at a certain temperature the material still will flow. Therefore, at still longer times viscous behaviour sets in and the compliance changes linearly with time.

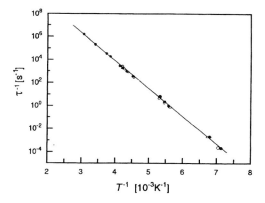

Fig. 20.8: Temperature dependence of the relaxation rates of the γ-process in PCHMA.

---

[m] Schwarzl, F.R. (1990), *Polymermechanik*, Springer, Berlin.

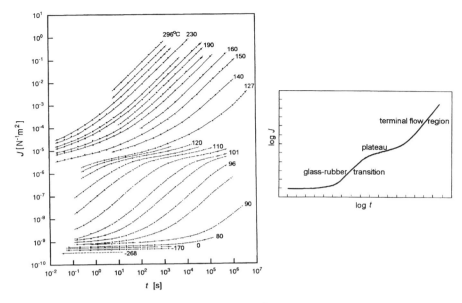

Fig. 20.9: Creep compliance of PS ($M_w = 3.85 \times 10^2$ kg/mol), as measured at the indicated temperatures (left) and the associated master curve (right).

Also relaxation experiments provide information about the visco-elastic properties. Similarly as for creep the response is measured at various temperatures within a limited time interval. The time-temperature equivalence principle is again used to construct a master curve. In Fig. 20.10 the results are shown for polyisobutylene[n] (PIB) using 298 K as a reference temperature while the insert shows the necessary shift function. Comparable information is extracted from the relaxation

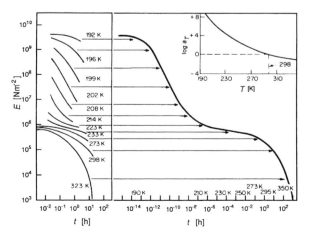

Fig. 20.10: Time dependence of the tensile modulus of PIB. Measurements at the indicated temperatures (left) and master curve, constructed at the reference temperature $T = 298$ K (right). The insert displays the applied shifts, i.e. the shift factor as a function of temperature.

---

[n] Castiff, E. and Tobolsky, T.S. (1955), J. Colloid Sci. **10**, 375.

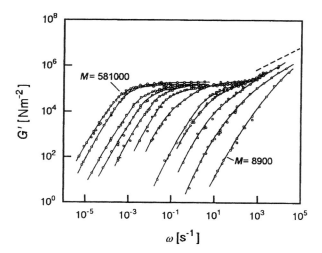

Fig. 20.11: Storage shear moduli measured for a series of fractions of PS with different molar
masses in the range $M = 8.9$ kg/mol to $M = 5.8 \times 10^2$ kg/mol.

curve as from the creep curve: glass-like behaviour at short times followed by a
transition to the rubbery plateau and eventually to viscous behaviour.

The molecular weight plays an important role in the width of the rubbery plateau.
In Fig. 20.11 the dynamic response for monodisperse PS[o] with varying molecular
weight is shown (Note that the order of appearance of the glass, rubbery and viscous
regions is reversed as compared with time-dependent measurements). For low
molecular weights hardly any plateau is observed while with increasing molecular
weight the width of the plateau increases. It can also be observed that the glass-rubber
transition is largely unaffected by the molecular weight, indicating that the chain
length between the entanglements is largely independent of the molecular weight. On
the other hand the terminal region strongly increases with increasing molecular
weight. Finally, the response in the low-frequency regime is largely independent of the
molecular weight, varies approximately as $G'(\omega) \sim \omega^2$ and is related to the behaviour
of flowing melts.

Chain motion

Since the visco-elastic behaviour is highly related to the viscosity we deal here
with the zero-shear viscosity η_0, i.e. $G''(0)$. It appears experimentally that the zero-
shear viscosity η_0 as a function of molecular mass M shows two regimes, as
illustrated[p] in Fig. 20.12. The crossover value is indicated as the *critical molar mass*
M_{cri}. The linear regions in the log-log plot can be characterised as a function of M by

$$\eta_0 \sim M^\nu$$

The exponent $\nu = 1$ for low M, while above M_{cri} it obeys $\nu \cong 3.4$. The value for M_{cri} is
dependent on the type of polymer but occurs for a chain length of about 300 to 700
main atoms. Although the steep increase in viscosity with molecular mass may seem

[o] Onogi, S., Masuda, T., Kitawaga, K. (1970), Macromolecules **3**,109.
[p] Berry, G.C. and Fox, T.G. (1968), Adv. Polym. Sci. **5**, 261.

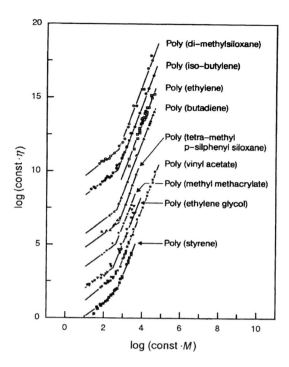

Fig. 20.12: The zero-shear viscosity η_0 of several polymers as a function of molar mass M. For clarity the position is shifted.

detrimental to polymer processing, fortunately at higher frequency the viscosity often decreases considerably, an effect that is known as *shear thinning*. Moreover, also with increasing temperature the viscosity decreases significantly, as indicated by the WLF equation. Since thermal degradation puts an upper limit to the processing temperature, it still holds that increased molar mass leads to increased processing difficulties. The dependence of the viscosity on chain length can be understood using models for the chain motion. We briefly describe the qualitative behaviour first and provide some details in Section 20.5.

It has to be realised that dynamics occur at different length and time scales. Going from a real polymeric chain with chemical structure to an equivalent chain with Kuhn segments effectively removes all the specific chemistry for a particular polymer. For the dynamic behaviour of polymers chains at a somewhat larger scale one usually distinguishes between the Rouse and reptation regime.

In the *Rouse regime*, first proposed by Rouse, it is assumed that the chains are not entangled and that can move more or less independently of the other chains in the melt. The chain is modelled as a chain of beads, which are connected with springs. These springs vibrate with a frequency dependent on the stiffness of the spring but are damped by the frictional forces exerted by the other chains. The vibrations can be analysed in terms of normal co-ordinates, similarly as for a crystal or a single molecule. These normal modes describe the overall vibrational behaviour of the chains. The time-dependent modulus is mainly influenced by the lowest frequency normal mode. Since the chains are supposed to be influenced only by the frictional

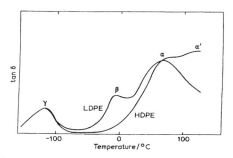

Fig. 20.13: Variation of tan(δ) with temperature for high- and low-density polyethylene (HDPE and LDPE, respectively).

forces, the Rouse theory predicts that the viscosity is linearly dependent on M or $\nu = 1$, in agreement with the experimental data (Fig. 20.12). The Rouse segments (bead and spring) can be chosen in principle independently of the Kuhn segments but a logical choice is to identify the two.

In reality, with increasing molar mass an increasing amount of entanglement occurs. The chains trap each other, thereby hampering the motion. The essence of the this regime is that the chain at the entanglement points can move only forward or backwards but in between can move in all directions. Moreover the chain contains a sufficiently large number of segments between the entanglements so that the subchain still can be described as Gaussian. De Gennes and Edwards proposed that the motion of the complete chain could be envisaged as the motion of a chain in a 'tube', representing the other chains. Because this motion resembles the motion of a snake it is called *reptation*. Effectively the chain thus can move more or less only along its length and this more restricted motion leads to an increased influence of the molecular mass. Using the reptation model the relation between the zero-shear viscosity and molar mass was found to be a power-law with exponent $\nu = 3$. Although this is not quite the experimental behaviour, the concepts involved are essential to understand the dynamical behaviour of polymers in the entangled regime. The elaboration to non-linear aspects is also capable of providing an explanation for $\nu \cong 3.4$.

Summarising, we expect that the dynamics at the largest and intermediate time scales are largely independent of the precise chemical nature but the response at the shortest time, dealing with motions within a Kuhn segment, is highly dependent of the nature of the polymer.

Mechanisms in partially crystalline materials

Due to the complex morphology of partially or semi-crystalline polymers, the elucidation is the relaxation behaviour of these materials is quite difficult. The most widely studied material is polyethylene, which we take as our guiding example[q] (Fig. 20.13). In polyethylene there are four transition regions designated α', α, β and γ. The γ-transition is very similar in both the high and low-density samples whereas the β-relaxation is virtually absent in the high-density polymer. The α and α' peaks are also somewhat different in the pre-melting region. Polyethylene normally is branched and this has an influence on the crystallinity, i.e. crystal size and crystal perfection and

[q] Flocke, H.A. (1962), Kolloid-Z. Z. Polym., **180**, 188.

therefore on the relaxation behaviour. Both the intensity of the α and α' peaks decrease with decreasing crystallinity, which suggests that these peaks are associated with motions in the crystals. With decreasing amount of branching the β peak eventually disappears which indicates that it is related to the relaxation of the branching points. The intensity of the γ peak increases with decreasing crystallinity, thus suggesting that it is associated with the amorphous phase and it has been tentatively assigned to a glass transition. More generally, assigning peaks in crystalline polymers to particular types of molecular motion is tricky and often matter for debate.

Herman Francis Mark (1895-1992)
Along with Hermann Staudinger and Wallace Carothers, Herman Mark can be credited as a cofounder of polymer science. In the 1920s his X-ray crystallographic studies of cellulose showed it to be made of giant molecules containing thousands of atoms, as Staudinger held. Mark also showed that most polymer molecules are made of flexible chains, while Staudinger had thought them to be rigid rods. The Mark-Houwink-Sakurada relationship describing the relationship between a polymer's solution viscosity and its molecular weight was another of his early discoveries. Escaping into Switzerland after the Nazis annexed Austria in 1938, Mark made his way first to Canada and then to the United States, where he joined the faculty of Brooklyn Polytechnic. There he established a strong polymer program, which included not only research but also the first undergraduate polymer education in the United States. To this day, very few American polymer chemists cannot trace their academic lineage back to Mark and Brooklyn Polytechnic.

20.5   Models for polymer visco-elasticity*

In this section we describe, in some detail, models for the dynamical behaviour as outlined above. The treatment is necessarily brief but several extensive treatments[r] exist. In order to describe models for these regimes we first address some basics of chain modelling. Thereafter the dynamics of disentangled chains are treated while entangled chains are deal with next. Finally, the connection to the (macroscopic) stress and viscosity is made. We follow here the approach as outlined by Doi (1996). We limit ourselves to linear, ideal chains.

---

[r] See e.g. Rubinstein and Colby (2003), Doi (1996), Strobl (1997) or Lin (2003).

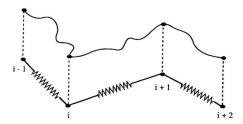

Fig. 20.14: Schematic representation of a part of the bead-spring (Rouse) model.

Chain basics

In Chapter 11 we learned that for an ideal (freely jointed) chain with N bonds and end-to-end vector $\mathbf{R} = \Sigma \mathbf{r}_n$ where $\mathbf{r}_n$ is the bond length vector for the nth bond, each taken with a length b so that $|\mathbf{r}_n| = b$, the mean square end-to-end length is given by

$$\langle \mathbf{R}^2 \rangle = N^{-1} \sum_n \sum_m \mathbf{r}_n \mathbf{r}_m = Nb^2$$

while the probability distribution appeared to be Gaussian and is given by

$$P(\mathbf{R}, N) = \left(\frac{3}{2\pi Nb^2} \right)^{3/2} \exp\left(\frac{-3\mathbf{R}^2}{2Nb^2} \right)$$

If we model a polymer as a chain of beads connected with harmonic springs by force constant K (Fig. 20.14), each spring represents a number of atoms and the particular choice determines the details of the behaviour. In any case each segment[s] in the chain has some flexibility and the potential energy of the chain Φ for the bead model is given by

$$\Phi = \frac{1}{2} K \sum_n^N (\mathbf{R}_n - \mathbf{R}_{n-1})^2 \tag{20.18}$$

If we assume that a Gaussian distribution describes the fluctuations of the bond length, we will have

$$p(\mathbf{r}) = \left(\frac{3}{2\pi b^2} \right)^{3/2} \exp\left(\frac{-3\mathbf{r}^2}{2b^2} \right) \tag{20.19}$$

If we write the position vector for the nth segment as $\mathbf{R}_n$, the distribution of bond vectors $\mathbf{r}_n = \mathbf{R}_n - \mathbf{R}_{n-1}$ is given by Eq. (20.19) and the probability distribution for the set of position vectors $\{\mathbf{R}_n\} \equiv (\mathbf{R}_0, \mathbf{R}_1, ..., \mathbf{R}_N)$ is proportional to

$$P(\{\mathbf{R}_n\}) = \left(\frac{3}{2\pi b^2} \right)^{3N/2} \exp\left(\frac{-3}{2b^2} \sum_n^N (\mathbf{R}_n - \mathbf{R}_{n-1})^2 \right)$$

The equilibrium state of the chain is also described by the Boltzmann equation given by $P(\mathbf{R}) \sim \exp(-U/kT)$, with as usual k the Boltzmann constant and T the temperature.

---

[s] It would be better to refer consistently to segments but usually the segments are called bonds.

Choosing the force constant as $K = 3kT/b^2$ these two distributions match and this provides an interpretation for the force constant.

The spatial distribution of the nth chain segment is described by the pair correlation function

$$g_n(\mathbf{r}) = \sum_m^N \langle \delta[\mathbf{r} - (\mathbf{R}_m - \mathbf{R}_n)] \rangle$$

where δ is the Dirac delta function. The (total) pair correlation function is given by the average over the individual functions and therefore reads

$$g(\mathbf{r}) = N^{-1} \sum_n^N g_n(\mathbf{r}) = N^{-1} \sum_n^N \sum_m^N \langle \delta[\mathbf{r} - (\mathbf{R}_m - \mathbf{R}_n)] \rangle$$

The Fourier transform of $g(\mathbf{r})$ is denoted as $g(\mathbf{q})$ and reads

$$g(\mathbf{q}) = \int g(\mathbf{r}) \exp(i\mathbf{q}.\mathbf{r}) \, d\mathbf{r} = N^{-1} \sum_n^N \sum_m^N \langle \exp[(\mathbf{q}.(\mathbf{R}_m - \mathbf{R}_n)] \rangle$$

For small $\mathbf{q}$ this expression can be expanded in a Taylor series with respect to $\mathbf{q}$ and we obtain

$$g(\mathbf{q}) = N^{-1} \sum_n^N \sum_m^N [1 - iq_i \langle (\mathbf{R}_m - \mathbf{R}_n)_i \rangle$$
$$- \frac{1}{2} q_i q_j \langle (\mathbf{R}_m - \mathbf{R}_n)_i \rangle \langle (\mathbf{R}_m - \mathbf{R}_n)_j \rangle + \cdots]$$

where the components of $\mathbf{R}_n$ are indicated by $(\mathbf{R}_n)_i$. For an isotropic vector we have

$$\langle r_i \rangle = 0 \qquad \text{and} \qquad \langle r_i r_j \rangle = \frac{r^2}{3} \delta_{ij}$$

with δ_{ij} the Kronecker delta function. Since the vector $\mathbf{R}_m - \mathbf{R}_n$ is assumed to be isotropic we may write

$$g(\mathbf{q}) = N^{-1} \sum_n^N \sum_m^N [1 - \frac{q^2}{6} \langle (\mathbf{R}_m - \mathbf{R}_n)^2 \rangle + \cdots] \equiv g(0) \left(1 - \frac{q^2}{3} R_g^2 + \cdots \right)$$

Here the *radius of gyration* R_g is defined as

$$R_g^2 = (2N^2)^{-1} \sum_n \sum_m \langle (\mathbf{R}_m - \mathbf{R}_n)^2 \rangle = N^{-1} \sum_n \langle (\mathbf{R}_n - \mathbf{R}_G)^2 \rangle$$

with the position of the centre of mass defined by

$$\mathbf{R}_G = N^{-1} \sum_n^N \mathbf{R}_n$$

For a Gaussian chain with $|n-m| \gg 1$, we have $\langle (\mathbf{R}_n - \mathbf{R}_m)^2 \rangle = |n-m| \, b^2$. Hence

$$R_g^2 = (2N^2)^{-1} \sum_n^N \sum_m^N |n-m| b^2 \cong \frac{b^2}{2N^2} \int_0^N \int_0^N |n-m| \, dn \, dm = \frac{Nb^2}{6}$$

where the last step can be made for large N. Similarly we can calculate for the Fourier transform for $\mathbf{q} \to 0$ that $g(\mathbf{q}) = N$ while for $qR_g \gg 1$ we have $g(\mathbf{q}) = 2N/q^2 R_g^2$. The *Ornstein-Zernike expression*

$$g(\mathbf{q}) = \frac{N}{1 + \frac{1}{2}R_g^2 \mathbf{q}^2}$$

provides a convenient interpolation formula, accurate to within 15% for all values of qR_g. The radius of gyration thus can be measured relatively easily with diffraction experiments.

To introduce the effect of random motion the theory of Brownian motion of particles is used. We employ here a 1D motion in the x-direction and write $v(t)$ for the distribution function of the velocity in that direction. The function is randomly fluctuating in time. The average of the product $\langle v(t_2)v(t_1)\rangle$ depends only on $|t_2-t_1|$ in the steady state and we can write

$$\langle v(t)v(0)\,\rangle = C(t)$$

where $C(t)$ is the velocity correlation function. At $t = 0$, $C = \langle v^2 \rangle = kT/m$ with m the mass of the particle. For $t \gg 0$, $C \to 0$. To make an estimate for C, we assume that the particle moving through a liquid with velocity V experiences a drag force $-\zeta v$ with ζ the friction coefficient. For a spherical particle with radius a moving in a liquid with viscosity η, the friction coefficient is given by Stokes law as $\zeta = 6\pi\eta a$. The equation of motion of the particle then becomes

$$m\dot{v} = -\zeta v$$

The solution is $v \sim \exp(-t/\tau)$ with $\tau = m/\zeta$ the relaxation time. In polymer dynamics the experimenting time is usually much larger than the relaxation time so that we consider the limit of $\tau \to 0$. The velocity correlation function can be written as

$$\langle v(t)v(0)\rangle = 2D\delta(t)$$

To interpret this expression we consider the displacement u of a particle during a time interval t given by

$$u(t) = \int_0^t v(t')\,dt'$$

and its average square $\langle u(t)^2 \rangle$ calculated as

$$\langle u(t)^2 \rangle = \int_0^t \int_0^t \langle v(t_1)v(t_2)\rangle\,dt_1 dt_2 = \int_0^t \int_0^t 2D\delta(t_2 - t_1)\,dt_1 dt_2 = 2Dt$$

Therefore, we see that D corresponds to a *diffusion coefficient*.

Introducing now the effect of a potential $\Phi(x)$ the particle also experiences a force $-\partial\Phi/\partial x$. Without Brownian motion the equation of motion would be

$$\frac{dx}{dt} = -\frac{1}{\zeta}\frac{d\Phi}{dx}$$

but with Brownian motion the velocity of the particle fluctuates due to a random force with probability function $g(t)$. Therefore, the equation of motion is

$$\frac{dx}{dt} = -\frac{1}{\zeta}\frac{d\Phi}{dx} + g(t) \tag{20.20}$$

which is often addressed as the *Langevin equation*. If the fluctuations are the same in the presence and absence of the potential, the mean and variance of $g(t)$ are given by

$$\langle g(t) \rangle = 0 \quad \text{and} \quad \langle g(t)g(0) \rangle = 2D\delta(t)$$

The equations for the Brownian motion can be solved numerically if we discretise the Langevin equation to

$$x(t + \Delta t) = x(t) - \frac{1}{\zeta}\frac{dU}{dx}\Delta t + \Delta G(t) \quad \text{with} \quad \Delta G(t) = \int_t^{t+\Delta t} g(t')\,dt'$$

Further it can be shown that if $x(t)$ satisfies Eq. (20.20), the probability equation $\psi(x)$ for $t \to \infty$ approaches

$$\psi(x) \sim \exp(-\Phi/D\zeta)$$

which coincides with the Boltzmann equation if

$$D = kT/\zeta \tag{20.21}$$

The latter relation is called the *Einstein relation*.

Disentangled chains

As before, the chain is modelled as a chain of beads connected with springs. If we now assume that the beads experience a drag force proportional to their velocity we can use the Langevin equation to describe its dynamics. Hence we have

$$\frac{d\mathbf{R}_n}{dt} = -\frac{1}{\zeta}\frac{d\Phi}{d\mathbf{R}_n} + \mathbf{g}_n$$

with ζ the friction coefficient for a bead and $\mathbf{g}_n$ the function describing the random fluctuations on bead n. Now employing Eq. (20.18) for the potential energy we obtain

$$\frac{d\mathbf{R}_n}{dt} = \frac{K}{\zeta}(\mathbf{R}_{n+1} + \mathbf{R}_{n-1} - 2\mathbf{R}_n) + \mathbf{g}_n \quad \text{for} \quad n = 1, 2, ..., (N-1) \tag{20.22}$$

$$\frac{d\mathbf{R}_0}{dt} = \frac{K}{\zeta}(\mathbf{R}_1 + \mathbf{R}_0) + \mathbf{g}_0 \quad \text{for} \quad n = 0 \tag{20.23}$$

$$\frac{d\mathbf{R}_N}{dt} = \frac{K}{\zeta}(\mathbf{R}_{N-1} - \mathbf{R}_N) + \mathbf{g}_N \quad \text{for} \quad n = N \tag{20.24}$$

If we define $\mathbf{R}_{N+1} \equiv \mathbf{R}_N$ and $\mathbf{R}_{-1} \equiv \mathbf{R}_0$, Eqs. (20.23) and (20.24) are included in Eq. (20.22). The dynamics of the chain are thus described by Eq. (20.22) and this model is addressed as the *Rouse model*.

Since the chain is supposed to be long, the beads can be assumed to be continuously distributed along the chain and n becomes a continuous variable. Writing $\mathbf{R}(n,t)$ for $\mathbf{R}_n(t)$ the Langevin equation becomes

$$\frac{\partial \mathbf{R}}{\partial t} = \frac{K}{\zeta}\frac{\partial^2 \mathbf{R}}{\partial n^2} + \mathbf{g}(n,t) \tag{20.25}$$

and Eqs. (20.23) and (20.24) become the boundary conditions $\partial\mathbf{R}/\partial n = 0$ for $n = 0$ and $n = N$. Introducing now the normal co-ordinates

$$\mathbf{X}_p(t) = N^{-1} \int_0^N \cos\left(\frac{p\pi n}{N}\right) \mathbf{R}(n,t)\, \mathrm{d}n \qquad \text{with} \quad p = 0, 1, \ldots$$

Eq. (20.25) can be written as

$$\frac{\partial \mathbf{X}_p}{\partial t} = -\frac{K_p}{\zeta_p} \mathbf{X}_p + \mathbf{g}_p \qquad \text{with} \tag{20.26}$$

$$\zeta_0 = N\zeta \quad \zeta_p = 2N\zeta \quad \text{and} \quad K_p = \frac{2\pi^2 K p^2}{N} = \frac{6\pi^2 kT p^2}{Nb^2}$$

The random force $\mathbf{g}_p(t)$ has zero mean and a variance given by

$$\langle [\mathbf{g}_p(t)]_\alpha [\mathbf{g}_q(t')]_\beta \rangle = 2\delta_{pq}\delta_{\alpha\beta}\delta(t-t')\frac{kT}{\zeta_p}$$

The correlation function for the normal co-ordinates can now be written as

$$\langle [\mathbf{X}_0(t) - \mathbf{X}_0(0)]_\alpha [\mathbf{X}_0(t) - \mathbf{X}_0(0)]_\beta \rangle = \delta_{\alpha\beta}\frac{kT}{\zeta_0}t \qquad \text{and} \tag{20.27}$$

$$\langle [\mathbf{X}_p(t)]_\alpha [\mathbf{X}_q(0)]_\beta \rangle = \delta_{pq}\delta_{\alpha\beta}\frac{kT}{K_p}\exp(-t/\tau_p) \qquad \text{with} \tag{20.28}$$

$$\tau_p = \frac{\zeta_p}{K_p} = \frac{\tau_1}{p^2} = \frac{\zeta N^2 b^2}{3\pi^2 kT}\frac{1}{p^2} \tag{20.29}$$

This function contains the dynamic behaviour of the chain. The relaxation time of mode $p = 1$, characteristic for the model, is proportional to M^2 and usually addressed as the *Rouse* (or *rotational*) *relaxation time* τ_R. The result includes the size of the Rouse unit b. Introducing $\langle \mathbf{R}^2 \rangle = Nb^2$, we see that $\tau_p = [(\zeta/b^2) \langle \mathbf{R}^2 \rangle^2]/3\pi^2 kT$. In order for τ_p to be a constant, it follows that ζ/b^2 should be independent of the precise choice of bead and spring. This is obviously true if the total friction coefficient ζ is proportional to the number of beads. This is not self-evident but appears to be true for amorphous polymers.

Let us now examine the behaviour of a chain in these terms and let us first discuss the centre of mass motion. The position of the centre of mass is given by

$$\mathbf{R}_G(t) = N^{-1} \int_0^N \mathbf{R}(n,t)\, \mathrm{d}n$$

and is equal to the first normal mode $\mathbf{X}_0(t)$. The mean square displacement then becomes

$$\langle [\mathbf{R}_G(t) - \mathbf{R}_G(0)]^2 \rangle = \frac{6kT}{\zeta_0}t = \frac{6kT}{N\zeta}t$$

From this result it follows that the centre of mass, i.e. the chain, diffuses with a diffusion coefficient

$$D_G \equiv \lim_{t\to\infty}\frac{1}{6t}\langle [\mathbf{R}_G(t) - \mathbf{R}_G(0)]^2 \rangle = \frac{kT}{N\zeta}$$

Since $D_G \sim N^{-1}$, the diffusion is inversely proportional to the molecular mass M.

Turning the attention now to the motion of the molecule as a whole we use the correlation function for the end-to-end vector $\mathbf{P}(t)$. From the normal co-ordinates we obtain for the correlation function for the end-to-end vector

$$\mathbf{P}(t) = \mathbf{R}(N,t) - \mathbf{R}(0,t) = -4\sum_p \mathbf{X}_p(t) \qquad \text{with} \quad p = 1, 3, 5, \dots$$

Employing the normal co-ordinate correlation function (20.28), the correlation function for the end-to-end vector becomes

$$< \mathbf{P}(t).\mathbf{P}(0) > = 16\sum_p \frac{3kT}{K_p} \exp(-t/\tau_p) = Nb^2 \sum_p \frac{8}{\pi^2 p^2} \exp(-tp^2/\tau_1)$$

Since the various terms decrease rapidly with increasing p, the first one contributes the most and the behaviour is approximately described by a single exponential with relaxation time τ_1. In fact the first three modes describe 90% of the mean-square end-to-end distance[t]. Since $\tau_1 \cong Nb^2/D_G$, the rotational relaxation time is also about the time required for the centre of mass to diffuse a distance comparable to the length of the chain.

For the internal motion of the chain we employ the mean square displacement of the nth segment given by

$$\phi(n,t) \equiv \langle [\mathbf{R}(n,t) - \mathbf{R}(n,0)]^2 \rangle$$

Using the displacement of the nth segment in terms of the normal co-ordinates

$$\mathbf{R}(n,t) = \mathbf{X}_0(t) + 2\sum_p^\infty \cos\left(\frac{p\pi n}{N}\right) \mathbf{X}_p(t)$$

the correlation function $\phi(n,t)$ can be calculated as

$$\phi(n,t) = 6D_G t + \frac{4Nb^2}{\pi^2} \sum_p^\infty \cos^2\left(\frac{p\pi n}{N}\right) \frac{1}{p^2}[1 - \exp(-tp^2/\tau_1)]$$

For $t \gg \tau_1$ the function $\phi(n,t) \cong 6D_G t$ and the displacement of the segments is determined by the centre of mass motion. For $t \ll \tau_1$ the summation can be replaced with an integral while the $\cos^2(p\pi n/N)$ term can be replaced by its average value ½. This results in

$$\phi(n,t) = 6D_G t + \frac{4Nb^2}{\pi^2} \int_0^\infty \frac{1}{2p^2}[1 - \exp(-tp^2/\tau_1)]$$

$$= \frac{2Nb^2}{\pi^2}\left[\frac{t}{\tau_1} + \left(\frac{\pi t}{\tau_1}\right)^{1/2}\right] \cong \frac{2Nb^2}{\pi^{3/2}}\left(\frac{t}{\tau_1}\right)^{1/2}$$

so that in this regime the average square displacement is proportional to $t^{1/2}$.

In conclusion in the disentangled regime Rouse theory predicts that

$$D_G \sim M^{-1} \qquad \text{and} \qquad \tau_R \sim M^2$$

---

[t] A simple dumbbell model, containing only two beads connected by a single spring, is sometimes used to describe the dynamics of a disentangled polymer chain. The result mentioned justifies its use.

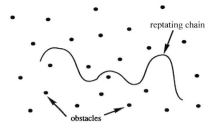

Fig. 20.15: Schematic representation of the model used in reptation.

Entangled chains

The dynamics of entangled chains are also often addressed, following de Gennes, as *reptation dynamics*. In this approach the dynamics are considered to be due to two different components. The first component is the wriggling of the chain in a tube or, equivalently, as the wriggling between entanglement points. This component corresponds to the Rouse part. The line connecting the entanglement points, representing the average of the wriggling motion, is known as the *primitive path*. The second component is the motion of the complete chain or the evolution of the primitive path, leading to a changing entanglement configuration. The second component corresponds to the reptation part (Fig. 20.15). In this image the tube only exists around a certain chain and ceases to exist as soon as the chain has left a certain part of space and is created as soon as the chain is entering a certain part of space.

Since the primitive path and the actual chain are both represented by a Gaussian coil with same end-to-end distance, we have

$$\langle \mathbf{R}^2 \rangle = Nb^2 = l_{\text{pp}} a_{\text{pp}} \tag{20.30}$$

where we introduced the length of the primitive path l_{pp} and the associated length a_{pp} characterising the stiffness of the primitive path as determined by the topology of the network. The changing entanglement configuration is envisaged as the motion of the chain through the tube along its primitive path, i.e. reptation. The associated diffusion coefficient D_{G} can be estimated from the Einstein relation

$$D_{\text{G}} = kT/N\zeta \tag{20.31}$$

where N is the number of Rouse segments and ζ is the friction coefficient for a single bead. The time necessary for a complete removal from the chain from its original tube is the characteristic time for this process and is given by

$$\tau_D = l_{\text{pp}}^2/D \tag{20.32}$$

The dependence on molecular mass is easily established as

$$\tau_D \sim \zeta N^3$$

using Eqs. (20.30) and (20.31). This simple model thus right away predicts an exponent $\nu = 3$ in the expression $\tau \sim N^\nu$. As briefly mentioned in Section 20.4, experimentally one observes $\nu \cong 3.4$. A further theoretical improvement can be obtained by incorporating processes related to neighbouring chains, like the release of

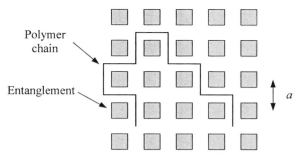

Fig. 20.16: Schematic of the lattice model of reptation.

constraints or tube deformation. This leads indeed to $v \cong 3.4$ (see e.g. Lin, 2003 or Rubinstein and Colby, 2003).

A more detailed but basic model for reptation is provided by a *lattice model*. In this model we position a chain of Z bonds on a lattice with spacing a (Fig. 20.16). The directions of the bond vectors are indicated by $\mathbf{u}_n$ with $n = 1, ..., Z$. In equilibrium these vectors will point in random directions. We assume that we may describe the process in time steps of Δt during which the head or tail of the chain advances by one spacing. If the head advances, all bonds vectors from 2 to Z change from $\mathbf{u}_n$ to $\mathbf{u}_{n-1}$ and $\mathbf{u}_1$ takes a random direction. In the same way if a tail advances, $\mathbf{u}_n$ changes to $\mathbf{u}_{n+1}$ and $\mathbf{u}_Z$ takes a random direction. This can be expressed as

$$\mathbf{u}_n(t + \Delta t) = \mathbf{u}_{n+\mathbf{n}(t)}(t)$$

where $\mathbf{n}(t)$ is a random variable taking either -1 or $+1$ and $\mathbf{u}_{-1}$ and $\mathbf{u}_{Z+1}$ take a random direction.

The parameters in the lattice model can be expressed in terms of Rouse model parameters. In the lattice model the mean square end-to-end distance is Za^2 while for the Rouse model the same quantity is Nb^2. Hence we may take $Z = Nb^2/a^2$. Similarly we may express Δt in terms of the centre of mass distance of the Rouse model with a diffusion constant $D = kT/N\zeta$. During a time t the average square distance moved by the chain along the tube $s(t)$ is given by $\langle s^2(t) \rangle = 2Dt$. On the lattice model the chain experiences $t/\Delta t$ random steps and with each step the chain moves by a distance a. Therefore $\langle s^2(t) \rangle = (t/\Delta t)a^2$ so that by combining we obtain

$$\Delta t = \frac{a^2}{2D} = \frac{a^2 N\zeta}{2kT}$$

The basic equations for lattice reptation in terms of the co-ordinates of the segments $\mathbf{R}_n$ then can be written as

$$\mathbf{R}_n(t + \Delta t) = \mathbf{R}_{n+\mathbf{n}(t)}(t)$$

where $\mathbf{R}_{-1}(t+\Delta t)$ and $\mathbf{R}_{Z+1}(t+\Delta t)$ are defined using the random unit vector $\mathbf{n}(t)$ by

$$\mathbf{R}_{-1}(t + \Delta t) = \mathbf{R}_0(t) - a\mathbf{n}(t) \qquad \text{and} \qquad \mathbf{R}_{Z+1}(t + \Delta t) = \mathbf{R}_Z(t) - a\mathbf{n}(t)$$

Proceeding in the same way as for the Rouse regime, we first discuss the centre of mass motion, which is given by

$$\mathbf{R}_G(t) = (Z+1)^{-1} \sum_0^Z \mathbf{R}_n(t)$$

For $\mathbf{n}(t) = 1$ we have

$$\mathbf{R}_G(t+\Delta t) = (Z+1)^{-1} \left[\sum_1^Z \mathbf{R}_n(t) + \mathbf{R}_Z(t) + a\mathbf{n}(t) \right]$$

$$= \mathbf{R}_G(t) + (Z+1)^{-1} \left[\mathbf{R}_Z(t) + a\mathbf{n}(t) - \mathbf{R}_0(t) \right]$$

$$= \mathbf{R}_G(t) + (Z+1)^{-1} \left[\mathbf{P}(t) + a\mathbf{n}(t) \right] \quad \text{with} \quad \mathbf{P}(t) = \mathbf{R}_Z(t) - \mathbf{R}_0(t)$$

the end-to-end vector. Similarly for $\mathbf{n}(t) = -1$ we can obtain

$$\mathbf{R}_G(t+\Delta t) = \mathbf{R}_G(t) - (Z+1)^{-1} \left[\mathbf{P}(t) + a\mathbf{n}(t) \right]$$

These two equations can be written as

$$\mathbf{R}_G(t+\Delta t) = \mathbf{R}_G(t) + \xi(t)\mathbf{f}(t) \quad \text{with} \quad \mathbf{f}(t) = (Z+1)^{-1} \left[\mathbf{P}(t) + a\mathbf{n}(t) \right]$$

The time correlation function becomes $\langle \mathbf{f}(t)\mathbf{f}(t') \rangle = \delta(t-t')\langle \mathbf{f}^2(t) \rangle$ so that we obtain

$$\langle [\mathbf{R}_G(t) - \mathbf{R}_G(0)]^2 \rangle = \frac{t}{\Delta t} \langle \mathbf{f}^2(t) \rangle$$

At equilibrium we also have $\langle \mathbf{P}^2(t) \rangle = Za^2$ and combining leads to

$$\langle \mathbf{f}^2(t) \rangle = (Z+1)^{-2} \left[\langle \mathbf{P}^2(t) \rangle + a^2 \right] = a^2 (Z+1)^{-1}$$

Finally the overall result is

$$\langle [\mathbf{R}_G(t) - \mathbf{R}_G(0)]^2 \rangle = \frac{t}{\Delta t} \frac{a^2}{Z} = \frac{2D}{Z} t$$

where $Z+1$ has been replaced by Z since $Z \gg 1$. The self-diffusion constant D_G reads

$$D_G = \frac{D}{3Z} = \frac{kT}{3N^2\zeta} \frac{a^2}{b^2}$$

The diffusion constant thus is predicted to be proportional to M^{-2}, in reasonable agreement with experimental data, the experimental exponent[u] for the entangled regime being about 2.3.

For the rotational motion we express the correlation function $\langle \mathbf{P}(t) \cdot \mathbf{P}(0) \rangle$ of the end-to-end vector $\mathbf{P}(t)$ in terms of the correlation function of the bond vectors $\mathbf{u}_n(t)$ given by $\psi_{n,m} \equiv \langle \mathbf{u}_n(t) \, \mathbf{u}_m(0) \rangle$ to obtain

$$\langle \mathbf{P}(t).\mathbf{P}(0) \rangle = a^2 \sum_n^Z \sum_m^Z \psi_{n,m} \tag{20.33}$$

After a time step Δt the vector $\mathbf{u}_n(t)$ becomes $\mathbf{u}_{n+1}(t)$ or $\mathbf{u}_{n-1}(t)$ and therefore $\psi_{n,m}$ obeys

$$\psi_{n,m}(t+\Delta t) = \frac{1}{2} \left[\psi_{n+1,m}(t) + \psi_{n-1,m}(t) \right] \tag{20.34}$$

Since $\mathbf{u}_0$ corresponds to a random vector $\mathbf{n}(t)$ we have

$$\psi_{0,m}(t) = \langle \mathbf{u}_0(t)\mathbf{u}_m(t) \rangle = \langle \mathbf{u}_0(t) \rangle \langle \mathbf{u}_m(t) \rangle = 0 \tag{20.35}$$

---

[u] See e.g. the data for hydrogenated polybutadiene at 175 °C: Lodge, T.P. (1999), Phys. Rev. **83**, 3218.

In the same way we obtain

$$\psi_{Z+1,m}(t) = 0 \tag{20.36}$$

Moreover at $t = 0$ we have the result

$$\psi_{n,m}(0) = \delta_{nm} \tag{20.37}$$

These equations form a set of difference equations. For the case of $Z \gg 1$ an analytical solution can be found if $\psi_{n,m}(t)$ is a sufficiently slow function of t and n. Eq. (20.34) then can be written as[v]

$$\psi_{n,m}(t + \Delta t) = \psi_{n,m}(t) + \frac{\partial \psi_{n,m}}{\partial t} \Delta t = \psi_{n,m}(t) + \frac{\partial \psi_{n,m}}{\partial t} \frac{a^2}{2D}$$

$$= \frac{1}{2}\left[\psi_{n+1,m}(t) + \psi_{n-1,m}(t)\right] = \psi_{n,m}(t) + \frac{\partial^2 \psi_{n,m}}{\partial n^2}$$

This leads to the differential equation

$$\frac{\partial \psi_{n,m}}{\partial t} = \frac{D}{a^2} \frac{\partial^2 \psi_{n,m}}{\partial n^2} \tag{20.38}$$

Eqs. (20.35), (20.36) and (20.37) become the boundary and initial conditions

$$\psi_{0,m}(t) = \psi_{Z+1,m}(t) = 0 \qquad \text{and} \qquad \psi_{n,m}(0) = \delta(n-m)$$

Solving these equations leads to

$$\psi_{n,m}(t) = \frac{2}{Z}\sum_p^\infty \sin\left(\frac{n\pi p}{Z}\right)\sin\left(\frac{m\pi p}{Z}\right)\exp\left(-\frac{tp^2}{\tau_{rep}}\right) \qquad \text{with} \quad \tau_{rep} = \frac{Z^2 a^2}{\pi^2 D}$$

where τ_{rep} is called the *reptation time*. Combining these results we obtain

$$\langle \mathbf{P}(t) \cdot \mathbf{P}(0) \rangle = Za^2 \psi(t) \qquad \text{with} \quad \psi(t) = \sum_p \frac{8}{\pi^2 p^2} \exp\left(-\frac{tp^2}{\tau_{rep}}\right) \tag{20.39}$$

and where the index p runs over all odd positive integers. Hence we have $\psi(0) = 1$ and with increasing t the function $\psi(t)$ decreases with a relaxation time τ_d. Using $D = kT/N\zeta$ and $Z = Nb^2/a^2$ the result is

$$\tau_{rep} = \frac{1}{\pi^2} \frac{\zeta N^3 b^4}{kTa^2}$$

so that the reptation time is proportional to M^3. This time can be measured with elastic measurements. As stated before, experimentally the exponent is observed to be 3.2 to 3.4 instead of the predicted value of 3.

Modulus and viscosity

In Chapter 18 we have seen that the steady-state viscosity is given by

---

[v] The same trick was used in Justification 11.1.

$$\eta = \int_0^\infty G(t)\, dt$$

Therefore it suffices to consider the time-dependent shear modulus. To that purpose we have to link first the macroscopic stress with the underlying structure and dynamics of the polymer and we do so by considering a set of particles in an incompressible viscous liquid. We take $\mathbf{R}_n$ as the position vectors of the nth particle and assume that the interaction between the particles can be described by a pair potential $\phi_{nm}(\mathbf{R}_n - \mathbf{R}_m)$. In order to calculate the stress σ_{ij} we consider a box with top and bottom area S and thickness L so that its volume $V = LS$. A plane parallel to the top side is our reference plane for the calculation of the stress. We recall that the stress component σ_{ij} denotes the i-component of the force on plane indicated by j and divided by the area of that plane. Since we have for the force between two particles $\mathbf{f}_{nm} = -\partial\phi_{nm}(\mathbf{R}_n - \mathbf{R}_m)/\partial\mathbf{R}_n$, we obtain

$$\hat{\sigma}_{ij} = S^{-1}\sum_{n,m}(\mathbf{f}_{nm})_i\, H[z-(\mathbf{R}_n)_j]H[(\mathbf{R}_m)_j - z] + \eta_{\text{sol}}d_{ij} - p\delta_{ij} \qquad (20.40)$$

Here H represents the step function (see Section 3.13) taking into account that the only contributions to the stress are when the particle m is above the plane, located at height z, and particle n is below that plane. The last two terms are due to the solvent (d_{ij} is the rate of deformation, η_{sol} is the viscosity of the solvent and p the pressure). Since this contribution is usually small, we further neglect it. For uniform flow, the stress σ_{ij} should be independent of the position in the box so that we may take the average over the box as

$$\sigma_{ij} = L^{-1}\int_0^L \hat{\sigma}_{ij}\, dz - p\delta_{ij}$$

Substitution of Eq. (20.40) and using

$$\int_0^L H[h-(\mathbf{R}_n)_j]H[(\mathbf{R}_m)_j - h]\, dz = (\mathbf{R}_m - \mathbf{R}_n)H(\mathbf{R}_m - \mathbf{R}_n) \equiv (\mathbf{R}_{mn})_j\, H[(\mathbf{R}_{mn})_j]$$

leads to

$$\sigma_{ij} + p\delta_{ij} = V^{-1}\sum_{n,m}(\mathbf{f}_{nm})_i(\mathbf{R}_{mn})_j\, H[(\mathbf{R}_{mn})_j]$$

$$= (2V)^{-1}\sum_{n,m}\left[(\mathbf{f}_{nm})_i(\mathbf{R}_{mn})_j\, H[(\mathbf{R}_{mn})_j] - (\mathbf{f}_{nm})_i(\mathbf{R}_{nm})_j\, H[(\mathbf{R}_{nm})_j]\right]$$

$$= -(2V)^{-1}\sum_{n,m}(\mathbf{f}_{nm})_i(\mathbf{R}_{nm})_j = -V^{-1}\sum_{n<m}(\mathbf{f}_{nm})_i(\mathbf{R}_{nm})_j$$

where $\mathbf{f}_{nm} = -\mathbf{f}_{mn}$, $\mathbf{R}_{nm} = -\mathbf{R}_{mn}$ and $H[(\mathbf{R}_{nm})_j] + H[(\mathbf{R}_{mn})_j] = 1$ are used. Since we assume that there are many particles in the box we may take the ensemble average

$$\sigma_{ij} = -V^{-1}\sum_{n<m}\langle(\mathbf{f}_{nm})_i(\mathbf{R}_{nm})_j\rangle - p\delta_{ij}$$

Now introducing the bead model we have only forces between beads, as given by[w]

---

[w] Here, and in the next equation, a comma is inserted between the indices n and $n+1$ just for clarity.

$$\mathbf{f}_{n,n+1} = \frac{3kT}{b^2} \frac{\partial \mathbf{R}(n,t)}{\partial n}$$

Because $\mathbf{R}_{n,n+1} = \partial \mathbf{R}_n / \partial n$ the stress becomes

$$\sigma_{ij} = \frac{c}{N} \frac{3kT}{b^2} \sum_n \langle \frac{\partial R_i(n,t)}{\partial n} \frac{\partial R_j(n,t)}{\partial n} \rangle - p\delta_{ij} \qquad (20.41)$$

where c is the number density of segments and N the number of segments in one polymer. We have used that cV/N polymers are present in the volume V. So far we have only considered forces between neighbouring beads along the chain and neglected entanglements. Although this appears to be a crude approximation, similarly as for the drag approximation of the beads, this approximation works reasonably well because the hydrodynamic interactions in amorphous polymers contribute not significantly. A slightly other approach to the calculation of the stress is via the fluctuation-dissipation theorem[x], for which we refer to the literature.

Now let us introduce the Rouse dynamics in this picture. This can be done as long as the size of the polymer is not larger than the characteristic size of the tube a. If we assume that the molecule is experiencing a flow field

$$\mathbf{v}(\mathbf{r},t) = \mathbf{d}(t).\mathbf{r}$$

the nth segment has increased its velocity by $\mathbf{d}.\mathbf{R}_n$. The Langevin equation becomes

$$\frac{\partial \mathbf{R}_n}{\partial t} = \frac{K}{\zeta} \frac{\partial^2 \mathbf{R}_n}{\partial n^2} + \mathbf{d} \cdot \mathbf{R}_n + \mathbf{g}(n,t) \qquad (20.42)$$

and introducing once more the normal co-ordinates results in

$$\frac{\partial \mathbf{X}_p}{\partial t} = -\frac{\mathbf{X}_p}{\tau_p} + \mathbf{d} \cdot \mathbf{X}_p + \mathbf{g}_p \qquad \text{with} \quad \tau_p = \zeta_p / K_p$$

Introducing also the normal co-ordinates in the expression for the stress yields

$$\sigma_{ij} = \frac{c}{N} \sum_p K_p \langle (\mathbf{X}_p)_i (\mathbf{X}_p)_j \rangle - p\delta_{ij} \qquad (20.43)$$

Specialising to a shear deformation in the x-direction we have

$$\frac{\partial (\mathbf{X}_p)_x}{\partial t} = -\frac{(\mathbf{X}_p)_x}{\tau_p} + \dot{\gamma}(\mathbf{X}_p)_x + (\mathbf{g}_p)_x \qquad (20.44)$$

$$\frac{\partial (\mathbf{X}_p)_y}{\partial t} = -\frac{(\mathbf{X}_p)_y}{\tau_p} + (\mathbf{g}_p)_y \qquad (20.45)$$

$$\frac{\partial (\mathbf{X}_p)_z}{\partial t} = -\frac{(\mathbf{X}_p)_z}{\tau_p} + (\mathbf{g}_p)_z \qquad (20.46)$$

with $\dot{\gamma}$ the shear rate. Defining the stress per mode $(S_p)_{xy}$ by $(S_p)_{xy} \equiv \langle [Xp(t)]_x [Xp(t)]_y \rangle$ we have

---

[x] Callen, H.B. and Welton, T. (1951), Phys. Rev. **83**, 34.

$$\sigma_{ij} = \frac{c}{N}\sum_{p} K_p(\mathbf{S}_p)_{ij}$$

For the calculation of $(\mathbf{S}_p)_{xy}$ Eqs. (20.44) and (20.45) are multiplied by $[\mathbf{X}p(t)]_y$ and $[\mathbf{X}p(t)]_x$, respectively, and the results are added. After averaging we obtain

$$\frac{d}{dt}(\mathbf{S}_p)_{xy} = -\frac{2}{\tau_p}(\mathbf{S}_p)_{xy} + \gamma\frac{kT}{K_p}$$

The solution of this equation is

$$(\mathbf{S}_p)_{xy} = \int_{-\infty}^{t} \frac{kT}{K_p}\exp\!\left[-2(t-t')/\tau_p\right]\dot\gamma(t')\,dt'$$

and substitution in Eq. (20.43) yields the final results for the stress becomes

$$\sigma_{xy}(t) = \int_{-\infty}^{t} G(t-t')\dot\gamma(t')\,dt' \qquad \text{with} \tag{20.47}$$

$$G(t) = \frac{c}{N}kT\sum_{p}^{\infty}\exp(-2t/\tau_p) = \frac{c}{N}kT\sum_{p}^{\infty}\exp(-2tp^2/\tau_1) \tag{20.48}$$

With this result the viscosity η becomes

$$\eta = \int_{0}^{\infty} G(t)\,dt = c\zeta Nb^2/36 \tag{20.49}$$

So we see that η is independent of the shear rate $\dot\gamma$ and proportional to the molecular mass M, which is the experimentally observed behaviour for not too long chains.

A similar exercise for reptation dynamics can be done yielding the results for high M polymers. We assume again that the number of bonds $Z \gg 1$. If the polymer segments are uniformly distributed along the tube, we have $\partial\mathbf{R}_n/\partial n = (L/N)\mathbf{m}_n$ with $\mathbf{m}_n$ a unit vector pointing in the direction of the tube at the position of segment n. From Eq. (20.41) we can obtain

$$\sigma_{ij} = \left(\frac{L}{N}\right)^2\frac{3kT}{b^2}\frac{c}{N}\sum_{n}[\mathbf{S}_n(t)]_{ij} - p\delta_{ij} \qquad \text{with} \quad [\mathbf{S}_n(t)]_{ij} = \langle(\mathbf{m}_n)_i(\mathbf{m}_n)_j - \frac{1}{3}\delta_{ij}\rangle$$

the orientation tensor of the primitive path of the tube.

For concreteness let us consider stress relaxation where at $t = 0$ a strain is applied to the material such that $\mathbf{r}' = \mathbf{E}\cdot\mathbf{r}$, which is kept constant for $t > 0$. We assume that the tube deforms affinely. Moreover, we assume that the tube returns to its original length Za after a time τ_R but that this time, since $Z \gg 1$ and therefore $\tau_R \ll \tau_{rep}$, can be neglected. From the above stress expression we obtain

$$\sigma_{ij} = \frac{3b^2kT}{a^2}\frac{c}{N}\sum_{n}[\mathbf{S}_n(t)]_{ij} - p\delta_{ij}$$

where $L = Za$ and $Z = Nb^2/a^2$ is employed. A part of the chain that was in the direction $\mathbf{m}$ before deformation will point in direction $\mathbf{m}' = \mathbf{E}\cdot\mathbf{m}/|\mathbf{E}\cdot\mathbf{m}|$ after deformation. Hence the initial value of the mode stress tensor $(\mathbf{S}_n)_{ij}$ will be given by

$$[S_n(t = +0)]_{ij} = \langle \frac{(\mathbf{E} \cdot \mathbf{m})_i (\mathbf{E} \cdot \mathbf{m})_j}{(\mathbf{E} \cdot \mathbf{m})^2} \rangle_0 - \frac{1}{3}\delta_{ij} \equiv Q_{ij}(\mathbf{E}) \qquad (20.50)$$

where $\langle \cdots \rangle_0$ represents the isotropic average defined by $\langle \cdots \rangle_0 = (4\pi)^{-1}\int \ldots d\mathbf{m}$. For a time $t > 0$ the chain moves gradually out of its tube but the changes in $(S_n)_{ij}$ will be due to reptation only since there is no external flow. Similar to Eq. (20.38), $(S_n)_{ij}$ will be determined by

$$\frac{\partial (S_n)_{ij}}{\partial t} = \frac{D}{a^2} \frac{\partial^2 (S_n)_{ij}}{\partial n^2} \qquad (20.51)$$

Since at the boundaries $n = 0$ and $n = N$ the new tube has an isotropic distribution, we must have $(S_n)_{ij} = 0$. Solving Eqs. (20.50) and (20.51) with these boundary conditions results in

$$(S_n)_{ij} = \sum_p^\infty \sin\left(\frac{np\pi}{N}\right) \exp\left(\frac{-tp^2}{\tau_{\text{rep}}}\right) \frac{4Q_{ij}(\mathbf{E})}{p\pi} \qquad \text{with} \quad p = 1, 3, 5, \ldots$$

The total stress becomes

$$\sigma_{ij} = \frac{3b^2 kT}{a^2} \frac{c}{N} Q_{ij}(\mathbf{E})\psi(t) - p\delta_{ij}$$

with $\psi(t)$ as defined in Eq. (20.39).

Specialising to a shear strain γ we obtain for $\gamma \ll 1$, $Q_{ij}(\gamma) = \gamma/5$, the stress reads

$$\sigma_{ij} = G(t)\gamma \qquad \text{with} \quad G(t) = \frac{3b^2 kT}{5a^2} \frac{c}{N} \psi(t) \qquad (20.52)$$

the stress relaxation function. Finally, the viscosity becomes

$$\eta = \int_0^\infty G(t)\, dt = \frac{\pi^2}{20} \frac{ckTb^2}{a^2} \tau_{\text{rep}} \qquad (20.53)$$

Because $\tau_{\text{rep}} \sim M^3$, the viscosity in the reptation regime is proportional to the third power of the molecular mass M. As stated before, experimentally an exponent of about 3.4 is observed. This slightly increased exponent can be rationalised by taking into account tube variations. The treatment can also be extended to general flow fields leading to

$$\sigma_{ij} = 5 \int_{-\infty}^t G(t - t') Q_{ij}[\mathbf{E}(t, t')]\, dt$$

For details of this more refined calculation and the general flow field we refer to the literature quoted.

Summarising, the dynamics of an entangled chain contains three contributions. At the highest frequency the motion at the smallest length, highly determined by the precise chemical nature of the polymer. At a somewhat larger time and length scale the Rouse dynamics, describing the motion between entanglements. Finally we have the reptation dynamics, which describes the diffusion of the complete chain. The latter two can be described by generic models, as outlined here, in which the chemical information is hidden in the Kuhn length.

John Douglas Ferry (1912-2003)

Born in Dawson, Canada, he obtained his bachelors degree from Stanford University in 1932 and his doctorate in 1935. He went to Harvard University in 1936 where he spent nine years in various capacities. In 1946 he was appointed assistant professor in the Chemistry Department at the University of Wisconsin, Madison and this association was a long and highly successful one. He was appointed full professor in 1947. Between 1959 and 1967 he became Chairman of the Department and from 1973 to his official retirement in 1982 he held the position of Farrington Daniels Research Professor. During his tenure Madison became an international centre of excellence in rheology and he supervised over 50 research students. Ferry's now classic book *Viscoelastic properties of polymers* was first published in 1961. It soon became a standard reference in the area of linear visco-elasticity and the third edition was published in 1980. He published over 300 research papers. Ferry was a highly successful and influential rheologist and received numerous honours during his career. However, he remained unassuming and charming.

20.6   Bibliography

Cottrell, A.H. (1953), *Dislocations and plastic flow in crystals*, Oxford University Press, Oxford.

Doi, M. (1996), *Introduction to polymer physics*, Clarendon, Oxford.

Evans, R.W. and Willshire, B. (1985), *Creep of metals and alloys*, Institute of Metals, London.

Evans, R.W. and Wilshire, B. (1992), *The theta approach to creep of structural ceramics*, Revs. Powder Metall. Phys. Ceram. **5**, 111-168.

Ferry, J.D. (1980), *Viscoelastic properties of polymers*, 3rd ed., Wiley, New York.

Findlay, W.N., Lai, J.S. and Onaran, K. (1976), *Creep and relaxation of nonlinear viscoelastic materials*, North-Holland, Amsterdam.

Frost, H.J. and Ashby, M.F. (1982), *Deformation mechanism maps*, Pergamon, Oxford.

Garofalo, F. (1965), *Fundamentals of creep and creep-rupture in metals*, McMillan, New York.

Krausz, A.S. and Eyring, H. (1975), *Deformation kinetics*, Wiley, New York.

Lin, Y.-H. (2003), *Polymer viscoelasticity*, World Scientific, New Jersey.

Poirier, J.-P. (1985), *Creep of crystals*, Cambridge University Press, Cambridge.

Rubinstein, M. and Colby, R. (2003), *Polymer physics*, Oxford University Press, New York.

Scherer, G. W. (1986), *Relaxation in glass and composites*, Wiley, New York.

Strobl, G. (1997), *The physics of polymers*, 2nd ed., Springer, Berlin.

Ward, I.M. (1983), *Mechanical properties of polymers*, 2nd ed., Wiley, Chichester.

Weertman, J. and Weertman J.R. (1983), *Mechanical properties, strongly temperature-dependent*, page 1309 in *Physical Metallurgy*, R.W. Cahn and P. Haasen, eds, 3rd ed., North-Holland, Amsterdam.

21

Continuum fracture

In the previous chapters elastic, plastic and visco-elastic deformation was described and the underlying molecular mechanisms were discussed providing descriptions at the micro-, meso- and macroscopic level. In all these cases the material remained intact and no cracks were present. In this chapter we deal with continuum fracture. After an overview the influence of stress concentration and the classical Griffith analysis are presented, followed by the description in terms of the stress intensity factor. The influence of limited plastic flow at the crack tip is discussed and we conclude with a thermodynamic version of fracture theory.

21.1   Overview

Fracture, being the large scale breaking of bonds, contains (at least) two important aspects. First, sufficient energy should be available and, second, a sufficiently high force must be provided for the process to occur.

Let us first consider energy and assume that sufficient energy is available for fracture to occur. The energy balance can be considered globally, i.e. on the scale of the structure tested, and this leads to what is called the *energy approach*. The mechanical energy that is provided from the outside and/or through the accumulated strain energy in the fracture process is converted to another form. First of all, this is energy related to surface formation but also the energy necessary for deformations associated with fracture. The latter may be quite substantial or even constitute the largest fraction of the energy. This conversion may essentially occur only at the crack tip in which case we speak of *brittle materials*. The basic mechanism is then transformation to the excess energy associated with the crack surface. The conversion may also occur in a small region (with respect to the size of the structure) ahead of or just after the crack tip where the process is called as *quasi-brittle* (in the jargon of inorganics) or as *small-scale yielding* (in the jargon of metals and polymers). The zone itself is usually addressed as *process-zone* for inorganics and as *plastic zone* for metals and polymers. In both cases a considerable amount of energy is used for the processes accompanying fracture in this zone. For inorganics these dissipation mechanism processes are, e.g. micro-cracking, crack bridging, phase transformations and the creation of surface roughness of the fracture plane. For metals and polymers plastic and/or visco-elastic deformation is the main deformation mechanism. The final possibility is, of course, that energy is dissipated globally, i.e. all over the structure. For inorganics this hardly occurs in tension but can happen in compression. The main dissipation mechanism is then microcracking. For metals and polymers this process can happen in tension as well as compression and the associated mechanism is plastic and/or visco-elastic deformation.

Now consider the force aspect and assume that a sufficiently high force is available to fracture a material. Eventually fracture means breaking of chemical bonds and bonds between atoms are generally quite strong, i.e. require a much higher stress than usually is applied. However, existing defects in a structure have a so-called stress

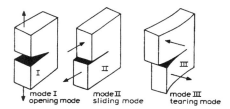

Fig. 21.1: The three different modes of crack opening: mode I or opening mode, mode II or sliding mode and mode III or tearing mode.

concentration effect, i.e. they amplify the relatively low globally applied stress to a high local stress at the crack tip so that bonds can break. This way of looking at fracture leads to the *stress intensity factor approach*. Although quite appealing in mechanistic terms, it appears difficult to generalise this approach to non-elastic materials. However, in many situations the structure globally can be considered to act elastically and in that case this approach still can be used. Due to this usefulness, fracture of metal structures in engineering situations is often discussed using the stress intensity factor approach.

In the remainder of this chapter we discuss some aspects of these ways of dealing with fracture. We do not discuss fracture with associated global deformation though, since this is a even more complex phenomenon. Sharp cracks can propagate in mode[a] I, mode II or mode III, as sketched in Fig. 21.1 for edge cracks. From the outset we limit ourselves to mode I or the *opening mode*, which is by far the most important one.

21.2   The energy approach

Generally the most satisfying physical principles are energy based. Therefore it is no surprise that also for fracture an energy-based approach is present. The originator of the energy approach was Griffith[b] who published his results in 1921. Griffith considered fracture as a *quasi-static thermodynamic process*. He considered a plate, as discussed by Inglis (see Section 21.3), but took the analysis further. Let us first assume that the plate is loaded at constant displacements and that these displacements remain at a fixed and prescribed value (displacement control) meaning that the plate is mechanically isolated. Now there are two energy[c] terms to consider when the crack extends. First, the change in *strain* or *deformation energy* F_{def} and, second, the energy associated with the increase in crack area, the so-called *fracture energy* F_{cra}. The crucial idea is that fracture occurs when strain energy can be exchanged to surface energy, i.e. when $dF/dA_{cra} = 0$, where $F = F_{def}+F_{cra}$ is the total (Helmholtz) energy and A_{cra} the crack area.

Let us first consider the strain energy. From the stress distribution as given by Inglis, Griffith calculated the elastic strain energy F_{def} stored in the plate as a function of the ratio a/ρ and took the limit for a sharp crack, i.e. $\rho \to 0$. The calculation is complex but the answer can be appreciated by considering the stress concentration in a thin plate having a Young's modulus E with central crack of length $2a$ loaded by a

---

[a] The labels I, II and III are also used for the principal values. Confusion is unlikely though.
[b] Griffith, A.A. (1921), Phil. Trans. Roy. Soc. London **A221**, 163.
[c] We recall that for isothermal conditions the relevant potential is the Helmholtz energy F (see Chapters 6 and 9).

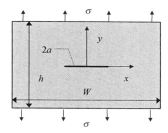

Fig. 21.2: Centrally cracked plate.

stress σ (Fig. 21.2). This plate is thus in plane stress. At the crack faces the stress σ_{yy} is zero but far removed from the crack it is unchanged and equal to σ. So, roughly a region with radius a around the crack is relieved of its elastic energy compared to a plate without crack. Hence the change in elastic energy is estimated to be $t(\pi a^2)(\sigma^2/2E)$, where t denotes the thickness of the plate and $\sigma^2/2E$ the strain energy density of the non-cracked plate. The exact calculation is given by twice this estimate, provided that the width of the plate $W \gg a$. The same result is obtained for a plate loaded in plane strain apart from an extra factor $(1-v^2)$. Hence the deformation energy of the cracked plate is

$$F_{\text{def}} = \frac{\sigma^2}{2E'}\left[(Wth) - 2\pi a^2 t\right] \tag{21.1}$$

where h is the height of the plate, $E' = E/(1-v^2)$ for plane strain and $E' = E$ for plane stress. Evidently F_{def} decreases when the crack extends.

Let us consider next the fracture energy. We distinguish between crack area and crack surface, which is twice the crack area[d]. So, for a central crack of length $2a$ in a plate of thickness t the crack area A_{cra} is by definition $A_{\text{cra}} = 2at$ and the crack surface is $2A_{\text{cra}} = 4at$. The energy necessary to form a crack U_{fra} is the product of the *specific fracture energy R* (by Griffith taken as the surface energy) and the magnitude of the crack surface $2A_{\text{cra}}$. The specific fracture energy R is assumed to be independent of the crack length $2a$. For the case of a central crack of length $2a$ in a plate of thickness t we thus have

$$F_{\text{fra}} = 2A_{\text{cra}}R = 4atR \tag{21.2}$$

Obviously F_{fra} increases when the crack extends. Equilibrium is obtained when $dF/dA_{\text{cra}} = 0$. Because the thickness t is constant, we have $dA_{\text{cra}} = 2tda$ leading to

$$\frac{\partial F}{\partial A_{\text{cra}}} = \frac{1}{2t}\frac{\partial F}{\partial a} = \frac{1}{2t}\frac{\partial}{\partial a}\left(F_{\text{def}} + F_{\text{fra}}\right) = 0 \quad \text{or} \quad \frac{\partial}{\partial a}\left(-F_{\text{def}}\right) = \frac{\partial}{\partial a}F_{\text{fra}} \tag{21.3}$$

So, the Griffith theory predicts that crack growth occurs whenever

$$\tfrac{1}{2}\pi\frac{\sigma^2}{E'}a = R \tag{21.4}$$

or, when the remote stress σ reaches its critical value,

---

[d] Note that this is completely analogous to the difference between the length of a river and the length of its banks.

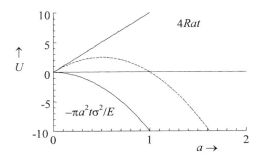

Fig. 21.3: The balance between elastic energy and fracture energy. The dotted line represents the total energy. The term Wth is skipped since it is independent of a and therefore the reference energy becomes zero.

▶ $$\sigma_{cri} \equiv S = \sqrt{\frac{2E'R}{\pi a}}$$ (21.5)

In many cases to distinguish applied stress σ from the *critical fracture stress* or *fracture strength* σ_{cri}, we indicate the latter with S. As long as $\sigma < \sigma_{cri}$ crack growth needs more energy than the body can deliver and further crack growth does not occur. If $\sigma > \sigma_{cri}$ the body delivers more energy than needed for crack extension and unstable crack growth occurs. In Fig. 21.3 the various contributions are indicated as a function of crack length, showing the balance between the elastic energy and fracture energy. For a certain crack of length a, fracture thus occurs when the applied stress σ is equal to the critical stress σ_{cri} given above. In that case sufficient energy is stored to extend the crack. So far it was assumed that experiment takes place under displacement control. Alternatively, if instead of the end displacements the forces are prescribed (load control) the result, i.e. Eq. (21.5), of the analysis remains the same. For other geometries similar results are obtained and the overall result can be written as

▶ $$\sigma_{cri} \equiv S = \frac{1}{\alpha}\sqrt{\frac{2E'R}{a}}$$ (21.6)

Generally the factor α depends on the crack length a. In the case of a wide plate with a central crack of length $2a$ ($W \gg a$) the factor α becomes $\alpha = \sqrt{\pi}$.

Problem 21.1*

Show that for linear elastic systems Eq. (21.5) is valid, irrespective of whether the experiment takes place under fixed displacement (fixed 'grip') conditions or constant load conditions.

Eq. (21.3) can be interpreted in a slightly different way. The term $\partial F_{fra}/\partial A_{cra}$ is interpreted as the resistance to fracture and is often indicated by G_{cri} ($= 2R$). The term $-\partial F_{ela}/\partial A_{cra}$ on the other hand is interpreted as the energy release rate and indicated by G. In this interpretation the fracture condition becomes

$$\frac{\partial F}{\partial A_{\text{cra}}} = -G + G_{\text{cri}} = 0 \qquad \text{or} \qquad G = G_{\text{cri}} \tag{21.7}$$

As soon as the strain energy that is 'released' equals the critical value G_{cri}, fracture occurs. This interpretation has its basis in the thermodynamics of cracked bodies. To see this consider the cracked plate as a linear elastic system in either load or in displacement control and we take for the energy again the Helmholtz energy F (in mechanics often denoted as potential energy Π, see Chapter 9).

Let us consider load control. For linear elastic systems loaded with a single, prescribed load P the corresponding end displacement u is given by $u = CP$, where the compliance C is given by $C = K^{-1}$ and K is the stiffness of the structure. The stiffness and compliance are dependent on the crack length $l = 2a$ and thus properly indicated by $K(l)$ and $C(l)$. Since u is the dependent variable it is also dependent on l, i.e. $u = u(l)$. In this case the deformation part of the Helmholtz energy F_{def} is given by

$$F_{\text{def}} = \tfrac{1}{2} K u^2 - Pu \tag{21.8}$$

where the first term is the strain energy in the system and the second the potential energy of the loading mechanism. Mechanical equilibrium entails $\partial F_{\text{def}}/\partial u = 0$, i.e. $u = P/K = CP$. The crack area $A_{\text{cra}} = lt$ is an internal variable[e], associated with the *energy release rate* G, given by

$$G = -\left. \frac{\partial F_{\text{def}}}{\partial A_{\text{cra}}} \right|_{P=\text{constant}} = \frac{1}{t} \frac{P^2}{2} \left. \frac{\partial C}{\partial l} \right|_{P=\text{constant}} \tag{21.9}$$

in which $u(l) = C(l)P$ was used because of equilibrium. If, instead of the load, the displacement u is prescribed (displacement control) the loading mechanism cannot perform work and we have now for the total Helmholtz energy

$$F_{\text{def}} = \tfrac{1}{2} K u^2 \tag{21.10}$$

and the energy release rate is found to be

$$G = -\left. \frac{\partial F_{\text{def}}}{\partial A_{\text{cra}}} \right|_{u=\text{constant}} = -\frac{1}{t} \frac{u^2}{2} \left. \frac{\partial K}{\partial l} \right|_{u=\text{constant}} \tag{21.11}$$

We thus can write either

▶ $$G = \frac{1}{t} \frac{P^2}{2} \left. \frac{\partial C}{\partial l} \right|_{P=\text{constant}} \qquad \text{or} \qquad G = -\frac{1}{t} \frac{u^2}{2} \left. \frac{\partial K}{\partial l} \right|_{u=\text{constant}} \tag{21.12}$$

as alternative expressions for the energy release rate. Relations (21.12) offer the possibility to determine the strain energy release rate for a particular system experimentally by measuring the compliance or the stiffness as a function of the crack length. Although derived for a special system, the relations are generally valid for linear elastic materials.

A few general comments on the above can be made. First, Griffith did his work on glass. His original but replotted data for glass with artificial defects of varying size

---

[e] This shows the reason to take A_{cra} as the important parameter since this choice leaves the Gibbs equation in its conventional form $\mathrm{d}F = -S\mathrm{d}T - P\mathrm{d}V - G\mathrm{d}A_{\text{cra}}$. The penalty is that in the equilibrium condition the energy release rate G_{cri} has to be taken as twice the fracture energy R, i.e. $G_{\text{cri}} = 2R$.

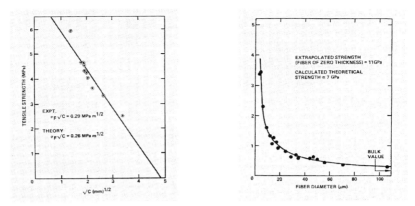

Fig. 21.4: Griffith's data for glass with defects and glass fibres without defects.

and data for glass fibres with varying diameter without artificial defects are shown in Fig. 21.4. The square root dependence is relatively well fulfilled for the glass with the defects. The fibre strength is extrapolated to diameter zero for an estimate of the theoretical strength resulting in 11 GPa. This corresponds nicely with model estimates (see Chapter 23).

The second remark is that fracture strength is a hybrid quantity, dependent on the intrinsic or material properties E, v and R and on the critical defect of length, which is an extrinsic property. Since the variation in Poisson's ratio is nearly always small, its influence is only minor and often neglected. Consequently a high fracture strength is realised by a high fracture energy, a high Young's modulus and a small critical defect. From these considerations it follows that polishing (removal of large flaws) yields stronger materials while abrading (introduction of large flaws) reduces the strength.

The third remark deals with the crack sharpness because it was assumed that the crack tip is sharp, i.e. $\rho \to 0$. The question how sharp a crack tip should be in order to comply with the model proposed is not clear a priori and will be treated in Chapter 23.

The fourth remark is that the fracture energy is not equal to the surface energy, as Griffith had in mind. A typical value for the surface energy is 1 J/m^2 while generally the fracture energy is much larger. This is due to the fact that not only new surface is created during fracture but, as mentioned before, also other mechanisms of energy dissipation occur. Amongst others we mention plastic deformation, phase transitions,

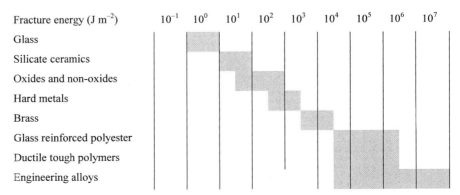

Fig. 21.5: Typical fracture energies for various materials.

micro-cracking, crack deflection and for composites fibre pullout. The magnitude of this energy is different for different types of materials and illustrated in Fig. 21.5. In Chapter 23 we discuss this aspect in some detail. Finally, we note that R needs not to be independent of the crack length.

A drawback of the Griffith approach is that details at the crack tip are not addressed. This is even recognised in continuum mechanics and therefore the stress and strain in the neighbourhood of the crack tip are specifically considered in the stress intensity approach, considered in the next two sections. Moreover, crack tip plasticity is not incorporated and this will be dealt in Section 21.5. Finally, in the Griffith approach fracture is always catastrophic while kinetic effects experimentally do occur. Section 21.8 deals with a thermodynamic approach, which can handle these effects.

For reference, Derby et al. (1992), Broek (1978) and Knott (1973) provide introductions to the energy approach to fracture. The book by Atkins and Mai (1985) is completely based on this approach and deals both elastic and elastic-plastic fracture.

Alan Arnold Griffith (1893-1963)
English engineer, educated in mechanical engineering at Liverpool University from 1911 to 1914, who was the father of fracture mechanics. He joined the Royal Aircraft Factory in Farnborough in 1915. Stress analysis of aerofoil sections of airscrew blading and other problems led to work with G.I. Taylor at Farnborough in 1917 on the use of soap films in solving the elastic torsion of hollow bars of irregular cross-section. It was for this work that he won the Thomas Hawksley Gold Medal of the Institution of Mechanical Engineers. After that he started to work on fracture and verified his theory by experiments on glass tubes subjected to internal pressure. The resulting paper entitled *Theory of rupture* is nowadays one of the most quoted ones in the history of science. After a glass-blower's torch set fire to his laboratory, the director, who did not know of his activities, found out what he did, stopped the research and transferred him to engine work. In 1939 he was invited to join Rolls-Royce. In the engine field he became an expert, who invented the multi-stage axial aero engine, the bypass engine and the use of jet-lift for vertical take-off. Due to the secret nature of his wartime work at Rolls-Royce, and later commercial confidentiality, his innovations in aeronautics are not widely known internationally.

21.3   Stress concentration

For quite some time fracture was described by a *maximum stress criterion*: if the stress reaches some critical value, fracture occurs. The background to this thought is simple. Experimentally it is observed that when a structure is mechanically loaded, it fails roughly at a certain, approximately constant stress. Moreover, stress

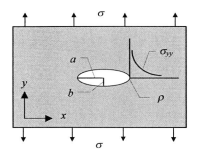

Fig. 21.6: Plate with an elliptical hole having a long axis a and a short axis b loaded remotely by a uniaxial stress σ.

concentrations occur at positions in the structure where holes or sharp corners are present and frequently indeed fracture occurs at these positions.

To illustrate the *stress concentration* effect we discuss the classical analysis by Inglis[f] who considered a plate with a central elliptical hole[g] (Fig. 21.6). The hole is considered to be small with respect to the size of the plate. If this plate is mechanically loaded, the stress (e.g. σ_{yy} in Eq. (21.13)) at the cross-section containing the hole is increased with respect to the remote loading stress σ. The stress is not distributed uniformly over the reduced cross-section but concentrated, primarily at the edges of the hole at the ends of the long axis and depending on the ratio of the long axis a and short axis b. At the edge and at the end of the long axis the radius of curvature[h] is given by $\rho = b^2/a$ and here σ_{yy} is given by

$$\blacktriangleright \qquad \sigma_{yy} = \sigma\left(1 + 2\sqrt{\frac{a}{\rho}}\right) = \sigma\left(1 + \frac{2a}{b}\right) \qquad\qquad (21.13)$$

which approaches $2\sigma(a/\rho)^{1/2}$ for a large ratio of a/ρ. If the ellipse reduces to a circle, $a = b$, and the stress concentration σ_{yy}/σ reduces to the well-known factor 3 for circular holes. At the end of the short axis a stress concentration in σ_{xx} is present but of a compressive nature. Note that for slender ellipses ($a/b \gg 1$) the stress concentration can be quite substantial, e.g. a ratio of $a/b = 10$ results in $\sigma_{yy}/\sigma = 21$.

The basic idea of the maximum stress criterion is that fracture occurs if $\sigma_{yy} = 2\sigma(a/\rho)^{1/2} = \sigma_{the}$, where σ_{the} denotes the theoretical strength. It appears that fracture is not adequately described by a maximum stress criterion. This, however, does not imply that stress concentration is unimportant. The result of the Inglis calculation carries over in a somewhat modified form for other geometries and always results in stress concentration for holes and corners, the amount of which is dependent on the sharpness of the geometry. The calculation of stress concentration is therefore important in the design of many structures. For a variety of structures the stress concentration factor is tabulated[i].

---

[f] Charles Edward Inglis (1875-1952). English scientist, gifted teacher who was head of the Department of Engineering of the University of Cambridge from 1919 to 1943. His most important contribution to engineering science was a treatise on the stresses in metal plates as a result of the presence of cracks.

[g] Inglis, C.E. (1913), Trans. Inst. Naval Architects **55**, 219. See also Knott, 1973.

[h] For a curve $y = y(x)$ the radius of curvature R at any point x is given by $R = [1+(dy/dx)^2]^{3/2}/d^2y/dx^2$.

[i] Peterson, R.E. (1974), *Stress concentration factors*, Wiley, New York.

Example 21.1

As an example of stress concentration we quote the expression for a strip of width W with a central hole of diameter d. For the direction normal to the tension the stress concentration factor is given by $\sigma_{max}/\sigma_{nom} = (3\beta-1)/(\beta+0.3)$ with $\beta = W/d$ and σ_{nom} the mean stress in the weakened section.

Problem 21.2*

The Airy stress function for a tensile loaded plate (Fig. 21.7) with a central hole is given by $\phi = \phi(r,\theta) = (Ar^2+Br^4+Cr^{-2}+D)\cos 2\theta$, where r denotes the distance and θ denotes the polar angle. The diameter of the hole is indicated by $2r_0$ and the remotely applied stress by σ. The boundary conditions are

$$\sigma_{rr}(r_0) = \sigma_{r\theta}(r_0) = 0$$

while from St. Venant's principle the stress field vanishes at large distance

$$\sigma_{xx}(\infty) = 0 \text{ and } \sigma_{yy}(\infty) = \sigma$$

Show that the stress distribution is given by

a) $\sigma_{rr} = \dfrac{\sigma}{2}\left[1-\dfrac{r_0^2}{r^2}+\left(1+\dfrac{3r_0^4}{r^4}-\dfrac{4r_0^2}{r^2}\right)\cos 2\theta\right]$, $\quad \sigma_{r\theta} = -\dfrac{\sigma}{2}\left(1-\dfrac{3r_0^4}{r^4}+\dfrac{2r_0^2}{r^2}\right)\sin 2\theta$

and $\sigma_{\theta\theta} = \dfrac{\sigma}{2}\left[1+\dfrac{r_0^2}{r^2}-\left(1+\dfrac{3r_0^4}{r^4}\right)\cos 2\theta\right]$

b) Show that the maximum stress concentration factor is 3 when the plate is loaded in tension while it is 1 when the plate is loaded in compression.

c) At what position do these maximum stresses occur?

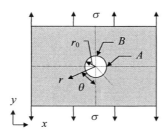

Fig. 21.7: A large plate with a central cylindrical hole.

21.4   The stress intensity factor approach

The energy approach as outlined in Section 21.2 considers the energetics of a system with a crack on a global scale and therefore tells us nothing about the situation near the crack tip itself. From Section 21.3 we know that at the edge of an elliptical hole the normal stress may be quite high and is, in fact, found to be infinitely high in the limit of an infinitely sharp crack $(a/\rho \to \infty$ in Eq. (21.13)). For fracture it is of course also important how (fast) the stress changes when going from the crack tip into the material because this determines how many bonds are loaded until breaking. With reference to Fig. 21.8 the question is how the stress depends on r and θ. For a wide plate with central crack of length $2a$ it was shown by Westergaard[j] that the stresses near the crack tip are given by

$$\sigma_{11} = K_1 (2\pi r)^{-1/2} \cos(\theta/2)[1 - \sin(\theta/2)\sin(3\theta/2)] + \cdots$$
$$\sigma_{12} = K_1 (2\pi r)^{-1/2} \cos(\theta/2)[\sin(\theta/2)\cos(3\theta/2)] + \cdots \qquad (21.14)$$
$$\sigma_{22} = K_1 (2\pi r)^{-1/2} \cos(\theta/2)[1 + \sin(\theta/2)\sin(3\theta/2)] + \cdots$$

For details of the derivation, see Justification 21.1. In these equations r and θ are the polar co-ordinates with respect to the crack tip. The dots indicate terms in second order of magnitude. The stresses decrease with increasing value of r and, obviously, diverge at the crack tip (as has been observed as well for the Inglis stresses when considered in the limit of $\rho \to 0$). We will come back to this divergence later. For a certain value of r the maximum tensile stress is straight ahead of the crack tip in the plane of the crack ($\theta = 0$). The factor K_I is the *stress intensity factor* for mode I defined by

$$K_1 = \lim_{r \to 0} \left[(2\pi r)^{1/2} \sigma_{22} \big|_{\theta=0} \right]$$

This factor characterises the stress field, that is to say the dominant part of the stress field immediately ahead of the crack tip. It appears that for any crack geometry under *plane conditions* loaded in mode I the stresses in the neighbourhood of the tip are given by Eq. (21.14) with K_I given by

▶ $$K_1 = \alpha\sigma\sqrt{a} \qquad (21.15)$$

where σ is the remotely applied stress. The shape factor α depends on the precise geometry and loading conditions of the crack. For the plate considered the factor α is $\sqrt{\pi}$. The factor α is calculated for many crack geometries and can be found in various data books[k]. The length parameter a characterises the crack length but it must be remembered that a is *not necessarily equal* to the physical crack length.

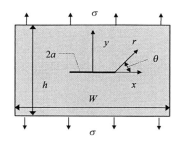

Fig. 21.8: Co-ordinate system for the stresses at the crack tip.

[j] Westergaard, H.M. (1939), Trans. ASME **61**, A49.
[k] Sih, G.C. (1973), *Handbook of stress intensity factor handbook*, Lehigh University, Bethlehem, PA; Rooke, D.P. and Cartwright, D.J. (1976), *Compendium of stress intensity factors*, Her majesty's stationary office, London.

Example 21.2

- Semi-infinite plate (width W) with edge crack (length a) with applied stress σ:

$$K_{\mathrm{I}} = a\sigma\sqrt{a} \qquad \text{with} \quad \alpha = 1.12\sqrt{\pi} \cong 1.99$$

- Semi-circular surface flaw (radius a) with applied stress σ:

$$K_{\mathrm{I}} = a\sigma\sqrt{a} \qquad \text{with} \quad \alpha = 1.12\sqrt{\pi}\,/(\pi/2) \cong 1.26$$

- Plate (width W) with central crack (length $2a$) with applied stress σ:

$$K_{\mathrm{I}} = a\sigma\sqrt{a} \qquad \text{with} \quad \alpha = \sqrt{\beta^{-1}\tan\pi\beta} \quad \text{and} \quad \beta = \frac{a}{W}$$

- Three-point bend bar (height h, width b, length l) with $l/h = 4$ and a notch of depth a:

$$K_{\mathrm{I}} = \alpha\frac{3Fl}{2bh^2}\sqrt{a} \quad \alpha = \frac{1.99 - \beta(1-\beta)(2.15 - 3.93\beta + 2.7\beta^2)}{(1+2\beta)(1-\beta)^{1.5}} \quad \beta = \frac{a}{h}$$

In principal stress terms the stress distribution near the crack tip is given by

$$\sigma_{\mathrm{I}}(r,\theta) = \frac{K_{\mathrm{I}}}{\sqrt{2\pi r}}\cos\frac{\theta}{2}\left(1 + \sin\frac{\theta}{2}\right) + \cdots$$

$$\sigma_{\mathrm{II}}(r,\theta) = \frac{K_{\mathrm{I}}}{\sqrt{2\pi r}}\cos\frac{\theta}{2}\left(1 - \sin\frac{\theta}{2}\right) + \cdots \tag{21.16}$$

The result for the third principal stress is dependent on whether the system is in plane stress or plane strain and using Poisson's ratio v given by

$$\text{plane stress}: \sigma_{\mathrm{III}} = 0 \quad \text{and} \quad \text{plane strain}: \sigma_{\mathrm{III}} = v(\sigma_{\mathrm{I}} + \sigma_{\mathrm{II}}) \tag{21.17}$$

The surface of the plate is not loaded in the z-direction. Therefore at the surface there is always a plane stress state and for a thin plate this stress state is also present in the interior of the plate. For a thick plate the situation is more complex. At the surface there is still the plane stress situation but in the middle a plane strain situation arises. A stress in the z-direction thus arises, dependent on v, as given by Eq. (21.17).

From the crack tip analysis also the displacements can be calculated and for plane strain are given by

$$u_1 = 2(1+v)\frac{K_{\mathrm{I}}}{E}\sqrt{\frac{r}{2\pi}}\cos\frac{\theta}{2}\left(1 - 2v + \sin^2\frac{\theta}{2}\right) \tag{21.18}$$

$$u_2 = 2(1+v)\frac{K_{\mathrm{I}}}{E}\sqrt{\frac{r}{2\pi}}\sin\frac{\theta}{2}\left(2 - 2v - \cos^2\frac{\theta}{2}\right)$$

Justification 21.1*

From the Airy stress function approach (Section 10.3) we recall that

$$\sigma_{11} = \Phi_{,22} \qquad \sigma_{22} = \Phi_{,11} \qquad \sigma_{12} = -\Phi_{,12}$$

where Φ is the Airy stress function. In this case a complex function $Z(z) = $ Re $Z + $ Im Z with $z = x+iy$ ($i = \sqrt{-1}$, Section 3.2) is used. The Westergaard function ψ is given by

$$\psi = \text{Re } \overline{\overline{Z}} + \text{Im } \overline{Z} \qquad \text{where} \qquad \frac{d\overline{\overline{Z}}}{dz} = \overline{Z} \qquad \frac{d\overline{Z}}{dz} = Z \qquad \frac{dZ}{dz} = Z'$$

From the stress expressions we obtain

$$\sigma_{11} = \text{Re } Z - y \text{ Im } Z' \qquad \sigma_{22} = \text{Re } Z + y \text{ Im } Z' \qquad \sigma_{12} = -y \text{ Re } Z'$$

For the mode I crack under biaxial stress Z is given by

$$Z = \sigma z \big/ \sqrt{z^2 - a^2}$$

which is analytic except for ($-a \le x \le a$, $y = 0$). Using a co-ordinate system with the origin at the crack tip z becomes $z+a$. The function Z generally must take the form $f(z)/z^{1/2}$ with $f(z)$ real, well-behaved and constant at the origin. The stresses σ_{22} and σ_{12} are then zero at the crack surface leading to stress free crack edges. The required value at the crack tip is denoted by K_I and thus $Z_{|z| \to 0|}$ $= K_I/(2\pi z)^{1/2}$. Taking polar co-ordinates with $z = r \exp(i\theta)$ the stresses become as given in Eq. (21.14). For a uniaxial stress a term $-\sigma$ has to be added to the stress σ_{11}, which is, however, has no effect on the singular terms. For an infinite plate the only characteristic length is the crack length a so that K_I must be of the form $K_I = c\sigma\sqrt{a}$. From the specific case of biaxial loading one obtains $c = \sqrt{\pi}$. From

$$\varepsilon_{\alpha\alpha} = \frac{1}{E}\left(\gamma\sigma_{\alpha\alpha} - \delta\nu\sigma_{\beta\beta}\right) \qquad \text{and} \qquad 2\varepsilon_{\alpha\beta} = \frac{\sigma_{\alpha\beta}}{G} = \frac{2(1+\nu)}{E}\sigma_{\alpha\beta} \quad (\alpha \ne \beta)$$

where $\gamma = \delta = 1$ for plane stress and $\gamma = 1-\nu^2$ and $\delta = 1+\nu$ for plane strain, respectively, the displacements for plane strain become

$$u_1 = \frac{1+\nu}{E}\left[(1-2\nu)\text{Re}\overline{Z} - y \text{ Im}Z\right] \qquad u_2 = \frac{1+\nu}{E}\left[2(1-\nu)\text{Im}\overline{Z} - y \text{ Re}Z\right]$$

which leads directly to Eq. (21.18).

With increasing stress at a certain moment K_I reaches a critical value, the *critical stress intensity factor* K_{Ic} or *fracture toughness* for short. The fracture criterion thus becomes

▶ $K_I \ge K_{Ic}$ (21.19)

Note that the quantity K_I characterises the stress field while the fracture toughness K_{Ic} is a material property[1] provided it has been determined in plane strain. Because $K_I = \alpha\sigma\sqrt{a}$ and the fracture strength S is the critical value of the applied stress σ, the fracture criterion can also be written as

▶ $$K_{Ic} = \alpha S\sqrt{a} \qquad (21.20)$$

Using the stresses and displacements as described above, the work to reach a certain stress intensity K_I can be calculated. For details, see Justification 21.2. However, the same result can be obtained from the Griffith analysis for a plane strain situation

$$G = -\frac{1}{t}\frac{\partial F}{\partial a} = \frac{\pi\sigma^2 a}{E'} = \frac{\left(\sigma\sqrt{\pi a}\right)^2}{E'} = \frac{K_I^2}{E'} \qquad (21.21)$$

where $E' = E/(1-v^2)$ for plane strain and $E' = E$ for plane stress. For other geometries the factor $\pi^{1/2}$ becomes the shape factor α. The above relationship is also valid at the moment the crack starts to grow resulting in

▶ $$G_{cri} = 2R = \frac{K_{Ic}^2}{E'} \qquad (21.22)$$

In linear elastic fracture mechanics (LEFM) there is thus a one-to-one relationship between the fracture energy R and the critical stress intensity factor K_{Ic}. Consequently the strength S can be expressed in terms of K_{Ic} or in terms of R and E and the expression reads

▶ $$S = \frac{1}{\alpha}\frac{K_{Ic}}{\sqrt{a}} = \frac{1}{\alpha}\frac{\sqrt{2E'R}}{\sqrt{a}} \qquad (21.23)$$

Eq. (21.21) provides a particular useful way to estimate the K_I-a relationship. From

$$K_I = \left[E'G(a)\right]^{1/2} \qquad (21.24)$$

the stress intensity factor can be calculated if the energy release rate is known. Since G can be calculated relatively easy as a function of crack length a using simple or sophisticated models and the accuracy can be improved systematically, this procedure provides a straightforward route for calculating and improving the expression for K_I.

Justification 21.2*

Consider an infinite plate with fixed ends containing a crack of size a and a cohesive zone of length δ and maximum allowed stress equal to the yield strength Y. The forces that are applied to the crack edge do work that is released as energy upon releasing the forces. It follows that

$$G = \lim_{\delta\to 0}\frac{2}{\delta}\int_0^\delta \frac{Yv}{2}\,dr$$

---

[1] In this respect the situation is quite comparable to the situation in plasticity where yielding occurs whenever the applied stress σ, a field quantity, equals the yield strength Y, a material property.

where the factor 2 is required because of the upper and lower edges and the factor ½ is introduced because the stresses increase from zero. The displacement v is, for the origin at the centre of the crack, given by

$$v = \frac{2\sigma}{E'}\sqrt{a^2 - x^2} = \frac{2K_1}{E'}\sqrt{(a - x^2/a)/\pi}$$

Since $x = r + a - \delta$ and neglecting second-order terms we can obtain

$$v = \frac{2K_1}{E'\sqrt{\pi}}\sqrt{2\delta - 2r + \frac{2r\delta}{a} - \frac{r^2}{a}} \cong \frac{2K_1}{E'\sqrt{\pi}}\sqrt{2(\delta - r)}$$

Because we also know that $Y = K_1/(2\pi r)^{1/2}$, we obtain after substituting,

$$G = \lim_{\delta \to 0} \frac{2K_1^2}{\pi E'\delta}\int_0^\delta \sqrt{\frac{1 - r/\delta}{r/\delta}}\, dr$$

which gives, after solving via $r/\delta = \sin^2\varphi$, $G = K_1^2/E'$, with $E' = E$ for plane stress and $E' = E/(1 - v^2)$ for plane strain[m].

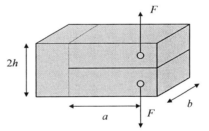

Fig. 21.9: Schematic of a DCB specimen.

Problem 21.3

Frequently the critical stress intensity factor or fracture toughness is measured using the double cantilever beam (DCB) test, as illustrated in Fig. 21.9. Neglecting the effect of the material before the load points, one can describe each beam by a cantilever with length a, clamped at the dotted line at the end of the crack. Using either the elastic energy per beam $F_{ela} = F^2a^3/6EI$ or the displacement per beam $u = Fa^3/3EI$, show that the stress intensity factor K_I is given by

$$K_I = Fa(bI)^{-1/2} = 2\sqrt{3}Fab^{-1}h^{-3/2}$$

where the moment of inertia is indicated by $I = bh^3/12$. The above expression for K_I for the DCB specimen is not particularly accurate. In particular, the estimate of the strain energy is based on the simplified modelling of Section 8.2. Moreover, the boundary condition 'clamped' is also not accurate. At the dotted line in Fig. 21.9 the structure is more likely to behave as an elastic hinge. However, this example clearly shows the procedure to be used.

[m] Eshelby, J.D (1968), ISI publ. **121**, 13.

The availability of α-factors for all kinds of crack geometries has contributed considerably towards the use of the stress intensity approach in the description for brittle fracture. In short, one proceeds as follows: determine for a certain crack geometry, e.g. in a test specimen designed in such a way that plane strain conditions prevail at the crack tip, the crack length and obtain the expression for the α-factor. From the critical stress at fracture σ_{cri} one obtains the value for the fracture toughness $K_{Ic} = \alpha \sigma_{cri}(a)^{1/2}$. Knowing the fracture toughness a prediction can be made for the critical stress at fracture (or strength) for any other geometry if the relevant α-factor for that structure is known. One catch is the prerequisite of brittle fracture. If yielding occurs but is strongly localised near the crack tip it is possible to correct for this and we will do so in the next section.

The stress intensity approach is the conventional one in fracture. Classical introductions are the books by Broek (1978) and Knott (1973). A more modern one is Miannay (1997). Kanninen and Popelar (1985) treat the topic on an advanced level.

21.5 Small scale yielding

Since the stress near the crack tip diverges one expects that yield always occurs in front of a crack. This is in fact true for most metals but not for inorganic materials. In the latter case the yield strength is generally quite high and the discrete nature of the solid cuts off the divergence of the crack tip stress (remember that real materials are not continua). Since the stress state ahead of the crack tip is dependent on the thickness of the plate (the thickness controls whether a plane stress state or plane strain state exists in the middle of the plate), not only the balance between the fracture toughness K_{Ic} on the one hand and yield strength Y on the other hand determines the failure behaviour but also the thickness. Assume a thin metal plate with relatively low yield strength so that the yield condition at the crack tip is reached before the fracture condition is reached. A zone around the crack tip starts to flow. The boundary of that zone is described by, say, the von Mises condition. Now consider a thick plate of the same material. Since in plane strain the third principal stress is not zero, the applied stress has to be increased much further before von Mises' criterion is met (Fig. 21.12). Consequently at the same stress the size of the plastically deformed zone is larger in plane stress than in plane strain. Since the energy spent in plastic deformation is proportional to the size of the zone, the specific fracture energy in plane stress will be higher. This implies that the strength of a thin plate with a certain defect size is higher than the strength of a thick plate because a larger fraction of the applied energy is needed for plastic work. It also implies that a certain thickness is required before the specific fracture energy is nearly fully determined by the plane strain state, that is before the specific fracture energy R (or fracture toughness K_{Ic}) as determined in an experiment can be considered as a material property. Generally, here and in the literature, whenever fracture energy is discussed, plane strain conditions are assumed unless stated otherwise.

Plastic zone and effective stress intensity

We limit ourselves initially to a plane stress situation, i.e. $\sigma_{III} = 0$, and assume elastic-perfect-plastic material behaviour for a wide plate (width W) with a central crack ($\alpha = \sqrt{\pi}$) of length $2a$ (Fig. 21.8). The shape and size of the plastic zone can be calculated from either the Tresca or the von Mises criterion using the stress ahead of

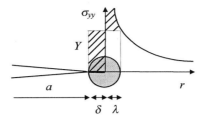

Fig. 21.10: The plastic zone ahead of a crack tip.

the crack tip as given by Eq. (21.16). Here we restrict ourselves to an estimate of the size. Straight ahead of the crack tip, i.e. along the line $\theta = 0$, and for $r = x-a$ the normal stress σ_{yy} is equal to σ_{II} and for an ideally brittle material we would have for $a < x < W/2$

$$\sigma_{yy}^{(ide)} = \sigma\sqrt{\frac{a}{2(x-a)}} + \tilde{\sigma}(x) \qquad (21.25)$$

The term $\tilde{\sigma}(x)$ was indicated in Eq. (21.16) by the dots because this term is very small in the vicinity of the crack. The total load P acting on the plate is

$$P = 2\sigma \int_0^{W/2}\sqrt{\frac{a}{2(x-a)}}\,dx + 2\int_0^{W/2}\tilde{\sigma}(x)\,dx \qquad (21.26)$$

Crack tip plasticity leads for a given load to larger displacements, that is to lower stiffness and Irwin[n], therefore, suggested that plasticity makes the plate behave as if the crack is longer than its actual physical size. So, one works with $a_{eff} = a+\delta$ as effective length[o] (Fig. 21.10). Moreover, within the plastic zone, i.e. the region $a < x < a+\delta+\lambda$, the magnitude of σ_{yy} is limited to the value of the yield strength Y. Therefore, it is assumed that actually the stress σ_{yy} is given by

$$\sigma_{yy} = Y \qquad\qquad\qquad a < x < a+\delta+\lambda$$
$$\sigma_{yy} = \sigma\sqrt{\frac{a_{eff}}{2(x-a_{eff})}} + \tilde{\sigma}(x) \qquad x > a+\delta+\lambda \qquad (21.27)$$

Note that the stress σ_{yy} only differs significantly from $\sigma_{yy}^{(ide)}$ in the neighbourhood of the crack tip. To determine the length parameters δ and λ one requires first that σ_{yy} be continuous across $x = a+\delta+\lambda$. Since $\tilde{\sigma} \cong 0$ at $x = a+\delta+\lambda$ we find

$$Y = \sigma\sqrt{\frac{(a+\delta)}{2\lambda}} \cong \sigma\sqrt{\frac{a}{2\lambda}} \qquad \text{or} \qquad \lambda = \frac{\sigma^2 a}{2Y^2} \qquad (21.28)$$

provided $\delta \ll a$. The next condition needed to determine δ and λ is that the total load on the system remains P. From

---

[n] Irwin, G.R. (1960), Proc. Sagamore Ordnance Materials Conf., Syracuse University, page IV-60, or (1958), Handbuch der Physik, vol. 6, Springer, Berlin, page 551.

[o] In the literature δ is often indicated by r_p and addressed as the *plastic zone radius*, the reason for which will become clear later.

$$P = 2\int_0^{W/2} \sigma_{yy}(x)\,\mathrm{d}x = 2\int_0^{W/2} \sigma_{yy}^{(\text{ide})}(x)\,\mathrm{d}x \tag{21.29}$$

one then finds that $\lambda \cong \delta$. Therefore, the plastic zone size $\delta + \lambda$ for a plane stress situation is

$$\delta + \lambda = 2\lambda = \frac{1}{\pi}\left(\frac{K_I}{Y}\right)^2 = \frac{\sigma^2 a}{Y^2} \tag{21.30}$$

where $\sigma = K_I/(\pi a)^{1/2}$ is the externally applied stress.

Because the plastic zone effectively increases the apparent crack length also the stress intensity factor is effectively increased. Using the relation

$$K_{\text{Ieff}} = \alpha\sigma\sqrt{a_{\text{eff}}} = \alpha\sigma\sqrt{a+\delta} \qquad \text{with} \quad \delta = \lambda = \frac{K_I^2}{2\pi Y^2} \tag{21.31}$$

and solving[p] for K_{Ieff} then leads to

▶ $$K_{\text{Ieff}} = \alpha\sigma\sqrt{a}\left(1+\frac{\alpha^2\sigma^2}{2\pi Y^2}\right)^{1/2} = K_I\left(1+\frac{\alpha^2\sigma^2}{2\pi Y^2}\right)^{1/2} \tag{21.32}$$

The effective stress intensity factor for a crack with a plastic zone is thus larger as the original stress intensity factor by a factor dependent on the ratio of applied stress and yield strength. The above approach fails of course if the correction δ becomes comparable to the physical crack length a. Another approach is then required which is outside the scope of this book. The strength S can also be estimated in this approximation. Applying Eq. (21.32) at failure and solving for S one obtains

$$\left(\frac{S}{Y}\right)^2 = \frac{\sqrt{1+\left(2\alpha^2/\pi\right)\left(S_a/Y\right)^2}-1}{\left(\alpha^2/\pi\right)}$$

where $S_a = K_{\text{Ieff}}/\alpha\sqrt{a} = S[1+(\alpha^2/2\pi)(S/Y)^2]^{1/2}$ is the strength the material would have with crack length a instead of $a+\delta$.

George Rankin Irwin (1907-1998)
Born in El Paso, Texas, he was educated at, first in English at Knox College in Galesburg and later in physics in which he obtained his Ph.D. at the University of Illinois, Urbana, in 1937.

---

[p] Using $\delta = K_{\text{Ieff}}^2/2\pi Y^2$ results in $K_{\text{Ieff}} = K_I(1-\alpha^2\sigma^2/2\pi Y^2)^{-1/2}$. This estimate is more in line with the spirit of self-consistency but not frequently used.

He joined the US Naval Research Laboratory, Washington DC, from which he retired as superintendent in 1967. During this period he researched penetration ballistics, combat damage to aircraft, the development of new armour materials and in particular the development of fracture mechanics. Thereafter he was full-time appointed at Lehigh University, Bethlehem, Pennsylvania, as Boeing professor of mechanics where he studied the fundamental and applied aspects of fracture mechanics relative to integration of this subject in engineering science at universities. During this period he co-operated with Paul Paris and founded the Journal of Engineering Fracture Mechanics. After his retirement from Lehigh in 1972 he was appointed as adjunct research professor and consultant at the University of Maryland. Irwin is recognized as the pioneer of modern fracture mechanics. He developed the scientific principles for understanding the relationships between applied stresses and cracks or other defects in metallic materials, starting around 1947 with the concept that fracture toughness should be measured in terms of resistance to crack propagation. Critical values of the stress intensity describing the onset of fracture, the onset of environmental cracking and the rate of fatigue crack growth were established later. As a consequence of Irwin's scientific work, fracture mechanics is now taught in many graduate schools and is an active field of R&D today. Irwin received many honours including the ASME Timoshenko medal (1986), was honorary member of numerous learned societies and published over 300 papers on fracture mechanics including its history.

Example 21.3

Consider a plate with a central crack of length $2a = 16$ mm, and width W, for which $W \gg 2a$ holds (Fig. 21.2). The yield strength is $Y = 1400$ MPa. The remotely applied stress is $\sigma = 350$ MPa. The plastic zone radius r_p becomes

$$r_p = \delta = \frac{1}{2\pi}\left(\frac{K_I}{Y}\right)^2 = \frac{1}{2\pi}\left(\frac{\sigma\sqrt{\pi a}}{Y}\right)^2 = \frac{1}{2\pi}\left(\frac{350(\pi 0.008)^{1/2}}{1400}\right)^2 \cong 0.25\,\text{mm}$$

The stress intensity factor K according to LEFM is $K = \sigma(\pi a)^{1/2} = 350(\pi 0.008)^{1/2} = 55.5$ MPa m$^{1/2}$. Due to the increase in effective length of the crack by the plastic zone, the effective stress intensity factor K_{Ieff} becomes

$$K_{\text{Ieff}} = K\left(1 + \frac{\alpha^2\sigma^2}{2\pi Y^2}\right)^{1/2} = 55.5\left(1 + \frac{1}{2}\left(\frac{350}{1400}\right)^2\right)^{1/2} = 56.4\,\text{MPa m}^{1/2}$$

In this case the difference is thus (almost) negligible. However, if the yield strength is much closer to the applied stress, say $Y = 385$ MPa, the difference cannot be neglected since r_p is now $r_p = \frac{1}{2\pi}\left(\frac{350(\pi\,0.008)^{1/2}}{385}\right)^2 \cong 3.3$ mm and

K_{Ieff} becomes $K_{\text{Ieff}} = 55.5\left[1 + \frac{1}{2}(350/385)^2\right]^{1/2} = 66.0$ MPa m$^{1/2}$. In this case the difference[q] is thus not negligible at all.

Problem 21.4

Calculate the plastic zone size for Al and Al$_2$O$_3$, using the data as given in Appendix B, and discuss the significance.

---

[q] Using the alternative expression (see footnote) yields $K_{\text{Ieff}} = K_I(1 - 350^2/2 \cdot 385^2)^{-1/2} = 72.4$ MPa m$^{1/2}$.

Plane stress versus plane strain and the transition

Let us now compare the size of the plastic zone for plane stress and plane strain. Substituting the stresses in the neighbourhood of the crack tip, as given by Eq. (21.16) and Eq. (21.17) in the expression for the plastic zone size as given by Eq. (21.30), and defining the plastic zone boundary via the von Mises criterion, results in an expression for the size as a function of polar angle θ, which reads

$$r_Y = \left[(1-2v)^2(1+\cos\theta) + \frac{3}{2}\sin^2\theta\right]\left(\frac{K_I}{Y}\right)^2 \frac{1}{4\pi}$$

$$= \frac{1}{4}\left[(1-2v)^2(1+\cos\theta) + \frac{3}{2}\sin^2\theta\right]r_p = f(\theta,v)r_p \qquad (21.33)$$

The plane stress situation can be described by taking $v = 0$ and $f(0,v) = 0.5$. The shape as determined by $f(\theta,v)$ is thus independent of Poisson's ratio. For plane strain on the other hand $v \neq 0$ and the result is dependent on the value for v. Taking a typical value $v = 1/3$ results in $f(0,1/3) = 1/18$, a factor of 9 different from the plane stress situation. On average the difference is much smaller. Presently it is now generally accepted that

▶ r_p(plane strain) $= \frac{1}{3}r_p$(plane stress)

Justification 21.3*

The plane strain plastic zone size is smaller than the plane stress size due to the fact that the effective yield strength in plane strain is larger than the uniaxial yield strength Y. A quantitative way is to express this is the plastic constraint factor p, defined by $p = \sigma_{max}/Y$, where σ_{max} is the maximum stress. Taking $\sigma_{II} = n\sigma_I$ and $\sigma_{III} = m\sigma_I$, the von Mises yield criterion can be written as

$$\left[(1-n)^2 + (n-m)^2 + (1-m)^2\right]\sigma_I^2 = 2Y^2 \quad \text{or equivalently} \qquad (21.34)$$

$$p = \frac{\sigma_I}{Y} = \left(1 - n - m + n^2 + m^2 - mn\right)^{-1/2} \qquad (21.35)$$

Using the principal stress expressions we have

$$n = (1-\sin \tfrac{1}{2}\theta)/(1+\sin \tfrac{1}{2}\theta) \quad \text{and} \quad m = 2v/(1+\sin \tfrac{1}{2}\theta)$$

For $\theta = 0$, $n = 1$ and $m = 2v$. For plane stress, $p = 1$. Similarly, for the typical value of Poisson's ratio of $v = 1/3$, for plane strain, $p = 3$. Since at the free surface of the crack tip a plane stress state exists, it follows that σ_{II} must be zero. Consequently even for a global plane strain state of the specimen at the crack tip itself $p = 1$. On average p is thus much smaller than 3. Irwin used $p = (2\sqrt{2})^{1/2} = 1.68$ and this implies

$$r_p = 2\pi^{-1}(K/pY)^2 \cong 6\pi^{-1}(K/Y)^2 \qquad (21.36)$$

The plastic zone is thus ~3 times smaller in plane strain than in plane stress.

Let us now discuss the transition from plane stress to plane strain with increasing thickness of the plate. Obviously, at the outer surface of a plate plane stress conditions

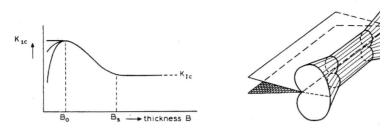

Fig. 21.11: Fracture resistance as a function of plate thickness and shape of the plastic zone over the plate thickness.

are present. If the plate is thin enough, the stress state will be approximately plane stress all over the thickness. With increasing thickness a transition to plane strain in the inner part of the plate takes place. However, a plane stress state remains at the crack surface. The shape of the plastic zone over the plate thickness is illustrated in Fig. 21.11. As long as the stress state is predominantly plane stress, the fracture energy increases with plate thickness due to the increasing energy dissipation in the plastic zone. At a certain thickness plane strain starts to dominate. Since the plastic zone size is much smaller for the plane strain state, also the specific energy dissipation for plane strain is much smaller. Hence with increasing thickness first the fracture energy increases and subsequently decreases and levels of to a constant value, as shown in the graph in Fig. 21.11. The graph also demonstrates that a measured value of K_{Ic} *can only be considered as a material property if the thickness of the plate is sufficiently large.*

Under what conditions then can the plate be considered to be fully in plane strain? Important are the thickness t and the crack parameter a. Because $(K_{Ic}/Y)^2$ defines an inherent length scale of the material, the critical thickness t must be a function of this length scale and the simplest way to obtain this is to take

$$t = C_t(K_{Ic}/Y)^2$$

with C_t some constant. A similar argument holds for the crack size a and leads to

$$a = C_a(K_{Ic}/Y)^2$$

Experiments show that both C_t and C_a should be ≥ 2.5. If the plane strain plastic zone radius is taken as $r_p = (K_{Ic}/Y)^2/6\pi$, it follows that for plane strain conditions to prevail

$$a, t \geq 50\, r_p$$

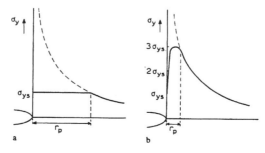

Fig. 21.12: Stress straight ahead of the crack tip for plane stress and plane strain. For r_p read $2r_p$.

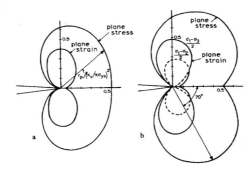

Fig. 21.13: The shape of the plastic zone according to the von Mises (a) and Tresca (b) yield criteria.

Since the stress field ahead of the crack tip is still supposed to be described by the elastic equations, a crack size larger than 50 r_p is not unreasonable. For many metals $C_t = C_a = 2.5$ is valid but there are exceptions and for tough polymers C_t and C_a may be much larger. For example, for high impact polystyrene plane strain is reached not before $C_t \cong 5.0$. The stress straight ahead of the crack is sketched in Fig. 21.12.

*Plastic zone shape**

Substituting the principal stresses as given by Eqs. (21.16) and (21.17) in either the Tresca criterion or the von Mises criterion the shape of the plastic zone can be calculated for the elastic-perfect-plastic material. We show without derivation only the results in Fig. 21.13. Although the extent of the plastic zone straight ahead of the crack is the same for both criteria, the shape is rather different. In fact both criteria do not describe the shape accurately. This is due to two effects: neglect of hardening and the neglect of the extra load bearing capacity required (similar to as described before). Incorporation of the latter is not straightforward in the general case. However, the rather different overall size for plane stress and plane strain is evident.

Problem 21.5

Show that the extent of the plastic zone as a function of θ for a material obeying von Mises' criterion is given by

$$r_Y(\theta, v) = \frac{1}{4\pi}\left(\frac{K}{Y}\right)^2\left[(1-2v)^2(1+\cos\theta) + \frac{3}{2}\sin^2\theta\right] = f(\theta, v)r_p$$

where v is the actual value of Poisson's ratio for plane strain and $v = 0$ for plane stress. Plot the result.

Problem 21.6

Show that the extent of the plastic zone as a function of θ for a material obeying Tresca's criterion is given by

$$r_Y(\theta, v) = \frac{1}{2\pi}\left(\frac{K}{Y}\right)^2\left[\cos\frac{\theta}{2}\left(1+\sin\frac{\theta}{2}\right)\right]^2 \qquad \text{for plane stress and}$$

$$r_Y(\theta,v) = \max\left\{\frac{1}{2\pi}\left(\frac{K}{Y}\right)^2\left[\cos\frac{\theta}{2}\left(1-2v+\sin\frac{\theta}{2}\right)\right]^2, \frac{1}{2\pi}\left(\frac{K}{Y}\right)^2\cos^2\frac{\theta}{2}\right\}$$

for plane strain where $\max(a,b)$ denotes the larger of a and b. Plot the result.

21.6   Alternative crack tip plastic zone ideas*

In this section we briefly discuss an alternative approach to the crack tip plastic zone, as presented by Dugdale[r]. In this approach also an effective crack, which is larger than the physical crack, is used (Fig. 21.14). A zone of length ρ in front of the crack is supposed to annihilate the stress singularity by carrying a stress equal to the yield strength Y, which tends to close the crack, i.e. the applied stress intensity factor K_σ due to the stress σ is compensated by the stress intensity K_ρ due to the zone stress or $K_\sigma = -K_\rho$. This allows us to determine ρ. The stress intensity due to a wedge force p is given by

$$K_A = \frac{p}{\sqrt{\pi a}}\sqrt{\frac{a+x}{a-x}} \qquad \text{and} \qquad K_B = \frac{p}{\sqrt{\pi a}}\sqrt{\frac{a-x}{a+x}} \tag{21.37}$$

In the case the wedge stress is distributed from a position s to a position a we have to integrate and the stress intensity becomes

$$K = \frac{p}{\sqrt{\pi a}}\int_s^a\left(\sqrt{\frac{a+x}{a-x}} + \sqrt{\frac{a-x}{a+x}}\right)dx = 2p\sqrt{\frac{a}{\pi}}\cos^{-1}\frac{s}{a} \tag{21.38}$$

For the Dugdale crack the integral has to be taken from $s = a$ to $a+\rho$ so that in the previous equation a has to be substituted for s and $a+\rho$ for a, meanwhile taking $\sigma = Y$. This results in

$$K_\rho = 2Y\sqrt{\frac{a+\rho}{\pi}}\cos^{-1}\frac{a}{a+\rho} \tag{21.39}$$

Equating this result to $K_\sigma = \sigma[\pi(a+\rho)]^{1/2}$ and solving for ρ yields

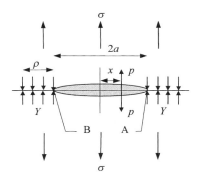

Fig. 21.14: The configuration of a Dugdale crack.

[r] Dugdale, D.S. (1960), J. Mech. Phys. Sol. **8**, 100.

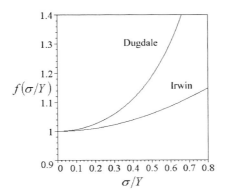

Fig. 21.15: A comparison between the Dugdale and Irwin plastic zone size correction crack using $K_{\text{Ieff}} = K_I f(\sigma/Y)$.

$$\rho = a\frac{1-C}{C} \quad \text{with} \quad C = \cos\frac{\pi\sigma}{2Y} \quad \text{or} \quad \rho \cong \frac{\pi^2\sigma^2 a}{8Y^2} = \frac{\pi K_I^2}{8Y^2} \qquad (21.40)$$

where in the second step $\cos x \cong 1 - x^2/2 + \cdots$ is used, neglecting higher order terms. This result can be compared with the Irwin result $2r_p = K^2/\pi Y^2$.

Duffy[s] proposed to use ρ as the plastic zone radius. In that case

$$a+\rho = a/\cos(\pi\sigma/2Y) = a\,\sec(\pi\sigma/2Y) \quad \text{and} \quad K_1 = \sigma[\pi a\,\sec(\pi\sigma/2Y)]^{1/2} \quad (21.41)$$

Fig. 21.15 compares this result with Eq. (21.32). We see that for $\sigma/Y > 0.2$ the difference is significant. Without derivation[t] we also quote the displacement at the end of the real crack $x = \pm a$ for the Dugdale model

$$\delta = \frac{8Y}{\pi E}a\ln\left[\sec\frac{\pi\sigma}{2Y}\right] \quad \left(\delta \cong \frac{\sigma^2\pi a}{YE} \quad \text{for} \quad \frac{\sigma}{Y} \ll 1\right) \qquad (21.42)$$

A model very similar to the Dugdale model is the Barenblatt model[u]. Instead of considering a macroscopic stress over the cohesive zone, he considered the atomic forces over that zone. He also assumed that the stress intensity from the cohesive zone cancels with the stress intensity due to the applied stress. It appears that the mathematical result is exactly the same as for the Dugdale model. Thus, although the mathematical outcomes are identical, the basic reasoning is rather different. Nevertheless, the two models are often mentioned in one stroke and referred to as the *Dugdale-Barenblatt* model.

Still another alternative method for calculating the size of the plastic zone (in mode III) and the associated displacements is the Bilby-Cottrell-Swinden (CBS) model. We refer to the literature[v] for details.

[s] Duffy, A.R. et al. (1969), page 159 in *Fracture*, vol. 1, Liebowitz, H., ed., Academic Press, New York.
[t] Burdekin, F.M. and Stone, D.E.W. (1966), J. Strain Anal. **1**, 145.
[u] Barenblatt, G.I. (1962), Adv. Appl. Mech. 7, 55.
[v] Bilby, B.A., Cottrell, A.H. and Swinden, K.H. (1963), Proc. Roy. Soc. **A272**, 304.

21.7   The J-integral*

The energy release rate approach is valid for linear elastic materials. For plastic materials obviously some generalisation is required. Some of the relevant ideas are due to Eshelby[w], Rice[x] and Cherepanov[y] and lead to the J-integral formulation. To that purpose they considered the two-dimensional deformation of a non-linear elastic but homogeneous body[z]. Neglecting body forces the potential energy is given by

$$\Pi(a) = \int W \, dA - \int_{\Gamma_t} t_i u_i \, dS \tag{21.43}$$

where A is the area of the body and Γ_t the contour on which the tractions t_i, which are supposed to be independent of the crack length a, are prescribed. The crack surfaces are supposed to be traction free. Differentiation yields

$$\frac{d\Pi(a)}{da} = \int \frac{\partial W}{\partial a} \, dA - \int_{\Gamma_0} t_i \frac{\partial u_i}{\partial a} \, dS \tag{21.44}$$

The contour for the second integral can be extended to the boundary of the body (counterclockwise from the lower crack face to the upper one) since $du_i/da = 0$ on Γ_u where the displacements are prescribed (Remember that $\Gamma_0 = \Gamma_t + \Gamma_u$). We introduce a co-ordinate system $X_i = x_i - a\delta_{i1}$ (with δ_{ij} the delta function) attached to the crack tip that moves along with the crack tip in x-direction. It follows that

$$\frac{d}{da} = \frac{\partial}{\partial a} + \frac{\partial X_1}{\partial a} \frac{\partial}{\partial X_1} = \frac{\partial}{\partial a} - \frac{\partial}{\partial X_1} = \frac{\partial}{\partial a} - \frac{\partial}{\partial x_1}$$

since $\partial X_1/\partial a = -1$ and $\partial/\partial X_1 = \partial/\partial x_1$. Hence we have

$$\frac{d\Pi(a)}{da} = \int \left(\frac{dW}{da} - \frac{dW}{dx_1} \right) dA - \int_{\Gamma_0} t_i \left(\frac{du_i}{da} - \frac{du_i}{dx_1} \right) dS \tag{21.45}$$

Further we have

$$\frac{\partial W}{\partial a} = \frac{\partial W}{\partial \varepsilon_{ij}} \frac{\partial \varepsilon_{ij}}{\partial a} = \sigma_{ij} \left(\frac{\partial u_i}{\partial a} \right)_{,j}$$

Because $\partial u_i/\partial a$ is kinematically admissible, the principal of virtual work allows us to write

$$\int_A \frac{\partial W}{\partial a} \, dA = \int_A \sigma_{ij} \left(\frac{\partial u_i}{\partial a} \right)_{,j} dA = \int_{\Gamma_0} t_i \frac{\partial u_i}{\partial a} \, dS \tag{21.46}$$

so that

$$-\frac{d\Pi}{da} = \int_A \frac{\partial W}{\partial x_1} \, dA - \int_{\Gamma_0} t_i \frac{\partial u_i}{\partial x_1} \, dS \tag{21.47}$$

---

[w] Eshelby, J.D. (1956), page 79 in *Solid state physics*, Seitz, F. and Turnbull, D., eds., Academic Press, New York.

[x] Rice, J.R. (1968), page 191 in *Fracture*, Liebowitz, H., ed., Academic Press, New York.

[y] Cherepanov, G.P. (1968), Int. J. Sol. Struct. **4**, 811.

[z] This implies that, if one wants to apply the results to plasticity, strictly speaking one limits oneself to proportional loading.

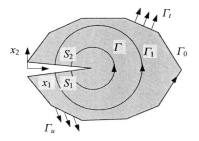

Fig. 21.16: Various *J*-integral contours.

Applying the divergence theorem this becomes

$$-\frac{\mathrm{d}\Pi}{\mathrm{d}a} = \int_{\Gamma_0}\left(Wn_1 - t_i\frac{\partial u_i}{\partial x_1}\right)\mathrm{d}S \tag{21.48}$$

with n_1 the normal in the *x*-direction.

Rice next considered an arbitrary contour Γ and defined

$$J = \int_{\Gamma}\left(Wn_1 - t_i\frac{\partial u_i}{\partial x_1}\right)\mathrm{d}S \tag{21.49}$$

This so-called *J-integral* can be shown to be path independent, given the conditions already stated, namely that body forces are absent and the crack surfaces are traction-free. If J_1 denotes the value of *J*-integral for another contour we can write

$$J_1 - J = \int_{X}\left(Wn_1 - t_i\frac{\partial u_i}{\partial x_1}\right)\mathrm{d}S \tag{21.50}$$

where the contour $X = \Gamma + \Gamma_1 + S_1 + S_2$ is closed by including (part of) the crack surfaces S_1 and S_2 on which $t_i = n_1 = 0$ (Fig. 21.16). Using the divergence theorem once more

$$J_1 - J = \int_{A}\left[\frac{\partial W}{\partial x_1} - \frac{\partial}{\partial x_j}\left(\sigma_{ij}\frac{\partial u_i}{\partial x_1}\right)\right]\mathrm{d}A = \int_{A}\left[\frac{\partial W}{\partial \varepsilon_{ij}}\frac{\partial \varepsilon_{ij}}{\partial x_1} - \sigma_{ij}\frac{\partial}{\partial x_1}\left(u_{i,j}\right)\right]\mathrm{d}A$$

$$= \int_{A}\left[\sigma_{ij}\frac{\partial \varepsilon_{ij}}{\partial x_1} - \sigma_{ij}\frac{\partial}{\partial x_1}\left(u_{i,j}\right)\right]\mathrm{d}A = 0 \tag{21.51}$$

Hence $J_1 = J$ and *J*-integral is independent of the path leading to

$$-\frac{\mathrm{d}\Pi}{\mathrm{d}a} = J \tag{21.52}$$

Therefore for a linear elastic material *J* and *G* are identical and we have

$$J = G = \frac{K_I^2}{E'} \tag{21.53}$$

For a link to yielding we consider again the Dugdale crack with a cohesive zone with stress σ, generally dependent on the separation $u = u^+ - u^-$ of the upper and lower prospective crack surfaces. For a contour around the cohesive zone $n_1 = t_1 = 0$ and therefore

$$J = -\int_0^\delta \left[\sigma \frac{\partial u^+}{\partial x_1} - \sigma \frac{\partial u^-}{\partial x_1} \right] dx_1 = \int_0^\delta \left[\sigma \frac{\partial u}{\partial x_1} \right] dx_1 = \int_0^\delta \sigma(u)\, du \qquad (21.54)$$

where δ is the crack tip displacement. In principle this is a general result, valid for any crack tip zone with cohesive forces. For a cohesive zone with constant stress equal to the yield strength Y, the result is

$$J = Y\delta \qquad (21.55)$$

For a small cohesive zone the deformation field deviates only marginally from the elastic field and therefore for a remote contour $J = G = K_I^2/E'$ while for the contour just enclosing the cohesive zone $J = Y\delta$. In view of the path independence the overall result for small scale yielding is

$$J = G = \frac{K_I^2}{E'} = Y\delta \qquad (21.56)$$

Noting that $K_I = \sigma(\pi a)^{1/2}$ and that for small scale yielding (SSY) $J_{\text{ssc}} = K_{Ic}^2/E' = S_a^2 \pi a/E'$ we obtain for the strength S using Eq. (21.42)

$$S = Y \frac{2}{\pi} \cos^{-1}\left[\exp\left(- \pi K_{Ic}^2 / 8aY^2\right) \right] = Y \frac{2}{\pi} \cos^{-1}\left[\exp\left(- \pi^2 S_a^2 / 8Y^2\right) \right] \qquad (21.57)$$

where S_a is given by $S_a = K_{Ic}/(\pi a)^{1/2}$.

To conclude, the J-integral, the energy release rate G, the stress intensity factor K_I and the crack tip opening displacement δ are all equivalent fracture parameters for small-scale yielding. For full-scale yielding, also the J-integral is applied. We refer to the literature (e.g. Kanninen and Popelar, 1985) for details. Finally, we note that, although the singularity at the crack tip is removed by process zone models as discussed above, a certain arbitrariness about the precise position of the crack tip is introduced.

John Douglas 'Jock' Eshelby (1916-1981)
Born in Puddington, Cheshire, he missed because of ill health his formal schooling from the age 13 and lived at the family home in north Somerset, where he learned instead from tutors. Through a contact with Sir Nevill Mott, he went early to the University of Bristol and obtained a first in physics there in 1937. During the Second World War he served first at the Admiralty and then in the Royal Air Force, where he reached the rank of squadron leader. He returned to Bristol in 1946 at a time when rapid advances were made in the theory of the deformation of crystals. Here he made his initial mark in dislocation theory. He obtained his Ph.D. in 1950 and two years later spent a year at the University of Illinois. Then followed some 10 years at

the University of Birmingham, a period in 1963 as visiting professor at the Technische Hochschule, Stuttgart, and then two years at Cambridge, where he became a Fellow and College Lecturer at Churchill College. In 1966 he went to the University of Sheffield, holding a readership and, from 1971, a personal chair in the theory of materials. His general field was the theoretical physics of the deformation, strength and fracture of engineering materials, and his principal interests were lattice defects and continuum mechanics. Though motivated by the desire to understand he kept a firm eye on application and had no time for useless erudition; like Gibbs his object was to make things appear simple by "looking at them in the right way". With a keen discrimination he selected those worthwhile difficult problems, which nevertheless had some chance of solution. He regarded himself as a modest "supplier of tools for the trade". His colleagues everywhere were always consulting him. Eshelby was elected a Fellow of the Royal Society in 1974, being "distinguished for his theoretical studies of the micromechanics of crystalline imperfections and material inhomogeneities". He made major contributions to the theory of static and moving dislocations and of point defects. By an elegant use of the theory of the potential he obtained some remarkable results on the elastic fields of ellipsoidal inclusions and inhomogeneities.

21.8 Fracture in anisotropic materials*

So far we have addressed only fracture in isotropic materials. Since usually engineering materials are polycrystalline, the influence of preferential orientation largely averages out so that the isotropic treatment is adequate. However, for fracture studies of single crystals an anisotropic treatment is necessary. Moreover also for laminates and highly oriented polymers such an anisotropic treatment required. We discuss only a few essentials.

Sih et al.[aa] have given a rather complete solution to the problems of displacement, stresses and stress intensity factors. Also Cherepanov (1979) deals with the matter. In their formulation the elastic compliance constants s_{ij} are used, as defined by

$$\varepsilon_i = s_{ij}\sigma_j$$

where the standard abbreviations $\varepsilon_1 = \varepsilon_{xx}$, $\varepsilon_2 = \varepsilon_{yy}$, ..., $\varepsilon_6 = \varepsilon_{xy}$ are used. For plane stress the only relevant constants are given by $i, j = 1, 2, 6$ while for plane strain the compliance constants r_{ij} become

$$r_{ij} = s_{ij} - \frac{s_{i3}s_{j3}}{s_{33}} \qquad (i,j = 1, 2, 6)$$

Using the complex variable formulation of plane and anti-plane problems to cracks the authors described both for plane stress and plane strain the elastic displacement and stress fields near the crack tip. In this formulation a parameter μ is involved which is given by one of the roots of the equation

$$a_{11}\mu^4 - 2a_{16}\mu^3 + (2a_{12} + a_{66})\mu^2 - 2a_{26}\mu + a_{22} = 0 \qquad (21.58)$$

Because the roots of this equation are either complex or purely imaginary, the roots can be expressed in pairs $(\mu_1, \mu_1{}^*)$ and $(\mu_2, \mu_2{}^*)$ where the asterisk denotes the complex conjugate. Comparable to the case for isotropic materials the energy release rate can be expressed in terms of the stresses and displacements. However, a treatment of angled cracks is prohibitively complicated and therefore one assumes normally planar crack growth. The expression for the energy release rate for mode I becomes

---

[aa] Sih, G.C., Paris, P.C. and Irwin, G.R. (1965), J. Fract. Mech. **1**, 189.

$$G = \lim_{\Delta a \to 0} \int_0^{\Delta a} \sigma_2(r,0) u_2 (\Delta a - r, \pi)\, dr$$

where the stress and displacement are given in terms of a polar co-ordinate system (r,θ) at the crack tip. The final result is

$$G = -\frac{K_I^2}{2\pi} a_{22} \; \mathrm{Im}\left(\frac{\mu_1 + \mu_2}{\mu_1 \mu_2} \right) \tag{21.59}$$

It appeared that in anisotropic bodies cracks are normally coupled, meaning that if two cracks are present with different orientation the result for energy release rate of the one crack is dependent on the one for the other crack. In orthotropic materials, however, $a_{16} = a_{26} = 0$ and there are only four independent constants which makes an analytical solution of Eq. (21.58) possible. In this case when the crack is in a plane of symmetry the three basic fracture modes are conveniently independent and the energy release rate for mode I cracks becomes

$$G = K_I^2 \sqrt{\frac{s_{11} s_{22}}{2}} \left[\left(\frac{s_{22}}{s_{11}} \right)^{1/2} + \frac{2 s_{12} + s_{66}}{2 s_{11}} \right]^{1/2} \tag{21.60}$$

with s_{ij} the elastic compliance constants. This expression reduces to the result for isotropic materials by substituting $a_{11} = a_{22}$ and $a_{66} = 2(a_{11}-a_{12})$. The authors emphasised that for self-equilibrating loads, the resulting stress intensity factors of anisotropic materials are identical to the one for isotropic materials and that the elastic anisotropy is only active through the elastic constants. This implies that the conventional formulation of fracture in terms of stress intensity factors remains valid for orthotropic materials. In particular the stress intensity factors as compiled in several handbooks for isotropic materials can also be used for orthotropic materials provided the proper elastic constants are used.

21.9   The thermodynamic approach*

In the previous sections we discussed the conventional (energy and stress) approach to fracture mechanics. It is, however, useful to consider a different approach, amongst other advocated by Rice[bb]. To that purpose we consider a crack as an internal variable (see Chapter 6), which is characterised by the crack surface A. Briefly reiterating from Chapter 6, we know that the dissipation function Φ is given by

$$\Phi = T\dot{S}^{(i)} = A_k^{(d)} \dot{a}_k \geq 0 \tag{21.61}$$

where T denotes the temperature, $\dot{S}^{(i)}$ the entropy production rate and $A_k^{(d)}$ the dissipative force associated with the kinematic variable a_k. The quasi-conservative force $A_k^{(q)}$ associated with a_k is given by $A_k^{(q)} = \partial F/\partial a_k$, where F denotes the Helmholtz energy. Since for internal variables it holds that the total force $A_k = A_k^{(q)} + A_k^{(d)} = 0$, we have

$$A_k^{(d)} = -A_k^{(q)} = -\frac{\partial F}{\partial a_k} \tag{21.62}$$

[bb] Rice, J.R. (1978), J. Mech. Phys. Solids **26**, 61.

For the case of a single crack the rate of the kinematic variable $\dot{a}_k = \dot{A}$, the rate of crack area extension, and $F = F_{def}+F_{cra}$. Restricting us to elastic solids F_{def} is given by Eq. (21.1), while F_{cra} is given by Eq. (21.2). Using the earlier introduced notation[cc] $G = -\partial F_{def}/\partial A$ and $2R = \partial F_{cra}/\partial A$ we obtain

$$\Phi = T\dot{S}^{(i)} = g\dot{A} \geq 0 \qquad \text{where} \quad g = G - 2R \qquad (21.63)$$

Since $T\dot{S}^{(i)} \geq 0$ we can have the following situations. If $g > 0$, $\dot{A} > 0$ and the crack extends. If $g < 0$, $\dot{A} < 0$ and the crack retards. In both cases the entropy increases. However, if the driving force $g = 0$, the crack growth rate $\dot{A}$ is not necessarily zero. The crack can grow or retard through a series of equilibrium states. Finally, if $\dot{A} = 0$, the Griffith criterion is regained.

Although discussions are often done in terms of the (internal) energy U where the dissipative force is given by $-\partial U/\partial a_k$ (at constant entropy S and kinematic variables a_j other than a_k) it is more appropriately done in terms of the Helmholtz energy where the dissipative force is equal to $-\partial F/\partial a_k$ (at constant temperature T and kinematic variables a_j other than a_k). It is clear that in normal laboratory conditions T and a_k are the independent variables and we will therefore favour the latter description.

Accepting this description we must now formulate a Helmholtz function, dependent on the conventional, external variables ε_{ij} and T as well on the internal variables a_k, i.e. $f = f(\varepsilon_{ij},T, a_k)$. The entropy, stress and crack driving force g are then given by respectively

$$s = -\frac{\partial f}{\partial T} \qquad \sigma_{ij} = \rho\frac{\partial f}{\partial \varepsilon_{ij}} \qquad A_k = -\frac{\partial f}{\partial a_k} \qquad (21.64)$$

In this connection it is convenient to use also the Gibbs energy, here denoted[dd] by h and given by the Legendre transform

$$h(\sigma_{ij}, T, a_k) = f - \sigma_{ij}\varepsilon_{ij} \qquad (21.65)$$

which leads to

$$s = \frac{\partial h}{\partial T} \qquad \varepsilon_{ij} = -\frac{\partial h}{\partial \sigma_{ij}} \qquad A_k = -\frac{\partial f}{\partial a_k} \qquad (21.66)$$

We also have to make a choice for the kinetic law for a_k. Here we simple use a limit function[ee] so that $\dot{a}_k = \lambda \,\partial\phi/\partial A_k$, where ϕ is a continuously differentiable single-valued limit function for which it holds that $\phi(A_k = 0) < 0$. Internal forces such that $\phi(A_k = 0) > 0$ are not obtainable and we have

$$\lambda \geq 0 \quad \text{if } \phi = 0 \text{ and } \dot{\phi} = 0 \qquad (21.67)$$

$$\lambda = 0 \quad \text{if } \phi < 0 \text{ or } \phi = 0 \text{ and } \dot{\phi} < 0$$

We can calculate the value of λ for any combination (A_k, ε_{ij}). Differentiating the driving force $A_k = -\partial f/\partial a_k$ we obtain

---

[cc] In the literature often $G_{cri} = R$ is used while we used in this chapter $G_{cri} = 2R$.
[dd] Normally one would use g but in the literature the crack driving force is also denoted by g. Therefore we use here h.
[ee] Of course, more complex kinetic laws can be used and this is done in Chapter 24.

$$\dot{A}_k = -\left(\frac{\partial^2 f}{\partial \varepsilon_{ij} \partial a_k} \dot{\varepsilon}_{ij} + \frac{\partial^2 f}{\partial a_k \partial a_l} \dot{a}_l\right) \qquad \text{with} \quad \dot{a}_k = \lambda \, \partial\phi/\partial A_k$$

and we are able to solve $(\partial\phi/\partial A_k)\dot{A}_k = 0$ to yield

$$\lambda = -\left(\frac{\partial^2 f}{\partial \varepsilon_{ij} \partial a_k} \frac{\partial \phi}{\partial A_k} \dot{\varepsilon}_{ij}\right)\left(\frac{\partial^2 f}{\partial a_k \partial a_l} \frac{\partial \phi}{\partial A_k} \frac{\partial \phi}{\partial A_l}\right)^{-1} \tag{21.68}$$

This expression gives the value for λ if $\phi = 0$ and for all other cases $\lambda = 0$.

Sufficient conditions for stability are that $f = f(\varepsilon_{ij}, T, a_k)$ and $\phi(A_k)$ are convex functions of their respective arguments. In this case this implies that a material element is stable if for the elastic response $\dot{\sigma}_{ij}\dot{\varepsilon}_{ij} > 0$ holds, which we assume is always satisfied. Moreover for inelastic response we have $\dot{\phi} = 0$ and $\lambda \geq 0$ for arbitrary $\dot{\varepsilon}_{ij}$. Global stability is insured if the matrix $\partial^2 f/\partial a_k \partial a_l$ (or $-\partial^2 h/\partial a_k \partial a_l$) is positive definite while local stability requires that

$$\left(\frac{\partial^2 f}{\partial a_k \partial a_l} \frac{\partial \phi}{\partial A_k} \frac{\partial \phi}{\partial A_l}\right) > 0 \qquad \left(\text{or} \quad \left(\frac{\partial^2 h}{\partial a_k \partial a_l} \frac{\partial \phi}{\partial A_k} \frac{\partial \phi}{\partial A_l}\right) < 0\right)$$

If these conditions cannot be satisfied, λ becomes unbounded and the material element is unstable, which can be identified as fast fracture.

Elastic fracture

To simplify the description as far as possible we consider[ff] an elastic material in the shape of a plate of width w loaded by a single tensile load p so that there is only one displacement u. The plate contains an edge crack of length a (equivalent to a central crack of length $2a$) perpendicular to the loading direction, which is small as compared with the width of the plate, i.e. $a \ll w$. We further consider the width of the crack constant so that we take $A = ta$ with t the (constant) thickness and a the crack length. We take further a unit thickness plate ($t = 1$). The Gibbs energy H then becomes

$$H = F - pu \tag{21.69}$$

with the equations of state (omitting the influence of temperature)

$$p = \frac{\partial F}{\partial u} \qquad g = -\frac{\partial F}{\partial a} \qquad u = -\frac{\partial H}{\partial p} \qquad g = -\frac{\partial H}{\partial a} \tag{21.70}$$

We consider for convenience a crack-free configuration first. For $a = 0$ we have the equation of state using E as the elastic modulus

$$p = \frac{\partial F}{\partial u} = Eu \tag{21.71}$$

and therefore a Helmholtz energy

$$F = \int_0^u p \, du = \frac{1}{2}Eu^2 \tag{21.72}$$

[ff] Carter, P. (1979), Eng. Fract. Mech. **11**, 441.

Expressing F as a function of p and using $H = F - pu$ one obtains for the Gibbs energy

$$H = F - pu = \frac{p^2}{2E} - p\frac{p}{E} = -\frac{p^2}{2E} \qquad (21.73)$$

Let us now consider the crack problem. The expression for the deformation part (as derived for the non-cracked solid) of the Helmholtz energy F_{def} and Gibbs energy H_{def} will change since F_{def} and H_{def} becomes a function of a. The potential energy Π is

$$\Pi = F_{def}(u,a) - pu \qquad (21.74)$$

From the elastic analysis of a cracked material we know that the energy release rate is

$$G = -\frac{d\Pi}{da} = \frac{\pi\sigma^2 a}{E} \qquad (21.75)$$

where $\sigma = p/w$. From a comparison of H_{def} and potential energy Π we see that $H_{def} = \Pi$. To obtain the total Gibbs energy H we have to add^{gg} to the deformation part H_{def} the fracture energy $H_{fra} = 2Ra$. Therefore,

$$g = -\frac{\partial H}{\partial a} = -\frac{\partial H_{def}}{\partial a} - \frac{\partial H_{fra}}{\partial a} = -\frac{\partial \Pi}{\partial a} - 2R = \frac{p^2 \pi a}{w^2 E} - 2R \equiv G - 2R \qquad (21.76)$$

with G the energy release rate for load control. Now we can integrate the equation of state $u = u(p,a)$ using the Maxwell condition $\partial u/\partial a = \partial g/\partial p$. Therefore, we have

$$du = \frac{\partial g}{\partial p} da = \frac{2p\pi a}{w^2 E} da \quad \text{or} \quad u = \frac{2p\pi}{w^2 E} \int a \, da + K_1 = \frac{p\pi a^2}{w^2 E} + K_1(p) \qquad (21.77)$$

From the boundary condition $u = p/E$ at $a = 0$ we obtain $K_1 = p/E$ and u becomes

$$u = \frac{p}{E}\left(1 + \frac{\pi a^2}{w^2}\right) \qquad (21.78)$$

From $u = -\partial H/\partial p$ we have $dH = -u \, dp$ or

$$H = \int u \, dp = -\frac{p^2}{2E}\left(1 + \frac{\pi a^2}{w^2}\right) + K_2(a) \qquad (21.79)$$

From the boundary condition $H = -p^2/2E$ at $a = 0$ we obtain $K_2 = Ca$ with C a constant, which we identify as $C = 2R$ using $H = 2Ra$ at $p = 0$. The expression for H as a function of its natural variables p and a thus becomes

$$H = -\frac{p^2}{2E}\left(1 + \frac{\pi a^2}{w^2}\right) + 2Ra \qquad (21.80)$$

Since $F = H + pu$ we finally obtain for the Helmholtz energy F as a function of its natural variables u and a

$$F = \frac{p^2}{2E}\left(1 + \frac{\pi a^2}{w^2}\right) + 2Ra = \frac{Eu^2}{2}\left(1 + \frac{\pi a^2}{w^2}\right)^{-1} + 2Ra \qquad (21.81)$$

from which we derive the equations of state as

---

^{gg} Strictly speaking the value of R can differ for constant (u,a) and (p,a). We neglect this difference.

$$p = \frac{\partial F}{\partial u} = \frac{Eu}{1 + \pi a^2/w^2} \quad \text{and} \quad g = -\frac{\partial F}{\partial a} = \frac{\pi E u^2 a}{w^2 (1 + \pi a^2/w^2)^2} - 2R \equiv G - 2R \quad (21.82)$$

with again G the energy release rate, but now for displacement control.

To complete the description we need the kinetic law or evolution equation. As indicated the simplest approach is a limit function $\phi(g)$ for which holds that

$$\dot{a} \geq 0 \quad \text{if} \quad g = G - 2R = 0 \quad \text{and} \quad \dot{g} = 0 \qquad (21.83)$$

$$\dot{a} = 0 \quad \text{if} \quad g = G - 2R < 0 \quad \text{or} \quad g = G - 2R = 0 \quad \text{and} \quad \dot{g} < 0$$

In load control we have $\dot{p} > 0$ and using Eq. (21.76) the rate of the crack driving force g becomes

$$\dot{g} = \frac{\pi p^2 \dot{a}}{w^2 E} + \frac{2\pi p \dot{p} a}{w^2 E} \qquad (21.84)$$

Using the equilibrium condition $\dot{g} = 0$ we obtain

$$\dot{a} = -2a\dot{p}/p \qquad (21.85)$$

and if $g = 0$ and $\dot{p} > 0$ we cannot satisfy both $\dot{g} = 0$ and $\dot{a} \geq 0$. This is due to the unstable behaviour of H (Eq. (21.80)) which results in $\partial^2 H/\partial a^2 = \pi p^2/a^2 E > 0$. In load control the fracture behaviour is thus always unstable. Using $g = 0$ in Eq. (21.76) and $\sigma = p/w$ yields

$$\sigma \sqrt{\pi a} = \sqrt{2ER} \equiv K_{\text{Ic}} \qquad (21.86)$$

which is the familiar fracture equation.

In displacement control we have $\dot{u} > 0$ and from Eq. (21.82) we obtain

$$\dot{g} = \frac{2\pi E a u \dot{u}}{w^2 (1 + \pi a^2/w^2)^2} + \frac{\pi E u^2 (1 - 3\pi a^2/w^2) \dot{a}}{w^2 (1 + \pi a^2/w^2)^3} \qquad (21.87)$$

In this case $\dot{g} = 0$ results in

$$\dot{a} = \frac{2a(1 + \pi a^2/w^2) \dot{u}}{u(1 - 3\pi a^2/w^2)} \qquad (21.88)$$

and $\dot{a} \geq 0$ for $a/w > (3\pi)^{-1/2}$. The same result is obtained from the stability condition $\partial^2 F/\partial a^2 > 0$ since

$$\frac{\partial^2 F}{\partial a^2} = \frac{\pi E u^2 (3\pi a^2/w^2 - 1)}{w^2 (1 + \pi a^2/w^2)^3} \geq 0 \qquad (21.89)$$

In Fig. 21.17 a plot is given of $u/\sqrt{w}$ versus a/w at $g = 0$ using Eq. (21.82) leading to

$$\frac{u}{\sqrt{w}} = \sqrt{\frac{2R}{E}} \frac{1 + \pi a^2/w^2}{\sqrt{\pi a/w}} = \frac{K_{\text{Ic}}}{E} \frac{1 + \pi a^2/w^2}{\sqrt{\pi a/w}}$$

from which it can be seen that the crack is unstable for $a/w \leq (3\pi)^{-1/2}$ and stable for $a/w > (3\pi)^{-1/2}$. For a displacement $u = u_1$ therefore a crack with length $a = a_1$ will grow unstably to length $a = a_2$ after which crack arrest occurs. This phenomenon is sometimes denoted as *pop-in*. Here it follows as a consequence of linear elastic fracture mechanics under displacement control.

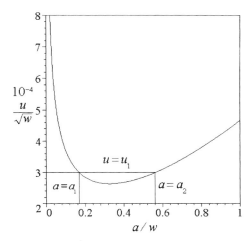

Fig. 21.17: Reduced displacement $u/\sqrt{w}$ versus reduced crack length a/w at the condition $g = 0$ for $K_{Ic}/E = 2\times10^{-4}$.

Elastic-plastic fracture

To introduce the complications due to plasticity we now assume that the material element behaves like an elastic-ideal-plastic material. Therefore, we have to introduce a displacement due to plasticity u_{pla} to the set of state variables. The fundamental equation thus becomes either $F = F(u,u_{pla},a)$ or $H = H(p,u_{pla},a)$. The equations of state become

$$p = \frac{\partial F}{\partial u} \quad q = -\frac{\partial F}{\partial u_{pla}} \quad g = -\frac{\partial F}{\partial a} \quad u = -\frac{\partial H}{\partial p} \quad q = -\frac{\partial H}{\partial u_{pla}} \quad g = -\frac{\partial H}{\partial a} \quad (21.90)$$

where q is the variable conjugated to u_{pla}. For elastic-ideal-plastic behaviour of a non-cracked element the equations of state for u and q become

$$u = \left.\frac{\partial H}{\partial p}\right|_{a=0} = \frac{p}{E} + u_{pla} \qquad q = -\left.\frac{\partial H}{\partial u_{pla}}\right|_{a=0} = p \qquad (21.91)$$

The yield surface is given by $q_0 = wY$ with Y the yield strength of the material and as before w the width of the element. The simplest extension of Eq. (21.78) for a cracked element that results in the above equation for u is

$$u = \frac{p}{E}\left(1 + \frac{\pi a^2}{w^2}\right) + u_{pla} \qquad (21.92)$$

If the other equations of state remain unchanged, i.e.

$$q(p,u_{pla},a) = p \qquad \text{and} \qquad g(p,u_{pla},a) = \frac{\pi a p^2}{w^2 E} - 2R \qquad (21.93)$$

they can be integrated to give the Gibbs energy

$$H = -\frac{p^2}{2E}\left(1 + \frac{\pi a^2}{w^2}\right) - p u_{pla} + 2Ra \qquad (21.94)$$

and by applying the Legendre transformation $F = H + pu$ this results in

$$F = \frac{E(u - u_{\text{pla}})^2}{2}\left(1 + \frac{\pi a^2}{w^2}\right)^{-1} + 2Ra \tag{21.95}$$

From this expression one obtains the equations of state

$$p = \frac{\partial F}{\partial u} = \frac{E(u - u_{\text{pla}})}{1 + \pi a^2/w^2} \qquad q = -\frac{\partial F}{\partial u_{\text{pla}}} = \frac{E(u - u_{\text{pla}})}{1 + \pi a^2/w^2} = p \qquad \text{and}$$

$$g = -\frac{\partial F}{\partial a} = \frac{\pi E(u - u_{\text{pla}})^2 a}{w^2(1 + \pi a^2/w^2)^2} - 2R \equiv G - 2R \tag{21.96}$$

where G is again the energy release rate.

We now have to consider the limit surface $\phi(q,g) = 0$, where $\phi(0,0) < 0$, for which tension is positioned in the first q-g quadrant. The kinetic equations are

$$\begin{bmatrix} \dot{u}_{\text{pla}} \\ \dot{a} \end{bmatrix} = \lambda \begin{bmatrix} \partial\phi/\partial q \\ \partial\phi/\partial g \end{bmatrix} \tag{21.97}$$

where $\lambda \geq 0$ if $\phi = 0$ and $\dot{\phi} = 0$, and $\lambda = 0$ otherwise. A single multiplier λ thus rules the kinetic behaviour in this description.

Let us now consider load control for which we have seen that the matrix of second derivatives of H with respect to the internal variables, denoted by d^2H, should be positive definite. For H as given by Eq. (21.94) this results in

$$d^2H = \begin{bmatrix} \partial^2 H/\partial u_{\text{pla}}^2 & \partial^2 H/\partial u_{\text{pla}}\partial a \\ \partial^2 H/\partial a\partial u_{\text{pla}} & \partial^2 H/\partial a^2 \end{bmatrix} \rightarrow \begin{bmatrix} 0 & 0 \\ 0 & \pi p^2/w^2 E \end{bmatrix}$$

which is positive semi-definite. Consequently, if $\phi = 0$ and $\dot{p} > 0$, the multiplier λ cannot be calculated and the element is unstable for all limit functions. Similarly for displacement control the matrix of second derivatives of F with respect to the internal variables, denoted by d^2F, should be positive definite which is true if

$$\partial^2 F/\partial u_{\text{pla}}^2 > 0 \qquad \partial^2 F/\partial a^2 > 0 \qquad (\partial^2 F/\partial u_{\text{pla}}^2)(\partial^2 F/\partial a^2) - (\partial^2 F/\partial a\partial u_{\text{pla}})^2 > 0$$

For F as given by Eq. (21.95) we have

$$\frac{\partial^2 F}{\partial u_{\text{pla}}^2} = \frac{E}{1 + \pi a^2/w^2} \qquad\qquad \frac{\partial^2 F}{\partial a\partial u_{\text{pla}}} = \frac{2\pi E a(u - u_{\text{pla}})}{w^2(1 + \pi a^2/w^2)^2}$$

$$\frac{\partial^2 F}{\partial a^2} = \frac{\pi E(u - u_{\text{pla}})^2(3\pi a^2/w^2 - 1)}{w^2(1 + \pi a^2/w^2)^3}$$

Therefore, we obtain

$$\frac{\partial^2 F}{\partial u_{\text{pla}}^2}\frac{\partial^2 F}{\partial a^2} - \left(\frac{\partial^2 F}{\partial a\partial u_{\text{pla}}}\right)^2 = -\frac{E^2(u - u_{\text{pla}})[\pi a^2/w^2(\pi + 1)]}{w^2(1 + \pi a^2/w^2)^4}$$

which means that d^2F is not positive definite. Hence stability for a general limit surface is not guaranteed.

Let us consider the simplest limit surface, a piecewise linear function given by

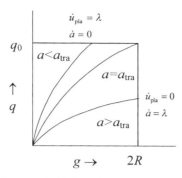

Fig. 21.18: Rectangular limit surface and various loading paths.

$$\phi_1 = q - q_0 = q - wY \qquad \text{and} \qquad \phi_2 = g = G - 2R \tag{21.98}$$

with $q \geq 0$, $g \geq 0$ and Y, G and R are the yield strength, energy release rate and fracture energy, respectively. For this limit surface a transition from pure plastic deformation to elastic fracture takes place for a crack length a_{tra} given by $a_{tra} = 2ER/\pi Y^2$. In Fig. 21.18 the loading path for three values of crack length a is indicated. We have for

- $a < a_{tra}$ failure by plastic flow with no crack growth,
- $a = a_{tra}$ failure by plastic flow and crack growth and
- $a > a_{tra}$ failure by crack growth with no plastic flow.

For $\dot{u} > 0$ we must consider the local stability condition

$$x = \begin{bmatrix} \dfrac{\partial \phi_\alpha}{\partial q} & \dfrac{\partial \phi_\alpha}{\partial a} \end{bmatrix} \begin{bmatrix} \dfrac{\partial^2 f}{\partial u_{pla}^2} & \dfrac{\partial^2 f}{\partial u_{pla} \partial a} \\ \dfrac{\partial^2 f}{\partial a \partial u_{pla}} & \dfrac{\partial^2 f}{\partial a^2} \end{bmatrix} \begin{bmatrix} \dfrac{\partial \phi_\alpha}{\partial q} \\ \dfrac{\partial \phi_\alpha}{\partial a} \end{bmatrix} \tag{21.99}$$

with ϕ_α given by Eq. (21.98) and determine whether the material is stable. For $a < a_{tra}$ failure occurs when $\phi_1 = 0$ and hence $\partial \phi_\alpha / \partial q = 1$ and $\partial \phi_\alpha / \partial g = 0$. For x we obtain

$$x = E/(1 + \pi a^2 / w^2)$$

Therefore, the conclusion is that when the limit surface is reached stable plastic deformation occurs but no crack growth. For $a > a_{tra}$ failure occurs when $\phi_2 = 0$ and therefore $\partial \phi_\alpha / \partial q = 0$ and $\partial \phi_\alpha / \partial g = 1$. Calculating again x the result is

$$x = \frac{\pi E (u - u_{pla})^2 (3\pi a^2 / w^2 - 1)}{w^2 (1 + \pi a^2 / w^2)^3}$$

Thus for these crack lengths stable crack growth occurs with no plastic deformation if $a/w > (3\pi)^{-1/2}$. If $a/w < (3\pi)^{-1/2}$ unstable crack growth occurs, followed by crack arrest, as described earlier. Summarising, unstable crack growth cannot occur unless

$$\frac{2ER}{\pi Y^2} < a < \frac{w}{\sqrt{3\pi}}$$

Stable crack growth is thus guaranteed if

$$2ER/\pi > w/(3\pi)^{1/2} \quad \text{or}$$

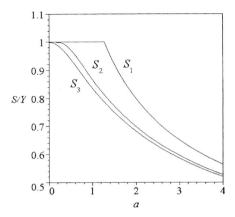

Fig. 21.19: Reduced failure stress versus crack length for the rectangular limit surface (S_1), the elliptical limit surface (S_2) and the Dugdale model (S_3), see Section 21.6.

$$(K_{\text{Ic}}/Y)^4 > \pi w^2/3$$

The failure stress behaviour of this model is sketched in Fig. 21.19.
    For a more realistic, e.g. an elliptic limit surface such as

$$\phi = \left(\frac{q}{q_0}\right)^2 + \left(\frac{g}{g_0}\right)^2 - 1 = 0 \tag{21.100}$$

with $q_0 = Y/w$ and $g_0 = K_{\text{Ic}}^2/2E$ we obtain using Eqs. (21.93) the failure stress S as

$$\pi^2 a^2 \left(\frac{S}{K_{\text{Ic}}}\right)^4 + \left(\frac{S}{Y}\right)^2 - 1 = 0 \tag{21.101}$$

and this failure stress is also indicated in Fig. 21.19. Differentiating Eq. (21.100) and using Eqs. (21.93) the kinetic equations for this model become

$$\begin{bmatrix} \dot{u}_{\text{pla}} \\ \dot{a} \end{bmatrix} = \lambda \begin{bmatrix} 2P/q_0^2 \\ 2\pi a P^2/Ew^2 g_0^2 \end{bmatrix} \tag{21.102}$$

Hence for all values of crack length a, failure occurs by a combination of flow and fracture. From a stability analysis very similar to the one before (of which we omit the details) it follows that under displacement control stable crack growth occurs when $(K_{\text{Ic}}/Y)^4 > \pi w^2/\sqrt{3}$. This stability condition thus deviates but marginally from the one derived for the rectangular limit surface.
    An explicit expression for the crack growth can be obtained as well using

$$\dot{a} = \lambda \frac{\partial \phi}{\partial g}$$

where $\lambda \geq 0$ if $\phi = 0$ and $\dot{\phi} = 0$ and $\lambda = 0$ otherwise. Solving $\dot{\phi} = 0$ as before yields

$$\dot{a} = -x^{-1} \left(\frac{\partial^2 f}{\partial u \partial u_{\text{pla}}} \frac{\partial \phi}{\partial q} + \frac{\partial^2 f}{\partial u \partial a} \frac{\partial \phi}{\partial g} \right) \frac{\partial \phi}{\partial g} \dot{u}$$

with x as defined in Eq. (21.99). This expression gives the crack growth rate if $\phi = 0$, $\dot{u} \geq 0$ and $x > 0$. If $\phi = 0$, $\dot{u} \geq 0$ and $x < 0$ crack growth is unstable. Otherwise $\dot{a} = 0$. Evaluating this expression we obtain

$$\frac{\dot{a}}{\dot{u}} = \frac{da}{du} = \frac{\pi a w \sigma E\left(1/Y^2 + 2\pi^2 a^2 \sigma^2/K_{\mathrm{Ic}}^4\right)}{\pi^2 a^2 \left[\left(\sigma/Y\right)^2 + 3\right] + \pi a^2 \left[\left(\sigma/Y\right)^2 - 1\right] + \left(K_{\mathrm{Ic}}/Y\right)^4}$$

Using this expression one can numerically calculate the complete (a,u) graph. It appears that these curves show either a continuous increase of a with increasing u in case of stable crack growth only or an increase in a with a jump with increasing u in case unstable crack growth is present. Having calculated the (a,u) graph one can eliminate the crack length a from Eq. (21.101) and obtain the load-displacement curve. The (p,u) curves similarly show, after the initial elastic behaviour, either a continuous decrease in P with increasing u for stable crack growth only or a decrease in P with a jump with increasing u in the case of unstable crack growth.

Obviously the model is limited but experimentally the (a,u) and (p,u) behaviour as described briefly above is observed for several materials. In principle the model can be extended both on the fracture and the flow side. For fracture one may wish to include multi-axial stress states, a more sophisticated kinetic law (limit surface) and the influence of plasticity on the energy release rate. For flow one may wish to introduce the stress over the section instead of the average stress the effect is work hardening. One way to deal with that is to extend the limit surface from $\phi(A_k)$ to $\phi(A_k, a_l)$. As we have seen this will destroy normality in stress space. However, so far the failure criterion itself remains a matter for experiment. Finally it will be clear that if a structure is designed to carry a particular stress σ (load control) one must have $\sigma < S = K_{\mathrm{Ic}}/(\pi a)^{1/2}$ and $\sigma < Y$. If the structure is subjected to a particular strain (displacement control) such that plasticity occurs, a high value of $K_{\mathrm{Ic}}/Y\sqrt{a}$ is required for stable crack growth to occur.

Microcracking

Similar approaches have been advocated for the description of the development of damage in brittle solids, relatively simple ones as well as more formal ones. In both cases microcracks, which are usually supposed to be non-interacting, are considered as damage and a descriptor of this microcracks is taken as the internal variable. In a simple approach the area of the microcracks is used as a scalar measure[hh] while more sophisticated approaches take (one or more) vectors or tensors as internal variable.

Here we limit our selves to description[ii] using a single damage vector **a** so that the Helmholtz energy f is again given by $f = f(\varepsilon_{ij}, T, a_k)$. As the kinetic law we employ a linear relation $\dot{\mathbf{a}} = \mathbf{L}(\mathbf{a}) \cdot \mathbf{A}(\mathbf{a})$ (see Chapter 6) where the second-order tensor **L** represents the proportionality between the growth rate $\dot{\mathbf{a}}$ and driving force **A**. For isotropic materials **L** reduces to $l\mathbf{I}$ with **I** the unit tensor and l a constant. This law

---

[hh] See e.g. Honein, E, Honein, T, Kestin, J and Herrmann, G. (1993), page 66 in *Nonlinear thermodynamic processes in continua*: ein gemeinsamer workshop der TU Berlin und des Wissenschaftskolleg zu Berlin, June 11-12, 1992, vol 3/61, W. Muschik and G. Maugin, eds.
[ii] Ván, P. (2001), J. Non-Equilib. Thermodyn. **26**, 167.

implies that cracks cannot heal, which for microcracking is a fair approximation[ij]. Neglecting again the temperature dependence and developing the Helmholtz energy to second order for an isotropic material we have in direct notation

$$f(\boldsymbol{\varepsilon}, \mathbf{a}) = (\delta + k_\delta \mathbf{a}^2) \operatorname{tr} \boldsymbol{\varepsilon} + (\mu + \tfrac{1}{2} k_\mu \mathbf{a}^2) \boldsymbol{\varepsilon} : \boldsymbol{\varepsilon} + \tfrac{1}{2} (\lambda + k_\lambda \mathbf{a}^2) \operatorname{tr}^2 \boldsymbol{\varepsilon} + \tfrac{1}{2} \alpha \, \mathbf{a}^2$$
$$+ \tfrac{1}{2} (\beta + k_\beta \operatorname{tr} \boldsymbol{\varepsilon}) \mathbf{a} \cdot \boldsymbol{\varepsilon} \cdot \mathbf{a} + \tfrac{1}{2} \gamma \, \mathbf{a} \cdot \boldsymbol{\varepsilon} \cdot \boldsymbol{\varepsilon} \cdot \mathbf{a}$$

(21.103)

The terms in this expression represent, respectively,

- the pressure independent (δ) and pressure dependent part (k_δ) of the hydrostatic energy,
- the usual elastic part with λ and μ the damage independent part and k_λ and k_μ the damage dependent part,
- a contribution due to the microcracks,
- a term ($\mathbf{a} \cdot \boldsymbol{\varepsilon} \cdot \mathbf{a}$) dealing with the crack opening and
- a term ($\mathbf{a} \cdot \boldsymbol{\varepsilon} \cdot \boldsymbol{\varepsilon} \cdot \mathbf{a}$) dealing with the rotation of the microcracks.

The stress is given in the usual way by

$$\boldsymbol{\sigma} = \frac{\partial f(\boldsymbol{\varepsilon}, \mathbf{a})}{\partial \boldsymbol{\varepsilon}} = (\delta + k_\delta \mathbf{a}^2) \mathbf{I} + (2\mu + k_\mu \mathbf{a}^2) \boldsymbol{\varepsilon} + (\lambda + k_\lambda \mathbf{a}^2) \operatorname{tr} \boldsymbol{\varepsilon} \mathbf{I}$$
$$+ (\beta + k_\beta \operatorname{tr} \boldsymbol{\varepsilon}) \mathbf{a} \mathbf{a} + k_\beta \mathbf{a} \cdot \boldsymbol{\varepsilon} \cdot \mathbf{a} \mathbf{I} + \tfrac{1}{2} \gamma (\mathbf{a} \boldsymbol{\varepsilon} \cdot \mathbf{a} + \mathbf{a} \cdot \boldsymbol{\varepsilon} \mathbf{a})$$

(21.104)

with $\mathbf{I}$ the unit tensor. The damage stress belonging to zero deformation is then

$$\boldsymbol{\sigma}_0(\mathbf{a}) = \boldsymbol{\sigma}(0, \mathbf{a}) = (\delta + k_\delta \mathbf{a}^2) \mathbf{I} + \beta \, \mathbf{a} \mathbf{a}$$

(21.105)

resulting in a total stress, using the fourth-order elastic constant tensor $\mathbb{C}$, represented by

$$\boldsymbol{\sigma}(\boldsymbol{\varepsilon}, \mathbf{a}) = \boldsymbol{\sigma}_0(\mathbf{a}) + \mathbb{C}(\mathbf{a}) \cdot \boldsymbol{\varepsilon}$$

(21.106)

Consequently the damage strain, remaining after unloading, is

$$\boldsymbol{\varepsilon}_0(\mathbf{a}) = -\mathbb{C}^{-1}(\mathbf{a}) \boldsymbol{\sigma}_0(\mathbf{a})$$

(21.107)

Eq. (21.104) can be considered as a generalisation of the classical idea where damage is connected to a reduction in load bearing area, see e.g. Kachanov (1986). In this case a second-order damage tensor $\mathbf{D}$ can be introduced using the elastic stiffness tensor $\mathbb{C}$ for every direction $\mathbf{n}$ via $\mathbf{n} \cdot \mathbb{C} \cdot \mathbf{n} = \mathbf{I} - \mathbf{D}$.

The stability of the material can be examined by evaluating

$$(\mathrm{d}\boldsymbol{\varepsilon}, \mathrm{d}\mathbf{a}) \cdot \mathrm{d}^2 f \cdot \begin{pmatrix} \mathrm{d}\boldsymbol{\varepsilon} \\ \mathrm{d}\mathbf{a} \end{pmatrix} = (\mathrm{d}\boldsymbol{\varepsilon}, \mathrm{d}\mathbf{a}) \cdot \begin{pmatrix} \dfrac{\partial^2 f}{\partial \boldsymbol{\varepsilon}^2} & \dfrac{\partial^2 f}{\partial \mathbf{a} \partial \boldsymbol{\varepsilon}} \\ \dfrac{\partial^2 f}{\partial \boldsymbol{\varepsilon} \partial \mathbf{a}} & \dfrac{\partial^2 f}{\partial \mathbf{a}^2} \end{pmatrix} \cdot \begin{pmatrix} \mathrm{d}\boldsymbol{\varepsilon} \\ \mathrm{d}\mathbf{a} \end{pmatrix} \geq 0$$

(21.108)

and by using Eq. (21.104) one can show that there are no values of material parameters that would result in a stable material for any strain and damage. Some necessary conditions can be calculated easily and are $\lambda > 0$ and $\mu + \lambda > 0$ for zero damage, which are the usual elastic conditions, and $\alpha > 0$ for deformation zero. For

---

[ij] In fact the so-called *Kaiser effect* is a firm experimental confirmation. The noise due to the extending microcracks can be detected in a preloaded sample only when the reloading arrives at the previously applied highest stress level. This is observed for several directions separately.

other conditions general relations are not available but in two dimensions we can derive that

$$-\frac{2\mu+\lambda}{2k_\mu+k_\lambda+k_\beta+\tfrac12\lambda}>\mathbf{a}^2 \qquad \text{if} \quad 2k_\mu+k_\lambda+k_\beta+\tfrac12\lambda<0 \qquad \text{and}$$

$$-\frac{4\mu}{4\mu+\gamma}>\mathbf{a}^2 \qquad \text{if} \quad 4k_\mu+\gamma<0$$

must hold for the material to be stable. These expressions restrict the possible values of the material coefficients and provide material dependent upper limits to the damage values.

The driving force for damage is given as usual by $\mathbf{A}=-\partial f/\partial\mathbf{a}$ and the kinetic law becomes

$$\dot{\mathbf{a}}=\mathbf{L}\cdot(\alpha+2k_\mu\boldsymbol{\varepsilon}:\boldsymbol{\varepsilon}+k_\lambda\text{tr}^2\boldsymbol{\varepsilon}+2k_\delta\text{tr}\boldsymbol{\varepsilon})\mathbf{a}+2(\beta+k_\beta\text{tr}\boldsymbol{\varepsilon})\mathbf{a}\cdot\boldsymbol{\varepsilon}+\tfrac12\gamma(\boldsymbol{\varepsilon}\cdot\boldsymbol{\varepsilon}\cdot\mathbf{a}+\mathbf{a}\cdot\boldsymbol{\varepsilon}\cdot\boldsymbol{\varepsilon})$$

Hence given the deformation $\boldsymbol{\varepsilon}=\boldsymbol{\varepsilon}(t)$ the damage growth rate can be calculated starting at damage $\mathbf{a}=\mathbf{0}$ using numerical integration and from that the resulting stress-strain curve. The overall behaviour is one that resembles the stress-strain behaviour of plastic solids and is therefore sometimes addressed as pseudo-plastic behaviour.

Also for microcracking a representation in terms of Gibbs energy h is often useful. From $h=f-\boldsymbol{\sigma}{:}\boldsymbol{\varepsilon}$ we obtain

$$h(\boldsymbol{\sigma},\mathbf{a})=(\delta'+k_\delta{}'\mathbf{a}^2)\,\text{tr}\,\boldsymbol{\sigma}+(\mu'+\tfrac12k_\mu{}'\mathbf{a}^2)\boldsymbol{\sigma}:\boldsymbol{\sigma}+\tfrac12(\lambda'+k_\lambda{}'\mathbf{a}^2)\,\text{tr}^2\boldsymbol{\sigma}$$
$$+\tfrac12\alpha\,\mathbf{a}^2+\tfrac12(\beta'+k_\beta{}'\text{tr}\,\boldsymbol{\sigma})\mathbf{a}\cdot\boldsymbol{\sigma}\cdot\mathbf{a}+\tfrac12\gamma'\mathbf{a}\cdot\boldsymbol{\sigma}\cdot\boldsymbol{\sigma}\cdot\mathbf{a} \tag{21.109}$$

In principle the material parameters in h should be the same as in f but since the second-order representation of the Helmholtz energy is an approximation, they may differ in practice, except for the parameters α, λ and μ which represent the pure elastic deformation and damage, respectively. In this case the strain becomes

$$\boldsymbol{\varepsilon}=-\frac{\partial h(\boldsymbol{\sigma},\mathbf{a})}{\partial\boldsymbol{\sigma}}=(\delta'+k_\delta{}'\mathbf{a}^2)\mathbf{I}+(2\mu+k_\mu{}'\mathbf{a}^2)\boldsymbol{\sigma}+(\lambda+k_\lambda{}'\mathbf{a}^2)\,\text{tr}\,\boldsymbol{\sigma}\mathbf{I}$$
$$+(\beta'+k_\beta{}'\text{tr}\,\boldsymbol{\sigma})\mathbf{aa}+k_\beta{}'\mathbf{a}\cdot\boldsymbol{\sigma}\cdot\mathbf{a}\mathbf{I}+\tfrac12\gamma'(\mathbf{a}\boldsymbol{\sigma}\cdot\mathbf{a}+\mathbf{a}\cdot\boldsymbol{\sigma}\mathbf{a}) \tag{21.110}$$

The damage strain belonging to zero stress becomes

$$\boldsymbol{\varepsilon}_0(\mathbf{a})=\boldsymbol{\varepsilon}(0,\mathbf{a})=(\delta'+k_\delta{}'\mathbf{a}^2)\mathbf{I}+\beta'\mathbf{aa} \tag{21.111}$$

Eq. (21.110) can be seen as a generalisation of the original Griffith idea where $\mathbf{a}$ represents a single crack of length a and the Gibbs energy is approximated by

$$h(\sigma,a)=-\frac{\sigma^2}{2E}\left(1+\pi a^2\right)+2Ra$$

This corresponds to the treatment of the elastic fracture given before.

Using this formalism Ván was able to show that with three material parameters used for fitting and the other ones calculated from known data or estimated, the biaxial failure envelope for Wombeyan marble could be represented as accurately as by several empirical failure criteria. Obviously also in this model several effects are not included. Kinetic effects and the fact that microcracking is initiated separately in different orientations are not included.

The literature of damage mechanics has become quite extensive and we refer to Krajcinovic (1996) and Lemaitre (1996) for further details. In fact on fracture theory and experiments as a whole an extensive literature exists and it is impossible to do justice to all approaches. For an overview we refer to Cherepanov (1998). The thermodynamic approach is one though which is firmly based on the physical principles of thermodynamics and stability and provides a general framework for elastic and elastic-plastic fracture as well as microcracking. Two evident drawbacks are present. The first is the failure criterion itself. The value of the fracture energy (or any other failure measure) has to be determined experimentally or to be modelled separately. This is, however, a drawback of all fracture theories. The second is the interpretation of the internal variable describing fracture. Various choices are possible and also here experiment has to provide the answer, which is the optimum choice. Again, also in other fracture theories usually some choice for the descriptor of the crack(s) has to be made, which has to be validated by experiment. Since both aspects depend on the mechanism of fracture, which in turn is related to the type and structure of the material, it is not surprising that thermodynamics alone cannot provide the answer[kk]. Finally, it has to be remarked that the approach as sketched in this section presents in a natural way the brittle-ductile transition. This transition of which the description is sometimes addressed as the two-criteria approach, is often used in engineering and is discussed in the next chapter.

21.10 Bibliography

Atkins, A.G. and Mai, Y-W. (1985), *Elastic and plastic fracture*, Ellis Horwood, Chichester.

Broek, D.J. (1978), *Engineering fracture mechanics*, 2nd ed., Sijthoff & Noordhoff, Alphen aan den Rijn.

Cherepanov, G.P. (1979), *Mechanics of brittle fracture*, McGraw-Hill, New York.

Cherepanov, G.P. ed. (1998), *Fracture*, Krieger, Malabar, Florida.

Derby, B., Hils, D. and Ruiz, C. (1992), *Materials for engineering*, Longman Scientific & Technical, Harlow, UK.

Kachanov, L.M. (1986), *An introduction to damage mechanics*, Martinus Nijhoff, Boston.

Kanninen, M.F. and Popelar, C.H. (1985), *Advanced fracture mechanics*, Oxford University Press, New York.

Knott, J.F. (1973), *Fundamentals of fracture mechanics*, Butterworths, London.

Krajcinovic, D. (1996), *Damage mechanics*, Elsevier, Amsterdam.

Lemaitre, J. (1996), *A course in damage mechanics*, Springer, Berlin.

Miannay, D.P. (1997), *Fracture mechanics*, Springer, New York.

[kk] Although it is claimed (Basaran, C. and Nie, S. (2004), J. Damage Mech. **13**, 205) that damage evolution can be described using the entropy production in the system only. In this paper, however, also a particular choice for the damage parameter is made, while global fracture is not addressed.

22

Applications of fracture theory

In this chapter some applications of fracture with the continuum point of view in mind are given. First, materials testing is discussed. This also includes the measurements of the defect size because defects like microcracks may initiate failure. Natural defects generally do not have all the same size but are statistically distributed. This leads to a statistical description for fracture in materials and structures where the defect size cannot be determined beforehand. Finally a brief discussion is given of those processes where fracture plays an important role, in particular to abrasive machining.

22.1 Materials testing

In Chapter 2 the description of the tensile test ended with fracture. We now describe briefly the measurement of strength itself and the relevant material's mechanical properties for the determination of strength.

The relevance of fracture theory

Fracture theory is highly relevant to prevent catastrophic events. Of the many large accidents due to fracture that can be quoted we just mention a few. On 15th January, 1919 a 27 m diameter and 15 m high tank located in Commercial Street, Boston, fractured thereby releasing about 7.5×10^6 l of molasses. The tank broke without any warning whatsoever. Twelve persons were killed and 40 injured. During World War II several new designed cargo ships broke at calm sea or even in dead calm weather in the harbour (S.S. Schenectady, 1943). Well-known is also the frequent failure of the troop ships, the so-called Liberty ships, during World War II. Famous also is the failure of the Tacoma Bridge on 7th November, 1940 due to its registration on film. More recently many oil tankers fractured. Another dramatic example is the loss of space shuttle *Challenger* during launching on 28th January, 1986 where a rubber ring became brittle by the low temperature involved. It was shown by Richard Feynman, one of the most famous American theoretical physicists, that a faulty design was involved. More recently, on 1st February, 2003 the space shuttle *Discovery* was lost on return to the atmosphere due to the impact of a piece of scrap during take-off.

Strength

In principle tensile testing can be done without undue complications for semi-ductile materials. A slight misalignment is usually not detrimental in the sense that unreliable data are obtained. Tensile testing is sometimes done for brittle materials in spite of the associated complications. In this case the most important aspect is to avoid bending stresses. Various solutions to this problem are in use, e.g. soft interlayers between the specimen and jig, air bearings for the application of the force or a cardanic coupling. At high temperature most of these solutions fail and extensive alignment has to be done.

For brittle materials normally the bend test is used. Although the measurement of bend strength data looks treacherously simple, many errors can be made. Aspects to consider are:

- application of simple beam theory,
- friction at the supports,
- local stresses at the supports,
- unequally applied moments and
- twisting and wedging of the specimen.

For sufficiently slender beams, beam theory is applicable. Free rollers are essential to avoid friction. The design of the test jig should be such that it is self-equilibrating and provide equal moments. Really planparallel specimens are to be used so that accurate preparation should is required. If all precautions of alignment and specimen preparation are made properly, an accuracy of 1 to 2% can be obtained. To minimize the influence of subcritical crack growth, usually the highest possible strain rate achievable on the testing machine at hand is used. Since a high humidity accelerates subcritical crack growth, the test jig is also often placed in a box through which dry N_2 or dry air is blown. Quinn and Morrell[a] have given an overview on bend testing.

Fracture toughness

Measurement of the fracture toughness K_{Ic} is possible by using the 3-(or 4-) point bend method (Fig. 8.5). A crack longer than any microstructural feature and of known length a is introduced in the specimen. From the load at fracture F and the dimensions the value of K_{Ic} is calculated using the equation[b]

$$K_{Ic} = \alpha \sigma_{cri} \sqrt{a} = \alpha \frac{3Fl}{2bh^2} \sqrt{a} \qquad (22.1)$$

For the meaning of the different parameters and the value of α, see Example 16.2 Note that the stress at failure σ_{cri} for a fracture toughness test is not equal to the strength S of a non-pre-cracked material. The value of the shape factor (also wrongly called compliance factor) α is well documented. Except for very deep cracks ($a/h > 0.95$) a loaded bend specimen is unstable. Once fracture occurs, enough energy is available to run the crack through the entire specimen.

Another possibility is the double cantilever beam (DCB) test (Fig. 16.9). In that case the expression for K_{Ic} is given by[c]

$$K_{Ic} = \frac{Fa}{bh^{3/2}} \left(3.47 + 2.32 \frac{h}{a} \right) \qquad (22.2)$$

Crack growth in this specimen is stable. Therefore this specimen also offers the possibility to measure the dependence of crack velocity on K_I.

Another convenient test specimen is the so-called 'double torsion' (DT) specimen (Fig. 22.1). The DT specimen is a neutral type specimen: the resulting stress intensity is independent of the length of the crack over the middle 50% of the length of the specimen. This makes crack length measurements in principle not necessary. Crack length values are nevertheless frequently measured for control purposes. A disadvantage of this specimen is that the crack front is not straight for which

---

[a] Quinn, G.D. and Morrell, R. (1991), J. Am. Ceram. Soc. **74**, 2037.
[b] Brown, W.F. and Srawley, J.E. (1966), ASTM-STP-410, page 12.
[c] Wiederhorn, S.M., Shorb, A.M. and Moses, R.L. (1968), J. Appl. Phys. **39**, 1569. Sometimes the width at the crack plane b' is taken smaller than the nominal width b. In that case $(b'b)^{1/2}$ replaces b.

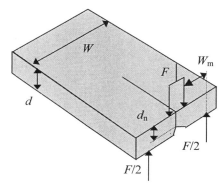

Fig. 22.1: Schematic of a DT specimen.

occasionally a correction factor is used. The relevant stress intensity equation (without crack front correction) is[d]

$$K_{Ic} = \left(\frac{3W_m^2(1+v)}{Wd^3d_n(1-1.26\,d/W)} \right)^{1/2} F \tag{22.3}$$

Elastic parameters

Although not strictly a fracture-related property the elastic parameters are of importance in fracture testing. Young's modulus E can be determined also from bending experiments by measuring not only the force but the deflection as well. From elementary beam theory the deflection δ at the midpoint of a 3-point bend and 4-point bend specimen is given by, respectively,

$$v_{3pb} = \frac{Fl^3}{48EI} \quad \text{and} \quad v_{4pb} = \frac{Fd}{48EI}(3l^2 - 4d^2) \tag{22.4}$$

where F is the applied load and d and l are the distances as indicated in Fig. 10.5. The moment of inertia I is given by $I = bh^3/12$ for a beam of width b and height h. For a cylindrical beam $I = \pi r^4/4$, where r is the radius of the beam. Although many testing machines provide an option for measuring the displacement of the crosshead, this measurement is insufficient for an accurate determination of Young's moduli for stiff materials. Direct measurement with a separate displacement transducer is a prerequisite.

Nowadays the elastic properties are usually determined by ultrasonic methods (see e.g. Pollard, 1977). The simplest is the pulse-echo method in which an ultrasonic pulse is transmitted in the material and received by the same transducer after reflection to the backside of a planparallel specimen. From the observed transmitting time and the thickness of the specimen the wave velocity can be calculated. The relations between the density ρ, longitudinal wave velocity v_{lon} and shear (or transverse) wave velocity v_{she} are

$$v_{lon}^2 = \frac{E}{\rho}\frac{1-v}{(1+v)(1-2v)} \quad \text{and} \quad v_{she}^2 = \frac{G}{\rho} = \frac{E}{\rho}\frac{1}{2(1+v)} \tag{22.5}$$

---

[d] D.P. Williams and Evans, A.G. (1973), J. Test. Eval. **1**, 264.

where G is the shear modulus. Inversion of these equations yields Young's modulus E and Poisson's ratio ν.

Another simple method is recording the (flexural) resonance frequency of a sample of known dimensions. Young's modulus can be calculated from the resonance frequency and the dimensions by the appropriate formulae. From vibration theory the natural frequencies of a beam of length l with a cross-sectional area A and moment of inertia I are given by

$$\omega_n = \left(\frac{\pi n}{l}\right)^2 \sqrt{\frac{EI}{\rho A}} \qquad \text{with} \quad n = 1, 2, \dots \tag{22.6}$$

An advantage of this method is that it can be used relatively easy at elevated temperature (see e.g. Fig. 11.7).

Defect size

The direct measurement of the critical defect size a using a non-destructive technique (NDT) is rather important. For the larger defect sizes (normal) ultrasonic or transmission X-ray techniques may be useful. For ductile materials like metals with a crack length of the order of millimetres NDT is routinely applied. NDT is much more difficult for brittle materials with much smaller defect sizes, say micrometres[e]. Potentially useful techniques are (high frequency) ultrasonic surface wave scattering, acoustic microscopy and X-ray tomography. For this purpose complicated and expensive equipment is necessary. For brittle materials up to now only in a limited number of cases some success has been met. Even in these cases the results are not too reliable. The problems associated with the determination of defects before failure form a major obstacle for further application of design by fracture mechanics in brittle materials. From a comparison of strength and fracture toughness, however, an estimate can be made for the defect size, assuming a certain flaw configuration (shape factor α). These estimates compare usually quite well with the size as determined from a fracture surface inspection after fracture (a 'post-mortem' investigation).

Example 22.1: Critical defects in ceramics

Flaws in ceramics can be divided into two categories. First, the fabrication related flaws and, second, the machining related flaws. Within the first category we can distinguish pores (Fig. 22.2), porous zones, inclusions, large grained areas and locally melted spots due to low melting eutectic compositions. In all cases improved processing can avoid these defects. Machining of ceramics is usually done by diamond grinding. This creates, first, a plastically deformed zone below the groove and, second, lateral, radial and median cracks. Because the median cracks have the largest extension they are considered to be the most risky ones. The consequence is that a circumferentially machined piece of material in general will have an

[e] Goebbels, K., Reiter, H., Hirsekorn, S. and Arnold, W. (1984), Sci. Ceramics **12**, 483; Biagi, E., Fort, A., Masotti, L., and Ponziani, L. (1995), Acoust. Imag. **21**, 667; Rice, R. W. (2001), Ceram. Trans. **122**, 105.

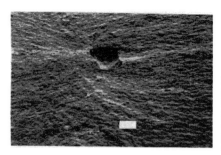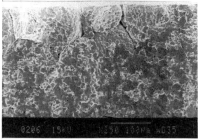

Fig. 22.2: Pore in barium titanate (bar = 50 μm) and non-adhering grain boundaries in Mn-Zn ferrite acting as fracture origin.

anisotropic strength. Machining defects, individually not having a critical size, may link by subcritical crack growth (see Chapter 24) to a critical defect.

22.2   Fracture in brittle structures

The basic aspects of fracture have been discussed in the previous chapter. Here we deal with brittle fracture in simple structures including test samples taking into account the intrinsic variability due to the distribution of defect sizes. In many materials the naturally occurring defects cannot be detected easily. Moreover, they will show a size distribution. This leads to a statistical description of fracture that can be used on the material as well as on the structure level.

Strength of brittle materials generally shows considerable scatter. From some consideration on fracture mechanics it becomes clear that this is due to the distribution in size for naturally occurring defects. After all, fracture mechanics teaches us that if the stress intensity factor of a defect (a field quantity characterising the effect of the applied stress on a defect) exceeds the fracture toughness (a material property), fracture occurs. The important part of the defect size distribution is the large size tail because the stress concentration factor is the largest at these defects if the stress variation over the structure is not too large. From $K_I = \alpha\sigma\sqrt{a}$, we see that fracture does not automatically occur at the largest flaw but generally it does. In this sense one can say that by means of a series of fracture experiments the tail of the defect size distribution is sampled. For this type of estimates, i.e. samples taken only from the tail of a distribution, so-called *extreme value statistics* are valid (see e.g. Castillo, 1988). For a sample of extreme values x taken from parent distribution bounded at the small size side and unbounded at the large size side, it can be shown that the most likely cumulative distribution function $F(x)$ is the so-called *Frechet distribution*, given by

$$F(x) = \exp\left[-\left(\frac{x}{x_0}\right)^{-n}\right] \qquad (22.7)$$

where $F(x)$ denotes the probability for x to occur, the parameter n characterises the width of the distribution and the parameter x_0, the *characteristic value*, the location of the distribution. If we apply this to the defect size distribution and combine the Frechet expression with the fracture mechanics equation

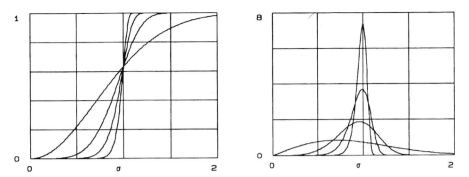

Fig. 22.3: Weibull graph and derivative for $m = 2, 5, 10$ and 20.

$$S = \frac{1}{\alpha}\sqrt{\frac{2RE}{(1-v^2)a}} = \frac{1}{\alpha}\frac{K_{Ic}}{\sqrt{a}}$$ (22.8)

we obtain for the strength distribution

$$W(S) = 1 - \exp\left[-\left(\frac{a}{a_0}\right)^{-n}\right] = 1 - \exp\left[-\left(\frac{S_0^2}{S^2}\right)^{-n}\right] = 1 - \exp\left[-\left(\frac{S}{S_0}\right)^{2n}\right]$$ (22.9)

which is a *Weibull*[f] *distribution* for the strength S with *Weibull modulus m* = $2n$ and *characteristic strength* S_0. The value of S_0 depends on the material and the specimen size. If the specimen is loaded with a stress σ, $W(S)$ is the probability that the specimen breaks before σ reaches the value S. Alternatively $1-W(S)$ is the probability that the specimen survives a load of S. The Frechet distribution (for the maximum in defect size a) thus becomes in a Weibull distribution (for the minimum in strength S). The (differential and cumulative) Weibull distribution for several values of m is shown in Fig. 22.3. From this plot it is observed that for small m-values like $m = 2$ or $m = 5$ the spread in strength can be as large as a factor of 2 and 1.5, respectively, with a long tail on the low strength side. For high m-values like $m = 20$, on the other hand, the distribution is much more narrow and symmetric.

The distribution of the strength values was originally described in a phenomenological way by the *weakest link model*, originated by Weibull[g]. The final result is similar to that described above. The route to arrive at the strength distribution is this way is as follows. The *survival probability* P' for a volume V, x times greater than a reference volume V_0, is given by

$$P'(V) = P'(V_0)^x$$ (22.10)

Writing $P'(V_0)^x$ in its exponential form, we define the *risk of rupture R* by

$$P'(V) = \exp[x \ln P'(V_0)] \equiv \exp(-R) \quad \text{or}$$ (22.11)

---

[f] Waloddi Weibull (1887-1979). Swedish engineer who published many papers on strength of materials, fatigue, rupture in solids, bearings, and of course, the Weibull distribution.
[g] Weibull, W. (1939), Ing. Vetenk. Akad. Proc. **153**, 151. Much easier accessible is: J. Appl. Mech. **18**, (1951) 293.

$$R = -\frac{V}{V_0} \ln P'(V_0) \qquad (22.12)$$

where we used that $x = V/V_0$. Since R is linear in V, $dR/dV = F(\sigma)$ is a function of the applied stress σ only. For an arbitrary volume the risk of rupture thus can be described quite generally by

$$R = \frac{1}{V_0} \int F(\sigma) dV \qquad (22.13)$$

where the integral is taken over the volume of the specimen. In the case of surface defects the integral should be taken over the area of interest. The general requirement for $F(\sigma)$ is that it is a positive non-decreasing function of σ, vanishing at a certain threshold stress σ_u not necessarily zero. Weibull postulated that the simplest function that satisfies the requirements for $F(\sigma)$, a power dependence, should be used:

$$F(\sigma) = \left(\frac{\sigma - S_u}{S_0} \right)^m \qquad (22.14)$$

where S_0 is a normalising factor, often addressed as *characteristic strength*, and m the Weibull modulus. The parameter S_u is the stress value below which no fracture occurs. Often one assumes that $S_u = 0$. The *failure probability*, $P = 1-P'$, of a structure in which a spatially inhomogeneous stress distribution is present is then given by

$$P = 1 - \exp\left[-\frac{1}{V_0} \int \left(\frac{\sigma - S_u}{S_0} \right)^m dV \right] \qquad (22.15)$$

In the case of a simple stress distribution the integral can be easily evaluated, e.g. for a tensile test where the stress σ is constant throughout the specimen. Writing S for the strength, the stress at failure, assuming $S_u = 0$, rearranging and taking logarithms twice yields

▶ $$\ln \ln \frac{1}{1-P} = m \ln S - m \ln S_0 + \ln \frac{V}{V_0} \qquad (22.16)$$

This relation can be used to determine the Weibull modulus for a series of n experimentally determined strength data. Obviously, to do so an estimate for P should be made. Various estimators for the failure probability are known, all based on an arranging the experimental strength values in order of increasing strength, i.e. $S_1 \le S_2 \le \dots \le S_n$. If i is the order number of the tested specimen and n the total number of tested specimens, for a large number of specimens P_i can be estimated as $P_i = i/n$. Using a limited number this estimator is no longer reliable and has to be adapted. Frequently used is the *mean rank estimator* and *median rank estimator*

$$P_i = \frac{i}{n+1} \qquad \text{and} \qquad P_i = \frac{i - 0.3}{n + 0.4} \qquad (22.17)$$

In a least-squares fitting of strength data, however, the *empirical estimator*

$$P_i = \frac{i - 0.5}{n} \qquad (22.18)$$

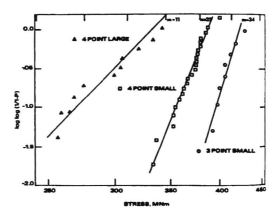

Fig. 22.4: Weibull plot of alumina with three different tests, indicating the difference in characteristic strength and modulus.

yields the least biased result with minimum variance[h]. Plotting $\ln \ln (1/(1-P_i))$ versus $\ln S_i$ results in a straight line if Weibull statistics are satisfied and the slope of this line is m. The higher the value of m, the less variation in fracture strength occurs. Values between 5 and 20 are common for (semi)-brittle materials. An example of strength measurement for various sizes of polycrystalline alumina samples is given in Fig. 22.4. For an estimate of the reliability of S_0 and m, use can be made of

$$s(m)/m = 1/\sqrt{n} \quad \text{and} \quad s(S_0)/S_0 = 1/m\sqrt{n}$$

where $s(x)$ and n denote the standard deviation of x and the total number of specimens, respectively. For the Weibull distribution the parameter S_0 determines the central location of the distribution while the parameter m determines its width. Frequently, however, for experimental data the mean strength $\langle S_{ten} \rangle$ and standard deviation $s(\langle S_{ten} \rangle)$ are given instead of the individual data points. The average strength $\langle S_{ten} \rangle$ is related to S_0 and m via $\langle S_{ten} \rangle = S_0(V_0/V)^{1/m}\Gamma(1/m+1)$ where Γ is the gamma function. An approximate relation between m, $\langle S_{ten} \rangle$ and $s(\langle S_{ten} \rangle)$ is

$$m \cong 1.2\langle S_{ten} \rangle / s(\langle S_{ten} \rangle)$$

Problem 22.1

Show that the use of the estimator $P_i = i/(n+1)$ always results in a smaller Weibull modulus m then the estimator $P_i = (i-0.5)/n$ by considering their dependence on the order number i.

Bend bars

The simple dependence on volume (or surface) as found for the tensile test is lost in more complicated stress distributions where the stress σ is no longer constant over

---

[h] Dortmans, L.J.M.G. and de With, G. (1991), J. Am. Ceram. Soc. **74**, 2293. Application of a weighting scheme with weights $w_i = (1-P_i) \ln(1-P_i)$, applied in order to correct for unequal weights due to the transformation S to $\ln S$, doesn't influence this result significantly.

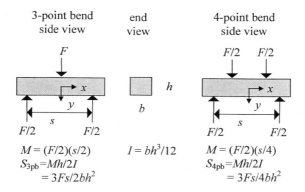

Fig. 22.5: Maximum outer fibre stress for a 3-point specimen S_{3pb} and a 4-point bend specimen S_{4pb}. M and I indicate the applied moment and moment of inertia, respectively.

the specimen. The same equation can be used, however, if we normalise the stress by the maximum stress S occurring in the structure.

$$\blacktriangleright \quad P = 1 - \exp\left[-\frac{1}{V_0}\left(\frac{S}{S_0}\right)^m \int \left(\frac{\sigma(\mathbf{r})}{S}\right)^m dV\right] = 1 - \exp\left[-\left(\frac{S}{S_0}\right)^m \frac{V_{eff}}{V_0}\right] \quad (22.19)$$

with the effective volume V_{eff} defined by

$$\blacktriangleright \quad \frac{V_{eff}}{V_0} = \frac{1}{V_0}\int \left(\frac{\sigma}{S}\right)^m dV \quad (22.20)$$

The quantity V_{eff}/V_0 is also known as the *stress volume integral*. The effective volume is thus always smaller than the real volume because $|\sigma(\mathbf{r})/S| \leq 1$. To assess the influence we consider bend configurations. For the 3-point and 4-point bend test (Fig. 22.5) the maximum occurring fibre stress is used for S. The stress distribution in a 3-point bend beam can be written as

$$\sigma = S_{3pb}\left(\frac{2y}{h}\right)\left(1 - \frac{2x}{s}\right)$$

where y is the distance in the depth as measured from the centre, x the distance in the length as measured from the centre and S_{3pb} represents the maximum occurring stress in a 3-point bend beam. At the neutral axis $\sigma = 0$ and the part above the neutral line loaded in compression is neglected. Moreover, we take only the tensile loaded surface between the loading rollers into consideration. Analogously an expression can be written for the strength S_{4pb} of a ¼-point loaded 4-point bend beam. This results for volume defects in the ratios

$$\frac{\langle S_{3pb}\rangle}{\langle S_{ten}\rangle} = \left[2(m+1)^2\right]^{1/m} \quad \text{and} \quad \frac{\langle S_{4pb}\rangle}{\langle S_{ten}\rangle} = \left[\frac{4(m+1)^2}{m+2}\right]^{1/m} \quad (22.21)$$

where $\langle S_{3pb}\rangle$, $\langle S_{4pb}\rangle$ and $\langle S_{ten}\rangle$ represent the average strength for 3-point, 4-point bending and tension, respectively. The corresponding expressions for surface defects are

$$\frac{\langle S_{3pb} \rangle}{\langle S_{ten} \rangle} = \left[\frac{2(m+1)^2}{f(m,b/h)} \right]^{1/m} \quad \text{and} \quad \frac{\langle S_{4pb} \rangle}{\langle S_{ten} \rangle} = \left[\frac{4(m+1)^2}{(m+2)f(m,b/h)} \right]^{1/m} \quad (22.22)$$

where

$$f(m,b/h) = 1 + [m/(1 + b/h)]$$

depends on the Weibull modulus m, the width b, and the height h of the beam. These expressions show that the average strength as measured with the various methods varies according to $\langle S_{ten} \rangle < \langle S_{4pb} \rangle < \langle S_{3pb} \rangle$. This is due to the decreasing volume under relatively high stress for this series of test geometries. The effect can be quite large. Consider the case of volume defects for the 3-point bend test. Assuming a Weibull modulus of 10, the ratio $\langle S_{3pb} \rangle / \langle S_{ten} \rangle$ is 1.73 and thus the tensile strength is significantly overestimated using a bend test (see also Fig. 22.4).

Problem 22.2

For an alumina a set of 3-point bend tests is done using bars of $0.5 \times 0.5 \times 5$ cm^3.
a) Calculate the Weibull modulus m and the characteristic strength S_0 using the strength data in the table shown. Use the mean rank and the empirical estimator in order to assess the effect of different probability estimators.

Table 22.1: Strength data S (MPa) for a 3pb test on alumina.

286	274	302	308	284
300	294	282	297	

b) Make an estimate of the standard deviation for m and S_0.
c) How large is the maximum allowable stress to keep the fracture probability $< 5\%$?
d) Calculate the maximum allowable stress for a fracture probability of 1% for a bar of size $5 \times 5 \times 50$ cm of the same material and preparation conditions.
e) What is the characteristic strength if the bars are tested in 4-point bending?
f) What is the characteristic strength if the bars are tested in tension?

---

*Plates**

Another frequently used geometry in materials testing is a plate. In this case also the simple dependency on volume or surface as obtained for the tensile test is lost since in this geometry the stress is again not constant over the specimen. For plates we briefly consider two loading situations as applied in the ball-on-ring test and the ring-

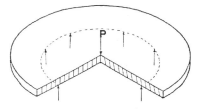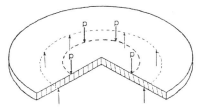

Fig. 22.6: The ball-on-ring and ring-on-ring test.

on-ring test (Fig. 22.6). While the ball-on-ring test can be considered as the axially symmetric equivalent of the 3-point bend test, the ring-on-ring test is the axially symmetric equivalent of the 4-point bend test. An advantage of plate specimens over bar specimens is that edge effects on the strength measurements are avoided. We follow here the analysis as provided by Scholten et al.[i]

The nominal fracture stress S_{bor} for the ball-on-ring test can be written as

$$S_{bor} = \frac{3(1+v)F}{2\pi t^2}\left[1+2\ln\left(\frac{a}{b}\right)+\left(\frac{1-v}{1+v}\right)\left(\frac{a}{R}\right)^2\left(1-\frac{1}{2}\left(\frac{b}{a}\right)^2\right)\right] \qquad (22.23)$$

where a is the support radius, b the effective contact radius, R the specimen radius, t the specimen thickness and v Poisson's ratio. For a loading ball the problem is to determine the proper value for b. It appears to be sufficient[j] to take the effective contact radius b equal to one-third of the plate thickness t. This expression describes the stress at the lower side of the plate just under the loading ball when the central ball is loaded with a force F.

For the ring-on-ring test the nominal fracture stress S_{ror} is equal to

$$S_{ror} = \frac{3(1+v)F}{2\pi t^2}\left[\ln\left(\frac{a}{b}\right)+\left(\frac{1-v}{1+v}\right)\left(\frac{a}{R}\right)^2\left(\frac{1}{2}-\frac{1}{2}\left(\frac{b}{a}\right)^2\right)\right] \qquad (22.24)$$

where a is the support radius (outer ring), b the loading ring radius (inner ring) and the other symbols have the same meaning as before. This expression describes the stress at the lower side of the plate within the area of the loading ring when the loading ring is loaded with a force F.

The stress integrals for these configurations can be calculated analytically but are not very accurate and therefore are usually calculated by numerical methods.

Arbitrary geometries and stress states*

In the examples above the stress integrals could be evaluated analytically. However, in case of more complex geometries and/or stress states this may be cumbersome or impossible. Fortunately the computation procedure is essentially straightforward and suited for numerical implementation. Using the Finite Element Method (FEM) the desired flexibility can readily be obtained. The strategy employed then can be summarized as follows:

- A standard FEM calculation is done to determine the stresses in the finite elements.
- Using these results and some information on the geometry of the elements the stress integrals can be evaluated numerically, e.g. by using Gauss integration. For this purpose several so-called postprocessors have been developed which use the results of a standard FEM program for start of the computation.

Several of these postprocessors[k] have been developed in the last decade such that all major commercial FEM programs can be used. For the calculation of failure probabilities the postprocessors essentially require three sets of input data, two of which are available through the FEM program (Fig. 22.7).

---
[i] Scholten, H., Dortmans, L., de With, G., de Smet, B. and Bach, P. (1992), J. Eur. Ceram. Soc. **10**, 33.
[j] De With, G. and Wagemans, H. (1989), J. Am. Ceram. Soc. **72**, C1538.
[k] Dortmans, L., Thiemeier, Th., Brückner-Foit, A. and Smart, J. (1993), J. Eur. Ceram. Soc. **11**, 17.

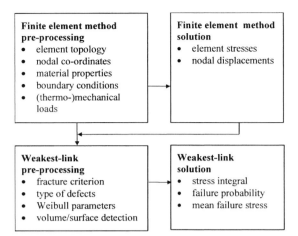

Fig. 22.7: Schematic overview of data sets and results for weakest-link failure probability calculation using the finite element method and a dedicated post-processor.

The first set of input data consists of data that are also the starting point for the conventional FEM analysis: element topology, nodal co-ordinates, material properties, boundary conditions and (thermo)-mechanical loads. Then the solver of the FEM program will produce nodal displacements and stresses at element level, which are the second set of input data for the weakest-link postprocessor. The third set of input data consists of: a choice for the appropriate fracture criterion in case of multiaxial loading, a choice for the relevant type of defects (volume, surface, values for α-factor relating to the crack geometry), values for the Weibull parameters m and S_0 and a description of which part of the geometry contributes to the volume or surface stress integrals. With respect to the material parameters it must be said that it is always tacitly assumed that the surface condition of test specimens and actual structure are identical. Obviously, this may not be the case.

Taking these sets of data the postprocessor will evaluate the stress integrals by numerical integration and produce a value for the failure probability of the component and/or the predicted mean failure stress. The main difficulty in using these codes is that sometimes an adapted FEM mesh is needed to get an accurate prediction for the failure probability. The reason for this can be found by considering the expression for the stress integral. As the stress σ in general is position dependent, the integrand σ/S can vary strongly throughout the component, especially for higher values of m or in case relatively large stress gradients are present. In these cases use of an adapted mesh may be required to increase the accuracy of the numerical integration process.

22.3   Design with brittle materials*

In recent years a great deal of attention has been paid to designing reliably with brittle materials. In this process many interrelated factors play a role. In order to illustrate the various issues relevant to the problem in Fig. 22.8 a metaphor, the reliability tree[1], is used. The various branches and twigs grow, dependent on each other but not necessarily knowing very much of one another. Often they tend to grow apart or to put it another way, workers in one field are not necessarily aware of the

---

[1] De With, G. (1992), J. Eur. Ceram. Soc. **9**, 337.

Fig. 22.8: The reliability tree illustrating various aspects in design with brittle materials.

developments in another field. The basic division in reliability is, as outlined in some detail in Section 22.4, between materials and design. For the materials side processing of improved and new materials, machining and non-destructive evaluation (NDE) are relevant. For the design side the first item is the stress distribution and its assessment using sufficient data and relevant failure criteria.

Mapping the most important factors on the strength equation, which relates the fracture toughness K, the residual stress Σ, the defect size a to the strength S, the distinction (Fig. 22.9) becomes perhaps more clear. The right-hand side deals with materials aspects such as new and improved materials and their associated properties but also with machining and non-destructive evaluation. It seems that NDE at present plays only a minor role and that cost-effective processing of existing materials is necessary. Really new materials have to follow a long road before they become accepted in structures. A proper characterisation of the machined surface is also required. The left-hand side deals with design as indicated by material data and failure criteria. The often-advocated statement 'design in compression' (because of the high compressive strength) must be considered as too simplistic. Further a carefully selected set of strength measurements, characterising the relevant effective stress for the multi-axial stress distribution in the structure, is highly desirable. In this book we dealt mainly with failure criteria, material's properties and data. However, for a reliable design the other factors should not be neglected. Creyke et al. (1982) present a pragmatic approach, while Mangonon (1999) provides practical rules for materials selection (for ceramics as well as other materials). Munz and Fett (1999) discuss the

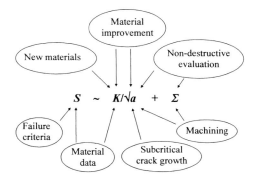

Fig. 22.9: The main influences on the strength S of ceramics: the fracture toughness K, the defect size a and the residual stress Σ.

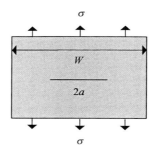

Fig. 22.10: Plate of width W with a central crack of length $2a$ remotely loaded by a stress σ.

selection of ceramic materials in relation to design in terms of the fundamental materials properties. Although for other materials the importance of the various aspects differs and even other aspects are relevant, e.g. plasticity for metals or visco-elasticity for polymers, similar considerations can be made. In particular the separation between materials and geometrical aspects remains valid.

22.4   The ductile-brittle transition

Failure of a structure can occur by global breaking (fracture) or by extensive plastic deformation (global flow). Whether one or the other failure mode is present depends on the material properties and the characteristic size of the defects in the structure. The term 'defect' does at the level of a complete structure like a ship, a reactor vessel or a storage container not only refer to cracks or inclusions but also to (functional) holes like doors, windows, manholes, ventilation slits, etc. Since fracture is usually a fast process as compared to plastic deformation and therefore more hazardous, it is from a safety point of view important to know already in the design stage whether a different shape may change the failure mode from brittle fracture to plastic deformation or the other way around. We illustrate this using a simple plate (Fig. 22.10) as an example.

Yield occurs when the maximum shear stress $\sigma^{(s)}$ equals the yield strength or

$$\sigma^{(s)} = \sigma_{\mathrm{I}} - \sigma_{\mathrm{III}} = k$$

where σ_{I} and σ_{III} denote the maximum and minimum principal stress, respectively, and k the yield strength in shear. In the case of brittle fracture the normal stress $\sigma^{(n)}$ on the crack plane reaches a critical value given by

$$\sigma^{(n)} = \frac{K_{\mathrm{Ic}}}{\alpha\sqrt{a}} \tag{22.25}$$

with, as before, K_{Ic} the fracture toughness and a the defect size. The answer to the question: flow or fracture? thus depends on whether the critical shear stress (or yield strength) or the critical normal stress (or fracture strength) is reached first. We use Mohr's circle to illustrate the situation. In plane stress

$$\sigma_{\mathrm{I}} = \sigma_{\mathrm{II}} = K_{\mathrm{I}}/\sqrt{2\pi r} \qquad \text{and} \qquad \sigma_{\mathrm{III}} = 0$$

This implies that with increasing stress Mohr's circle expands but remains 'attached' to the origin (Fig. 22.11). In plane strain though,

$$\sigma_{\mathrm{I}} = \sigma_{\mathrm{II}} = K_{\mathrm{I}}/\sqrt{2\pi r} \qquad \text{and} \qquad \sigma_{\mathrm{III}} = \nu(\sigma_{\mathrm{I}} + \sigma_{\mathrm{II}})$$

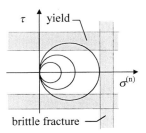 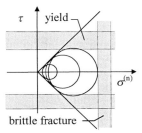

Fig. 22.11: Mohr's circles showing competition between flow and fracture in plane stress (left) and plane strain (right).

In this case the Mohr circle not only expands but also shifts away from the origin (Fig. 22.11). So, we see that depending the magnitude of k, K_{Ic}, a, plane stress or plane strain, the plate may fail by plastic flow or by brittle fracture. We call this phenomenon the ductile-brittle transition because for fixed k (or Y) and K_{Ic} one can by altering the 'defect' parameter a change from ductile to brittle failure. In the remainder of this section a simple way to deal with the ductile-brittle transition is discussed more quantitatively.

To that purpose we consider a plate with unit thickness and width W. The plate contains a central crack of length $2a$ and is remotely loaded by a force per unit thickness F. Hence the applied stress $\sigma = F/W$ (Fig. 22.10). For the moment we take the plate wide with respect to the crack length. In fracture the material has a constant fracture energy R while in flow it shows elastic-perfect-plastic behaviour with yield strength Y.

Let us first discuss fracture using the classical Griffith analysis. We have seen that, considering a balance between elastic energy and fracture energy, fracture occurs when the strain energy release rate $G \geq G_{cri} = 2R$. We have also seen that it has become customary to rephrase this equation a bit by introducing the fracture toughness $K_{Ic} = (2E'R)^{1/2}$, where E' is the appropriate elastic modulus given by $E' = E/(1-v^2)$ for plane strain conditions and by $E' = E$ for plane stress conditions. As usual, E denotes Young's modulus and v Poisson's ratio. So an equivalent description is given by $K_I \geq K_{Ic}$, where K_I is the stress intensity factor given by $K_I = (E'G)^{1/2} = \alpha\sigma\sqrt{a}$. The shape of the crack (for a certain loading condition) is reflected in the shape factor, which for $a/W \ll 1$ and an ideally sharp crack, as is the case here, is given by $\alpha = \sqrt{\pi}$. Hence fracture occurs when

$$\sigma \geq \frac{1}{\alpha}\sqrt{\frac{2E'R}{a}} = \frac{1}{\alpha}\frac{K_{Ic}}{\sqrt{a}} \qquad (22.26)$$

The fracture strength, i.e. the stress at fracture, thus decreases with increasing crack size. We also note that fracture is a process requiring an amount of energy proportional to the crack surface and requires R J/m^2.

For other configurations similar considerations apply. Consider, for example, again a plate of the above-mentioned material with unit thickness but now of width W comparable to the central crack of length $2a$, again loaded remotely with a force per unit thickness F (Fig. 22.10). The stress intensity factor K_I for crack in such a plate is dependent on the plate width W and is given to a good approximation by

$$K_I = \left[\frac{W}{a}\tan\left(\frac{\pi a}{W}\right)\right]^{1/2}\sigma\sqrt{a} \equiv \alpha\left(\frac{a}{W}\right)\sigma\sqrt{a} \tag{22.27}$$

For small values of a/W, the shape factor approaches $\sqrt{\pi}$, the value for an infinite plate, as it should do. For an edge crack in a plate, where a is now the crack length itself, the same shape factor applies but with an extra factor 1.12. In all cases, fracture occurs when $K_I \geq K_{Ic}$. Essentially only the shape factor α is dependent on the geometry of the problem.

Let us now direct our attention to flow. For overall plastic deformation (plastic collapse) the applied stress σ has to surpass the uniaxial yield strength Y of the material. In the simplest approach and considering the same structure the load is carried homogeneously by the non-broken part of the plate. Hence if $\sigma W/(W-2a) > Y$, flow will occur. More generally we may write that flow occurs when $\beta\sigma > Y$ or when $\sigma \geq \beta^{-1}Y$. Generally the plastic collapse stress can be determined by limit analysis (see Chapter 14). We note that global flow is a process proportional to the volume of the structure and requires an energy $Y^2/2E'$ J/m^3.

Next we define the dimensionless applied stress σ^*, the dimensionless fracture stress σ_{fra}^* and the dimensionless crack parameter a^* by

$$\sigma^* = \frac{\sigma}{Y} \qquad \sigma_{fra}^* = \frac{K_{Ic}}{\alpha\sigma\sqrt{a}} \qquad a^* = \frac{a}{(E'R/Y^2)} \tag{22.28}$$

So, flow occurs whenever σ^* reaches its critical value for flow

$$\sigma_{pla}^* = 1/\beta \tag{22.29}$$

first. On the other hand, brittle fracture occurs if σ^* first reaches

$$\sigma_{fra}^* = \frac{1}{\alpha}\sqrt{\frac{2}{a^*}} \tag{22.30}$$

A crossover from plastic collapse to brittle fracture occurs when the fracture strength equals the flow strength, i.e. when $\sigma_{fra}^* = \sigma_{pla}^*$ or when

$$\frac{1}{\alpha}\left(\frac{2ER}{a}\right)^{1/2} = \frac{Y}{\beta} \tag{22.31}$$

Hence the crossover occurs at a crack length $a^* = 2(\beta/\alpha)^2$ or, equivalently, at $a =$

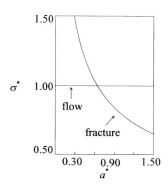

Fig. 22.12: Fracture and flow for a wide plate with a central crack of length $2a$ ($\alpha = \sqrt{\pi}$).

$2(\beta/\alpha)^2 ER/Y^2$. For the wide plate $a^* = 2/\pi = 0.64$. The next step is to plot this information in a diagram of stress versus crack length. Fig. 22.12 shows plots of the critical values of σ^* for flow and fracture as a function of the crack parameter for a wide plate ($\beta = 1$). Two regimes can be recognised. For $a^* < 2/\pi$, flow results in the lowest critical stress and thus the structure will flow. For $a^* > 2/\pi$, fracture results in the lowest critical stress and thus the structure will fracture.

Summarising in general terms, since fracture occurs at $\sigma_{fra} = \alpha^{-1}(2E'R/a)^{1/2}$ and flow occurs at $\sigma_{pla} = Y/\beta$ a transition between fracture and flow occurs when $\sigma_{fra} = \sigma_{pla}$ or when $a = (\beta/\alpha)^2 2E'R/Y^2$.

The result for the plate is qualitatively the same as for a material containing defects. It thus appears that for failure processes a critical length C is present given by:

$$\blacktriangleright \qquad C = \frac{\beta^2}{\alpha^2}\frac{2E'R}{Y^2} \qquad\qquad\qquad (22.32)$$

If the characteristic defect in the system is smaller than the critical value, flow will occur. For a characteristic defect length above that value, fracture will result. Most interesting is the fact that the value for C is a product of two factors: one (β^2/α^2) representing the geometrical features of the process and one $(2ER/Y^2)$ representing the material properties. This implies that both the material properties and the geometry (design) can be used for optimisation of the structure at hand.

Of course, in the transition region neither the treatment of the brittle part nor of the plastic part of the model is entirely correct. In particular the brittle part needs adjustment, e.g. by applying small scale yielding fracture mechanics and the use of an appropriate equivalent stress. This, however, does not change the basic procedure and leads to only to relatively small differences in predictions in the transition region between fracture and flow due to the different values of the shape factor effect α^{-2}. Also the plastic part needs some adjustment since the stress is not homogeneously distributed over the smallest cross-section and hardening may be present. Again, this does not change the basic procedure. Therefore one can say that for any mechanically loaded structure a balance between fracture and flow is present. In assessing the structural integrity of a structure both failure mechanisms should be considered before predictions can be made and one should be aware that a transition of failure mechanism can take place. In the case compressive stresses play a significant role, buckling should be considered in a similar way as done here for flow and fracture. We also considered here only simple structures loaded in load control. For displacement control in detail different results will occur. However, the essential result, namely that a critical length is present, which is constructed from a factor dependent on the materials properties only and one dependent on geometry only, remains. Both Chell (1979) and Atkins and Mai (1985) provide a more detailed review, while Atkins and Mai also illustrates the concepts by applying them to abrasive processes.

*The R-6 model**

In this section we show a few improvements leading to an often-used practical model. From Chapter 21 we know that the J-integral J and crack opening displacement δ are connected via the yield strength Y as expressed by $J = Y\delta$ so that we have

704 22 Applications of fracture theory

$$J = Y\delta = Y\left(\frac{8Y}{\pi E'} a \ln \sec \frac{\pi \sigma}{2Y}\right) = \frac{8Y^2}{\pi E'} a \ln \sec \frac{\pi \sigma}{2Y}$$

Noting that $K_I = \sigma(\pi a)^{1/2}$ and that for small scale yielding (SSY)

$$J_{ssc} = K_{Ic}^2/E' = S_a^2 \pi a/E'$$

we obtain for the strength S

$$S = Y \frac{2}{\pi} \cos^{-1}\left[\exp\left(\frac{-\pi K_{Ic}^2}{8aY^2}\right)\right] = Y \frac{2}{\pi} \cos^{-1}\left[\exp\left(\frac{-\pi^2 S_a^2}{8Y^2}\right)\right]$$

where S_a is, as in Chapter 21, given by $S_a = K_{Ic}/(\pi a)^{1/2}$. When $S/Y \ll 1$ the conventional LEFM result is obtained but as $a \to 0$, $S \to Y$ instead of $S \to \infty$. Hence this approach displays in a natural way the transition from fracture to flow. From the same expressions we have as well

$$\frac{J}{J_{ssc}} = \frac{8}{\pi^2}\left(\frac{Y}{\sigma}\right)^2 \ln\left[\sec\frac{\pi\sigma}{2Y}\right] \cong 1 + \frac{\pi^2}{24}\left(\frac{\sigma}{Y}\right)^2 \tag{22.33}$$

where the second step is a series development to first order. The term $(\pi^2/24)(\sigma/Y)^2$ is about 18% smaller than the Irwin first-order correction expressed by $\frac{1}{2}(\sigma/Y)^2$. At fracture we have

$$\sigma = S \quad \text{and} \quad J = J_{cri} \tag{22.34}$$

and defining K_{cri} via

$$J_{cri} = K_{cri}^2/E' \tag{22.35}$$

we can also write

$$(J_{cri}/J_{ssc})^{1/2} = K_{cri}/K_{Ic} \tag{22.36}$$

Therefore, we can determine K_{cri} for intermediate scale yielding when S is measured in a fracture test from

$$\left(\frac{J_{cri}}{J_{ssc}}\right)^{1/2} = \frac{K_{cri}}{K_{Ic}} = \frac{Y}{S}\left[\frac{8}{\pi^2}\ln\left(\sec\frac{\pi S}{2Y}\right)\right]^{1/2} \tag{22.37}$$

where the yield strength Y and the crack length a are known. The latter is required since $K_{Ic} = S(\pi a)^{1/2}$. Vice versa, if J_{cri} (or K_{cri}) is known, S can be predicted.

Eq. (22.33) is one way to describe[m] failure incorporating both linear elastic fracture ($S/Y \gg 1$) and (relatively) large scale yielding ($S/Y \to 1$). If the yield strength is replaced by the plastic collapse stress σ_{col} (see Chapter 14), the resulting failure assessment curve is denoted as R-6 curve.

A useful representation of that curve is obtained by plotting the relative stress $S_{rel} = \sigma/\sigma_{col}$ and the relative stress intensity $K_{rel} = K_I/K_{cri}$. The R-6 curve in terms of the relative co-ordinates is given in Fig. 22.13. It reads

$$K_{rel} = \left[\frac{8}{\pi^2 S_{rel}^2}\ln\sec\left(\frac{\pi S_{rel}}{2}\right)\right]^{-1/2}$$

---

[m] Harrison, R.P. Loosemore, K. and Milne, I. (1996), CEGB report R/H/R6, Central Electricity Generating Board, UK.

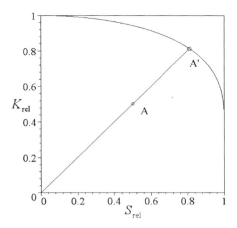

Fig. 22.13: The R-6 curve in relative co-ordinates. The line OAA′ indicates the increase in stress at constant crack size *a*.

Since both the relative strength S_{rel} and the relative stress intensity K_{rel} are proportional to the applied stress, the distance from the origin to a point on the failure curve is also proportional to the stress. For a crack of fixed length the point (K_{rel}, S_{rel}) is displaced along a ray through the origin when the applied stress is changed and failure occurs when the point reaches the R-6 curve. Hence, the failure stress is given by $f\sigma$, where f is the safety factor given by the ratio OA′/OA (Fig. 22.13).

A detailed comparison with FEM calculations for various plane strain configurations shows that a universal failure curve does not exist but that the differences between various calculations are relatively small. It appears that the R-6 curve approximates a lower bound for various structures (with some exceptions) and thus provides a useful design tool. To account for variability of materials properties sometimes an overall reduction by 15% is applied to the failure curve.

So far, using the two-criteria approach, it has been assumed that the material behaves elastic-ideal plastic, i.e. no work hardening occurs. Therefore the yield strength is the only material parameter. Because for work hardening materials this results in conservative estimates, it is customary to replace the yield strength Y with the flow strength $Y' = \frac{1}{2}(Y+S)$ where, as usual, S denotes the strength. Furthermore, in plane strain the restriction of the out-of-plane deformation puts a constraint, which effectively increases the yield strength with about 15%. In plane strain thus it is customary to use as flow strength $Y'' = 1.15Y'$ to incorporate this effect. Of course, this is an approximate procedure and for materials that exhibit appreciable work hardening this procedure is questionable. More sophisticated procedures, taking explicitly into account work hardening, exist for which we refer to the literature, e.g. Kanninen and Popelar (1985), for further details. Finally, it must be said that the experimental evidence suggests that the fracture criterion should be written as $J_{cri} = K_{cri}^2/E' = mY\delta$ where m is a dimensionless constant with a value between 1 and 2, depending on structural geometry, the degree of stress triaxiality and work hardening capacity. However, in spite of these shortcomings, the failure assessment diagram is appealing because of its simplicity.

Problem 22.3

In this section we initially used $K_I = \left[(W/a)\tan(\pi a/W)\right]^{1/2}\sigma\sqrt{a}$ as the expression for the stress intensity. Another approximate expression for the stress intensity factor K for a finite width plate, resulting from the Dugdale analysis, is $K = \sigma(\pi a)^{1/2}\cos^{-1/2}(\pi a/2W)$. Calculate K according to Eq. (22.27) and according to the above expression for $W = 50$ mm and $2a = 16$ mm, using an applied stress of $\sigma = 350$ MPa. Plot both expressions and comment on the differences.

Problem 22.4

Consider a steel plate with Young's modulus $E = 200$ GPa, Poisson's ratio $v = 0.3$, fracture toughness $K_{Ic} = 150$ MPa m$^{1/2}$ and yield strength $Y = 800$ MPa. The plate has a width $W = 0.2$ m and a length that can be considered as infinite. The plate is loaded in tension along the length. The plate thickness is such that it is fully in plane stress. In this plate centrally a crack is introduced with a length of $2a$. The shape factor for this plate is given in Problem 22.3.
a) Sketch the critical fracture stress as a function of the crack size.
b) Sketch the critical yield stress as a function of the crack size.
c) Discuss how the plate will behave under tensile loading as a function of the crack size.
d) How will the behaviour change if the width can be considered as infinite?
e) Discuss the influence of effect of the plastic zone on the deformation behaviour.

Problem 22.5*

Show that for a plate with width W failure occurs according to the R-6 approximation at $a/W = 0.66$ when $\sigma/Y = 0.25$ and $\sigma(\pi W)^{1/2}/K_{cri} = 0.75$.

22.5   Fracture in processes

Fracture is often a nuisance since it leads to non-functioning of systems but it can also be used in processes to obtain shapes of materials or surface conditions that are required. A well-known example is glass cutting. Glass cutting is actually running a crack along a guiding scratch. Amongst the other useful processes there are milling, erosion, grinding, lapping and polishing. To quote Bilby, fracture can be put at work. *Milling* is the process in which small particles are made smaller by breaking them in a mill. The most frequently used mill is a *ball mill*, a rotating vessel filled with the powder and *milling stones*. The latter are small pieces of hard, abrasion resistant materials usually in the shape of balls or short cylinders. During rotation the balls impact upon each other with the powder particles in between. The impact results in high stresses in the particles, which lead to fracture. During the impact a competition between fracture and flow is present. Quite generally the transition length between fracture and flow was shown to be proportional to ER/Y^2. Hence there is a lower limit that can be obtained by milling. This limit is obviously material dependent but typically about 0.1 to 1 μm. The milling process is abundantly used in the process industry and in powder technology.

Abrasive machining is a collective noun for processes in which mechanically loaded abrasive grains are involved, with the purpose to remove a certain amount of material from a component in a controlled way. In the field of abrasive machining one can distinguish three basic modes, based on the difference in the degree of freedom of an abrasive grain during machining.

- *Free Abrasive Machining* (FAM). The abrasive particles are fully free. Powder (or grit) blasting, also called erosion, is an example. *Erosion* is a process where an air stream containing small abrasive particles is blown against a workpiece. The impact of the abrasive particles lead to several types of cracks, which in turn lead to break-out of small material pieces. The efficiency of the process typically is 5%, i.e. to remove 5 gram of material 100 gram powder has to be used. Advantages are that it is possible to produce rather detailed structures and that the price is relatively low.
- *Contained Abrasive Machining* (CAM). In this process the abrasive particles are more or less contained between the workpiece and the tool. Polishing, lapping and wire sawing are examples. In *lapping* and *polishing* the small abrasive particles are free to move between the workpiece and a counter plate. Polishing and lapping are similar except that polishing is generally a more gentle process. Extremely flat surfaces can be obtained by polishing, in particular if combined with chemical action. Dependent on the material characteristics, an average roughness below 1 nm is possible.
- *Bonded Abrasive Machining* (BAM). In this category the abrasive particles are fully bonded in a matrix. Examples are grinding, sawing and drilling. These are the most frequently used techniques for ceramics machining. In *grinding* a wheel containing many abrasive particles is rotated with slight contact with a workpiece so that the scratches that arise deform the surface and/or cause breakout of small pieces of the material. Grinding can be done with the rotation axis either parallel to the surface or perpendicular to the surface.

In all these processes brittle cracking as well as plastic deformation is important. While cracking is related to the fracture toughness, plasticity is related to the yield stress, thus the hardness. The phenomena during abrasive machining are determined, apart from the above-mentioned material parameters, also by the machining forces, machining rate, temperature and environment (coolant) and type of abrasive. Dependent on these parameters either plastic deformation or brittle fracture may dominate. The environment can play an important role too. In particular in drilling and polishing proper use of additives can ease the process or yield results difficult to obtain otherwise.

While for fracture during milling the process description is generally in statistical terms, for erosion, lapping and grinding fracture mechanics based models are proposed. We discuss some of them in the next sections.

22.6   Bonded abrasive machining*

The most frequently applied techniques in the category bonded abrasive machining are grinding and sawing with diamond grinding wheels or sawing blades. In both cases the basic mechanism is the same. Detailed information can be found in various proceedings (Schneider and Rice, 1972, Hockey and Rice, 1979, Subramanian and Komanduri, 1985).

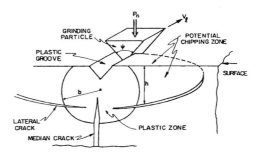

Fig. 22.14: Schematic of the interaction process during grinding. Below a particle with load P and velocity v, a plastic zone with radius b is present. The depth of lateral cracks is denoted by h and it is assumed that the lateral and median cracks originate approximately at this depth.

Grinding can be considered as repeatedly grooving with single diamond grains, bonded in a wheel (Fig. 22.14). With increasing load during grooving, plastic deformation occurs and median cracks appear while during unloading lateral cracks appear. The lateral cracks are responsible for the material removal while the median cracks act as mechanical flaws. For cracking a threshold load exists. Below this threshold only plastic deformation takes place while above also cracking occurs. For coarse grinding fracture is dominating. In this case the co-operation of the cracks developed by different diamonds may result in considerable amount of chipping (Fig. 22.15). An extensive description of the damage is given by Marshall et al. (1983).

The grinding wheels consist of a binder (a metal like brass or a plastic resin) in which the diamond grains are dispersed. In case of a metal wheel diamonds can also be bonded galvanically, e.g. by Ni. Dependent on the requirements for the workpiece, one chooses the size of the wheel, concentration of the diamonds, type of binder and machine settings. The latter are the width of the wheel b, the circumferential speed of the wheel v_s, the cutting depth a, and the feed rate v_w (Fig. 22.16). Also the choice between single and multiple pass grinding must be considered. The choice of parameters is still often determined empirically in spite of the fact that some guidance from theory can be obtained. The tangential force F_t and normal force F_n determine the (effective) coefficient of friction for grinding. Detailed analysis of grinding is complicated because of the many, interacting parameters, such as the type of grinding

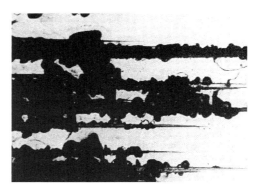

Fig. 22.15: Co-operation of separate cracks leading to chipping during grinding of a Mn-Zn ferrite, as observed by optical microscopy during first contact (with a wheel speed of $v_s = 75$ m/s).

wheel, the stiffness of the grinding machine, the cooling liquid and, of course, the material to ground. Since the whole grinding system is a closed loop mechanical system, vibrations, related to the stability (stiffness and damping of the equipment, play an important role.

Before using a wheel on a grinding machine, it is usually 'trued'. The trueing process is required for removing initial eccentricity of the wheel to a level less than the average grain size in the wheel. This is done by grinding over a coarse grained, porous brittle material ('dressing' stone), for which often carborundum is used. In use grinding wheels do 'load', which means that the surface of the wheel becomes smeared and the abrasive grains do not protrude sufficiently, resulting in a loss in efficiency and quality during grinding. Therefore they have to be 'dressed' regularly, that is manipulated in such a way that adhering material is removed and the grains protrude sufficiently again. This is normally also done by grinding a dressing stone.

Improving quality

In normal grinding procedures a relatively broad distribution of diamond grit sizes in the grinding wheel is used. This results in a wide range of loads, creating widely different lateral crack sizes. Since also widely different median crack sizes originate, the strength in principle degrades and sometimes becomes more variable. The decrease in strength is frequently compensated though by the compressive stresses introduced by the plastic deformation. Moreover, since the grinding operation can be considered as a closed loop mechanical process, the stiffness of the grinding machine, which controls the amount of vibrations, is also of importance. In conventional grinding the specific material removal rate is always achieved by using small cutting depth and high feed rates. High cutting depth and low feed rate also enable the same amount of material to be removed. This process is called creep feed grinding. The major difference between these two processes is the size of contact area between the grinding wheel and workpiece (Fig. 22.16). In case of creep feed grinding the contact length is essentially larger than in conventional grinding. Consequently the material removal rate is distributed over a large number of cutting edges (grains) so that the

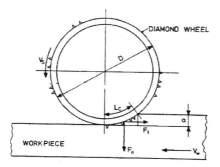

Fig. 22.16: Grinding wheel in contact with workpiece, indicated are the radius and circumferential speed of the wheel R and v, respectively, the cutting depth a, the feed rate v. The normal and transverse forces are F_n and F_t, respectively. The difference in contact situation for conventional grinding (high feed rate v, low cutting depth a) and for creep feed grinding (low feed rate v, high cutting depth a) is essentially the larger contact area in case of creep feed grinding resulting in larger cutting forces.

depth of cut of the individual grains is smaller. The creep feed process has considerable advantages since a higher strength with improved reproducibility and surface quality can be obtained. The greater number of grains in contact entails a rise in cutting force, however, requiring a stiffer machine. Due to the decreased chip thickness (see next section) higher cutting speeds are possible. Optimising this process can lead to double removal rates meanwhile increasing strength from 15 to 50%.

In view of the above-mentioned mechanisms, in a number of cases extremely stiff machines with carefully prepared and dressed grinding wheels are used[n]. This results in so-called *damage free* or *ductile grinding* which may increase the strength of the ground product considerably as compared with normal grinding, however, accompanied by a considerable increase in cost. An example can be found in 3-point bend strength of sialon, ground conventionally and 'damage free'. For the former type a strength of 482 MPa was reported while for the latter an increase to 875 MPa was observed.

The above-mentioned improvements do have a price. Mentioned are the high grinding forces and therefore the need for an extremely stiff machine. Also important are the high wear rate of the grinding wheel or insufficient removal of the debris, which can lead to instability of the process and the difficulty to use finer grained abrasives.

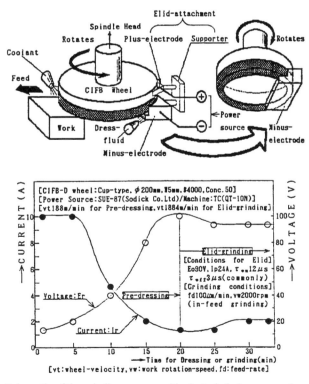

Fig. 22.17: Schematic of the grinding process with electrolytic in-process dressing (upper). Electrical behaviour during predressing and grinding (lower).

---

[n] Shore, P. (1990), Brit. Ceram. Soc. Proc. 46, page 189 in *Advanced engineering with ceramics*, Morrell, R., ed., The Institute of Ceramics, Stoke-on-Trent.

A recent development[o] is grinding with electrolytic in-process dressing (ELID). This process, illustrated in Fig. 22.17, avoids the problems mentioned above and has the additional advantage of increased quality. A power source, such as used in electro-discharge machining or in an electrochemical apparatus, is added to a conventional grinding machine. A metal bonded diamond-grinding wheel becomes the positive electrode and a negative electrode, usually made from copper or graphite, is positioned opposite to the wheel. In the clearance between grinding wheel and negative electrode electrolysis occurs by supply of an electrical current by which the metal of the wheel is oxidised. Thereby the protrusion of the diamond grains on the surface of the wheel is kept effectively constant. The process is as follows: after a precision trueing process, the wheel is predressed electrolytically (Fig. 22.17), typically for 10 to 15 min. In this process proper protrusion of the grains is achieved. During predressing the oxide generated covers the wheel, thereby increasing its resistance. Consequently the current drops and the voltage rises to the maximum of the power source. During grinding the oxide layer is partly removed by the grinding action but restored by the electrolysis so that a steady state arises. In the steady-state process continuously new grains emerge resulting in constant average protrusion of grains, greatly improving the grinding ability and removing the problem of 'loading'.

The method can be used for coarse, fine and mirror surface quality grinding. It is important to note that no special equipment is required and a variety of conventional grinding machines can be used due to the reduced grinding forces. Frequently cast iron (possibly fibre reinforced) diamond wheels are used, but other metals will do as well. The diamond grain size that can be used ranges from around a 100 μm (conventional quality resulting in $R_{max} \cong 500$ nm) via a few micrometre (fine quality resulting in $R_{max} \cong 50$ nm) to submicrometre (mirror quality resulting in $R_{max} \cong 10$ nm). In the latter case hydrostatic bearings are recommended.

Finally it should be mentioned that it has been tried to reduce the brittleness of ceramics by heating the surface so that turning can become possible. Heating by flame as well as lasers has been pursued. At present these developments seems to be applicable in a laboratory environment only.

Classical approach

In view of the many individual events that comprise the grinding process, modelling of grinding is almost inevitably dealing with an average approach. Grinding requires energy and therefore we discuss in this section the classical, energy-based approach. The specific (grinding) energy e is given by $e = W/Z$, where W is the power delivered by the equipment. This power is consumed by the interaction of the grinding wheel and workpiece and is thus on average equal to $W = F_t v_s$. The rate of volume removal Z can easily be calculated by $Z = abv_w = F_t v_s/e = F_n v_s/H$, where a is the depth of cut, b the width of the wheel, v_w the feed rate and H the grinding hardness. Combination yields

$$F_t = W/v_s = eZ/v_s = eb(av_w/v_s) \equiv ebh_{eq} \quad \text{and} \quad F_n = ZH/v_s = Hbh_{eq}$$

The *equivalent cutting depth* h_{eq} is a parameter combining the machine settings. The equivalent cutting depth represents the average thickness of a layer of debris continuously spread over the surface of the grinding wheel. Because of this, h_{eq} is also

[o] Ohmori, H. and Nakagawa, T. (1990), Annals CIRP **39**, 329.

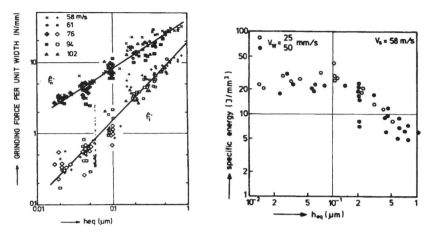

Fig. 22.18: Relation between cutting forces Fn and F_t and equivalent cutting depth h_{eq} for SrFe$_{12}$O$_9$ (left) and specific grinding energy e and equivalent cutting depth h_{eq} for Mn-Zn ferrite (right).

called the *continuous chip thickness*. It is often used for comparative purposes. For coarse grinding a typical value of h_{eq} is 1 μm while for fine and ultrafine grinding h_{eq} is typically about 0.01 and < 0.001 μm, respectively.

The forces are thus supposed to be proportional h_{eq} and the specific energy e to be independent of h_{eq}. in Fig. 22.18. In Fig. 22.18 (left)[p] the cutting forces for machining polycrystalline SrFe$_{12}$O$_9$ are shown. The fact that the cutting forces for various velocities fit well to a single line indicates that h_{eq} indeed can be used as a single parameter for various settings. In Fig. 22.18 (right) the grinding energy e for Mn-Zn ferrite is given, which appears to depend on h_{eq}. This behaviour can be rationalised in terms of single grooves. Roughly at low values of h_{eq} the material removed corresponds to that of the grooves while at high h_{eq} the effect of chipping becomes important, thereby increasing the removal rate Z. This is equivalent to lower effective value of e. Indeed the amount of chipping for Mn-Zn ferrite is much larger than for SrFe$_{12}$O$_9$. This is directly related to the value for the fracture toughness K for both materials: $K \cong 1.1$ MPa m$^{1/2}$ for Mn-Zn ferrite and $K \cong 2.1$ MPa m$^{1/2}$ for SrFe$_{12}$O$_9$. However, it is clear that a more elaborate grinding model, i.e., one that goes beyond simple grooving and takes the detailed interaction between grinding wheel and workpiece into account, is required.

Modern approach

A number of drawbacks could be raised to the classical approach to grinding. The existence of many energy-dissipating mechanisms, which are difficult to quantify such as losses in the bearings, is one of the most important ones. The classical approach can be compared with weighing the captain of a ship via weighing the ship with and without captain. The modern approaches therefore consider the plastic deformation and cracking associated with a single scratch without considering the total energy

[p] Broese van Groenou, A. and Brehm, R. (1979), page 61 in *The science of ceramic machining and surface finishing II*, Hockey, B.J., Rice, R.W., eds. NBS-SP 562, Nat. Bureau Standards, Washington DC.

balance. Most researchers start from single scratch (or indentation) results considering only the normal forces and without considering the interaction between scratches or indentations. It can be rationalised that the effect of lateral forces is minor. Moreover, fortunately, the effect of crack-crack interaction as judged from indentations appears too small[q]. These modern approaches have also their drawbacks. We mention in particular the complexity of the description of the unit event (a groove or an indentation) and statistical aspects of the process. In the following the modelling as given by Evans and Marshall[r] is used.

The basic operation for grooving is shown in Fig. 22.14. For material removal the lateral cracks are the most important ones. As mentioned a threshold below which no fracture occurs exists. For static indentation the threshold $P*$ is dictated by hardness H, fracture toughness K, and elastic modulus E according to

$$P* = \xi(K^4/H^3)f(E/H)$$

where the function $f(E/H)$ and the dimensionless constant ξ depend upon the type of crack. For lateral cracks $f(E/H)$ appears to be a slowly varying function of E/H and $\xi f(E/H)$ is approximately 2×10^5. For the median crack system, a similar analysis is in good agreement with experiment. In case of moving indentation (a scratch or groove) the stress is modified in such a way that the tensile stresses at the surface increase (see Appendix E) which results in a lower threshold load for fracture. The threshold load for various inorganic materials is low, typically in the order of a few tenth of a Newton. In many cases the loads on individual (diamond) grains is well above this threshold.

The lateral cracks develop due to the residual stress created by the indentations (or scratches). The maximum extension is thus realized when the load is removed. The cracking behaviour was modelled by regarding the plastically deformed zone as a precompressed spring, which drives the crack extension. This spring is considered to be a thin elastic plate, clamped at its outer edge to a rigid substrate. For axisymmetric indentation the lateral crack length c is given by (see Appendix E)

$$c = c_L\left[1-(P*/P)^{1/4}\right]^{1/2}$$

where P is the peak load during particle penetration. It has been shown that, for loads sufficiently above the threshold, the crack length for a two-dimensional lateral crack system accompanying the indentation is approximately given by

$$c = c_L \cong \alpha_1(E/H)^{2/5}P^{5/8}/H^{1/8}K^{1/2} \tag{22.38}$$

where α_1 is a material-independent constant depending on indenter shape that can be calibrated using materials with known fracture data. For a groove a comparable result is obtained, namely

$$c = c_L \cong \alpha_2(E/H)^{3/5}P^{5/8}/H^{1/8}K^{1/2} \tag{22.39}$$

where α_2 is another material-independent constant. The only difference is in the exponent for E/H but since the range of that value is limited, this factor is considered to be of minor importance.

[q] Buijs, M. and Martens, L.A.A.G. (1992), J. Am. Ceram. Soc. **75**, 2809.
[r] Evans, A.G. and Marshall, D.B. (1981), page 439 in *Fundamentals of friction and wear of materials*, D.A. Rigney, ed., Am. Soc. Metals, Metals Park, Ohio, USA.

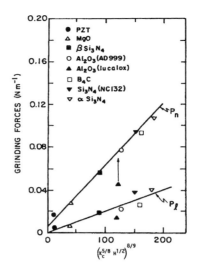

Fig. 22.19: Normal and longitudinal grinding forces for various materials showing the dependence on $K^{-1/2}H^{5/8}$.

Using this expression an estimate can be made for the material removal rate Z. The material removal rate Z_i by the passage of each particle is determined by the crack length c, the depth of lateral fracture h and the lateral velocity v_l according to

$$Z_i = 2hcv_l \tag{22.40}$$

The depth of lateral fracture h latter scales with the plastic zone radius b, which in turn relates to the plastic deformation width a by

$$b \cong a(E/H)^{2/5} \tag{22.41}$$

where the indentation width a is given by

$$a^2 \cong P_n/H \tag{22.42}$$

where P_n is the normal force. Substituting Eqs. (22.38), (22.41), (22.42) in Eq. (22.40) yields for the volume removal rate

$$Z_i \cong 2hcv_1$$

$$= \alpha_3 \left(\frac{E}{H}\right)^{4/5} \frac{P^{9/8}}{K^{1/2}H^{5/8}} v_1 \tag{22.43}$$

where P denotes again the peak load and α_3 a material-independent constant.

The predictions as given above by this model appear to be reasonably verified by the experiments available. In particular, neglecting the dependence on $(E/H)^{4/5}$, the dependence of the grinding forces on the factor $(K^{-1/2}H^{5/8})^{8/9}$ appears to be reasonably satisfied, as shown in Fig. 22.19 for a variety of inorganics. The analysis shows the importance of the fracture toughness K and hardness H. It can be further analysed that only a small influence of thermal effects is anticipated except for materials with a large thermal expansion coefficient or small specific heat. For forces more close to the fracture threshold a more complete analysis taking into account the microstructure of the material is required.

22.7 Contained abrasive machining*

Polishing and lapping are the most well known contained abrasive machining processes. In both processes material removal can take place by mechanical action only and the removal process may be assisted by chemical reactions. Polishing and lapping are similar except that polishing is generally a more gentle process.

One can distinguish between several polishing techniques. Classical is the procedure using a cloth on a flat plate and an abrasive powder dispersed in a fluid. The hardness of the abrasive is normally higher than that of the workpiece. Common abrasives are diamond, alumina, silicon carbide and boron carbide. A useful variant is the polishing by ultrafine quartz particles dispersed in an alkaline solution. This is an example of chemical polishing.

In practice polishing is done in a number of stages. Later stages use smaller particle size and include chemical effects. In some case an abrasive material is used which degrades in use so that an in-situ multistage polishing process is obtained. In the optical industry pumice is used as an abrasive material for this process.

Newer polishing techniques are float polishing and soft grain polishing. In float polishing the workpiece is mounted upside down on an upper plate. A much larger lower plate provided with grooves rotates slowly below the workpiece. The whole set-up is immersed in polishing fluid. This fluid is agitated by the rotation and the particles hit the workpiece. Thereby they remove only the outer atoms of the workpiece. A surface roughness of a few nanometre can be realised. In soft grain polishing the workpiece is moved over a tin plate covered with a dispersion of Fe_2O_3 particles (typical size 1 to 7 nm) in a fluid. In this case the hardness of the abrasive is smaller than of the workpiece. By the normal forces the particle adheres to and possibly penetrates slightly the workpiece. The following particles remove this primary particle again, together with some of the material of the workpiece. The process may be assisted by chemical action of the fluid, possibly dependent on pH. Even more so than for grinding, polishing (and lapping) is based on experience and different people obtain different results with the same material[s]. Very low roughness values can be obtained, however, by proper procedures (Fig. 22.25).

Ultrasonic machining also belongs in the category contained abrasive machining. In this technique the abrasive particles are agitated by ultrasonic energy provided by a so-called sonotrode. Highly localised material removal can be realised with this technique, e.g. it is used to drill small, deep holes.

Finally also wire sawing belongs to this category. In the technique a piece of material is cut by abrasive particles in a slurry which are agitated by a moving wire. It is a relatively 'soft' technique in the sense that only limited damage occurs. In some cases the abrasive particles are bonded to the wire for quicker material removal. Apart from changing to a bonded abrasive process, these wires, however, quickly deteriorate by wear.

Modelling of lapping

Buijs and Korpel-van Houten[t] have put forward an approach for lapping similar to the one by Evans and Marshall for grinding. Again only the normal forces are considered and no lateral interactions are allowed. The relationship between wear rate

---

[s] Clinton, D.J. (1987), *A guide to polishing and etching of technical and engineering ceramics*, Stoke-on-Trent, The Institute of Ceramics.

[t] Buijs, M. and Korpel-van Houten, K. (1993), J. Mater. Sci. **28**, 3014.

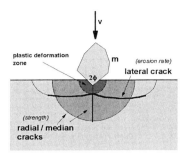

Fig. 22.20: Schematic of the indentation process. Shown are the plastic zone and median crack
formed during loading (left) and the median cracks formed during unloading (right).

Z on the one hand and load P and velocity v on the other hand has been known for
quite some time to be linear[u] (Preston's law): $Z = \alpha P v$. Modelling should, apart from
explaining the linear relationship, also provide an expression for the coefficient α.

In this case the indentation is used as the basic event. As indicated in Fig. 22.20 an
indentation is accompanied by a plastic zone, lateral cracks and median cracks. Above
the threshold load for cracking the length of the lateral cracks c in an indentation is
given according to Marshall et al. by

$$c = \alpha_0 \frac{E^{3/8}}{K^{1/2} H^{1/2}} P_i^{5/8}$$

where α_0 is a material-independent constant, P_i the normal load on the particle and H,
E and K have their usual meaning. The depth of the lateral cracks h is given by

$$h = \alpha_1 \frac{E^{1/2}}{H} P_i^{1/2}$$

where α_1 is another constant, depending on the shape of the abrasive particle. The

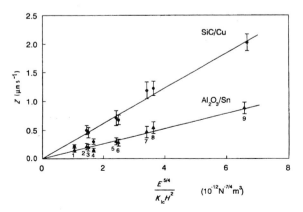

Fig. 22.21: Wear rate for various glasses as function of $E^{5/4}/KH^2$ for the abrasive system SiC
(abrasive)/Cu (lapping plate) and Al_2O_3 (abrasive)/Sn (lapping plate).

---

[u] The Evans and Marshall analysis indicates actually $Z = \alpha P^{9/8} v$ but there P refers to the load on a single
particle.

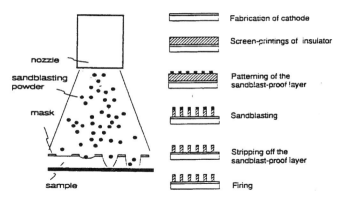

Fig. 22.22: Schematic of the erosion process (left) and process of producing ribs with small dimensions with erosion (right). The thickness of the electrode is 20 μm while the height and width of the ribs are 200 and 50 μm, respectively. These structures can be realized with a pitch of 150 μm.

volume of workpiece removed per indentation event V_i is thus given by

$$V_i = \pi c^2 h = \alpha_2 \frac{E^{5/4}}{KH^2} P_i^{7/4}$$

The total removal rate Z is evidently the sum over all indentation events and should proportional to $E^{5/4}/KH^2$. For various glasses tested this was indeed the case (Fig. 22.21). The slope of the line Z versus $E^{5/4}/KH^2$ indicates the difference in abrasive system. Moreover, the linear dependence on load and velocity was also confirmed[v]. It thus appears that lapping, as illustrated here for the case of glass, can be described with the concept of lateral fracture, originating from indentation by the individual grains during the lapping process.

22.8   Free abrasive machining*

In free abrasive machining the particles are entirely free. With a nozzle an air stream containing the abrasive particles are blown against the workpiece. Dependent on the velocity and particle type the technique can be used for cleaning surfaces or machining. In the latter case in many cases the workpiece is provided with a mask so that the technique can be used e.g. to drill holes in plates and for contouring plates. Other names in use for the process are erosion and sand (grit) blasting.

It appears to be possible to produce rather detailed structures by free abrasive machining[w]. Barrier ribs for plasma display panels (Fig. 22.22) provide an example. A thick film glass layer of 200 μm is deposited by the Doctor Blade process and dried on an electrode printed/burned out glass substrate. A photoresistive lacquer is laminated on this plate and patterned by ordinary photolithographic processing. This is more precise than directly providing the pattern by screen-printing. The lacquer acts as a mask for the erosion process. Typical erosion processing conditions are 25 μm glass powder at a pressure of 0.5 to 1.0 bar. Uniform ribs with a spacing of 500 down to 150 μm, a height of 100 to 200 μm and a width of 150 to 50 μm can be produced. The erosion rate and the selectivity can be influenced by heat treatment of the printed layer

[v] Buijs, M. and Korpel-van Houten, K. (1993), Wear **166**, 237.
[w] See e.g. Terao, Y., Masuda, R., Koiwa, I., Sawai, H. and Kanamori, T. (1992), SID **92** Digest, 724.

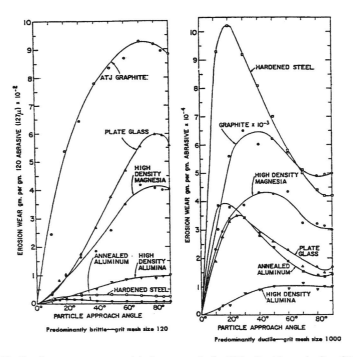

Fig. 22.23: Erosion wear versus particle impact angle for SiC grit with a velocity of 150 m/s.

and a proper choice of abrasive powder. While heat-treating changes the mechanical behaviour of the printed layer, the removal rate for materials with different hardness is different for abrasive powders with different hardness.

A relatively new development is the replacement of the particle containing air stream by a water stream (without particles!). In this case the water acts as abrasive material as well as transport medium. Applications are drilling and cutting of plates.

An essentially different behaviour is observed for brittle and ductile substrates. While brittle materials (e.g. graphite, glass, alumina) show the largest wear rate at normal incidence, ductile substrates (e.g. steel, magnesia, aluminium) often show a maximum in wear rate at non-normal incidence (Fig. 22.23). Moreover, the shape of the particles plays an important role since the threshold for fracture during indentation depends on the angularity. Also the particle flux is important. While at low flux the effect of the various impacting particles is additive and thus the wear rate increases with flux, at high flux the particles hinder each other resulting in a decreased wear rate.

Modelling of the erosion process is again based on indentation fracture mechanics. An extensive review has been given by Evans[x] and Ruff and Wiederhorn[y]. Here also similar use as before is made of indentation fracture mechanics. All considerations lead to power law expressions in E, K, H, ρ, r and v, where ρ, r and v denote the particle density, radius and velocity, respectively, while the other parameters have their usual meaning. It appears difficult, however, to establish the validity of the various exponents experimentally.

[x] Evans, A.G. (1979), Treat. Mater. Sci. Tech. **16**, 1.
[y] Ruff, A.W. and Wiederhorn, S.H. (1979), Treat. Mater. Sci. Tech. **16**, 69.

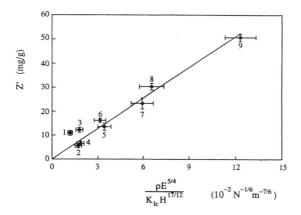

Fig. 22.24: Erosion removal rate Z' as function of the material parameters E, H, K and ρ. The solid line is least-squares fit constrained through the origin.

An estimate for the material parameter dependence can be made as follows. Modelling the erosion process as a series of quasistatic indentations, once more use can be made of the expression for the volume of work piece removed per indentation

$$V_i = \alpha_2 E^{5/4} P_i^{7/4} / KH^2$$

For erosion the load acting on the particle increases during the impact event and reaches its maximum value at maximum penetration depth. The load can estimated from $H = \alpha_3 P/d^2$, where α_3 is a constant dependent on the geometry of the indenting particle and d the depth of penetration. The maximum penetration during penetration d_m can be obtained by equating the fraction of kinetic energy W_{kin} that is lost during the impact of the particle on the workpiece with the plastic work W_{pl}

$$W_{pl} = \int_0^{d_m} P(z)\, dz = W_{kin} = \frac{1}{2} m_p v_n^2 (1 - e^2)$$

where v_n is the normal component of the velocity, m_p the mass of the particle and e the coefficient of restitution. Substitution of $P(z) = Hz^2/\alpha_3$ and integration leads to

$$d_{max} \sim H^{-1/3}$$

Substituting back in $P = Hd^2/\alpha_3$ yields $P \sim H^{1/3}$. Consequently the removal rate Z, i.e. the thickness of the layer removed per unit time, for erosion is given by

$$Z \approx E^{5/4} / KH^{17/12}$$

Frequently the removal rate is expressed in mass of workpiece removed per mass of abrasive used Z' and this results in

$$Z' \approx \rho E^{5/4} / KH^{17/12}$$

where ρ is the density of the workpiece. As shown by Buijs[z] this expression leads to good agreement with experiment for a range of glasses studied (Fig. 22.24).

---

[z] Buijs, M. (1994), J. Am. Ceram. Soc. 77, 1676.

Nowadays more extensive numerical simulations of erosion are available in which the full details of the indentation cracks and the surface topology are taken into account[aa].

22.9   Characterising finished products*

Geometrical control includes the verification of the various dimensions, roughness, non-planarity, eccentricity, etc. In many cases only a sample is tested and, when found to be satisfactory, the lot is approved. If the sample is not satisfactory, frequently another, larger sample is taken, from which a more reliable judgement can be made, before disposing of the whole lot of products.

While dimensions eccentricity and non-planarity are easily determined, roughness is somewhat trickier. Only a few remarks will be made here. An extensive review is given by Thomas (1982). Roughness measurements are usually done with a profilometer, which registers the height of a component along a certain track by tracing it with a stylus. The conventional parameter for the roughness is the centre line average (CLA) or R_a-value defined by

$$R_a = \left(\sum_i |z_i| \right) / n$$

where n is the number of points on the centre line at which the profile deviation z_i is measured. The centre line is defined as the line, which divides the profile in such a way that the areas above and below that line are equal. Another frequently used parameter is the root mean square deviation (RMS) or R_q-value, defined by

$$R_q = \left[\sum_i (z_i)^2 / n \right]^{1/2}$$

Although the definitions are simple enough, the tricky part is in the evaluation. Here two factors play a role. First, various kinds of filters are used, either in the measuring device or during data processing, which supposedly eliminate edge effects due to the finite track length registered. However, this filtering can have a large influence on the value of R_a and R_q obtained. Typically, the R_a and R_q values are reduced by 10 to 50% by applying filtering[bb]. Secondly, the centre line is frequently determined by applying such a filter repeatedly. This procedure can also cause a

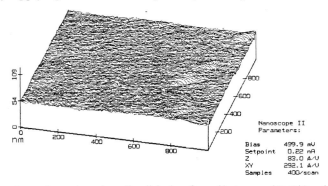

Fig. 22.25: Topology of an extremely well polished surface of hot-pressed BaTiO$_3$ with a grain size of ~0.3 μm and 99.9% relative density, as established by AFM. The R_a value is only 0.83 nm.

[aa] Verspui, M.A., de With, G., Corbijn, A. and Slikkerveer, P.J. (1998), Wear **233-235**, 436.
[bb] de With, G. and Corbijn, A.J. (1992), *Roughness evaluation of ceramic and glass surfaces*, unpublished work.

severe bias. A fairly simple but suitable approach seems to be to determine the centre line by a least-squares fitting of the entire profile by a suitable low order polynomial and to avoid further filtering as much as possible. It should be remarked that this type of measurement does not adequately characterize the surface morphology. Quite different morphologies can correspond to the same R_a or R_q values. Finally we mention the maximum roughness, R_{max}, being the maximum difference in height over the track measured. This parameter is much more sensitive to sampling.

Although contactless measurements of the surface roughness are possible for quite a while, through the invention of the scanning tunnelling microscope (STM) and atomic force microscope (AFM) and in particular the confocal microscope, they have become much easier. In Fig. 22.25 the topography of a polished surface of a hot-pressed $BaTiO_3$ with a grain size of about 0.3 µm, as determined by STM, is displayed. This figure illustrates the results that can be obtained by polishing and its representation by the AFM[cc].

Apart from size, the machined material also has changed in surface properties. In particular a residual surface stress has developed which can have a considerable magnitude. A direct way to measure these stresses is by its effect on the strength of a test piece. From the strength of an as-machined (stress containing) test piece and an annealed (stress free) test piece the effective residual stress can be evaluated.

Another way is to use indentation mechanics and measure the length of the cracks accompanying a hardness indentation. If the material parameters are known, the crack measurements in principle directly deliver the (effective) stress over the crack. In case these parameters are unknown, a comparison with annealed, that is stress free, material may be necessary. If the orientation dependence of the stress is important, it may be wise to use a Knoop instead of a Vickers indentation.

A non-destructive way is to use X-ray diffraction with the so-called $\sin^2\psi$-technique[dd]. Essentially in this technique the lattice plane distances are measured in various directions so that the strains in the those directions in the surface due to the stress can be calculated, given the stress free lattice constants. It should be noticed that, since the penetration depth of the X-ray beam is limited, this technique essentially measures the residual stress at the surface of the machined component. The effect on the strength, however, is determined by the residual stress profile over the critical defect. Therefore, the magnitude of the residual stress as determined from strength measurements on annealed and machined surfaces may be quite different from the one as determined from $\sin^2\psi$-measurements.

22.10 Bibliography

Atkins, A.G. and Mai, Y-W. (1985), *Elastic and plastic fracture*, Ellis Horwood, Chichester.

Castillo, E. (1988), *Extreme value theory in engineering*, Acadamic Press, Boston.

Chell, G.G. (1979), *Elastic-plastic fracture mechanics*, page 67 in *Developments in fracture mechanics*, G.G. Chell, ed., Applied Science Publishers, London.

[cc] de With, G. (1996), *Process control in the manufacture of ceramics*, page 27 in Materials Science and Technology, vol. 17A, R.J. Brook, ed., VCH, Weinheim.
[dd] Eigenmann, B., Scholtes, B. and Macherauch, E. (1989), page 249 in *Joining ceramics, glass and metal*, W. Kraft, ed., DGM Verlag.

Creyke, W.E.C., Sainsbury, I.E.J. and Morrell, R. (1982), *Design with non-ductile materials*, Applied Science Publishers, London.

Hockey, B.J., Rice, R.W. (1979), *The science of ceramic machining and surface finishing II*, NBS-SP 562, Nat. Bureau Standards, Washington DC.

Kanninen, M.F. and Popelar, C.H. (1985), *Advanced fracture mechanics*, Oxford University Press, New York.

Mangonon, P.L. (1999), *The principles of materials selection for engineering design*, Prentice Hall, Upper Saddle River, NJ.

Marshall, D.B., Evans, A.G., Khuri Yakub, B.T., Tien, J.W. and Kino, G.S. (1983), *The nature of machining damage in brittle materials*, Proc. Roy. Soc. London **A385**, 461.

Munz, D. and Fett, T. (1999), *Ceramics: mechanical properties, failure behaviour, materials selection*, Springer, Berlin.

Pollard, H.F. (1977), *Sound waves in solids*, Pion, London.

Schneider, S.J. and Rice, R.W. (1972), *The science of ceramic machining and surface finishing*, NBS-SP 348, Nat. Bureau Standards, Washington DC.

Subramanian, K. and Komanduri, R., eds. (1985), *Machining of ceramics materials and components*, Am. Soc. Mech. Eng., New York.

Thomas, T.R. (1982), *Rough surfaces*, Longman, London.

23

Structural aspects of fracture

In the previous chapters, the basics of continuum fracture and the application to structures and processes were discussed. In this chapter, the structural background of fracture is dealt with. We discuss first the theoretical strength. Subsequently we deal with various general aspects. After an overview of structural aspects for the various material types, we finally discuss some of these aspects in some detail.

23.1   Theoretical strength

Previously we referred occasionally to the theoretical strength σ_{the}. To estimate the theoretical strength of solids, we consider a simple lattice through which a crack is formed. The model is essentially elastic and has been used to estimate the theoretical strength of alkali halides, metals, oxides and covalently bonded materials. It is often attributed to Gilman[a] but in fact Polanyi[b] and Orowan[c] basically used it as early as 1921 and 1934, respectively. We largely follow the discussion as given by Kelly and McMillan (1986).

As has been discussed before, the lattice potential energy is complex and depends on the positions of all atoms. Frequently the pair potential approximation is invoked. While for lattice dynamics the second derivatives of the potential at the equilibrium position of all the atoms are the important features of the potential, in fracture we need a much larger part of the potential energy. In a 2D representation, we need the potential energy curve up to the inflection point of this curve. However, since the potential energy surface is specific for each substance, it is difficult to extract a general view on fracture from considerations using these energy surfaces. Therefore, a drastic approximation to a simple energy surface is made.

We assume (Fig. 23.1) that the attractive stress $\sigma(x)$ between two lattice planes can be approximated by a sine function from r_0, the equilibrium interplanar distance, up to $r_0 + \lambda$, where λ is the range on which the attractive forces differ significantly from zero, i.e.

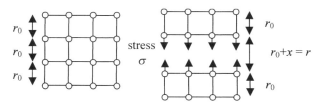

Fig. 23.1: Theoretical strength. Unloaded state (left) and separation of lattice planes (right).

[a] Gilman, J.J. (1959), page 193 in *Fracture*, Averbach, B.L., Felbeck, D.K., Hahn, G.T. and Thomas, D.A., eds., J. Wiley, New York.
[b] Polanyi, M. (1921), Z. Phys. **7**, 323.
[c] Orowan, E. (1934), Z. Krist. A **89**, 327.

$$\sigma = \sigma_0 \sin \frac{\pi x}{\lambda} \qquad \text{where} \qquad 0 < x \equiv r - r_0 < \lambda \tag{23.1}$$

Further we assume, similarly to the case of shear deformation, that all bonds are broken simultaneously (Fig. 23.1). For small displacement from the equilibrium interplanar spacing, the material should behave linearly elastic. Consequently, the strain $\varepsilon = x/r_0$ is proportional to the stress

$$\left.\frac{d\sigma}{d\varepsilon}\right|_{x=0} = \frac{d\sigma}{dx}\left(\frac{d\varepsilon}{dx}\right)^{-1}\bigg|_{x=0} = \left.\frac{\sigma_0 \pi}{\lambda} \cos\left(\frac{\pi x}{\lambda}\right) r_0\right|_{x=0} = \frac{\sigma_0 \pi r_0}{\lambda} = E \tag{23.2}$$

where E is the appropriate Young's modulus. The full expression for the attractive stress is then given by

$$\sigma = \frac{E\lambda}{\pi r_0} \sin \frac{\pi x}{\lambda} \tag{23.3}$$

The theoretical strength σ_{the} is the amplitude of the sine and thus given by $\sigma_{the} = \sigma_0 = E\lambda/\pi r_0$. The theoretical strength σ_{the} is related to the surface energy γ since for the creation of two surfaces of unit area 2γ J/m^2 is required. Assuming that all energy is converted to surface energy only and that no other dissipating mechanisms operate, we may approximate the fracture energy R by the surface energy γ to obtain

$$2\gamma = \int_{r_0}^{\infty} \sigma \, dr = \sigma_{the} \int_0^{\lambda} \sin \frac{\pi x}{\lambda} \, dx = 2\sigma_{the} \frac{\lambda}{\pi} = \frac{2\sigma_{the}^2 r_0}{E} \tag{23.4}$$

This yields expressions for the theoretical strength σ_{the} and the surface energy γ

$$\blacktriangleright \qquad \sigma_{the} = \sqrt{\frac{\gamma E}{r_0}} \qquad \text{and} \qquad \gamma = \frac{E}{r_0}\left(\frac{\lambda}{\pi}\right)^2 \tag{23.5}$$

Solids with a high elastic modulus and surface energy are thus the prime candidates for strong materials. Taking reasonable values, $\gamma = 1$ J/m^2, $r_0 = 0.2$ nm and $E = 200$ GPa, results in the estimate $\sigma_{the} \cong 32$ GPa $= E/6.3$, corresponding to a theoretical fracture strain of $\varepsilon_{the} = \sigma_{the}/E \cong 0.16$. A comparison for various materials is given in Table 23.1. Generally quite high values, of the order of several tens of GPa's, are thus obtained.

The choice of λ largely dictates the extent of agreement that is reached with experiment. Gilman[d] and others state that good agreement between the theoretical and

Table 23.1: Estimates for the theoretical strength for various materials.

Material	Direction	E (GPa)	γ (J/m^2)	$(E\gamma/r_0)^{1/2}$ (GPa)
Ag	$\langle 111 \rangle$	121	1.13	24
Ag	$\langle 100 \rangle$	44	1.13	16
Ni	$\langle 100 \rangle$	138	1.73	37
NaCl	$\langle 100 \rangle$	44	0.25	6.3
MgO	$\langle 100 \rangle$	245	1.20	37
SiO$_2$	–	73	0.56	16
Si	$\langle 111 \rangle$	188	1.20	32

experimental values of the surface energy is reached when λ is taken to be equal to the mean radius of atoms or ions in the surface plane. A more elaborate analysis[e] indicates that in this case γ is overestimated with a factor of 1.4 to 2, however.

For single crystals, the choice of λ is more or less straightforward. For polycrystalline materials, however, it is not clear what choice should be made for λ^2/r_0.

A slightly different way to estimate the surface energy is by bond energy counting. It has been shown in the literature[f] that the surface energy can be estimated as $\gamma \cong 0.25 N_A^{-1/3} V_{mol}^{-2/3} \Delta H(0)$, where N_A denotes Avogadro's constant, V_{mol} the molar volume and $\Delta H(0)$ is the atomisation energy at 0 K. For purely brittle materials like silica, diamond or boron carbide, this estimate yields reasonably accurate results.

Justification 23.1

Some simple arguments using a simple bond pair model can be advanced to rationalise the expression $\gamma \cong 0.25 \ N_A^{-1/3} V_{mol}^{-2/3} \Delta H(0)$. Consider a lattice in which only nearest-neighbour interactions are present. Furthermore, the co-ordination number is z, and we assume that the bond energy B is temperature independent. In this approach the heat of sublimation ΔH_S can be estimated as

$$\Delta H_S = \tfrac{1}{2} N_A z B$$

where N_A denotes Avogadro's number. Now consider the (111) plane in a FCC lattice where $z = 12$. There are six nearest neighbours in the plane, three above and three below. Hence after cleavage, three bonds are broken yielding two surfaces. Hence the surface energy per atom $3B/2 = 3 \Delta H_S/(2N_A \times 12 \times \tfrac{1}{2}) = \Delta H_S/4N_A$. For the number of atoms per unit surface N, we take the estimate $N \cong (N_A/V_{mol})^{2/3}$, where V_{mol} is the molar volume, so that the final result becomes

$$\gamma = N \Delta H_S/4N_A = \tfrac{1}{4} \Delta H_S V_{mol}^{-2/3} N_A^{-1/3}$$

The numerical difference between ΔH_S and $\Delta H(0)$ can be neglected.

From the above discussion, it is clear that the theoretical strength far exceeds the actual strength. The origin is obviously the presence of defects of one kind or another in the material[g]. Only a few materials exist for which these defects are (nearly) absent and the theoretical strength indeed is approached. *Optical fibres* as used in telecommunication provide one example. Freshly drawn glass fibres are defect free and to keep that condition they are already coated with a polymeric coating during cooling. Strength values up to 15 GPa, equivalent to about $E/5$, are possible for fibres with diameter in the range of 10-100 µm. Another example is provided by *whiskers*, small elongated single crystals containing only one screw dislocation (at the cylinder

---

[d] Gilman, J.J. (1959), page 193 in *Fracture*, Averbach B.L., Felbeck, D., Hah, G.T. and Thomas D.A., eds., Wiley, New York.
[e] McMillan, N.H. and Kelly, A. (1972), Mater. Sci. Eng. **10**, 139.
[f] de With, G. (1984), J. Mater. Sci. **19**, 457.
[g] Note the similarity between the arguments for the existence of dislocations in plasticity and for the existence of defects in fracture.

axis) leading to almost perfect single crystalline materials. Also some polymers can reach a strength value close to theoretical, e.g. for polydiacetylene a strength of about 1.4 GPa, equivalent to $E/40$, at a fibre diameter of about 40 µm has been obtained. Generally, however, nearly all materials contain defects and they cause considerable stress concentration, as discussed Chapter 21. Roughly speaking one can say that at the crack tip one of two situations can be present. One situation is that the yield strength is reached, as in most metals and polymers, leading to relatively blunt crack tips with limited stress concentration. The other situation is that the theoretical strength is reached, as in most inorganic materials, leading to atomically sharp crack tips with considerable stress concentration. In either case, the macroscopic strength is reduced considerably as compared with theoretical strength.

Problem 23.1

Show that the theoretical strength is estimated as $\sigma_{the} \cong 31$ GPa $\cong E/6$ by using $\gamma = 1$ J/m^2, $r_0 = 0.2$ nm and $E = 200$ GPa. Also show that the theoretical fracture strain is given by $\varepsilon_{the} \cong 0.16$.

Problem 23.2

Consider a parabolic stress-strain curve, given by $\sigma = Ar^2 + Br + C$ with r_0 as the position of the theoretical strength σ_{the}, instead of a sine curve for the stress-displacement of two lattice planes.
a) Show that $A = -\sigma/r_0^2$, $B = -2Ar_0$ and $C = 0$.
b) Show that the surface energy γ is given by $\gamma = 4\sigma_{the}r_0/6$.
c) Show that the theoretical strength σ_{the} is given by $\sigma_{the} = (3/2)^{1/2}(\gamma E/r_0)^{1/2}$.

Problem 23.3

For a low molecular weight 'van der Waals' material, the intermolecular potential can be described with the Mie potential

$$V(r) = \frac{nm}{n-m}V_0\left[\frac{1}{n}\left(\frac{r_0}{r}\right)^n - \frac{1}{m}\left(\frac{r_0}{r}\right)^m\right]$$

where r is the intermolecular distance and r_0 is the equilibrium intermolecular distance. Often for the exponents $n = 12$ and $m = 6$ is taken, in which case one speaks of a Lennard-Jones potential. Assume that the structure shows simple cubic packing and that only nearest-neighbour interactions have to be taken into account.
a) Show that for this potential Young's modulus of the material in the [100] direction is given by $E = nmV_0/r_0^3$ and for $n = 12$ and $m = 6$ by
$E = 72V_0/r_0^3$.
b) Show that the stress to separate two (100) lattice planes over a distance r is

equal to $\sigma(r) = \dfrac{E}{n-m}\left[-\left(\dfrac{r_0}{r}\right)^{n+1} + \left(\dfrac{r_0}{r}\right)^{m+1}\right]$.

c) Show that for materials that can be described by this potential, fracture occurs at

$$r = r_{\text{fra}} = \left(\frac{n+1}{m+1}\right)^{\frac{1}{n-m}} r_0.$$

d) Calculate the strain at fracture $\varepsilon_{\text{fra}} = (r_{\text{fra}} - r_0)/r_0$.

e) Show that the theoretical strength σ_{the} is equal to

$$\sigma_{\text{the}} = \frac{E}{6} \left[-\left(\frac{m+1}{n+1}\right)^{\frac{n+1}{n-m}} + \left(\frac{m+1}{n+1}\right)^{\frac{m+1}{n-m}} \right]$$

and thus for $n = 12$ and $m = 6$ to $\sigma_{\text{the}} = E/26.8$.

f) Discuss whether these low molecular weight materials will reach this strength or not.

Problem 23.4

Provide the argument why the difference between ΔH_S and $\Delta H(0)$ can be neglected.

Problem 23.5*

Verify for fused silica (amorphous SiO_2) the surface energy γ on the basis of broken bond counting using the data from Table 23.2 for the density ρ, the bond energy B and the molecular weight M. Compare the results with the experimental value of the fracture R as determined from the fracture toughness K_{Ic}, Young's modulus E and Poisson's ratio v. Compare the result also with surface energy of SiO_2. Silanol (Si–O–H) surfaces have a surface energy of 0.13 J/m^2 while for siloxane (Si–O–Si) surfaces a value of 0.26 J/m^2 generally is measured.

Table 23.2: Data for fused silica.

Theory		Experiment	
ρ (g/cm^3)	2.2	K_{Ic} (MPa m$^{1/2}$)	0.75
B (kJ/mol)	420	E (GPa)	72
M (g/mol)	60	v (-)	0.16
γ (J/m^2)	4.1	R (J/m^2)	3.8

23.2   Some general fracture considerations

In this section, we discuss several general aspects relevant for fracture phenomena. We limit ourselves to strength reduction (associated with crack tip sharpness), the nature of the fracture energy, to the temperature dependence of fracture and stress localisation.

Strength reduction and crack tip sharpness

In the previous section, we learned that the value for the theoretical strength is quite high but that in practice this value is hardly ever realised for the actual strength and we indicated that defects are responsible for this effect. When an external stress σ

is applied to a material containing a defect of size a and radius ρ, the stress is locally concentrated. If we equate the Inglis expression (Section 21.3) for the crack tip stress

$$\sigma_{tip} = 2\sigma \, (a/\rho)^{1/2}$$

to the estimate for the theoretical strength

$$\sigma_{the} = (E\gamma/r_0)^{1/2}$$

we obtain

$$\sigma = (E\gamma\rho/4r_0a)^{1/2} \tag{23.6}$$

The sharpest possible crack would be one for which $\rho \cong r_0$ and this leads for the strength S to

$$S = (E\gamma/4a)^{1/2}$$

Using the same data as before ($\gamma = 1$ J/m^2 and $E = 200$ GPa) combined with a defect size of $a = 2$ μm results in $S \cong 160$ MPa $= E/1250$. A strong reduction in the real strength S as compared with the theoretical strength $\sigma_{the} = E/6.3$ is thus the result. Let us consider the topic of crack tip sharpness a bit further, as promised in Section 21.2. Comparing Eq. (23.6) with the Griffith expression, in which we again take the surface energy γ as the only contribution to the fracture energy R so that $R = \gamma$,

$$\sigma = (2E\gamma/\pi a)^{1/2}$$

we see that both equations coincide if $\rho = 8r_0/\pi$, which corresponds to about 3 bond distances. Hence for a crack tip radius $\rho < 8r_0/\pi$, a crack can be considered as sharp and the Griffith equation $\sigma = (2E\gamma/\pi a)^{1/2}$ applies. If, on the other hand, the crack tip radius $\rho > 8r_0/\pi$, the modified Inglis equation $\sigma = (E\gamma\rho/4r_0a)^{1/2}$ applies and the crack cannot be considered as sharp. Since the estimate for the theoretical strength is only approximate, the numerical value for the estimate of the crack tip radius characterising the transition from a sharp to a blunt crack tip should not be taken too literally.

The estimates as discussed above are reasonable for completely brittle materials. Generally that encompasses materials such as glass and fully dense inorganics with covalent bonds. However, for brittle metals and even for other brittle inorganics, the estimates for the crack size using the approximation $R = \gamma$ yields unrealistic values. Therefore other contributions to R arise and in the next paragraph we discuss the influence of crack tip plasticity.

The nature of fracture energy

In the previous paragraph we limited the contributions to the fracture energy to surface energy only. It appeared experimentally that the Griffith equation can be applied for materials that show plasticity and this led Irwin and Orowan to the use of a fracture energy, defined as the sum of the surface energy γ and the energy U_{pla}, necessary for plastic deformation, i.e. $R = \gamma + U_{pla}$. Since $U_{pla} \gg \gamma$, the fracture energy is approximated by $R \cong U_{pla}$. This formulation, however, masks the nature of the fracture energy and we elaborate a bit further along the line as given by Weertman[h]. We recall that fracture starts when for a volume crack of length $2a$ subject to a stress σ the energy release rate G equals a critical value G_{cri}, related to the fracture energy R.

---

[h] Weertman, J. (1978), Acta Met. **26**, 1731.

For a linear elastic solid $G = -\partial U_{ela}/\partial a$. For a plate with a central crack, small as compared to the width of the plate, the energy release rate per unit thickness of plate[i] is given by

$$G = \alpha \pi \sigma^2 a / \mu \qquad (23.7)$$

where μ is the shear modulus and $\alpha = 1 - v$. The energy required for unit extension of the crack per unit plate width is

$$U_{sur} = 4\gamma \qquad (23.8)$$

considering surface energy only for the moment ($R = \gamma$). Equating those two expressions at the fracture stress S yields the Griffith equation

$$S = \sqrt{\frac{4\mu\gamma}{\pi\alpha a}} \qquad (23.9)$$

Let us now introduce crack tip plasticity. The energy required for unit extension of a single crack tip per unit plate width due to plastic deformation is given by

$$U_{pla} = \beta \sigma^2 a / \mu \qquad (23.10)$$

where β is a numerical factor dependent on the exact shape of the stress-strain curve. The value of β cannot be larger than $\pi\alpha/2$, otherwise the plastic work term would be larger than the elastic energy release rate. The form of this expression can be justified on dimensional grounds as well as on the expectation that it is of the same order of magnitude as the total plastic work done divided by the plastic zone lengths when an undeformed material that contains a crack of a fixed length is loaded by a stress σ. The fracture energy R can be obtained from noting that

$$G = U_{sur} + 2U_{pla} \equiv 4R \qquad \text{or} \qquad \frac{\alpha\pi\sigma^2 a}{\mu} = 4\gamma + \frac{2\beta\pi\sigma^2 a}{\mu}$$

at the fracture stress S. This results in

$$S = \sqrt{\frac{4\mu\gamma}{a}\frac{1}{\pi\alpha - 2\beta}} \qquad (23.11)$$

The plastic work is thus $U_{pla} = 4\beta\gamma/(\pi\alpha-2\beta)$ leading to a fracture energy

$$R = \gamma + \frac{U_{pla}}{2} = \gamma\left(1 + \frac{2\beta}{\pi\alpha - 2\beta}\right) = \gamma\left(\frac{\pi\alpha}{\pi\alpha - 2\beta}\right) \qquad (23.12)$$

We see that the fracture energy is proportional to the surface energy by a factor that can rapidly increase, dependent on the value of the ratio $x = \pi\alpha/2\beta$.

A few comments are in order. First, the plastic energy as expressed by Eq. (23.7) may not be taken as the fracture energy since it is dependent on both the applied stress σ and crack length a. Moreover this expression makes it difficult to explain the influence of the environment on fracture since the plastic work is generally non-sensitive to the environment. This environmental effect is well established

---

[i] We use a slightly generalised expression covering all three fracture modes. The parameter $\alpha = (1-v)$ for mode I and II cracks and $\alpha = 1$ for mode III cracks. For mode I, we thus have $2\mu/(1-v) = E/(1-v^2)$, consistent with our earlier formulation.

experimentally though. Because the value of γ is dependent on the environment, Eq. (23.12) can explain this influence via γ. Second, it has been shown by Weertman that using the stress intensity approach instead of the energy approach as used here, an identical result is obtained for small-scale yielding. Unfortunately solutions for a growing crack at present are only partial ones so that the value of β cannot be determined exactly. It can be estimated only by calculating the plastic work done in a zone of a stationary crack. Third, similar arguments can be given for other crack tip effects, e.g. crack plane roughness, microcracking and phase transformations, and as long as an equation similar to Eq. (23.10) can be proposed, the nature of R can be interpreted similarly.

The temperature dependence of fracture behaviour

Let us now discuss the effect of temperature. From the theoretical considerations given in the previous section, an estimate can be made for the temperature dependence of the fracture toughness K_{Ic} for completely brittle materials ($R = \gamma$). The temperature dependence of the fracture energy γ can be estimated[j] by differentiating γ as given in Eq. (23.5). Assuming λ to be independent of temperature, we obtain

$$\frac{1}{\gamma}\frac{d\gamma}{dT} = \frac{1}{E}\frac{\partial E}{\partial T} - \frac{1}{r_0}\frac{\partial r_0}{\partial T} = \frac{1}{E}\frac{\partial E}{\partial T} - \alpha \tag{23.13}$$

where α is the linear thermal expansion coefficient. Since the relation between the fracture toughness K_{Ic} and the surface energy γ can be described by

$$K_{Ic}^2 = \frac{2E\gamma}{1-v^2} \tag{23.14}$$

we obtain, differentiating once more, neglecting the temperature dependence of Poisson's ratio v and substituting Eq. (23.13),

$$\frac{1}{K_{Ic}}\frac{dK_{Ic}}{dT} = \frac{1}{E}\frac{\partial E}{\partial T} - \tfrac{1}{2}\alpha \tag{23.15}$$

From the temperature dependence of K_{Ic} and the thermal expansion coefficient α, the relative decrease in γ with T thus can be estimated. For covalently bonded materials, a small decrease in γ is predicted as has been confirmed in a number of cases, e.g. B_4C as demonstrated in Problem 23.6.

However, pushing the analysis as far as this may be actually too far. One would expect consistency of results using various approaches. Using the expression $\gamma \cong 0.25\,N_A^{-1/3}V_{mol}^{-2/3}\Delta H(0)$ instead of $\gamma \cong (E/r_0)(\lambda/\pi)^2$ leads to different results. This can be shown by approximating the atomisation energy $\Delta H(0)$ by the sublimation energy U_{sub} as given by $U_{sub} = 9V_{mol}K/mn$ (Eq. (11.9)), where K indicates the bulk modulus, and m and n are the exponents for the Lennard-Jones potential. Further noting that $K = E/3(1-2v)$ and neglecting the temperature dependence of the Poisson ratio v and taking only into account the temperature dependence of Young's modulus E, leads, after some analysis, to

---

[j] de With, G. (1984), J. Mater. Sci. **19**, 457.

Fig. 23.2: Typical stress-strain curve for rubber.

$$\frac{1}{\gamma}\frac{d\gamma}{dT} = \frac{1}{E}\frac{\partial E}{\partial T} + \alpha \qquad (23.16)$$

in direct contrast to Eq. (23.13). In fact one can expect this. Generally it holds that approximating a curve by another, which describes the overall behaviour reasonably, the behaviour of the derivative of the original function using the approximation is not necessarily well described. Different approximations can yield different answers. Since the surface energy is here approximated in two ways, different temperature derivatives result. Since the major contribution is arising from Young's modulus, the numerical difference for the materials considered is limited, though.

Stress localisation

Fracture mechanics is based on stress concentration and this leads to either fracture because the theoretical strength is reached or to yielding because the yield strength is reached. However, it is possible to avoid stress concentration to some extent because in some cases the constitutive behaviour of the material helps. A crack localises the stress in the material in the neighbourhood of a defect due to limited deformation locally and therewith enhances the local stress considerably. If this stress concentration can be avoided by deforming as much as possible in the neighbourhood of the defect, the stress concentration is reduced. Rubber gives the clearest example. Due to the shape of the stress-strain curve (Fig. 23.2), it is possible to stretch this material considerably without introducing a high stress. Therefore the elastic energy stored is kept low. In a rubber with a crack, frequently insufficient energy is stored to propagate the crack, in spite of the fact that the fracture energy of rubbers is low. The fracture energy can be accounted for reasonably well by the total energy of broken bonds per unit fracture area and in this sense rubber is a brittle material. Most interesting is that many biological materials, such as skin, exhibit a similar stress-strain curve as rubber.

Unfortunately, most synthetic materials do not exhibit the same characteristics in the stress-strain curve as rubbers. In those cases one has to increase the fracture energy and/or introduce smaller defects. The next section provides an overview.

Problem 23.6

For B_4C the molar volume $V_{mol} = 2.19{\times}10^{-5}$ m³/mol, the atomisation energy at 300 K $\Delta H(300) = 3.14{\times}10^6$ J/mol and Young's modulus $E = 461$ GPa while its temperature coefficient is given by $(1/E)(\partial E/\partial T) \cong -2.6{\times}10^{-5}$ K⁻¹. The thermal expansion coefficient is $\alpha = 4.3{\times}10^{-6}$ K⁻¹.
a) Show that the surface energy can be estimated as $\gamma \cong 11.8$ J/m² and its relative decrease with temperature as $(1/\gamma)(d\gamma/dT) \cong -3.0{\times}10^{-5}$ K⁻¹.

b) Compare these data with the experimentally determined fracture toughness $K_{Ic} = 3.7$ MPa m$^{1/2}$ and its relative decrease with temperature $(1/K_{Ic})(\mathrm{d}K_{Ic}/\mathrm{d}T) \cong 0$ K^{-1} up to 1500 K.

23.3   Overview of effects

From fracture mechanics, it has become clear that a better performance of a structure can be realised by either improving the mechanical behaviour of the material via increasing the fracture energy^{k} R or lowering the defect size a or by improving the design via lowering the stress concentration thereby lowering the strain energy release rate G. In this section, we consider only the materials aspects. Herzberg (1989) provides an elaborate general overview.

From the expression for strength

$$S = \frac{1}{\alpha} \frac{K_{Ic}}{\sqrt{a}} = \frac{1}{\alpha} \frac{\sqrt{2E'R}}{\sqrt{a}}$$

one easily deduces that a high strength can be reached by increasing the fracture energy R, increasing the elasticity parameter E' or decreasing the defect size a. An increase in fracture energy can be realised in various ways, as discussed briefly in the following sections. Also the stiffness can be considerably influenced by micro-structural manipulation, as discussed in Chapter 12. These parameters can be optimised by proper processing. Also minimisation of the defect size is a processing aspect but the defect size can be considerably enlarged by improper handling, in particular for brittle materials. We now briefly discuss the various material classes. In the next sections further details are presented.

In many metals and some inorganic materials, the presence of dislocations can result in macroscopic plastic deformation. Of course, the proper conditions, such as slip system availability, sufficient Frank-Read sources, etc., must be fulfilled. It will be clear by now that the large amount of energy that can be spent in plastic deformation is used for either the dislocation movement or for the elastic energy associated with each dislocation. Because the dislocation density can reach considerable values during crack propagation, large values for the fracture energy can be obtained, e.g. 10^4 to 10^6 J/m^2. This should be compared with the surface energy, which is of the order of 1 J/m^2.

In most inorganic materials the fracture energy is low, typically 10 to 100 J/m^2. This relatively small increase above the surface energy (typically 1 J/m^2) is generally due to various factors. We mention crack surface roughness, crack branching, microcracking ahead of the crack tip and crack bridging after the crack tip. Of these mechanisms microcracking and crack bridging seem to be the most important ones. Note by the way that if crack bridging is important, the (implicit) condition of stress-free crack flanks in the Griffith analysis is not fulfilled. A less general mechanism is energy dissipation due to a phase transformation. This has been explored in particular for partially stabilised ZrO$_2$, leading to a fracture energies of about 400 J/m^2.

In polymers, as discussed before, two types of bonds are present: the strong covalent bonds and the weak secondary bonds. The influence of the secondary bonds

---

^{k} Since in general the fracture energy R contains not only the surface energy γ, as assumed in the discussion of the theoretical strength, we use from now on again R.

has already been discussed. In the case of long chain polymers, the chain may cross the primary crack and the covalent bonds become stressed, leading to an orientation more or less perpendicular to the crack plane. These stressed chains form fibrils leading to the formation of crazes. The fibrils can carry load until they are stretched fully. Thereafter the chain breaks and the crack propagates through the material. The fracture energy is expected to be proportional to the chain length because the longer the chains the larger the probability that they cross the crack plane. In the case of polystyrene, this has been observed experimentally up to molecular weights of 10^5 g/mol.

23.4 Inorganic materials*

In the part dealing with elasticity and plasticity, we have seen a profound influence of the microstructure on the properties of inorganic materials. The same is true for fracture. For inorganics we deal with single crystals, monophase polycrystals and multiphase materials/ceramic matrix composites, respectively. Thereafter we briefly discuss temperature effects and fracture maps. An early review is made by Davidge (1979). Lawn (1993), Wachtman (1996) and Green (1998) provide an overview as well but from rather different perspectives.

Single crystals

Single crystals often show cleavage, i.e. fracture dominated by crystal structure. This leads to a preferential fracture plane, which is usually constant within a series of iso-structural crystals but whose degree of perfection may differ considerably. For example, the halite or rock salt structure (NaCl) shows (100) cleavage while the fluorite structure (CaF_2) and sphalerite structure (ZnS) show (110) and (111) cleavage fracture, respectively. Some inorganic crystals do not show a preferential fracture plane, e.g garnets. The cleavage planes do not necessarily correspond to the habitus planes (outer planes as observed for grown crystals), although sometimes the two are the same as in the cases of NaCl (100) and CaF_2 (111). Bradt[1] has considered the various models that have been proposed for cleavage, amongst others the models that state that the cleavage planes are parallel to unit cell planes, planes of low bond density, planes with low modulus or planes with low surface energy. He came to the conclusion that cleavage could be well described by fracture mechanics using the proper data for single crystals. Normally the cleavage toughness is quite low, typically (far) less than 1 MPa m$^{1/2}$. We follow his discussion to some extent.

Brittle fracture models generally reasonably describe the fracture behaviour of single crystalline inorganics and the influence of plasticity is limited, in particular for low temperature and fast deformation rates[m]. The fracture energy can reasonably be interpreted as an unrelaxed surface energy. Nevertheless a limited influence of plasticity can be present as shown for the relatively soft rock salt-type crystals, including LiF and MgO. For these materials an increase in fracture energy with yield strength is observed for which several interpretations are given. One interpretation is referring to the Cottrell mechanism of intersecting slip bands nucleating a new cleavage crack ahead of the blunted, existing crack. Another interpretation is

[1] Bradt, R.C. (1997), page 355 in *George R. Irwin Symp. on Cleavage Fracture*, Chan, K.S., ed. Minerals, Metals and Materials Soc., Warrendale PA.
[m] Pratt, P.L. (1980), Metal Sci. **14**, 163.

Table 23.3: Fracture toughness data for some cubic single crystals (data from Bradt).

Crystal	Plane	K_{Ic} (MPa m$^{1/2}$)
LiF	{100}*	0.50
	{110}	0.70
	{111}	1.50
GaP	{100}	0.73
	{110}*	0.65
	{111}	0.81
Si	{100}	0.95
	{110}	0.90
	{111}*	0.82
MgAl$_2$O$_4$	{100}*	1.18
	{110}	1.54
	{111}	1.90

* Commonly observed cleavage plane.

discussing the effect in terms of a mechanism of dislocation focusing. In this mechanism screw dislocations are emitted and via double cross-slip on parallel slip planes redirected towards the crack tip.

For the harder crystals the effect of plasticity is even more limited and only occurs at elevated temperature. For example, it has been shown for fracture of single crystal alumina[n] (Al$_2$O$_3$) at room temperature by TEM that at the crack tip no dislocations are emitted so that one can speak of a truly brittle process. From fracture toughness measurements for spinel (MgAl$_2$O$_4$) as a function of temperature, it appears that ductility becomes only important above about 900 °C.

Even for cubic single crystals the fracture response is anisotropic, as can be expected since the surface energy and elastic modulus are both anisotropic. Rather large differences in fracture toughness can be present for different crystallographic orientations of the fracture plane. Table 23.3 provides some examples for cubic single crystals. Apart from the difference in intrinsic fracture energy, also the nature of the fracture plane is of importance, in particular the electrical neutrality of a plane. Some crystallographic planes are non-neutral, e.g. the {111} planes for the NaCl structure, the {100} and {110} for CaF$_2$ structure and the {100} and {111} for ZnS structure. Non-neutral planes normally lead to serrated fracture planes. E.g. for alumina the basal (C) or {0001} plane is non-neutral and $R(0001) \cong 22$ J/m^2 is measured for a highly serrated fracture surface. This value is to be compared with the theoretical non-relaxed surface energy of about 6.1 J/m^2. For the prism (M) or $1\bar{1}00$ plane $R(1\bar{1}00) \cong 11.4$ J/m^2 was obtained, to be compared with a calculated value of about 6.7 J/m^2 for the non-relaxed surface energy. This plane is relatively smooth and in this case, according to Bradt, possibly limited crack tip plasticity and/or twinning adds to the fracture energy. In contrast, for the rhombohedral (R) or $10\bar{1}1$ plane $R(10\bar{1}1) \cong 6.5$ J/m^2 results, to be compared with the theoretical value of about 6.4 J/m^2, so that in this case there is a truly brittle fracture.

Influence from the environment is also clearly evident. A humid atmosphere leads generally to lower fracture energy. For example, consider the spinel MgAl$_2$O$_4$ (see also Section 8.10) for which similar anisotropic behaviour is observed as for alumina. Experiments show that $R(100) \cong 3.6$ J/m^2, $R(110) \cong 4.1$ J/m^2 and $R(111) \cong 4.9$ J/m^2.

---

[n] Lawn, B.R., Hockey, J.R. and Wiederhorn, S.H. (1980), J. Mater. Sci. **15**, 1207.

For the 100 orientation invariably a serrated fracture plane results, which is attributed to the stronger decrease in surface energy due to water adsorption for {111} and {110} planes as compared with the {100} at plane so that the crack preferentially proceeds along these other planes with as average direction {100}. Subcritical crack growth occurs as well. For this effect we refer to Chapter 24.

Monophase materials

For monophase polycrystalline inorganic materials, the influence of the microstructural parameters is much stronger on the fracture energy R than on Young's modulus E. For a fully dense material, it was thought for a long time that the fracture energy should be equal to the (thermodynamic) surface energy and initial experiments seemed to confirm this. Nowadays it is clear that this is not true. Obviously bonds have to be broken to form a fracture surface but in many cases the fracture velocity is so high that surface relaxation only occurs after the crack has already passed. A simple calculation of the energy necessary to break the bonds is in covalently bonded materials therefore a good estimate. Fused silica is discussed in Problem 23.5. The cleavage data for single crystals also provide evidence for this interpretation.

Fracture energy is calculated per projected unit area but a fracture surface is not flat. This may increase the available surface by about 40% in the case of monolithic materials but considerably more in the case of composites. Moreover, limited plastic deformation at the crack tip as well as subsidiary cracking may occur (Fig. 23.3). It is therefore presently impossible to estimate the fracture energy accurately and we have to resort to measurements. Typically the fracture energy is 10-100 J/m^2. In the following, we discuss the influence of several phenomena.

Porosity has a severe influence on the fracture energy. One could think on stereological arguments that the decrease in R is proportional to the amount of porosity present. In many cases, however, fracture is preferential through the pores. Rice[°] has made plausible that in that case the dependence on porosity becomes exponential

$$R = R_0 \exp(-cP)$$

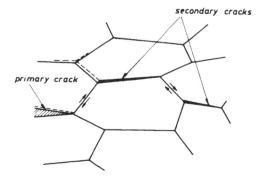

secondary cracks

primary crack

Fig. 23.3: A network of secondary cracks (or process zone) accompanying a main crack.

[°] Rice, R.W. and Freiman, S.W. (1977), page 800 in *Ceramic Microstructures '76*, R.M. Fulrath, J.A. Pask, eds., Westview Press, Boulder.

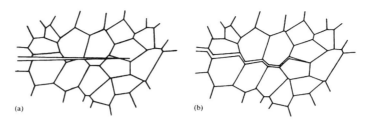

(a)   (b)

Fig. 23.4: Transgranular and intergranular fracture.

where R_0 is the fracture energy for the fully dense ceramic, c is an exponent dependent on the shape of the pores, typically 1 to 2, and P is the porosity. Obviously the limiting behaviour for $P \to 0$ is incorrect. In those few cases, where porosity was the only parameter varied[p], reasonable agreement was observed over the first tens of percent of porosity.

Preferred crystallographic orientation, in inorganics often introduced for functional reasons, can also be important for the fracture energy. In general cylinder symmetry applies and for the direction perpendicular to the axis the value of R becomes smaller than the value for the random polycrystal while it becomes larger for the other, parallel direction. For example, the fracture toughness of Sr-hexaferrite is about 1 MPa m$^{1/2}$ for fracture planes perpendicular to the axis and about 3 MPa m$^{1/2}$ for fracture parallel to the axis[q]. The value for the random polycrystal is about 1.5 MPa·m$^{1/2}$. A relatively large difference is thus observed for the various directions.

While grain size is unimportant (except in special cases) for Young's modulus E, it is rather important for the fracture energy R. Unfortunately the issue is confused by the transition from intergranular to transgranular fracture (Fig. 23.4), which occurs with increasing grain size. In many cases a mixed fracture mode occurs. For cubic crystals, the value of R is more or less independent of the grain size unless a transition occurs from intergranular to transgranular. For non-cubic materials, the anisotropy in thermal expansion coefficient (and Young's modulus) results in stresses at the grain boundaries. These stresses may increase the value of R. If the grains become sufficiently large, microcracks can arise resulting in diminished fracture energy (and strength!). The size of this effect is largely dependent on the amount of anisotropy present. In the case of the heavily anisotropic $MgTi_2O_5$ first an increase in fracture energy is present, followed by a decrease at still larger grain size[r]. For the much less anisotropic Al_2O_3 a more or less constant fracture energy is observed[s]. In this case for all grain sizes fracture was intergranular due to the presence of a limited amount of porosity at the grain boundaries. The results of various experiments are sometimes rather conflicting. This is due to the above-mentioned transition from intergranular to transgranular fracture as well as the fact that different methods yield slightly different quantities, e.g. the bend test delivers the fracture initiation energy while the double cantilever beam test gives the fracture propagation energy. The effect of the grain size on the fracture energy as measured with a particular method is often not very large,

[F] Evans, A.G. and Tappin, G. (1972), Proc. Brit. Ceram. Soc. **20**, 275.
[q] Iwasa, M., Liang, C.E. Bradt, R.C. and Nakamura, Y. (1981), J. Am. Ceram. Soc. **64**, 390.
[r] Kuszyk, J.A. and Bradt, R.C. (1973), J. Am. Ceram. Soc., **56**, 420.
[s] Simpson, L.A. (1974), *Fracture Mechanics of Ceramics 2*, R.C. Bradt, D.P.H. Hasselman and F.F. Lange, eds., Plenum, New York.

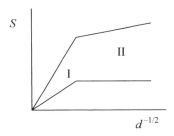

Fig. 23.5: The strength as a function of grain size at constant other parameters. For large grain size (area I) the strength decreases strongly as a function of the grain size, possible weakly dependent on the surface quality. For small grain size (area II) the strength is roughly constant, independent of the grain size with the level determined by the quality of the machining processes.

e.g. for the earlier-mentioned $MgTi_2O_5$ about a factor of 2. In some case the grain shape has a significant influence[t].

In the case of intergranular fracture, there can be a significant influence of segregation of impurities to the grain boundaries. For example, in Al_2O_3 ceramics Ca segregates strongly to the grain boundaries. With increasing Ca content the fracture energy decreases strongly in the case of intergranular fracture while for transgranular fracture R remains approximately constant[u].

Finally the grain size interferes with the defect size. We distinguish between material inherent defects and machining induced defects. If the machining damage is larger than the grain size, the strength is strongly influenced by the machining processes. In a number of cases, it has been shown that the defects due to machining are more or less constant in size, independent of the grain size[v]. In these cases the strength is thus also approximately independent of the grain size, at least if other parameters do not change. In those cases where the machining defects are smaller than the grain size, the defect size is controlled by the average grain size d and it holds that

$$S \sim \frac{1}{\sqrt{d}}$$

where S is the strength. Generally a small grain size and high machining quality are beneficial for strength (Fig. 23.5) but a small scatter is only achieved if both grain size distributions and machining damage are homogeneously distributed.

Multiphase materials

Multiphase materials and composites differ only slightly in nature. Generally a multiphase material contains more than one phase as a consequence of the chemical composition and thermal treatment. Composites on the other hand are usually deliberately produced using a mixture of matrix and reinforcing raw materials in the pre-densification stage.

---

[t] Lange, F.F. (1974), J. Am. Ceram. Soc. **57**, 84.
[u] De With, G. (1981), J. Mater. Sci. **16**, 841 and Jupp, R.S., Stein, D.F. and Smith, D.W. (1980), J. Mater. Sci. **15**, 96.
[v] Rice, R.W. (1974), page 323 in *Fracture Mechanics of Ceramics 1*, R.C. Bradt, D.P.H. Hasselman and F.F. Lange, eds., Plenum, New York.

The effect of second phases acts via two mechanisms, which we discuss subsequently: the normal mechanism and the microcrack/transformation toughening mechanism.

In the case of the normal mechanism, a second phase is present with a higher K_{Ic} (and usually also higher E) as the matrix material. Typical examples are porcelain, glass-ceramics, particle reinforced glass and grinding wheel materials. During cooling of the two-phase material stresses occur because generally the thermal expansion coefficient of the matrix α_{mat} is different from the value for the inclusion α_{inc}. We have to distinguish between two cases: $\alpha_{inc} \cong \alpha_{mat}$ and $\alpha_{inc} \neq \alpha_{mat}$.

In the case of the normal mechanism with $\alpha_{inc} \cong \alpha_{mat}$, a crack will be deflected at the interface if $E_{inc} > E_{mat}$. A rougher fracture surface is obtained. An increase of surface area and thus of the fracture energy by a factor of about 2 should be realizable. In glass with a high volume fraction of Al_2O_3 spheres an increase of about 5 is realized, however. This extra increase is due to pinning of the crack front between particles, comparable to the pinning of a dislocation at an impurity. The increase of the fracture energy appears to be increasing with increasing inclusion size. Moreover, it is necessary that the inclusions are larger than the matrix grains. If it holds that $E_{inc} < E_{mat}$ the crack will pass through the particles and an increased fracture energy is only realized if the fracture energy of the inclusions R_{inc} is higher than the fracture energy of the matrix R_{mat}. In systems with $\alpha_{inc} \cong \alpha_{mat}$ at low volume fractions the defect size is determined by the matrix. At high volume fractions the defect size is limited by the distance between the inclusions and a higher fracture energy as well as strength is realized. The size of the inclusions, however, should be as small as possible to prevent fracture initiation from the inclusions themselves.

In the case of the normal mechanism with $\alpha_{inc} \neq \alpha_{mat}$, a crack only circumvents the inclusions if $\alpha_{inc} > \alpha_{mat}$ and $R_{inc} > R_{mat}$. Apart from the earlier-mentioned effects of roughening and pinning, a third effect occurs. The matrix is in compression and therefore a higher tensile stress has to be applied for extending a crack. The risk exists that, though the fracture energy increases, the defect size increases more, resulting in a lower strength. For systems with $\alpha_{inc} \neq \alpha_{mat}$ the size of the inclusions should be as small as possible to prevent microcracks. In particular, radial cracks that occur when $\alpha_{inc} < \alpha_{mat}$ are detrimental because they easily link to form a macro-defect. For both systems fracture energy and defect size thus increase with increasing inclusion size. The optimum choice for the inclusion size is therefore dependent on whether high fracture energy, high strength or a reasonable compromise between the two is required.

Let us now discuss the second mechanism, the microcrack and transformation toughening mechanism. In both the phase transformation in ZrO_2 from tetragonal to monoclinic occurring at about 1000 °C is relevant. With this phase transformation a volume increase of about 1.4% is associated. If one introduces ZrO_2 particles in a matrix the phase transformation from tetragonal to monoclinic can be wholly or partly avoided through the constraining action of the matrix. The effect on the main crack can be twofold. In the microcrack mechanism the phase transformation from tetragonal to monoclinic of the ZrO_2 occurs during cooling of the composite material. This results in compression in the matrix and these residual stresses in the matrix can cause microcracking in the matrix in the neighbourhood of the crack tip if the main crack extends. The total amount of energy spent increases because of the formation of the many microcracks. In the transformation toughening mechanism, the ZrO_2 remains

Table 23.4: Thermo-mechanical data for barium titanate.

	Below T_c (tetragonal)	Above T_c (cubic)
Unit cell volume (nm^3)	0.639	0.644
α (10^{-6} K^{-1})	6.5	9.8
E (GPa)	124$^{\#}$ (138*)	120
K_{Ic} (MPa m$^{1/2}$)	1.3	1.0

#: Constant electric field. *: Constant electric polarization.

in the tetragonal form and transforms if a crack passes and if the main crack extends, the ZrO_2 phase transformation is induced in the neighbourhood of the crack tip, thereby again dissipating energy. Tetragonal ZrO_2 has been used in various matrices, e.g. Al_2O_3 and Si_3N_4, but also with cubic ZrO_2. In the latter case, part of the material is tetragonal and part is cubic and referred to partially stabilised zirconia (PSZ). The expected increase in strength occurs only when the particles are well dispersed through the matrix. If no proper dispersion is realized, there is a large chance of introducing larger defects that counteracts the increase in toughness. In principle, all materials with such a phase transformation can be used. There is, however, a limited choice due to the absence of practical materials but it must be said that nowadays zirconia ceramics with an average strength of about 1000 MPa and a Weibull modulus of about 20 are commercially available.

Temperature effects

In Section 23.2 the temperature dependence of rather pure, brittle monolithic materials was discussed. For inorganic materials containing a second phase a quite different behaviour for R is observed. Take debased alumina as an example. At relatively low temperature the alumina grains as well as the glassy phase behave elastic. The overall behaviour is thus a slight decrease of R with temperature. At somewhat higher temperature the glassy phase becomes viscous. The energy dissipation that is required for the viscous deformation is reflected in the fracture energy and the value for R increases. At still higher temperature the viscosity of the glassy phase becomes so low that energy dissipation due to viscous deformation during deformation can be neglected. Moreover the connectivity between the grains

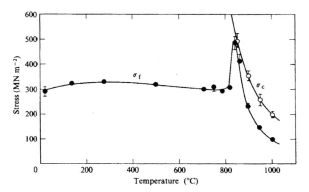

Fig. 23.6: Temperature dependence of the strength σ_f and compressive flow strength σ_c for a 95% alumina.

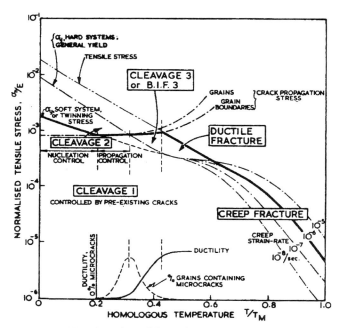

Fig. 23.7: Overview of the various fracture mechanisms.

has diminished so that a strong decrease of R and strength S results. An example is given in Fig. 23.6.

In case a phase transformation takes place, the elastic modulus as well as the fracture energy are generally different above and below the phase transformation temperature. Other physical and thermal properties change as well. As an example, take $BaTiO_3$, which shows a transition at $T_c = 125$ °C. From the data as shown in Table 23.4 it is clear that substantial change in α, E and $R = K_{Ic}^2/2E$ occurs.

During grinding, usually residual compressive stresses are introduced. These stresses may be relieved by a high temperature treatment and thus result in lower strength. Although the depths of these stresses generally amount only 10 to 20 μm, the influence on strength may be considerable. For alumina, ground with 126 μm diamond, a strength reduction of 20% from 335 MPa to 270 MPa has been reported after annealing at 1000 °C for 2 hours.

Finally it should be noted that, apart from a temperature dependence of properties, also the continuous modification of defect populations by the service conditions, such as corrosion and/or erosion, could occur. In general these processes lead to a larger defect size. Also the prolonged high service temperature may influence the defect population. At present, it is not clear how to handle these effects in a practical though reliable way.

Fracture mechanism maps

If sufficient data on the fracture behaviour of ceramics are known, it is possible to plot the strength as a function of the temperature and grain size. Frequently the normalized strength S/E where E denotes Young's modulus and relative temperature T/T_{mel} with T_{mel} the melting point are used. In such a plot, various regimes with the prevailing mechanism or fracture mode are indicated. These maps are indicated as

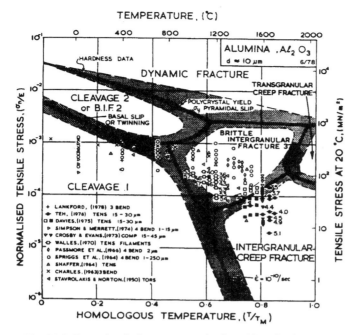

Fig. 23.8: Example of a fracture map: alumina with grain size 10 μm.

fracture mechanism maps. In Fig. 23.7, an overview of the possible fracture mechanisms is given, a brief description[w] of which is given below.

- *Cleavage I.* For 'true' brittle fracture, the behaviour is governed by the Griffith equation with a flaw size equal to the size of the pre-existing defect.
- *Cleavage II.* If pre-existing defects are small the stress can reach a level to initiate slip or twinning. The failure is still governed by the Griffith equation but the flaw size is generally proportional to the grain size.
- *Cleavage III.* As the temperature rises, the flow stress falls until plasticity or creep precedes failure, which may be, nevertheless by cleavage. Significant plastic strain (1-10%), sufficient to blunt pre-existing defects, raises effectively the fracture energy. In all three cases, fracture may either or transgranular (cleavage) or intragranular (brittle intragranular fracture or BIF).
- *Ductile failure.* This appears when general plasticity without cleavage is sufficient to permit large strains (10-100%). Failure occurs by nucleation, plastic growth and coalescence of voids, either pre-existing or nucleated.
- *Creep fracture.* Creep becomes dominant for most ceramics above about $0.5T_{mel}$. The morphology of fracture is similar to that of ductile failure but the deformation mechanism has changed. Failure is time dependent. Fracture may be either transgranular or intragranular. For ceramics usually transgranular creep fracture prevails because cleavage persists up to temperatures at which creep starts.
- *Rupture.* This appears when ductility becomes so great that significant necking occurs. The process is accompanied by dynamic recrystallisation or recovery. This process is rarely occurring for ceramics.

---

[w] Ashby, M.F. (1979), page 1 in *Fracture mechanics*, Smith, R.A. ed., Pergamon, Toronto.

- *Dynamic fracture.* This is the regime where initial loading must be described in terms of propagation of elastic waves through the material roughly $\dot{\varepsilon} = 10^6$ s.

Generally for oxide ceramics the cleavage I regime is rather large, extending to about $0.5T_{mel}$. In Fig. 23.8, the fracture map for fine-grained alumina is presented. For alumina all mechanisms discussed except rupture do appear on the map. For covalently bonded ceramics the cleavage I regime is even extending further to 0.6-$0.7T_{mel}$. They retain their strength up to high temperature and show no real ductility. It must be said though that data from two different sources can differ considerably; see e.g. the maps for hot-pressed Si_3N_4 by Gandhi et al.[x] and Quinn et al.[y]

The transition from one mechanism to another is not always clear. Moreover, unfortunately, the mechanical behaviour of ceramics is highly dependent on the microstructure so that for one compound several fracture maps are possible. Finally, the exact positions of the boundaries depend on the strain rate applied. The shading on the field boundaries indicates these effects. Therefore this representation of fracture is rather useful for educational purposes but renders it less useful for quantitative design.

23.5   Metals*

Similarly as for inorganics we deal first with single crystal and polycrystal monophase metals and thereafter with multiphase metals. Where possible, similarities are indicated. A more extensive reference is Dieter (1976), which we follow here on several points. We also refer to Broek (1978) and Herzberg (1989).

Single crystals

Fracture of single crystal metals is studied but limitedly. For a long time *Sohncke's law*, which states that fracture occurs when the resolved normal stress reaches a critical value (CNRS), was accepted although it was not based on extensive experimental evidence. As can be expected, the experimental results strongly depend on temperature, purity, heat treatment and orientation. Here we discuss only two typical examples, the brittle (semi)-metal silicon (Si) and the ductile metal zinc (Zn). Similar as for inorganics the fracture behaviour is anisotropic. For Si the amount of anisotropy present is minor as can be expected for a cubic material with the diamond structure, although cleavage preferentially occurs for the (111) plane. The theoretical strength has been estimated[z] as 7 GPa. The fracture energy has been measured by several techniques[aa] and data for some important crystallographic planes are given in

Table 23.5: Fracture data for Si single crystals.

	K_{Ic} (MPa m$^{1/2}$)	CRNS (MPa)
Si (001)	0.91-0.95	1.9
Si (011)	0.94-1.19	3.0
Si (111)	0.6*, 0.82-1.23	12.0

Polycrystalline Si 0.94 MPa m$^{1/2}$. *: Estimated from cleavage energy.

---

[x] Gandhi, C. and Ashby, M.F. (1979), Acta Met. **27**, 1565.
[y] Quinn, G.D. and Wirth, G. (1990), J. Eur. Ceram. Soc. **6**, 169-177.
[z] Petersen, K.E. (1982), Proc. IEEE. **70**, No. 5. Fitzgerald, A.M., Iyer, R.S., Daushardt, R.H. and Kenny, T.W. (2002), J. Mater. Res. **17**, 683.
[aa] Ericson, F. et al. (1988), Mater. Sci. and Eng. A **105/106**, 131.

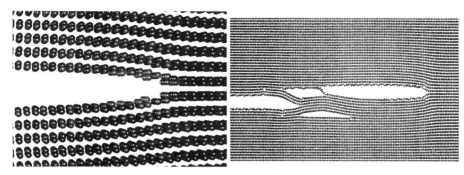

Fig. 23.9: Simulation of stable fracture in Si at $G \cong 4R$ (left) showing a smooth fracture surface and unstable fracture at $G \cong 10R$ (right) showing branching and a multiple fracture plane.

Table 23.5. Si becomes ductile at about 900 °C and then shows {111} slip in well-developed band for stresses above about 10 MPa.

In view of its applications and relatively simple structure, it is no wonder that simulations of the fracture process have been made for Si. From these simulations[bb] it appeared that for stable fracture at relatively low deformation a crack travelling through the material can be atomically sharp, leaving perfectly flat surfaces behind. Fig. 23.9 displays a crack travelling at 1.8 km/s through a piece of Si, strained so that $G \cong 4R$. At high strain the crack becomes unstable because the energy flowing to the crack tip is too great. The crack is no longer able to behave in a smooth way, but advances by blunting, emitting dislocations, jumping planes, branching and creating rough surfaces. Fig. 23.9 shows also a crack travelling at approximately 3.6 km/s using $G \cong 10R$. It cannot go any faster. Higher strain would just mean more damage.

For Zn, however, which has a hexagonal structure, there is a strong preference for cleavage along the basal plane. The critical normal resolved stress (CNRS) for several purities of Zn single crystals is given in Table 23.6. The influence of the amount of impurities again can be noticed. Even at room temperature there is an influence of plasticity[cc]. Although the fracture energy as calculated from a conventional fracture analysis yields only ~0.51 J/m^2 at 298 K, a value somewhat below the accepted surface energy of about 0.6 to 0.8 J/m^2, it was nevertheless inferred on the basis of dislocation etch pit density that plasticity was important. It was concluded that the surface energy was ~0.08 J/m^2, in good agreement with the experimental value of at 77 K. At the intermediate temperature of 209 K a value of ~0.36 J/m^2 was obtained. Hence the surface energy of Zn as estimated from the fracture experiments is well below the thermodynamic surface energy. This is in contrast to other materials,

Table 23.6: Cleavage data for Zn single crystals.

	Plane / Temperature	CRNS (MPa)
Zn (0.03% Cd)	0001 / −185 °C	1.9
Zn (0.13% Cd)	0001 / −185 °C	3.0
Zn (0.53% Cd)	0001 / −185 °C	12.0

Data from Dieter (1976).

[bb] Hauch, J.A., Holland, D., Marder, M. P. and Swinney, H.L. (1999), Phys. Rev. Lett. **82**, 3823.
[cc] Bilello, J.C., Dew-Hughes, D. and Pucino, A.T. (1983), J. Appl. Phys. **54**, 1821.

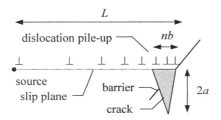

Fig. 23.10: Schematic view of microcrack formation at a dislocation pile-up at a barrier.

including W, WC, Si, several compound semiconductors and ionic crystals, where a good agreement exists between the cleavage and thermodynamic surface energy when plastic relaxation is taken into account. The reason for this anomaly is not known. Detailed information about the flow and fracture of Zn is available[dd].

Polycrystals

Turning now to polycrystalline metals, we note that, while for inorganic materials flaws are generally already present before mechanical loading, in semi-brittle metals the loading process itself[ee] via local plastic deformation can also introduce defects. Therefore we have to consider the creation of microcracks. One frequently occurring mechanism is local plastic deformation resulting in dislocation pile-ups. This mechanism is strongly influenced by the presence of secondary phase particles. On the one hand, if the particles are penetrable by dislocations, planar slip will occur and relatively large pile-ups will result. This leads to high stresses, relatively easy creation of microcracks and therefore to brittle fracture. If, on the other hand, the particles are impenetrable and finely dispersed, slip is strongly reduced leading to a reduced number of dislocations in the pile-ups. Moreover, once a crack is formed, it will be forced to bow around these particles thereby increasing the fracture energy. Hence fine, well-dispersed particles can lead to an increased toughness. This process is relevant for mild and low-carbon steels. Particles more ductile than the matrix can also improve ductility if they are large enough to yield before large pile-ups are created. Another possible mechanism is the creation of microcracks at the intersection of (110) slip planes.

For a semi-quantitative model of microcrack nucleation, we again make use of the dislocation pile-up concept as introduced in Section 16.5. The configuration is sketched in Fig. 23.10. On the slip plane of length L a shear stress squeezes the dislocations together. At some critical stress the dislocation at the head of the pile-up is pushed so close together that they coalesce[ff] in a microcrack of the width nb and the length $2a$. If the stress at the head is not relieved by plastic deformation the stress at the pile-up is given by

[dd] Stofel, E.J. (1962), *Plastic flow and fracture of zinc single crystals*, Ph.D. thesis, California Institute of Technology, Pasadena.

[ee] Although more rare, flaws can be introduced in inorganics by the loading process as well. Mechanisms similar to those as in metals can occur in those ceramics that are made from materials, which are ductile in their single crystal state, e.g. MgO. Flaws may also be generated in pristine glass: see Bouten, P.C.P. and de With, G. (1988), J. Appl. Phys. **64**, 3890.

[ff] Stroh, A.N. (1958), Adv. Phys. **6**, 418.

$$\sigma = \beta\tau_{\text{eff}}(L/r)^{1/2} \equiv \beta(\tau_{\text{res}}-\tau_{\text{obs}})(L/r)^{1/2}$$

with the effective stress $\tau_{\text{eff}} = \tau_{\text{res}} - \tau_{\text{obs}}$ where τ_{res} is the resolved stress and τ_{obs} the friction stress due to obstacles on the glide plane. The factor β is close to unity and r is the distance from the tip of the pile-up. Microcrack nucleation occurs when this stress equals the theoretical strength given by

$$\sigma_{\text{the}} = (E\gamma/r_0)^{1/2}$$

so that, solving for τ_{eff}, we have

$$\tau_{\text{eff}} = \frac{1}{\beta}\left(\frac{E\gamma r}{Lr_0}\right)^{1/2} = \frac{1}{\beta}\left(\frac{2(1+v)G\gamma r}{Lr_0}\right)^{1/2} \cong C\left(\frac{2G\gamma}{L}\right)^{1/2} \quad \text{with} \quad C = \frac{(1+v)^{1/2}}{\beta}$$

where the last step can be made assuming that $r \cong r_0$. The number of dislocations in the pile-up is given by (Section 16.5)

$$n = \alpha\pi\tau_{\text{eff}}L/2Gb \tag{23.17}$$

with, as before, $\alpha = 1$ for screw and $\alpha = 1/(1-v)$ for edge dislocations. Eliminating L from these equations results[gg] in

▶ $\qquad \tau_{\text{eff}}\, nb = (C^2\alpha\pi)\gamma \qquad$ or $\qquad (\tau_{\text{res}}-\tau_{\text{obs}})\, nb \cong 2\gamma \tag{23.18}$

From this result it follows that a microcrack will form when the work done by the applied stress τ_{res} by producing a displacement nb equals the work done in moving dislocations over a distance nb against the obstacle stress τ_{obs} and the work done for producing new fracture surfaces 2γ. Remarkably this expression does not contain the crack length $2a$ and the crack grows as long as the dislocation source continues to supply dislocations in the pile-up. Since only shear forces are involved, microcrack nucleation can occur in tension as well as compression. For propagation, however, a tensile stress is required.

There is experimental evidence that the most difficult step in propagation of deformation induced microcracks is breaking through a barrier. Petch[hh] observed that for brittle fracture in iron and steel the tensile strength S could be expressed by

$$S = S_0 + k_S D^{-1/2}$$

with the parameters S_0 and k_S, analogous to the Hall-Petch expression for the yield strength. To interpret this expression we recast Eq. (23.18), using $\sigma \cong 2\tau_{\text{eff}}$, to

$$\sigma\, nb \cong 4\gamma$$

and, assuming that the dislocation source is in the middle of the grain so that $L \cong D/2$ meanwhile substituting Eq. (23.17), we obtain

$$\sigma\, \alpha\pi\tau_{\text{eff}}D \cong 16\gamma G \qquad \text{or} \qquad \sigma(\tau_{\text{res}}-\tau_{\text{obs}})D \cong 8\gamma G$$

In the last step $\alpha\pi \cong 2$ is used. Since microcracks form when the applied stress equals the yield strength we may use the Hall-Petch relation $\tau_{\text{res}} = \tau_{\text{obs}} + k_Y D^{-1/2}$ and obtain

$$S\, k_Y D^{1/2} \cong 8\gamma G \qquad \text{or} \qquad S \cong 8\gamma G/k_Y D^{1/2}$$

[gg] Cottrell, A.H. (1958), Trans. Metall. AIME **212**, 192.
[hh] Petch, N.J. (1953), J. Iron Steel Inst. London, **174**, 25.

This expression represents the stress to propagate a microcrack of length D, i.e. the strength S, and shows that it is proportional to $D^{-1/2}$.

The above analysis shows that the dependence on grain size is again via an inverse square root. In fact, this behaviour generally can be expected since the intrinsic length scales L of a polycrystalline metal are the Burgers vector length b and grain size D. Since the above considerations dealing with the application of a stress σ are essentially of a continuum nature, the Burgers vector length should disappear. This leaves the grain size as the only relevant length scale. Since fracture occurs whenever the stress intensity as given by $\sigma L^{1/2}$ reaches a critical value, we easily obtain the scaling law $S \sim L^{-1/2}$ for the strength S.

For polycrystalline monophase metals mechanisms to enhance the fracture energy are grain size control (to influence the yield strength), work hardening, solid solution hardening or alloying (both to influence the yield strength and hardening behaviour). The effects of grain control (as described by the Hall-Petch relation), work-, solid solution- and dispersion hardening on the yield strength have already been discussed in Sections 16.3, 16.7 and 16.8. Other mechanisms leading to toughening and/or strengthening often introduce different phases. We just mention fibre reinforcement, which is used to stiffen a metal, which in turn leads to strengthening and is briefly discussed in Section 23.7. Dispersion hardening is another mechanism to control the toughness. We limit ourselves here to some brief remarks about dispersion hardening.

Many metals contain a distribution of the particle sizes ranging from small particles up to 50 nm (precipitates, relevant for the yield strength), intermediate size particles of 50 to 500 nm (serving as grain growth inhibitors and/or improving yield strength) and large size particles from 0.5 to 50 μm (no purpose in some materials but for wear resistance and/or hardness improvement in others). The largest particles are the most important ones with respect to fracture toughness. These large particles fail easily causing voids in the highly stretched area in front of the main crack. For an estimate of the size of this area, we need the displacement of the crack edges, the crack tip opening displacement (CTOD). It can be shown that the displacement v perpendicular to the crack plane for a crack in the LEFM description is given by $v = 2\sigma(a^2-x^2)^{1/2}/E$, where x is the running co-ordinate along the crack. For small-scale yielding the plastic zone correction has to be added, leading to $v = 2\sigma[(a+r_p)^2-x^2]^{1/2}/E$. The CTOD $2v$ is then

$$2v(a) = 4\sigma[(a+r_p)^2-a^2]^{1/2}/E \cong 4\sigma(2ar_p)^{1/2}/E$$

since $r_p \ll a$. Using the plastic zone size expression $r_p = (K_{Ic}/Y)^2/2\pi = \alpha^2\sigma^2a/\pi Y^2$, with α the shape factor in $K_I = \alpha\sigma\sqrt{a}$, we obtain

$$2v = 4K_{Ic}^2/\pi EY$$

The extension δ of the highly stretched area before the actual crack tip is assumed to be a fraction of $2v$, say $\sim\frac{1}{3}$ or for convenience $\pi/8$, so that we have

$$\delta \cong K_{Ic}^2/2EY$$

Rice and Johnson[ii] assumed that cracking can proceed if the particle spacing s equals the size of the heavily stretched zone, i.e. if $\delta = s$. This leads to

---

[ii] Rice, J.R. and Johnson, M.A. (1970), page 641 in *Inelastic behaviour of solids*, Kanninen, M.F. et al., eds. McGraw-Hill, New York.

$$K_{Ic} = (2YEs)^{1/2}$$

Estimating the average particle distance from a simple cubic packing of particles with the size d and the number density n, the volume fraction $f = \pi d^3 n/6$. The volume available per particle $V = s^3$ is $1/n$ and thus

$$s^3 = \pi d^3/6f \quad \text{or} \quad s = d(6f/\pi)^{-1/3}$$

Combining leads to

$$K_{Ic} = (6f/\pi)^{-1/6}(2YEd)^{1/2}$$

The 1/6-th power dependence on f has indeed been observed[ii]. The dependence on particle size has not been confirmed though, while experimentally K_{Ic} decreases with increasing Y, contrary to the model prediction. The matter is thus not clear.

We conclude with two remarks. First, similarly as for inorganics, fracture maps have been produced for all kinds of metals and alloys. Since the dependence on the microstructure is usually less pronounced as for the case of inorganics, they are somewhat more practical. Their use as convenient representations of fracture remains, of course. Second, we note that another important aspect of the analysis of brittle fracture in metals, similar to inorganics, is fractography. This part of fracture methodology has become quite important for the elucidation of the failure of materials in complex structures, both from an engineering pint of view as well as a legal point of view. Hull (1999) has provided a recent introduction.

Constance Fligg Elam (1894-1995)
Later known as Constance Tipper and an undergraduate at Newnham, she was one of the first women to take the Natural Sciences Tripos, in 1915. On graduating she joined the National Physical Laboratory and then went to the Royal School of Mines. While employed by the School of Mines she worked with G.I. Taylor on the deformation of crystals under strain. She moved to Cambridge in 1929 and became a research fellow at Newnham. Cambridge University appointed her a Reader in 1949. From that time she was a full member of the Faculty of Engineering and the only woman to hold office in the otherwise all male department. She used the second SEM that was ever built, which was produced by a team headed by C.W. Oatley in the Dept. of Engineering for the examination of metallic fracture faces. Her speciality was in the strength of metals, and the way in which this affected engineering problems. Her major contribution was, during the Second World War, investigating the causes of brittle fracture in Liberty Ships. Professor J. Baker, Head of Engineering at that time, was asked to launch an investigation into the reasons why these ships were breaking, so he brought Constance Tipper in as the technical expert. Her experience was summarised in the book *The Brittle Fracture Story* (1962, CUP).

[ii] Hahn, G.T. and Rosenfield, A.R. (1973), 3rd ICF Conf. 1, PL111-211.

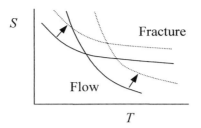

Fig. 23.11: The transition from brittle to ductile at strain rate $\dot{\varepsilon}_1$ (—) $\dot{\varepsilon}_2 > \dot{\varepsilon}_1$ (---).

23.6   Polymers*

In general the fracture strength of polymers is low as compared with metals and ceramics. Since many phenomena are relevant for the discussion of the fracture behaviour of polymers, we will refrain from an elaborate discussion but only indicate the most salient features. A general introduction of the fracture of polymers is given by Ward (1983) whil an early review is provided by Andrews[kk]. For further details we refer to Kinloch and Young[ll], Williams[mm] and Kausch[nn].

The fracture mode for thermosetting polymers is typically brittle and during fracture the covalent bonds in the cross-linked network break. For thermoplastic materials both ductile failure with large fracture strain ε_{fra} but low strength S and brittle failure with small fracture strain ε_{fra} but higher strength S occurs. The transition from ductile to brittle is favoured by low temperature (Fig. 23.11), high strain rate, the presence of sharp notches and large thickness. As an example we mention polymethyl methacrylate (PMMA), which is totally brittle at 4 °C ($\varepsilon_{fra} \cong 0.03$, $S \cong 80$ MPa), shows limited ductility at room temperature and slightly above ($\varepsilon_{fra} \cong 0.05$, $S \cong 50$ MPa) and becomes fully ductile at about 60 °C ($\varepsilon_{fra} \cong 1.3$, $S \cong 15$ MPa). Generally there is no correlation between the brittle ductile transition temperature and the temperature of the glass transition or the secondary transitions. This may not be too unexpected since the relaxations are detected at small strains whereas the brittle ductile transition occurs at relatively large strain and is affected by external factors such as the presence of notches.

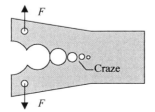 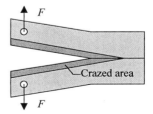

Fig. 23.12: Schematic of the fracture plane during (left) and after (right) failure of a glassy polymer.

[kk] Andrews, E.H. (1968), *Fracture in polymers*, Oliver and Boyd, London.
[ll] Kinloch, A.J. and Young, R.J. (1983), *Fracture behaviour of polymers*, Applied Science, New York.
[mm] Williams, J.G. (1984), *Fracture mechanics of polymers*, Ellis Horwood, Chichester.
[nn] Kausch, H.-H. (1985), *Polymer fracture*, Springer, Berlin.

The fracture behaviour of glassy polymers can be described by fracture mechanics along the usual lines. We recall that one of the features is an increased strength with decreasing flaw size. However, this rise in strength is not unlimited and when the flaw size is reduced below a certain value, the strength becomes independent of the flaw size. This size is about 1 mm for PS and 0.07 mm for PMMA at room temperature. It appears that the 'natural' flaw size responsible for this behaviour is due to crazing, which often precedes fracture in lightly cross-linked glassy polymers.

Crazes are regions of highly localised yielding leading to the formation of small and interconnected voids. Between these voids bridges of fibrils exist in which the molecules are oriented. These bridges elongate with increasing stress, meanwhile dragging the molecules out of the bulk, and finally break, whereby the voids coalesce (Fig. 23.12, left). Experimentally it appears that cross-tie fibrils connecting the main parallel fibrils over the craze are present. In the absence of cross-tie fibrils one would expect a blunt crack tip, while experimentally a parabolic shape is observed. A considerable amount of energy can be absorbed during this crazing fracture process, e.g. in PMMA (polymethyl methacrylate) 200 to 400 J/m^2 and in PS (polystyrene) 100 to 2000 J/m^2. After failure the fracture surface is covered with the remainders of the crazes (Fig. 23.12, right). In summary, a craze is a kind of process-zone, which can carry load and, similar to a plastic zone in metals, increases the effective crack length. Although via this mechanism a considerable amount of energy can be dissipated, it usually does not contribute significantly to overall deformation in view of its localised nature. Since a craze can act as a crack initiator, this explains their frequent association with brittle fracture.

An empirical criterion for crazing has been introduced in Section 13.4 and the role of the mean tensile stress was emphasised, consistent with the fact that crazing primarily occurs at stress concentrators in the structure, such as notches and inclusions. Here we discuss briefly some relevant structural considerations. It appears that polymers with a high entanglement density show more easily crazing than polymers with a low entanglement density. During fibril formation entanglements are lost and the higher the entanglement density, the more energy is dissipated. This leads to a higher crazing stress σ_{cra} and in such a case overall ductility may be favoured. Secondary relaxations are probably also strongly influence the competition between crazing and overall ductile deformation. It has been shown[oo] that for PC (polycarbonate) shifting the β-relaxation to higher temperature via a chemical modification indeed leads to embrittlement, while otherwise it deforms in a homogeneous way. This is consistent with the fact that polymers that show easily crazing, e.g. PS and PMMA, typically have high transition temperatures for the secondary relaxations.

To estimate the contribution of crazing to the fracture toughness, the craze was modelled[pp] as an orthorhombic material showing linear elasticity. Referring to Fig. 23.13 the maximum width of the craze is indicated by v_m while Δ represents its length. The extension over the craze is given by

$$v = v_{max}[1-(1/\lambda)] \tag{23.19}$$

where λ is the stretch. Since the stress S at the face of an isolated craze with the bulk is approximately constant over its length, the craze can be modelled as a Dugdale

[oo] Chen, L.P., Yee, A.F. and Moskala, E.J. (1999), Macromolecules **32**, 5944.
[pp] Brown, H.R. (1991), Macromolecules **24**, 2752.

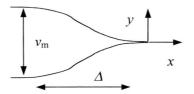

Fig. 23.13: Schematic showing the dimensions of a craze model.

crack. The fracture energy R is then given by $R = Sv_m$. Locally the driving force for crack extension g is given by $g = vW$ where $W = S^2/2E_2$ represents the energy density with E_2 the modulus of the fibril parallel to its length. Here we assume that g is used only for a relatively small area of the craze near the crack tip that has a strain significantly different from either a stretched, uncracked craze (in front of the crack) or a completely relaxed craze (after the crack tip has passed). The next step is to realise that a craze is highly anisotropic and therefore the driving force g should be described by the fracture mechanics of anisotropic materials (Section 21.8) yielding

$$G = K_I^2 \sqrt{\frac{s_{11}s_{22}}{2}} \left[\left(\frac{s_{22}}{s_{11}} \right)^{1/2} + \frac{2s_{12} + s_{66}}{2s_{11}} \right]^{1/2} \tag{23.20}$$

where the s_{ij}'s are the elastic compliances. In view of the really strong anisotropy of the fibril, we have $s_{22} \ll s_{11}$, $s_{66} \cong s_{11}$ and $s_{12} \cong 0$ and we may approximate Eq. (23.20) by

$$g = k_I^2 \sqrt{E_1 E_2} \tag{23.21}$$

where E_1 and E_2 are the moduli normal and parallel to the fibril direction, respectively. From these preliminaries one easily obtains

$$k_I = \left[\left(\frac{E_1}{E_2} \right)^{1/2} \frac{v_{max}}{2} \right]^{1/2} S \tag{23.22}$$

For the stress along the crack, we have from fracture mechanics $\sigma_{22} = k_I/(2\pi r)^{1/2}$ with r the distance from the crack tip. The stress here is the stress in the fictional continuous material that has the same average elastic properties as the craze. The stress in the y-direction in actual craze fibrils is larger than this by a factor λ. The mean stress at the fibril closest to the crack tip can be approximated as the stress at $D/2$ with D the fibril diameter. The maximum fibril stress thus becomes

$$\sigma_{fib} = \lambda k_I / \sqrt{\pi D} \tag{23.23}$$

From Eqs. (23.22) and (23.23) we obtain

$$\sigma_{fib} = \lambda S \left(\frac{v_{max}}{2\pi D} \right)^{1/2} \left(\frac{E_1}{E_2} \right)^{1/4} \tag{23.24}$$

Combining Eqs. (23.19) and (23.24) with $R = Sv_m$ we obtain a relation between the fibril stress σ_{fib} and the fracture energy

$$R = \frac{\sigma_{fib}^2 2\pi D}{S\lambda^2}\left(\frac{E_2}{E_1}\right)^{1/2}\left(1-\frac{1}{\lambda}\right) \tag{23.25}$$

Since the fracture of a glassy polymer requires the fracture of the molecular chains, it is reasonable to assume that the force to break a craze fibril is the product of the force to break a polymer chain f_{pol} and the number of effectively entangled chains in the fibril. If we denote the real density of the chains by Σ, we have $\sigma_{fib} = \Sigma f_{pol}\lambda$ so that the fracture energy becomes

$$R = \frac{\Sigma^2 f_{pol}^2 2\pi D}{S}\left(\frac{E_2}{E_1}\right)^{1/2}\left(1-\frac{1}{\lambda}\right) \tag{23.26}$$

The above expression relates the macroscopic fracture energy to the to the density of entangled strand and the force necessary to break a strand. Given sufficient information the strength of a single polymer chain can be estimated.

For PMMA the fracture energy $R = 600$ J/m^2, the fibril stress $S = 70$ MPa, the stretch $\lambda = 3$ while the density Σ is estimated as 2.8×10^{17} m^{-2}. This results in the molecular fracture force $f_{pol} = 3.8\times10^{-9}(E_1/E_2)^{1/2}$ N. For an estimate of the ratio of the moduli, we use structural information from low angle electron diffraction. This technique teaches us that for PMMA the fibrils are oriented with an angle θ of about $9°$ with the principal stress direction. We assume E_1/E_2 to vary with $\tan^2\theta$ so that we have $E_1/E2 = 0.025$. The final estimate for f_{pol} is thus 1.4×10^{-9} N. Since independent experimental estimates for f_{pol} are between 2.5×10^{-9} and 12×10^{-9} N, the agreement is fair.

For semi-crystalline polymers at a temperature between T_g and T_m also a significant amount of local deformation occurs, akin to crazing. It is usually assumed that the deformation initiates in the amorphous regions between the lamellae. These regions deform and cavitations occur until their size becomes comparable to that of the lamellae. After that the lamellae themselves provide the material for the fibrils. Therefore a high entanglement density and high degree of connectivity between the amorphous regions and the lamellae, i.e. high molecular weight, are expected to enhance the fracture toughness.

To enhance the fracture energy for easily crazing polymers such as PS, one strategy is to promote crazing via the introduction of stress concentrators. Since the stiffness of rubbers is largely different from that of PS, one way is to introduce rubber particles in the PS matrix. At each of these particles crazing can occur and the overall amount of energy dissipated during fracture increases.

Similar to inorganics, brittle polymers may exhibit subcritical crack growth, empirically often described by a power law in the crack growth rate $\dot{a}$ and reading

$$K = A\dot{a}^n \qquad \text{or} \qquad R = A'\dot{a}^{2n}$$

with K the fracture toughness, R the fracture energy and A and A' are the temperature dependent material parameters. The exponent n is quite low as compared with inorganics, e.g. $n \cong 0.07$ for PMMA. The modelling as described in Section 24.5 is also applicable here. The mechanism involved seems to be the breakdown of the craze at the tip of a growing crack. In tough polymer, such as PC, general yielding as well as (multiple) crazing takes place at the crack tip, leading to a more complex situation.

Finally, we note that, although rubbers do show extensive elongation upon deformation, their fracture behaviour is essentially brittle, as will be made clear. The fracture behaviour can be described by fracture mechanics, although rubber scientists developed their own jargon and usually employ only the fracture energy R, by them called tearing energy due to usual mode of measuring. The fracture energy R empirically depends again on the rate of crack propagation $\dot{a}$ via a power as given above. Also for rubbers the exponent n is small, e.g. $2n \cong 0.25$ for styrene-butadiene rubber (SBR). At room temperature at normal deformation rates the values for R are typically in the range 10^3 to 10^5 J/m^2, which is much higher than the energy required for just bond breaking. This high value is due to the large deformation at the crack tip and the associated visco-elastic relaxation when the crack propagates. At low deformation rate these losses can be reduced significantly and R approaches the bond breaking energy. The 'intrinsic' energy necessary for fracture is thus just the bond breaking energy and that is why a rubber can be called brittle. In filled rubbers crack propagation occurs in discrete jumps, making an analysis much more difficult.

23.7   Composites*

Although many materials can be designated as composites, we will be brief and limit ourselves to a few remarks about uniaxially oriented fibre reinforced materials and laminate-like materials. Hull and Clyne (1996) have provided an excellent introduction to composite materials.

For fibre-reinforced materials the fracture process proceeds as follows. At low load the matrix fractures. Thereafter the fibres (partially) debond and elongate. At still higher load the fibres themselves fracture. Because the fibres are not exactly in the plane of the crack fibre pull-out has to take place. Next to the strain energy in the fibres this pull-out usually forms a significant part of the fracture energy. Considerable increases in fracture toughness and strength can be realised. Table 23.7 provides an example of a C-fibre reinforced glass matrix material[qq]. Since pull-out is important, chemical surface modification of the fibres is often employed to optimise the bonding between matrix and fibres in such a way that pull-out becomes possible.

We consider a simple model for an uniaxially oriented fibre composite and deal with uniaxial loading only. First, we assume the fibres to be continuous. The load P will be carried by the fibres and matrix and we have $P = P_{mat}+P_{fib}$. From compatibility considerations the strain in the fibre ε_{fib} and matrix ε_{mat} are equal to the total strain ε. Since the load is given by $P = \sigma A$, where A denotes a cross-sectional area, we obtain

$$\sigma = PA = P_{fib} + P_{mat} = \sigma_{fib} A_{fib} + \sigma_{mat} A_{mat} \tag{23.27}$$

where σ, σ_{fib} and σ_{mat} denote the strength of the composite, fibre and matrix, respectively. The area fractions of the matrix and fibres equal their volume fractions so that $A_{fib} = \phi_{fib}A$ and $A_{mat} = \phi_{mat}A$. This leads to

$$\sigma = \sigma_{fib}V_{fib} + \sigma_{mat}V_{mat} = \sigma_{fib}V_{fib} + \sigma_{mat}(1-V_{fib}) = E_{fib}\varepsilon_{fib}V_{fib} + E_{mat}\varepsilon_{mat}V_{mat} \tag{23.28}$$

where in the last step use is made of Hooke's law. For the uniaxial composite we have for the Young's modulus $E = \phi_{fib}E_{fib} + \phi_{mat}E_{mat}$. Therefore, we obtain

[qq] Philips, D.C., Sampbell, R.A.J. and Bowen, D.H. (1972), J. Mater. Sci. 7,1454.

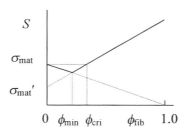

Fig. 23.14: Composite strength as a function of volume fraction fibres ϕ_{fib}.

$$\frac{P_{fib}}{P_{mat}} = \frac{E_{fib}\varepsilon_{fib}\phi_{fib}}{E_{mat}\varepsilon_{mat}\phi_{mat}} = \frac{E_{fib}\phi_{fib}}{E_{mat}\phi_{mat}} \tag{23.29}$$

As expected, the load is carried by the fibres and matrix in proportion to their modulus and volume fraction.

If the matrix is no longer reacting elastically but plastically, we may assume that the stress at the matrix equals the stress σ_{mat}' corresponding to the fracture strain of the fibre. For an elastic ideal plastic material, this would be the uniaxial yield strength Y while for other materials this value depends on the precise shape of the stress-strain curve. We thus have

$$\sigma = \sigma_{fib}\phi_{fib} + \sigma_{mat}'(1-\phi_{fib}) \tag{23.30}$$

Note that to obtain strengthening we must have $\sigma > \sigma_{mat}$ so that the volume fraction fibres must be larger than a critical value $\phi_{cri} = (\sigma_{mat}-\sigma_{mat}')/(\sigma_{fib}-\sigma_{mat}')$, as indicated in Fig. 23.14. Further for small value of ϕ_{fib} the strength is not given by Eq. (23.30) because insufficient fibres are present to restrain the elongation of the matrix and the fibres are rapidly stressed to their points of fracture. However, the matrix will carry part of the load and if we assume that this occurs when all fibres are fractured, the composite strength will be given by $\sigma \geq \sigma_{mat}(1-\phi_{fib})$. Hence the strength of the composite starts to follow Eq. (23.30) when $\phi_{min} = (\sigma_{mat}-\sigma_{mat}')/(\sigma_{fib}+\sigma_{mat}-\sigma_{mat}')$, also indicated in Fig. 23.14.

For discontinuous fibres the composite strength is smaller than the ideal value indicated by Eq. (23.28). It appears that the load from the matrix to the fibres is transferred by the shear stress τ_{rz} acting along the interface between matrix and fibre but mainly near the end of the fibres over a length z. For a fibre with a circular cross-section and a radius r and experiencing a stress σ_{zz} we obtain

$$\sigma_{zz}\pi r^2 = \tau_{rz}2\pi rz \qquad \text{or} \qquad \sigma_{zz} = 2\tau_{rz}z/r \tag{23.31}$$

At $z = 0$, $\sigma_{zz} = 0$. When σ_{zz} reaches the fibre strength σ_{fib}, the fibre either breaks or deforms plastically. This occurs at the critical transfer length $z = l_{cri}/2$. Substituting

Table 23.7: C-fibre reinforced glass.

	E (GPa)	S (MPa)	R (J/m^2)	K_{Ic} (MPa m$^{1/2}$)
glass	60	100	4	0.7
composite (50 vol.%)	193	700	5000	44

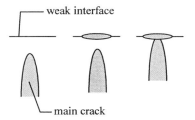

Fig. 23.15: The Cook-Gordon mechanism showing the approaching of the main crack and the T-shaped stopper.

these results in Eq. (23.31), we obtain with τ the interfacial shear strength

$$l_{cri} / r = \sigma_{fib} / \tau \tag{23.32}$$

To realise a stress equal to the strength at the centre of a fibre, the fibre aspect ratio l_{cri}/r must increase with increasing fibres strength and decreasing shear strength. For a fibre with a critical length l_{cri}, the average stress $\overline{\sigma}_{fib}$ in the fibre is equal to $\sigma_{fib}/2$. For longer fibres, the stress remains at σ_{fib} in the midsection so that the average stress becomes

$$\overline{\sigma}_{fib} = (\sigma_{fib} l - \sigma_{fib} l_{cri} / 2)/l = \sigma_{fib}(1 - l_{cri} / 2l) \tag{23.33}$$

The average strength thus increases with increasing fibre length and approaches σ_{fib} for $l \gg l_{cri}$. Substitution in Eq. (23.30) yields

$$\sigma = \sigma_{fib}\left(1 - \frac{l_{cri}}{2l}\right)\phi_{fib} + \sigma_{mat}{'}(1 - \phi_{fib}) \tag{23.34}$$

which describes the strength of a composite with discontinuous fibres dispersed in a plastically deforming matrix.

For a fibre with a length l, less than a critical fibre length l_{cri} for fracture, the fibres will show pull-out instead of fracture. The energy U_{fib} associated with the pull-out of single fibre can be calculated from the shear stress τ acting along the interface directed in the z-direction as

$$U_{fib} = \int_0^z 2\pi r \tau z \, dz = \pi r \tau z^2 \tag{23.35}$$

The energy is maximised when the pull-out length is $z = \tfrac{1}{2} l_{cri}$. Using Eq. (23.32) and realising that for n fibres per unit area we have $n\pi r^2 = A_{fib} = \phi_{fib}A$, we obtain for the fracture energy due to pull-out

$$R = \frac{nU_{fib}}{A} = \frac{n\pi r \tau z^2}{A} = \frac{n\pi r \tau l_{cri}^2}{4A} = \frac{n\pi r^3 \tau (\sigma_{fib}/\tau)^2}{4A} = \frac{n\pi r^2 r \sigma_{fib}^2}{4A\tau} = \frac{\phi_{fib} r \sigma_{fib}^2}{4\tau} \tag{23.36}$$

The fracture energy thus decreases with decreasing fibre radius and emphasises that the fibres should not be too thin. It also decreases with increasing shear strength of the interface, indicating once more its importance. The fibre pull-out mechanism is the main mechanism for the fracture energy of wood. Finally, we note that for brittle matrices with proper control of the interface in this way considerable toughening can be realised. For fibres with a length larger than the critical length l_{cri}, also the energy

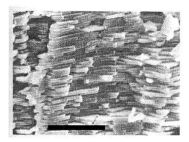

Fig. 23.16: The fracture surface of a shell showing the individual $CaCO_3$ crystals due to the Cook-Gordon mechanism leading to high tortuosity and fracture energy. The bar indicates 100 μm. The fracture direction is from bottom to top.

associated with relaxation of the broken fibres must be added. If these fibres relax over a characteristic length l_{rel}, this energy is $\phi_{fib}l_{rel} \cdot \sigma_{fib}^2/2E_{fib}$. Table 23.7 provides an example.

The fracture energy of non-ductile crystalline materials and amorphous low molecular weight materials is intrinsically low and to enhance it several mechanisms have been introduced. Fibre reinforcement is one of them and the idea of low strength interfaces was introduced. The same idea can be used for other mechanisms. In all considerations on crack propagation so far, we considered only crack propagating in a direction in its own plane, obviously since the stress concentration is the most severe in that direction. Moreover, we assumed the fracture energy to be isotropic. If we can introduce fracture planes with low fracture energy but in another direction then the main crack, say perpendicular to it, the following can happen. If the crack plane deflects in that low fracture energy direction in spite of the lower stress concentration in that direction, the crack tip for the main crack is effectively blunted leading to much lower stress concentration and thus to higher allowable stress. This causes the crack to stop. Moreover, since the deflected crack covers a significantly larger area, the macroscopic fracture energy is increased. The weak interface thus acts effectively as a crack stopper (Fig. 23.15). This so-called *Cook-Gordon mechanism* occurs in many natural materials, e.g. wood and shells (Fig. 23.16). It should be clear that this mechanism crucially depends on the ratio of the fracture energies in the main and deflected direction. The Cook-Gordon effect has been used to advantage to reinforce and toughen materials by Clegg[π] using laminar materials. For cracks perpendicular to the weak interfaces considerable toughness has been realised.

23.8   Bibliography

Broek, D. (1978), *Elementary engineering fracture mechanics*, 2nd ed., Sijthoff & Noordhoff, Alphen aan den Rijn.

Davidge, R.W. (1979), *Mechanical behaviour of ceramics*, Cambridge University Press, Cambridge.

Derby, B., Hils, D. and Ruiz, C. (1992), *Materials for engineering*, Longman Scientific & Technical, Harlow, UK.

Dieter, G.E. (1976), *Mechanical metallurgy*, McGraw-Hill Kogakusha, Tokyo.

[π] Clegg, W.J. (1999), Nature **286**, 1097.

Green, D.J. (1998), *An introduction to the mechanical properties of ceramics*, Cambridge University Press, Cambridge.

Herzberg, R.W. (1989), *Deformation and fracture mechanics of engineering materials*, 3rd ed., Wiley, New York.

Hull, D. (1999), *Fractography*, Cambridge University Press, Cambridge.

Hull, D. and Clyne, T.W. (1996), *An introduction to composite materials*, 2nd ed., Cambridge University Press, Cambridge.

Kelly, A. and McMillan, N.H. (1986), *Strong solids*, 3rd ed., Clarendon, Oxford.

Lawn, B.R. (1993), *Fracture of brittle solids*, Cambridge University Press, Cambridge.

Wachtman, J.B. (1996), *Mechanical properties of ceramics*, Wiley, New York.

Ward, I.M. (1983), *Mechanical properties of polymers*, 2nd ed., Wiley, Chichester.

Young, R.J. (1981), *Introduction to polymers*, Chapman and Hall, London.

24

Fatigue*

Fracture theory and its applications were dealt with in the previous chapters. In this chapter fatigue is discussed. After a brief indication of the importance of fatigue, we discuss first metal fatigue, both the classical and more modern theory. As before we deal with macroscopic and structural aspects. Thereafter the behaviour of inorganics and polymers is discussed although to a lesser extent for the latter category.

24.1   The S-N curve: the classical approach for metals

Fatigue is the fracture of a material after a certain time when loaded at a stress lower than the instantaneous strength. This load can be constant (static fatigue), periodic (cyclic fatigue) or random. Fatigue occurs in metals, inorganics and polymers. While for metals static fatigue is rare, it is common for inorganic materials. For both metals and inorganics cyclic fatigue can occur although the nature of the process is quite different. Fatigue is important in many structures. Some sources, e.g. Dieter (1976), state that 90% of all failures is due to fatigue. It will be clear that even if the 90% is exaggerated, fatigue is a highly relevant failure mechanism.

The impact of fatigue

Fatigue occurs in many materials and in many applications. Due to the many, complex and interacting factors involved, reliable predictions require extensive and time-consuming testing of the particular material used and therefore the topic is not very popular amongst researchers. As stated in the main text, the phenomenon is extremely important. The most well-known example in daily life is probably the failure of a paperclip after repetitive bending, as frequently done in offices and elsewhere. Machine parts that are designed for many load cycles are typically found in turbines, engines, railway wheels and axles, airplanes, ships at high sea, etc. Nevertheless also many structures for which cyclical loading is limited and therefore not specially designed with respect to fatigue fail by this mechanism. Toys provide a daily life example but also for doorknobs, water tap handles and similar household items as well as for technical items such as tools, fatigue is often the responsible failure mechanism.

In this introductory section, we discuss the general picture of fatigue as encountered for metals. For the moment we limit ourselves to cyclic fatigue. In cyclic fatigue we can distinguish various situations (Fig. 24.1). The applied stress can be zero on average ($\sigma_{ave} = 0$) or have a finite value ($\sigma_{ave} \neq 0$). Furthermore the amplitude of the stress (σ_{amp}) is relevant. Of course, in many practical situations the loading cycles are not purely sinusoidal but we restrict ourselves for the moment to this type of loading. Usually the experimental results are expressed in the parameters R and A, defined by

▶ $$R = \sigma_{min} / \sigma_{max} \qquad \text{and} \qquad A = \sigma_{amp} / \sigma_{ave} \qquad (24.1)$$

where $\sigma_{max} = \sigma_{ave} + \sigma_{amp}$ and $\sigma_{min} = \sigma_{ave} - \sigma_{amp}$, respectively. A completely reversed, average stress zero cycle is thus characterised by $R = -1$ and $A = \infty$.

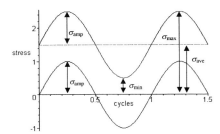

Fig. 24.1: Sinusoidal cycles with ($\sigma_{ave} = 0$, $\sigma_{amp} = 1$) and ($\sigma_{ave} = 1.5$, $\sigma_{amp} = 1$).

The usual way to represent fatigue data is in the so-called *Wöhler*[a] curve[b]. To obtain this curve, the strength S_N is measured in a rapid loading test after the specimen has experienced N load cycles for a certain combination of R and A. This is done for a range of cycle numbers. Next, S_N is plotted against the logarithm of the number of cycles N. A typical curve for ferro and non-ferro metals is given in Fig. 24.2. In the case of most ferro-metals a *fatigue* (or *endurance*) *limit*, here indicated with F, is often observed. This implies that below this stress the material has an infinite lifetime. The value of F for steel is typically $0.35S$ to $0.50S$. For non-ferro metals this limit is usually absent, Ti being an important exception. In this case the endurance limit is defined as the stress amplitude, which the specimen can endure for an arbitrary numbers of cycles, say 10^7 or 10^8. Experimentally the S_N-N relation follows often the *Basquin law*[c] $S_N = SN^{-a}$ with a in the range of 0.05-0.12.

The S_N-N curve is normally determined with a limited number of specimens, e.g. 10. There is a considerable spread in the results though and one should actually speak of the number of cycles at a certain failure probability p. Since there is some experimental evidence, the spread is often discussed in terms of a normal distribution of the strength S_N with the logarithm of the number of cycles $\ln(N)$, the so-called lognormal distribution. However, other distributions, such as the Weibull distribution, are

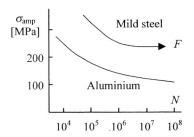

Fig. 24.2: Typical S_N-N or fatigue curve for ferro- and non-ferro metals.

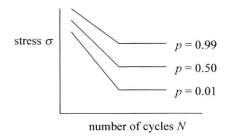

Fig. 24.3: A schematic of the S_N-N curve for various failure probabilities p.

---

[a] A. Wöhler (1819-1914). German engineer whose main work on fatigue of metals was carried out with a view to remedy the occurrence of annoying fractures of railway axles.
[b] In the literature, it is often addressed as the *S-N* curve. We will use S for the (zero cycle) tensile strength and S_N for the strength after N cycles and thus refer to the S_N-N curve. The applied stress is indicated by σ.

used as well. In order to estimate the probabilities mentioned reasonably, many more than ten specimens are required, of course. A schematic example of the variability is shown in Fig. 24.3, indicating in particular the spread in the fatigue limit. It should be remarked that, like in the case of brittle fracture, variability is intrinsic, i.e. dependent on material characteristics and not solely on experimental uncertainty.

Orowan[d] presented a simple model that can explain the existence of a fatigue limit. To this purpose we consider a linear hardening material without Bauschinger effect (Fig. 24.4), which is cyclically taken between the upper strain limit ε_∞ and the lower strain limit $-\varepsilon_\infty$. In tension, after initial elastic deformation with Young's modulus E until the yield strength σ_0, plastic deformation occurs up to the stress σ_1 at which the maximal upper strain $\varepsilon_\infty = \sigma_1/E$ is reached. In the following, compressive part of the cycle elastic relaxation occurs up to a stress $-\sigma_1$, after which again plastic deformation occurs up to $-\sigma_2$. At this stress, the maximum strain $-\varepsilon_\infty$ is reached. With an increasing number of cycles the elastic part of the cycles becomes larger and larger and the hysteresis loop becomes smaller and smaller until only elastic deformation occurs between $\sigma_\infty = E\varepsilon_\infty$ and $-\sigma_\infty = -E\varepsilon_\infty$. If the stress σ_∞ is larger than the instantaneous strength S the material fractures. On the other hand, if the stress σ_∞ is smaller than the strength S, a fatigue limit is present.

In fatigue usually a large number or cycles, say $N > 10^5$, is involved. The fatigue process normally occurs at relatively low stress with respect to the instantaneous strength. At relatively high stress the number of sustainable cycles is much lower, say $N < 10^4$, and one speaks of *low cycle fatigue*. This part of fatigue is highly determined by the instantaneous strength and diameter reduction of the specimen. An empirical approach to describe the transition between high and low cycle fatigue runs as follows. The total strain ε is the sum of the elastic strain $\varepsilon^{(e)}$ and the plastic strain $\varepsilon^{(p)}$. For short lives (at relatively high stress) fatigue is controlled by plasticity, i.e. by

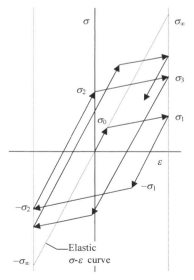

Fig. 24.4: Stress-strain curve for a material in cyclic fatigue between two strain limits.

[c] Basquin, O.H. (1910), Proc. ASTM, **10**, 625.
[d] Orowan, E. (1939), Proc. Roy. Soc. London A **171**, 79.

strain, while for long lives (at relatively low stress) it is strength controlled, i.e. by stress. For both regimes the experimental strain amplitude data can be described reasonably well by $\Delta\varepsilon = CN^{-a}$, with the parameters a and C, of course, different for different materials. For the high cycle regime the exponent, indicated by b, is small. The exponent varies a bit for various materials, say in the range from 0.05 to 0.12. A typical value is $c = 0.08$. This low c-value could be expected since for pure elastic deformation we would have $\Delta\varepsilon^{(e)} = S/E$, where, as before, S and E denote the (tensile) strength and Young's modulus, respectively. The weak N-dependence allows for a gradual degradation of the maximum allowable elastic strain with the number of cycles. In fact we will write for the elastic strain amplitude $\Delta\varepsilon^{(e)}/2$,

$$\Delta\varepsilon^{(e)}/2 = \sigma_{amp}/E = (\sigma_f'/E)(2N)^{-b}$$

where σ_f' is the fatigue strength coefficient, approximately equal to the true strength σ_f. It appears that for low cycle fatigue the exponent, for this regime indicated by c, varies but little for various materials and is about 0.5 to 0.7. Here we write

$$\Delta\varepsilon^{(p)}/2 = \varepsilon_f'(2N)^{-c}$$

sometimes called the *Coffin-Manson law*[e]. Here ε_f' is the fatigue ductility coefficient, approximately equal to the true ductility ε_f.

The total strain range is given by $\Delta\varepsilon = \Delta\varepsilon^{(e)} + \Delta\varepsilon^{(p)}$ and therefore one obtains for the maximum alternating strain amplitude[f]

▶ $$\Delta\varepsilon/2 = (\sigma_f'/E)(2N)^{-b} + \varepsilon_f'(2N)^{-c} \tag{24.2}$$

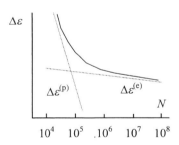

Fig. 24.5: The transition in fatigue life.

Table 24.1: Typical fatigue data for Al- and Fe-alloys (steel).

Metal	S (MPa)	Y (MPa)	F (MPa)	σ_f' (MPa)	ε_f'	b	c
Al 100	90	34	34	193	1.80	0.106	0.69
Al 2024	483	~360	138	1103	0.22	0.124	0.59
Fe 1015[*]	455	~250	240	827	0.95	0.110	0.64
Fe 4340[#]	1260	1170	670	1655	0.73	0.076	0.62

* Annealed. # Quenched and tempered at 538 °C.

---

[e] Coffin, L.F. (1954), Trans ASME **76**, 931 and Manson, S.S. (1954), *Behavior of materials under condition of thermal stress*, NACA report 1170, Cleveland, Lewis Flight Propulsion Laboratory.
[f] It is conventional to take the number of stress reversals $2N$ instead of the number of cycles N.

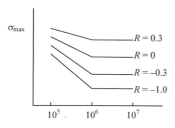 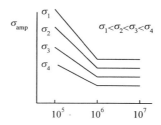

Fig. 24.6: Influence of the parameter R (left) and the mean stress σ_{ave} (right) on the fatigue behaviour of metals. On the right-hand side, the increasing mean stresses are indicated by σ_1, σ_2, etc.

as illustrated in Fig. 24.5. The number of cycles where the contributions of $\Delta\varepsilon^{(e)}$ and $\Delta\varepsilon^{(p)}$ are equal is denoted as *transition fatigue life*. If one accepts the values of b and c as given, fatigue behaviour can be predicted from a tensile test since this test provides the values of S, E and $\varepsilon^{(f)}$. We see that for high stress (strain) cycles ductile alloys are the best while strong alloys are better for the region of low stress (strain). In the transition region, typically at $\varepsilon \approx 0.1$ and $N \sim 10^3$-10^5, there is no preferred type of alloy. Some typical data are given in Table 24.1.

24.2   Influence of average stress, load fluctuations and multi-axiality

Most of the work on fatigue of metals has been done for completely reversed loading ($R = -1$ and $A = \infty$). Limiting us to metals exhibiting a clear fatigue limit, the influence of the value for R and σ_{ave} as observed experimentally is schematically represented in Fig. 24.6. From the left part of this figure, it can be seen that with increasing R, equivalent to increasing σ_{ave}, the value for the fatigue limit increases. Plotting the σ_{amp} data against the number of cycles is also done in Fig. 24.6. From the right part of this figure, one can see that with increasing mean stress σ_{ave}, in the figure indicated by σ_1, σ_2, etc., the allowed alternating stress σ_{amp} decreases.

One way to summarise the data is via the *Goodman diagram* (Fig. 24.7), i.e. a plot

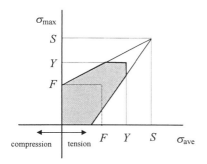 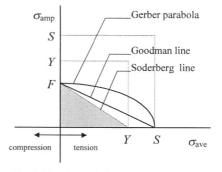

Fig. 24.7: The Goodman diagram showing the stress range as a function of σ_{ave}. The grey area indicates infinite life with the yield strength Y as the limiting stress.

Fig. 24.8: The Soderberg diagram showing σ_{amp} as a function of σ_{ave}. The grey area indicates infinite life.

of the limiting range of stress from σ_{min} to σ_{max} versus σ_{ave}. At $\sigma_{ave} = 0$, $\sigma_{min} = -F$ and

$\sigma_{max} = F$ while at $\sigma_{ave} = S$ the range $\sigma_{max} - \sigma_{min} = 0$. In the absence of data, Goodman[g] considered that the area between the lines connecting $(0,F)$ and (S,S) and $(-F)$ and (S,S) could be considered as safe. Thus note that as σ_{ave} becomes more tensile the allowable range for σ_{amp} is reduced until it becomes zero at $\sigma_{amp} = S$. For a more conservative estimate, stresses larger than the yield strength Y should be also excluded. From this figure, an estimate can easily be made for the allowable σ_{amp} for a given σ_{ave}.

Another way to represent this behaviour, sometimes addressed as the *Soderberg plot*, is shown in Fig. 24.8. In this plot the alternating stress σ_{amp} is plotted versus the mean stress σ_{ave}. As said, Goodman suggested to make an estimate of the fatigue limit as a function σ_{ave} by drawing a straight line between the momentary strength S to F, the fatigue limit for $R = -1$ and $A = \infty$. Alternatively, Gerber[h] used a parabola. Both lines are indicated in Fig. 24.8. The Goodman and Gerber lines, usually addressed as *failure envelopes*, can be represented by

$$\frac{\sigma_{amp}}{F} + \left(\frac{\sigma_{ave}}{S}\right)^x = 1 \quad \text{or} \quad \sigma_{amp} = F\left[1 - \left(\frac{\sigma_{ave}}{S}\right)^x\right] \tag{24.3}$$

where $x = 1$ and $x = 2$ for the Goodman and Gerber lines, respectively. Also said, to be a little more conservative, the applied stress can be limited to the yield strength Y (as was done by Soderberg[i]) so that the triangle above the yield strength is also declared a finite life area. This behaviour similarly can be represented by

$$\frac{\sigma_{amp}}{F} + \frac{\sigma_{ave}}{Y} = 1 \quad \text{or} \quad \sigma_{amp} = F\left(1 - \frac{\sigma_{ave}}{Y}\right) \tag{24.4}$$

Obviously, other empirical relations are possible. Summarising, it may be said that the Soderberg relation provides a conservative estimate for most engineering alloys, the Goodman relation is quite closely followed by brittle materials while the Gerber relation is generally good for ductile alloys.

In practice many structures are subjected to load fluctuations and not to constant amplitude and frequency loads. A number of procedures have been developed to reduce non-regular load patterns into a sequence of regular patterns. Essentially these procedures should take care of two points. First, counting of the various types of cycles and, second, accumulation of damage for various sequences.

The counting of cycles for a random load pattern has become a topic in its own right and only a few remarks will be made. Empirical methods of counting the number of cycles can be counted, like the *rainflow method* (see Herzberg, 1989 or Suresh, 1991), have been devised. Nowadays the frequency spectrum of the mechanical loading can be analysed relatively simple by Fourier analysis. This procedure yields the distribution of amplitudes for each frequency that occurs during loading, i.e. the load spectrum, and the effects of the various sine waves in the spectrum can be added. The interaction of various waves, i.e. the difference in effect of a large amplitude wave followed by a small amplitude one and vice versa, is not taken into account though (see the next section) in both methods.

---

[g] Goodman, J. (1899), *Mechanics applied to engineering*, Longmans, London.
[h] Gerber, H. (1874), Z. Bayer. Arch. Ing.-Ver. **6**, 101.
[i] Soderberg, C.R. (1930), Trans. ASME **52**, APM-52-2, 13.

For the accumulation of damage, we consider stress cycles with a stress above the fatigue limit as an *overstress* cycle. If a specimen is loaded in time with n_i overstress cycles of varying magnitude and N_j represents the number of cycles (life) at a particular overstress level, a linear cumulative damage rule, also frequently denoted as the *Palmgren-Miner rule*[j], can be used. It assumes that the total life of a component can be estimated by adding the percentage of life consumed at each overstress cycle and thus reads

▶ $\sum_i n_i / N_i = 1$ (24.5)

It neglects entirely the order of different loading cycles, which do have a significant influence. Although many deviations have been observed and many modifications proposed, none of these modifications seem to be essentially better and thus have not gained wide acceptance.

Example 24.1

Suppose that for a metal the fatigue curve is linear, i.e. follows Basquin's law, as described by

$$\log N = 14\left[1 - (\sigma / S)\right]$$ (24.6)

and that we apply n_1 load cycles at $\sigma = 0.6S$. Let us estimate how many cycles the steel can withstand if we increase the load to $\sigma = 0.7S$. From Eq. (24.6) we calculate for the load $\sigma = 0.6S$, $N_{0.6S} = 3.98 \times 10^5$, while for the load $\sigma = 0.7S$ we have $N_{0.7S} = 1.58 \times 10^4$. According to the Palmgren-Miner rule we obtain for the allowable number of of load cycles n_2 at $\sigma = 0.7S$

$$\frac{n_1}{N_{0.6S}} + \frac{n_2}{N_{0.7S}} = \frac{1 \times 10^5}{3.98 \times 10^5} + \frac{n_2}{1.58 \times 10^4} = 1 \quad \text{or} \quad n_2 = 1.18 \times 10^4$$

So far we have addressed only uni-axial loading, which is not as restrictive as it appears at first sight since cracks, even in multi-axial loading, grow mainly perpendicular to the largest tensile principal stress. However, we will make a few remarks about multi-axial failure envelopes (see also Broek, 1978 and Derby et al, 1992). Generally the equations proposed are empirical generalisations of the uni-axial ones. For example, Gough[k] performed fatigue tests in pure bending (with stress σ) and in torsion (with stress τ). The fatigue limits for these tests are indicated by F and T, respectively. For combined tests with stress components σ and τ, he proposed the failure envelope

$$\left(\frac{\sigma}{F}\right)^2 + \left(\frac{\tau}{T}\right)^2 = 1$$

---

[j] Palmgren, A. (1924), Z. Ver. Deutscher. Ing. **68**, 339 and Miner, M.A., (1945), J. Appl. Mech. **12**, A159.
[k] Gough, H.J. (1949), Proc. Inst. Mech. Eng. London, **160**, 417.

Similarly, Sines[1] generalised the Goodman equation to

$$\frac{\sigma_{vM}*}{F} + \frac{3\sigma_m}{S} = \frac{(3J_2')^{1/2}}{F} + \frac{J_1}{S} = 1$$

where $\sigma_{vM}* = (3J_2')^{1/2}$ is the von Mises equivalent stress, $\sigma_m = \frac{1}{3}J_1$ the mean stress and S is the (zero cycle) tensile strength. As usual, J_2' and J_1 represent the second invariant of the deviatoric stress tensor and the first invariant of the full stress tensor, respectively. As can be expected, in view of the remarks made on their uni-axial counterparts, these approaches have met variable success. More details about the topics of this section can be found in the books by Dieter (1976), Herzberg (1989) and Suresh (1991).

Problem 24.1

In Fig. 24.9, a schematic representation of the fatigue behaviour of a particular steel is shown for three failure probabilities p. The data have been gathered for a tensile-loaded bar using the conditions $R = \sigma_{min}/\sigma_{max} = -1$ and $A = \sigma_{amp}/\sigma_{ave} = \infty$, where σ_{max} is the maximum applied stress, σ_{min} the minimum applied stress, σ_{ave} the average applied stress and σ_{amp} is the amplitude stress.

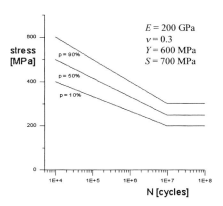

Fig. 24.9: Schematic fatigue behaviour.

a) How many cycles will the steel survive with 50% probability using a maximum stress $\sigma_{max} = 300$ MPa?
b) What is the failure probability p for a stress of 300 MPa after 10^6 cycles?
c) Construct the Goodman diagram for 10^5 cycles and $p = 50\%$. Hatch the area for allowed loading, conservatively assuming that the yield strength Y may not be exceeded.
d) If an average stress $\sigma_{ave} = 300$ MPa is applied, what is the maximum allowable amplitude stress σ_{amp}?

Problem 24.2

A metal has a linear fatigue curve starting at $(S, \log N = 0)$ and passing through $(S/3, \log N = 7)$.
a) Determine the expression for the fatigue curve.
b) Estimate how many cycles at $\sigma = 0.8S$ the metal can still endure after 3×10^4 cycles at $\sigma = 0.7S$.

24.3   Fracture mechanics: the modern approach

Although the above outlined approach is still used to a significant extent, another, more modern, approach is based on fracture mechanics. It is based on the crack growth per cycle da/dN (Fig. 24.10) and the data are usually plotted as a da/dN versus ΔK curve, where the stress intensity factor $K = \alpha\sigma\sqrt{a}$ and $\Delta K = K_{max} - K_{min}$ the range of the stress intensity factor, associated with σ_{min} and σ_{max}. During each cycle the crack grows until a stress intensity is obtained at which catastrophic fracture occurs. Three regions can be often discerned. In region I initiation occurs and the microstructure, surface finish, etc. are quite important. Below a certain threshold ΔK_{thr} no crack growth occurs. If the crack is well enough developed region II starts and here the crack growth is largely determined by the applied stress. Finally, in region III the stress intensity approaches the fracture toughness K_{Ic}, the crack accelerates and final fracture occurs. The fracture mode will vary with material and service conditions.

The most widely used kinetic equation relating the crack growth per cycle da/dN to the range in the stress intensity factor $\Delta K = K_{max} - K_{min}$ in region II, often called the *Paris-Erdogan equation*[m] (1962), reads[n]

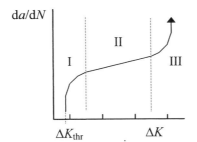

Fig. 24.10: Typical da/dN - ΔK curve.

Table 24.2: Fatigue data for several metals.

Material	S (MPa)	Y (MPa)	R	N	ΔK^* (MPa m$^{1/2}$)	ΔK_{thr} (MPa m$^{1/2}$)
Mild steel	325	–	−1.0	–	–	6.4
	430	230	0.74	3.3	6.2	3.8
Mar. steel	2010	–	0.67	3.0	3.5	2.7
18/8 steel	685	230	−1.0	–	–	6.0
		–	0.74	3.1	6.3	4.1
Copper	220	–	−1.0	–	–	2.7
		–	0.80	3.9	4.3	1.3
60/40 Brass	330	–	−1.0	–	–	3.1
		–	0.72	3.9	4.3	2.6
Titanium	540	440	0.6	4.4	3.1	2.2
Nickel	440	–	−1.0	–	–	5.9
		–	0.71	4.0	8.8	3.6

ΔK^* denotes the ΔK for $da/dN = 10^{-6}$ mm/cycle.

---

[m] The origin of this equation is not clear. See Paris, P. and Erdogan, F. (1963), J. Basic Eng. **85**, 528.
[n] Note that is again an example of an empirical equation which should be written differently, e.g. as $da/dN = C(\Delta K/K_{Ic})^n$. In the form quoted above the dimension of C depends on the value of n.

▶ $$\frac{da}{dN} = C\Delta K^n \tag{24.7}$$

The parameters C and n are assumed to be material parameters. Since the volume of the damaged zone is proportional to the square of the plastic zone size r_p and r_p itself is proportional to K_I^2, the exponent n is frequently assumed to be $n = 4$ for fully unloading cycles. In practice n varies between 1 and 6 approximately, depending on the metal, the average stress level σ_{ave} and environment. Some typical data are given in Table 24.2.

Obviously the bend-over from region I to II and II to III are not included in the Paris expression. It is simply supposed to be valid in the range of $K_{thr} < K_I < K_{Ic}$. Forman et al.[o] (1967) argued that $da/dN \rightarrow \infty$ if $K_I \rightarrow K_{Ic}$ and using $R = \sigma_{min}/\sigma_{max}$ they proposed

$$\frac{da}{dN} = \frac{C\Delta K^n}{(1-R)K_{Ic} - \Delta K} \tag{24.8}$$

in order to take into account both regions II and III. Similarly for the incorporation of the transition from region I to II, Donahue et al.[p] proposed

$$\frac{da}{dN} = C\left(\Delta K - \Delta K_{thr}\right)^n \tag{24.9}$$

Relations taking both aspects into account also exist, e.g. the relation proposed by Priddle[q]

$$\frac{da}{dN} = C\left(\frac{\Delta K - \Delta K_{thr}}{K_{Ic} - K_{max}}\right)^n \tag{24.10}$$

where ΔK_{thr} is not a material constant but may depend on R. Unfortunately not a great deal of data is available for these more extensive equations. Many more exist but usually they are not based on sound mechanical considerations.

For variable amplitude fatigue, another important aspect is that an occasional high stress will introduce a relatively large plastic zone, which in subsequent normal cycles retards the crack growth. Crack retardation is complex and depends on the amount and order of the overloads. A frequently used simple model is due to Wheeler[r] who assumes that the compressive residual stress at the crack tip due to the overload is the main cause of retardation. In his approach, the retarded crack growth rate $(da/dN)_{ret}$ for variable amplitude fatigue is related to the normal crack growth rate $(da/dN)_{nor}$ for constant amplitude fatigue according to

$$\left(\frac{da}{dN}\right)_{ret} = \phi\left(\frac{da}{dN}\right)_{nor} = \left(\frac{r_{p,i}}{\lambda}\right)^m \left(\frac{da}{dN}\right)_{nor} \tag{24.11}$$

[o] Forman, R.G., Keary, V.E. and Engle, R.M. (1967), J. Basic Eng. Trans. ASME **89**, 459.
[p] Donahue, R.J., Clark, H.M., Atanmo, P., Kumble, R. and McEvily, A.J. (1972), Int. J. Fract. Mech. **8**, 209.
[q] Priddle, E.K. (1976), Int. J. Pressure Vessels and Piping **4**, 89.
[r] Wheeler, O.E. (1972), J. Basic Eng. **94**, 181.

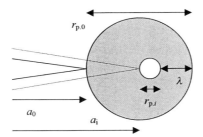

Fig. 24.11: Schematic of the Wheeler retardation model.

In the factor $\phi = (r_{p,i}/\lambda)^m$, the terms $r_{p,i}$ denote the plastic zone size at the ith overstress cycle and λ is the distance from the current crack tip to the greatest prior elastic-plastic boundary due to a previous overload (see Fig. 24.11), i.e. $\lambda = a_0 + r_{p,0} - a_i$. If $a_i + r_{p,i} \geq a_0 + r_{p,0}$, the factor ϕ is per definition $\phi = 1$. The exponent m is an empirical constant dependent on material and type of loading spectrum. Wheeler reports $m = 1.43$ for D6ac steel and $m = 3.4$ for the alloy Ti-06Al-4V. It appears that retardation is of significant influence on fatigue crack growth and has to be taken into account for a realistic crack growth prediction. Broek (1978) and Herzberg (1989) have given examples of such calculations. Several other models have been proposed in the literature but the Wheeler model has gained some acceptance since his model has led to fairly good predictions using a cycle-by-cycle integration although it does take into account multiple overloads and negative loads.

Problem 24.3

A steel plate of 75 by 100 cm and a thickness $t = 12.5$ mm contains a central through-the-thickness crack with a length $a = 2.5$ mm. The plate is loaded in tension with an applied stress σ along the long side. The material data are Young's modulus $E = 200$ GPa, Poisson's ratio $v = 0.3$, the fracture toughness $K_{Ic} = 50$ MPa m$^{1/2}$ and the yield strength $Y = 1000$ MPa. The fatigue behaviour can be represented by da/d$N = C\,\Delta K^n$ with $C = 5 \times 10^{-15}$ m/(MPa m$^{1/2}$)^n and $n = 3$.

a) Discuss whether this plate can be considered to be in plane strain?
b) Calculate the critical strain energy release rate.
c) Calculate the maximum allowable crack length for $\sigma = 100$ MPa.
d) Calculate the constant C for the proper form of the crack growth equation
   da/d$N = C(\Delta K/K_{Ic})^n$.
e) Estimate the number of cycles the plate will survive during cyclical tensile loading between 0 and 100 MPa using the Paris equation.

24.4   Structural aspects of metal fatigue

Having discussed the basic phenomenological aspects, we now turn to some structural aspects. We discuss the initiation and propagation mechanism and thereafter the influence of the surface conditions.

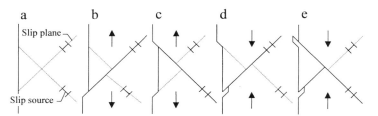

Fig. 24.12: Mechanism for the formation of slip band extrusion and intrusions.

Initiation and propagation

The fatigue process in metals roughly can be divided into four parts:

- *Crack initiation.* This is the first damage that occurs and nucleates at discontinuities, be it at the surface (scratches or steps) or in the interior (inclusions or grain boundaries). It can be removed largely by thermal treatment if necessary.
- *Slip band crack growth.* This represents the increase in length of the initial crack on crystal planes with a high shear stress. This is referred to as stage I growth.
- *Crack growth on planes of high tensile stress.* This represents the growth of the well-developed crack in a direction perpendicular to the highest tensile stress. This is referred to as stage II growth.
- *Final fracture.* This occurs when the crack has become long enough to satisfy the momentary fracture criterion, i.e. $K_I = K_{Ic}$.

We discuss the various stages briefly in the following paragraphs.

An important phenomenon during stage I growth is the occurrence of extrusion and intrusions at the surface of a specimen. Cottrell and Hull[s] have given an explanation, which is shown schematically in Fig. 24.12. Two slip band systems are required to explain the phenomenon. During tensile loading in the cycle (Fig. 24.12 (b) and (c)) two steps originate at the surface. During compression an intrusion appears at one step and an extrusion at the other (Fig. 24.12 (d) and (e)). By repetition of this mechanism a microcrack forms. At some point a macrocrack is developed which on its turn can develop via stage II growth. Of course slip forwards and backwards does not always occur at exactly the same lattice plane so that a band of slipped material grows during the process (Fig. 24.13). The fracture surface does not reveal clear marks of the process.

During stage II growth, crack growth occurs via a mechanism of plastic deformation at the crack tip, which alternately blunts and sharpens the crack tip (Fig. 24.14). At the beginning of a cycle the crack tip is sharp (Fig. 24.14 (a)). Slip at 45°

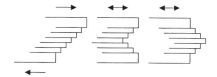

Fig. 24.13: Formation of extrusions of certain width by slip on various lattice planes.

---

[s] Cottrell, A.H. and Hull, D. (1957), Proc. Roy. Soc. **A242**, 211.

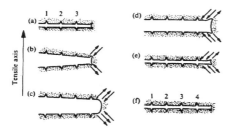

Fig. 24.14: Blunting and sharpening of the crack tip during stage II growth.

planes to the crack tip occurs during the tensile loading (Fig. 24.14 (b)) which blunts the crack tip. During further loading the crack grows (Fig. 24.14 (c)). During the compressive part of the cycle the slip direction reversed (Fig. 24.14 (d)). The fracture surfaces approach each other until they touch and the crack sharpens again (Fig. 24.14 (e)). The crack plane is again becoming perpendicular to the direction of the tensile stress. During each cycle this mechanism repeats itself, the result being that a *striation pattern* develops on the fracture surface that can be examined by microscopy. This model is due to Laird[t]. It must be said, though, that care should be exercised in the interpretation of the striations. It has been shown that the striation distance does not always correspond to the crack growth per cycle. Moreover, the environment plays a role in the development of the striations.

In summary, during stage I a fracture surface without any clear marks is present. On the contrary, during stage II a fracture surface with clear ripples develops, corresponding with the cyclic loading. Finally, if the crack has reached its critical length, i.e. $\alpha\sigma\sqrt{a} = K_{\mathrm{I}} \rightarrow K_{\mathrm{Ic}} = (2ER)^{1/2}$, catastrophic failure occurs. This process has been discussed in Chapter 21.

Influence of the surface

Damage initiates preferentially at those points where stress concentration occurs, primarily at micro-defects. Therefore the surface plays a major role in fatigue since stress concentration at surface defects is the largest. Fatigue is thus influenced by

- *Notches.* Stress concentration is larger at a notch than at a smooth surface. It is thus important to take this aspect into account in the design stage. Various empirical approaches have been forwarded for the incorporation of the effect of

Table 24.3: Lifetime for SAE 3130 steel and $R = -1$ and $\sigma_{\mathrm{ave}} = 160$ MPa.

Machining	R_{a} (µm) *	Life time
Lathe	2.7	24×10^3
Partially hand polished	0.16	91×10^3
Hand polished	0.13	137×10^3
Ground	0.18	217×10^3
Ground/polished	0.05	234×10^3

* R_{a} indicates the centre line average.

[t] Laird, C. (1967), page 131 in *Fatigue crack propagation*, ASTM-STP 415, Philadelphia, ASTM.

notches on fatigue life. We refer to the literature.

- *Surface roughness.* Also at micro-defects due to surface roughness stress concentration occurs. Increasing roughness leads to decreasing lifetime. An example is provided in Table 24.3.
- *Coatings.* Coatings generally have a large influence on the fatigue behaviour. For example, a hard Cr coating deteriorates the fatigue resistance while a soft Cd coating shows little or no influence.
- *Residual stress.* Residual stress, due to welding or surface machining like shot peening, has to be added to the applied stress. Since shot peening, i.e. the bombardment of the surface with small metallic spheres, introduces a compressive residual stress, the fatigue limit may be enhanced by as much as a factor of 2. The process is routinely used in industry.
- *Corrosion.* Corrosion creates pits at which again stress concentration occurs.
- *Fretting.* During contact of two surfaces with relatively small displacement damage arises. The process is called *fretting* and leads also to stress concentration.

From the above list, it is clear that the exact prediction of the fatigue lifetime is a complex problem involving many aspects.

24.5   Fatigue in inorganics: subcritical crack growth

If fracture behaviour of brittle, inorganic materials were governed entirely by fracture mechanics as dealt with in Chapter 21, the problem would have been already quite complicated. Unfortunately this is not true and fatigue plays a role as well, although in this connection it is usually denoted as slow or *subcritical crack growth* (SCG). The essence of this process is that even at stress intensities below the critical one, flaws can grow slowly. This means that when a material is stressed below its momentary strength, it still will fail after the time necessary for the flaw to grow to its critical size at that particular stress level. The rate of slow crack growth is dependent on the temperature and atmosphere. As to be expected, increased temperature results in an enhanced growth rate, typically described by an Arrhenius-like behaviour. For oxides, water from the environment enhances the slow crack growth rate. In case the material is submerged in water, increasing pH will generally also lead to an enhanced growth rate. The general dependence of crack growth on the stress intensity applied is shown in Fig. 24.15. Three different regimes can be distinguished. In regime I the crack velocity is determined by the reaction of H_2O at the crack tip while in regime II

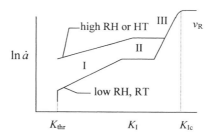

Fig. 24.15: Schematic of the relation between the crack velocity $\dot{a}$ and the applied stress intensity K. The influence of relative humidity RH is indicated as well. K_{thr} indicates the possible existence of a stress corrosion limit. Increased temperature and enganced humidity have a similar effect. The maximum crack velocity at catastrophic failure is the Rayleigh wave velocity v_R.

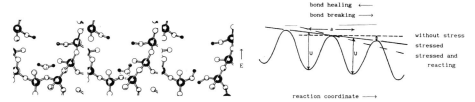

Fig. 24.16: Schematic of the reaction of H_2O with the Si–O–Si bonds at the crack tip (left) and the corresponding energy-reaction path diagram (right).

the diffusion of H_2O to the crack tip is rate determining. In regime III various mechanisms, dependent on the particular material, operate. For non-oxides, the subcritical crack growth is normally of lesser importance at room temperature, although exceptions do occur.

The influence of H_2O on the slow crack growth rate[u] can be visualized as follows (Fig. 24.16). In the stress-free situation a bond at the crack tip has equal probabilities for healing and breaking due to the equal reaction barriers for the forward and reverse reactions. Applying a stress changes this situation to a lower barrier, and thus a higher probability, for bond breaking than for bond healing. This is the basic subcritical crack growth mechanism. The energy of the system after breaking of the bonds can be lowered further by the addition of OH to the Me and H to the O–Me part of the broken Me–O–Me bond. Bond healing becomes more difficult. Other polar liquids react in a similar way (Fig. 24.17). The shape and size of the molecule, however, can prevent reactions at the crack tip itself.

The power-law formalism

At low stress intensities, a limit to slow crack growth, usually called stress corrosion limit, can be present. Its existence has been verified for just a few materials. In the case of glass, it has been estimated as 0.17 the value of K_{Ic}. For lifetime calculations the dependence of the crack growth rate $\dot{a}$ on stress intensity in regime I (and III) is often taken as

$$\dot{a} = A \left(\frac{K_I}{K_{Ic}} \right)^n \tag{24.12}$$

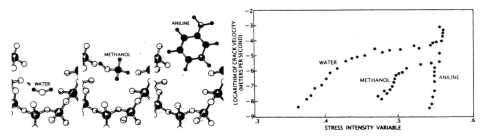

Fig. 24.17: Subcritical crack growth rates of glass in different liquid environments. The increasing size of the reactant renders approach to the crack tip more difficult (left) and has a marked influence on the growth rate (right).

---

[u] Michalske, T.A and Bunker, B.C. (1987), Sci. Am. **257**, 79.

Taking into account only region I the lifetime of a stressed material t_{fin} can thus be estimated by

$$t_{fin} = \int_{a_{ini}}^{a_{fin}} \dot{a}^{-1} \, dt \tag{24.13}$$

Here a_{ini} and a_{fin} indicate the initial and final crack length, respectively. Separation of variables meanwhile using $K_I = Y\sigma a^{1/2}$ leads to

$$\int_0^{t_{fin}} \sigma^n \, dt = \int_{a_{ini}}^{a_{fin}} \frac{K_{Ic}^n}{AY^n} \frac{da}{a^{n/2}} \tag{24.14}$$

If we take Y as constant, integration of the right-hand side results in

$$\frac{2K_{Ic}^2}{Y^2 A(2-n)} \left[(S)^{n-2} - (S*)^{n-2} \right] \equiv B\left[(S)^{n-2} - (S*)^{n-2} \right] \tag{24.15}$$

where $K_{Ic} = YS*a_{ini}^{1/2} = YSa_{fin}^{1/2}$ are used. Here S and $S*$ denote the strength and *inert strength*, i.e. the strength under conditions of no subcritical crack growth, respectively. The result of the left-hand side of Eq. (24.14) depends on the applied stress profile. For a constant stress $\sigma = S$, the integral becomes $S^n t$ so that yields the lifetime t_{fin}

$$S^n t_{fin} = B\left[(S)^{n-2} - (S*)^{n-2} \right] \quad \text{or} \quad t_{fin} \cong B\left(\frac{S*}{S} \right)^n \frac{1}{(S*)^2} \tag{24.16}$$

where the last step can be made since $n \gg 1$ and $S* > S$. Long lifetimes are thus achieved by a high value of K_{Ic}, a small initial defect size a_{ini}, corresponding to a high inert strength $S*$, and a high, respectively low value for the SCG parameters n and A.

Similarly an estimate can be made for the strength at a given constant stress rate $\dot{\sigma}$. Using $\sigma(t) = \dot{\sigma} t$ and indicating the fracture stress by S, we obtain for the left-hand side of Eq. (24.14) $(\dot{\sigma} t)^{n+1}/\dot{\sigma}(n+1) = S^{n+1}/\dot{\sigma}(n+1)$ so that

$$\frac{S^{n+1}}{\dot{\sigma}(n+1)} = B\left[(S)^{n-2} - (S*)^{n-2} \right] \quad \text{or} \quad S^{n+1} \cong \dot{\sigma}(n+1)B(S*)^{n-2} \tag{24.17}$$

where the same approximation is made as before.

The value of n can vary considerably. For oxides a low value is 10 while values above 50 seldom occur. For bulk fused silica $n \cong 40$, while for fused silica fibres $n \cong 20$. For non-oxides generally higher n-values are reported. In the few cases where n has been determined for region III, values of $n \cong 100$ result.

The SCG parameters n and A are to be determined by experiment. This can be done in two ways. The first is making use of a macroscopic crack whose extension is followed. The second method makes use of the 'natural' defects.

In the macroscopic crack approach the 'double torsion' (DT) specimen (Fig. 22.1) is frequently used. An initial crack is made. After rapid loading to about 90% of the force necessary for catastrophic fracture, the load relaxation is monitored. As mentioned before, for the DT specimen the resulting stress intensity is independent of the length of the crack over a large part of the length of the specimen. This makes crack length measurements during load relaxation not necessary. From the relaxation curve and the length of the curve at the beginning or end of the load relaxation experiment, the entire crack velocity-stress intensity diagram can be determined. The previously mentioned stress intensity equation is used. The DCB specimen (Fig. 21.9) can be used also but in this case monitoring the crack length during relaxation is necessary. This generally is also true for other type of specimens.

The second method uses the microscopic, 'natural' defects in bending specimens and let them grow under an applied constant stress, smaller than the momentary strength. The value of n can easily be calculated from Eq. (24.16) using measurements of the lifetime $t_{fin}(S)$ at different stress levels S, according to

$$\frac{t_{fin}(S_1)}{t_{fin}(S_2)} = \left(\frac{S_2}{S_1}\right)^n$$

where S_1 and S_2 are two different stress levels. Another possibility using the 'natural' defects is to apply different loading rates. At slower loading rates, there is more time for the existing defects to grow, resulting in a lower strength. From Eq. (24.17), it is clear that the strength values at different stressing rates $\dot{\sigma}_1$ and $\dot{\sigma}_2$ are related by

$$\frac{\dot{\sigma}_1}{\dot{\sigma}_2} = \left(\frac{S_1}{S_2}\right)^{n+1}$$

where S_1 and S_2 are the fracture stresses at the rates $\dot{\sigma}_1$ and $\dot{\sigma}_2$. The calculation of the parameter A from these types of experiments is also possible. It requires, however, inert strength values for the material, the measurement of which is not always that easy.

Activated complex theory

From the discussion on the crack growth mechanisms, it is clear that a propagating crack should be described by 'activated complex' theory instead of arbitrarily taking a power law in the stress intensity. This theory (see Chapter 8 or e.g. Krausz and Eyring, 1975 or Krausz and Krausz, 1988) provides

$$\dot{a} = p^{\eta} \frac{kT}{h} l_0 \left[\exp\left(-\Delta U_{for}/kT\right) - \exp\left(-\Delta U_{bac}/kT\right)\right] \tag{24.18}$$

where, as usual, k and h denote the Boltzmann and Planck constant, respectively[v]. Further l_0 and p denote the elementary crack growth step and pressure (activity) of the surface-active component, respectively. Finally, ΔU_{for} and ΔU_{bac} indicate the potential energy for crack growth (forward crack growth) and crack healing (backward crack growth), respectively. Usually the backward reaction is neglected and we do so here. In this case for the rate equation an exponential relation is obtained

$$\dot{a} = p^{\eta} \frac{kT}{h} l_0 \exp\left(-\Delta U_{for}/kT\right) \equiv \dot{a}_0 \exp\left(-\Delta U_{for}/kT\right) \tag{24.19}$$

We omit further the subscript 'for'. In a stress-free configuration only the interatomic forces determine the barrier height ΔU. However, here we apply a stress and this effect should be taken into account. Since we do not know the stress dependence, the usual thing to do is to develop the expression for the barrier height in a Taylor series.

---

[v] In this section, R is the fracture resistance and we use the Boltzmann constant k in activated complex expressions in order to avoid confusion.

Immediately the question pops up: developed with respect to which parameter? Different authors use different parameters but generally they fall in two categories. The first is development with respect to the local stress at the crack tip and the second is with respect to the global driving force for crack growth. We deal briefly with both.

Let us start with the local stress approach. In this case we write

$$\Delta U = U_0 - \frac{\partial \Delta U}{\partial \sigma_{loc}} \sigma_{loc} + \frac{1}{2!} \frac{\partial^2 \Delta U}{\partial \sigma_{loc}^2} \sigma_{loc}^2 + \cdots \cong U_0 - \frac{\partial \Delta U}{\partial \sigma_{loc}} \sigma_{loc}$$

The zero-order term U_0 is usually referred to as the *activation energy* U_{act} and the first derivative $\partial \Delta U / \partial \sigma_{loc}$ is known as the *activation volume* V_{act}. Therefore we have

$$\dot{a} = \dot{a}_0 \exp\left(\frac{-U_{act} + V_{act}\sigma_{loc}}{kT} \right) \tag{24.20}$$

Before we proceed with this approach we mention another representation, which is sometimes used. If Eq. (24.20) is combined with the Inglis expression $\sigma_{loc} = \sigma(1+2(a/\rho)^{1/2}) = \sigma(1+2x)$, which for long defects $a \gg \rho$ can be approximated by $\sigma_{loc} \cong 2\sigma x$, and $K_1 = Y\sigma a^{1/2}$, the above formula can also be written as

$$\dot{a} = \dot{a}_0 \exp\left(\frac{-U_{act} + bK_I}{kT} \right)$$

where $b = 2V_{act}/Y\rho^{1/2}$. Remembering that fracture occurs when the local stress σ_{loc} equals the theoretical strength σ_{the}, we obtain $b = V_{act}\sigma_{the}/K_{Ic}$. This equation was used by Wiederhorn[w] for fused silica resulting in $b' = N_A b = 0.216$ m$^{5/2}$/mol, corresponding to $V_{act}' = N_A V_{act} = 8.1$ cm^3/mol using $K_{Ic} = 0.75$ MPa m$^{1/2}$ and $\sigma_{the} = 20$ GPa.

For the stress representation[x] we obtain after substitution of the (full) Inglis expression

$$\dot{a} = \dot{a}_0 \exp\left(\frac{-U_{act} + \sigma V_{act}(1+2x)}{kT} \right) \equiv \dot{a}_0 \exp\left(\frac{-U_{act} + \sigma V_{act}}{kT} + 2Q \right)$$

with $Q = \sigma x V_{act}/kT$. After integration (keeping in mind that $(a_{ini}/\rho_{ini})^{1/2} \ll (a_{fin}/\rho_{fin})^{1/2}$) the failure time t_f is given by

$$t_{fin} = \frac{\rho}{\dot{a}_0 \sigma^2} \left(\frac{kT}{V_{act}} \right)^2 (1+2Q) \exp\left[\frac{U_{act} - V_{act}\sigma(1+2x)}{kT} \right] \tag{24.21}$$

The lifetime thus depends largely on the applied stress σ and the factor $x = (a/\rho)^{1/2}$. Now we have to distinguish between two cases. First, the process proceeds always at minimum energy and in that case the defect radius ρ is constant. Second, the process proceeds at zero-energy change. In this case, it can be shown that the defect radius ρ is proportional to σ^{-2}. We can write therefore

$$\ln t_{fin} = \frac{U_{act}}{kT} - \frac{V_{act}}{kT}(1+2Q) - q \ln\sigma + 2 \ln T - \ln\dot{a}_0 + \ln(1+2Q) + P$$

[w] Wiederhorn, S.M. Fuller jr, E.R. and Thomson, R. (1980), Metal Sci. **14**, 450.
[x] Bouten, P.C.P. and de With, G. (1985), J. Appl. Phys. **64**, 3890.

with $P = \rho k^2/2V_{act}^2$ for $q = 2$ when ρ is constant and $P = \gamma E(k/2V_{act})^2$ for $q = 4$ when energy is constant. The slope of a plot of $\ln t_f$ versus σ yields the apparent activation volume V_{app}

$$V_{app} = -kT\frac{\partial \ln t_{fin}}{\partial \sigma} = V_{act}(1+2x) + \frac{kT}{\sigma}\left(q - \frac{2Q}{1+2Q}\right) \qquad (24.22)$$

The last term varies between kTq/σ and $kT(q-1)/\sigma$, and since q is either 2 or 4 between 1 and 4, it is only weakly varying. The primary influence on the apparent activation volume is through the term $x = (a/\rho)^{1/2}$.

Let us see how this result compares with power-law representation. Calculating the crack growth exponent n results in

$$n \equiv -\frac{\partial \ln t_f}{\partial \ln \sigma} = \frac{V_{act}\sigma}{kT}(1+2x) + \frac{kT}{\sigma}\left(q - \frac{2Q}{1+2Q}\right) \qquad (24.23)$$

The last term is identical to the last one of Eq. (24.22). It thus appears that n is directly related to V_{act} and is primarily dependent on the variable x and on the stress σ. This resolves the discrepancy in n-values for fused silica as measured with bulk glass ($n \cong$ 40) and glass fibres ($n \cong 20$): It appears to be due to the high stress levels in the case of fibres. Using $V_{act}' = 8.1$ cm^3/mol, as obtained from bulk silica fracture experiments which yield $n \cong 40$, and using a fracture stress of 5 GPa for silica fibres we obtain for $x > 1$ the vale $n \cong 20$, practically independent of the value of x. In view of the fact that the same value for V_{act}' for highly defective surfaces (bulk fused silica) and pristine surfaces (fused silica fibres) can be used, these results are rather satisfactory. Moreover these results are also coherent. The failure strain of silica fibres is ~0.06 under ambient conditions to 0.15-0.20 in vacuum or inert environment. A rough estimate of V_{act}' is thus ~$0.2V_{mol}$ or $\cong 6$ cm^3/mol, comparing favourably to the observed value of 8.1 cm^3/mol, which is roughly 0.3 of the molar volume $V_{mol} = 27.3$ cm^3/mol.

The crack driving force approach runs slightly different. Crack growth is enhanced by the energy release rate G (related to the applied stress intensity K_I via $G = K_I^2/E$) but retarded by the fracture resistance[y] R (related to the surface energy γ via $R = 2\gamma$ if we neglect all other effects associated with fracture apart from surface creation). Therefore in this case ΔU is developed with respect to $g = G-R$. It starts again from activated complex theory. To first order we can write for the barrier height

$$\Delta U_{for} = U_{for} - \alpha(G - R) \qquad \text{and} \qquad \Delta U_{bac} = U_{bac} + \alpha(G - R)$$

respectively, with $\alpha = \partial \Delta U_{for}/\partial g$ the *activation area*. Usually the backward reaction is neglected and we do so here. In this case the rate equation can be written as

$$\dot{a} = p^n \dot{a}_0 \exp[\alpha(G - R)/kT] \qquad \text{with} \qquad \dot{a}_0 = \frac{kT}{h}l_0 \exp(-U_{for}/kT)$$

Using $G = K_I^2/E$ and $R = K_{Ic}^2/E$, the driving force for crack extension can be written as

---

[y] In Chapter 22 the fracture criterion $G = G_{cri} = 2R$ was formulated, making the factor 2 for the two surfaces explicitly visible. Here we use $G = G_{cri} = R$ in accordance with the literature.

$$G - R = R\left(\frac{K_I^2}{K_{Ic}^2} - 1\right)$$

yielding for the rate equation

$$\dot{a} = \dot{a}_0 p^n \exp\left[-\beta\left(1 - \frac{K_I^2}{K_{Ic}^2}\right)\right] \qquad \text{with} \quad \beta = \frac{\alpha R}{kT}$$

Since $\exp(-ax) \cong (1-x)^a$ for $x \ll 1$ and $a \gg 1$, we can write

$$\dot{a} = \dot{a}_0 p^n \left(\frac{K_I}{K_{Ic}}\right)^n \qquad \text{with} \quad n = 2\beta \qquad\qquad (24.24)$$

providing some background for the power law discussed in the previous paragraph.

This approach has been amplified by the explicit incorporation of adsorption[z]. As indicated, the environment influences the fracture energy R and this is described by the *Gibbs adsorption equation* (see Chapter 6) reading

$$\frac{d\Delta\gamma}{d\ln z} = -kT\Gamma$$

Here $\Delta\gamma = \gamma - \gamma_0$ denotes the change in surface energy upon adsorption, z the activity of the active species in the environment (usually water) and Γ is the excess surface concentration. For the low pressures involved, the activity can be safely replaced by the partial pressure p and therefore we can write

$$\Delta\gamma = -kT\Gamma_{max} \int \frac{\theta}{p} dp$$

where $\theta = \Gamma/\Gamma_{max}$ is the relative surface coverage, dependent on the partial pressure. The surface coverage θ has to be determined experimentally and we use here the *Langmuir isotherm* with the parameter b reading

$$\theta = \frac{bp}{1 + bp}$$

This leads after integration to

$$R = R_0 - 2kT\Gamma_{max} \ln(1 + bp) = R_0\left[1 - \phi\ln(1 + bp)\right] \qquad (24.25$$

where $\phi = 2kT\Gamma_{max}/R_0$ and $R_0 = 2\gamma_0$ indicates the fracture energy for the clean surface. The use of more realistic adsorption isotherms leads rapidly to rather complex expressions.

The combination of these effects can decrease the strength in two ways. First, by decreasing the fracture resistance from R_0 to R via adsorption. Thereto we define the *inert toughness* $\tilde{K} = (ER_0)^{1/2} = Y\tilde{\sigma}a_{ini}^{1/2}$ with $\tilde{\sigma}$ the inert strength and the *adsorption-controlled toughness* $\hat{K} = (ER)^{1/2} = Y\hat{\sigma}a_{ini}^{1/2}$ with $\hat{\sigma}$ the adsorption-controlled strength. Second, by enhanced subcritical crack growth via the action of reactive species at the crack tip leading to an increased crack length from a_{ini} to a_{fin} as

---

[z] Donners, M.A.H., Dortmans, L.J.M.G. and de With, G. (2000), J. Mater. Res. **15**, 1377.

described by the relation $\widetilde{K} = (ER_0)^{1/2} = Y\sigma a_{\mathrm{fin}}^{1/2}$ with the strength σ. The combination of these two effects leads to the expression for the experimental toughness $\hat{K} = (ER)^{1/2} = Y\sigma a_{\mathrm{fin}}^{1/2}$. Since we have many labels already, we omit here the subscripts 'I' and 'Ic' for K.

Using the same procedure as before we now obtain

$$
\begin{aligned}
\dot{a} &= p^n \dot{a}_0 \exp\left\{ \frac{\alpha R_0}{kT} \left[\frac{K^2}{\widetilde{K}^2} - 1 + \phi \, \ln(1 + bp) \right] \right\} \\
&= p^n \dot{a}_0 \exp\left[-\beta \left(1 - \frac{K^2}{\widetilde{K}^2} \right) \right] \exp\left[\ln(1 + bp)^{\beta\phi} \right] \quad \text{with} \quad \beta = \alpha R_0 / kT \\
&\cong p^n \dot{a}_0 \left(\frac{K}{\widetilde{K}} \right)^{2\beta} (1 + bp)^{\beta\phi}
\end{aligned}
\tag{24.26}
$$

Introducing the adsorption-controlled toughness $\hat{K} = \widetilde{K}[1 - \phi \, \ln(1 + bp)]^{1/2}$ (which is the one usually measured) we obtain

$$
\dot{a} \cong a_{\mathrm{kin}} a_{\mathrm{ads}} [1 - \phi \, \ln(1 + bp)]^{1/2} \left(\frac{K}{\hat{K}} \right)^n \quad \text{with} \quad a_{\mathrm{kin}} = \dot{a}_0 p^n \text{ and } a_{\mathrm{ads}} = (1 + bp)^{\beta\phi}
$$

where the factors a_{kin} and a_{ads} account for kinetic and adsorption effects, respectively.

The above-discussed model was tested on Mn-Zn ferrite and the experiments have shown that at low partial pressure the crack growth process is limited by adsorption while at higher pressure thermal activation is limiting. Moreover, good agreement was obtained between $\hat{K}$ as determined from single edge notched beam experiments using the model discussed and as measured directly with the double torsion technique. It appears experimentally that for many oxides[aa] the kinetic factors are controlling. However, occasionally an exception is observed. For example, Sr-hexaferrite[bb] does not exhibit subcritical crack growth at all. The strength, however, when measured in water decreases significantly as compared when measured in air and the fracture toughness shows a corresponding decrease. A theoretical estimate of the decrease in fracture energy due to adsorption corresponds well with the experimental data. It thus appears that for this material adsorption is controlling the fracture process for all partial pressures. The environmental effect discussed above is not only limited to oxides but can also occur in nitrides and other compounds, though usually less extensive. The effect is also not exclusively due to water but can also be caused by alcohols and other polar and/or polarizable liquids.

In conclusion, it will be clear that the prediction of lifetimes of ceramic components using experimental data is far from a routine exercise although the basic mechanism for relatively simple microstructures seems to be clear. Apart from the relatively strong influence of the precise type of material yielding a somewhat limited reliability of the SCG parameters, many other drawbacks are present. Extension of the

---

[aa] Van der Laag, N.J., Dortmans, L. and de With, G. (2004), in press.
[bb] De With, G. (1984), Silicates Industrielles **9**, 185.

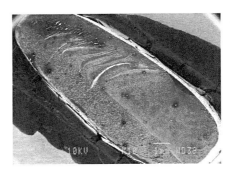

Fig. 24.18: Fracture surface of a grip of a pair of scissors, showing the area broken by fatigue (approximately the lower one third) and by brittle fracture (approximately the upper two third).

uni-axial data to multi-axial data is just one of the (not discussed) problems involved. More information can be found e.g. in Lawn (1993).

24.6   Fatigue in polymers

Although polymers are generally considered as more forgiving with respect to failure than either inorganics or metals, nevertheless the failure mechanism is often defect initiation and initial growth by fatigue, followed by catastrophic brittle fracture. This process is common for all kinds of polymers as used for utensils. Fig. 24.18 shows an example of a fracture surface of a grip of a pair of scissors on which the fatigue and brittle part of the fracture clearly can be discerned.

The phenomenology of polymer fatigue is similar to that of metals. Typically two or three regions can be distinguished (Fig. 24.19) at room temperature. Region I is only present for those polymers, which do show substantial crazing, like polystyrene (PS) while it is absent for polymers not exhibiting crazing, like polycarbonate (PC). In region II the Paris-Erdogan law is often followed but the exponent can be quite large and ranged up to 20. Similar to the case of metals deviations occur, such as the presence of a threshold for low stress intensity amplitude and an increased crack velocity when the stress intensity amplitude approaches the fracture toughness. Also the influence of the mean stress is not represented. Williams[cc] attempted to model fatigue failure using the Dugdale plastic zone model. In each cycle the craze stress on

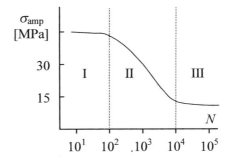

Fig. 24.19: Typical fatigue curve for polymers showing region I due to crazing.

[cc] Williams, J.G. (1980), J. Mater. Sci. **12**, 2525.

the one side if the plastic zone is reduced while on the other side it increases. This leads to

$$\frac{da}{dN} = \beta\left(K^2 - K_{lc}^2\right)$$

This expression shows a good fit to the data for PS over a wide range of temperatures. Semi-crystalline materials (e.g. Nylon 66) are more fatigue resistant than amorphous, ductile polymers (e.g. PVC or PC), which on their turn are more resistant than brittle, amorphous materials (e.g. PS or PMMA)[dd]. Generally with increasing molecular weight M, da/dN decreases[ee], approximately as $da/dN \sim 1/M$. For example, for PS an increase by a factor of 5 in M results in a factor of 10 increase in lifetime. The interpretation of parameters in equations like the Paris equation for polymers is difficult.

As shown in Fig. 24.18 the fracture surface of polymers shows two regions. One is a region of smooth, slow crack growth surrounding the fracture initiation site, sometimes showing striations. The other is a rough region corresponding rapid growth. In polymer science, the term striations strictly applies to the advance of the crack per cycle. Another discontinuous growth process results in so-called discontinuous growth bands. They correspond to the discontinuous growth of the fatigue crack after many cycles. The usual interpretation is that a craze is formed at the fatigue crack tip and that the fatigue crack propagates along the craze matrix interface and then along the craze mid part until arrest, resulting in band on the fracture surface (Fig. 24.20). The discontinuous growth bands thus represent the position of the crack tip that has been blunted, advanced and arrested repeatedly.

An important part in polymer fatigue is the local heating due the hysteresis that can lead to local melting. In that case failure is assumed to occur by viscous flow. The energy dissipated in each cycle $\dot{U}$ is given by

$$\dot{U} = \pi f J''(f,T)\sigma^2$$

where f is the frequency, J'' is the loss compliance and σ is the peak stress. If heat losses are neglected this leads to the temperature rate

$$\dot{T} = \dot{U}/\rho c_p = \pi f J''(f,T)\sigma^2/\rho c_p$$

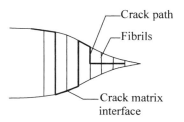

Fig. 24.20: Discontinuous growth bands at the tip of a fatigue crack in polymers.

[dd] Hertzberg, R.W., Manson, J.A. and Skibo, M. (1975), Polymer Eng. Sci. 15, 252.
[ee] Kim, S.L., Skibo, M., Manson, J.A. and Hertzberg, R.W. (1977), Polymer Eng. Sci. 17, 194.

with ρ the density and c_p the specific heat. From this equation the major factors influencing the temperature rate can be discerned. With increasing temperature first of all the elastic modulus decreases so that larger deflections result. When the apparent modulus has decreased to 70% of the original modulus, the specimen is usually considered to have failed. With increasing temperature also J'' increases, in particular in the vicinity of the glass transition temperature. This increases the temperature rate and leads to increased local heating and thus to shorter lifetime. Further, both with increasing stress and increasing frequency the temperate rate increases, again leading to a decreasing lifetime. Since heat losses become smaller with increasing specimen size, the lifetime of polymer specimens decreases with increasing thickness. Tests with intermittent rest periods in which the heat can flow away lead to significantly longer lifetime. Accordingly the linear cumulative damage rule cannot be applied.

24.7   Bibliography

Broek, D.J. (1978), *Engineering fracture mechanics*, 2nd ed., Sijthoff & Noordhoff, Alphen aan den Rijn.

Derby, B., Hils, D. and Ruiz, C. (1992), *Materials for engineering*, Longman Scientific & Technical, Harlow, UK.

Dieter, G.E. (1976), *Mechanical metallurgy*, 2nd ed., McGraw-Hill Kogakusha, Tokyo.

Herzberg, R.W. (1989), *Deformation and fracture mechanics of engineering materials*, 3rd ed., Wiley, New York.

Krausz, A.S. and Eyring, H. (1975), *Deformation kinetics*, Wiley-Interscience, London.

Krausz, A.S. and Krausz, K. (1988), *Fracture kinetics of crack growth*, Kluwer, Dordrecht.

Lawn, B.R. (1993), *Fracture of brittle solids*, Cambridge University Press, Cambridge.

Suresh, S. (1991), *Fatigue of materials*, Cambridge University Press, Cambridge.

25

Perspective and outlook*

In this final chapter I discuss the perspective of the present knowledge and outlook on possible future development of thermomechanics. For this a brief discussion is given on science and engineering in general and on its multidisciplinary and related aspects. After that I follow the order as used everywhere else, i.e. elasticity, plasticity, visco-elasticity and fracture with attention for the macroscopic, microscopic and mesoscopic aspects. Again I try to divide the attention even over the material categories of metals, inorganics and polymers. I do not attempt to be complete or to provide only those topics that are accepted troublesome by everybody in the respective fields. Much more this is a personal account of many interesting features that are around in thermomechanics, illustrated with some historical remarks, and a number of the problem areas that I have encountered during the last few years and I think that are interesting.

25.1   Science and Engineering

One of the questions that might come up after reading the contents of this book is probably: are we talking about science or engineering? Let us define what generally is understood to be science or engineering. In brief *engineering* deals with the question *how to make things*, i.e. materials and/or structures. *Science* is much more concerned with the question *why things behave as they do*. There seems to be roughly consensus about the following[a]. Science uses experiments and observations to support or falsify theoretical edifices. These edifices are abstract and their purpose is to answer why? In engineering the edifices are more descriptive and prescriptive containing information on how to make things. In both engineering and science all the accumulated knowledge is used and thus they are interrelated through the common *knowledge pool*. Another important aspect is the *time horizon*. Science works in principle without a definite time horizon since it postulates hypotheses and theories and these are, at least in principle, under constant testing, accepted if correct and corrected otherwise. On the other hand, engineering must provide solutions within a certain time frame, generally implying that there is a focus on the use of methods and materials available today. In some cases engineering is ahead of science, a typical example being metallurgy. In this position engineering poses the same questions as nature does and, in this particular case, metal physics has to provide the answers. In other cases science offers opportunities to engineering, of which a good example is microelectronics. Here the basic principles provide all sorts of options and the challenge for engineering is to realise these options. In all cases there is generally a great deal of collaboration between the two and it will be clear that the boundary between them is vague[b].

---

[a] Hoddeson, L., Braun, E., Teichmann, J. and Weart, S. (1992), *Out of the crystal maze*, Oxford University Press, New York.
[b] This short elaboration is incapable of representing the complex relation between science and engineering and violates the well-known statement of Albert Einstein: 'Simplify as far as possible but

Now then, is thermomechanics as defined in the preface and introduction part of science or part of engineering? I think that not many people will be offended if I say that I feel that it is somewhere in the middle, so that you may call it an *engineering science*. Thermomechanics has many rather basic aspects either partially solved, such as the fundamental dissipation law or laws, or not solved at all, such as the description of the collective motion of internal structural features in materials and their effect on the macroscopic behaviour. On the other hand, there are also many applications of theories and models in daily (engineering) life in which much of the framework as described here is used. Elastic calculations may be the most frequently used and most well-known example.

Let us now have a look at some aspects of science and engineering that relate to the multidisciplinary nature of our field. Thereafter we throw a glimpse on materials versus design and overview the various fields within thermomechanics.

Multidisciplinarity and all that

We have seen in this book that the field of thermomechanics in the definition as used here contains many aspects, e.g. from thermodynamics, mechanics, crystallography, chemical physics and physical chemistry (whatever the difference between the latter two may be). Although multidisciplinarity is generally advocated nowadays, this inevitably results in problems with the terminology that is used in different fields, i.e. the matter of *semantics*. Similar phenomena are addressed differently in different fields or really different phenomena are addressed in the same way in two different fields. Obviously this does not add greatly to a better or easier mutual understanding. I just need to remind the reader of the fact that within the domain of thermomechanics often a very sloppy use is made of words such as stiff, strong and brittle. This provides newcomers (and likely also more experienced workers) in a field more difficulties to grasp the ideas when reading or studying different sources than necessary. Also on a much more fundamental level, semantic problems cannot be neglected[c]. The discussion among various thermodynamicists on the very basis of their discipline is at least partly due to semantics[d]. Therefore even within a field as 'limited' as thermomechanics, a clear vocabulary has not to be taken for granted. Although this topic may not seem to be of basic importance, it is often a serious stumble block. In particular, since many problems require a multidisciplinary view, a common language is a must for a general, accepted description of the models to be used. We have seen this in other fields as well. An example not far from the present field is that of cybernetics. In the introduction of his well-known book *Cybernetics: or control and communication in the animal and the machine* (1961), Norbert Wiener explicitly refers to the hampering of progress by the absence of a common terminology.

---

no further'. A great deal more can be said including the relation between academic and industrial science, see e.g. Ziman, J.M. (2000), *Real science*, Cambridge University Press, Cambridge.
[c] Kestin, J.A. (1993), J. Non-Equilib. Thermodyn. **18**, 360.
[d] A fake but true common citation (according to Maugin, G.A. (1999), *The thermomechanics of irreversible behaviors*, World Scientific, Singapore) is: So many thermodynamicists, so many thermodynamics!

Same phenomena, different names

Throughout the history of science independent discoveries and developments have taken place. This often leads to a different terminology for essentially the same phenomena. Within the realm of thermomechanics a clear example is the description of the temperature dependence of the viscosity. In the world of inorganics, in particular glasses, the Vogel-Fulcher-Tammann (VFT) equation is used while in the polymer world the same equation is usually addressed as the Williams-Landel-Ferry (WLF) equation. Both equations are based on the same, relatively simple free volume model.

Mathematics is increasingly used as the common language in which one expresses his or her ideas in science and engineering. Mathematics not only provides a *universal language* but also makes clear that solutions to problems as found in one area can be transferred into another, once it is clear that the mathematical outline of the problem is the same. In this way mathematics greatly contributes to the common knowledge pool. May be the most striking example in this respect is that of differential equations. The same equation occurs in different areas, e.g. the Laplace equation in diffusion and thermal conduction. This does not mean that every scientist has to be a mathematician, far from that, but he should be capable of explaining his thoughts to mathematicians. Of course, the reverse remark can and has to be made for the mathematician. He should be willing and capable to understand the problems put forward by non-mathematicians. Otherwise, one can expect more limited progress. I again refer to the introduction of *Cybernetics* by Wiener mentioned earlier, where the connection between mathematics and biology was stressed. Finally mathematics often provides clarity. For example, the distinction between force and energy was for a long time not very clear for our ancestors and only became fully clear after both quantities were expressed as mathematical entities with energy conserved and force expressed as a derivative. A well-known paper by Helmholtz[e] was highly instrumental in this process.

The last topic to discuss deals with the nature of ideas and models. There are two types of models and ideas, distinctly different from each other, which have their place in both science and engineering. Some ideas start from a small base and spread in depth and scope. Such ideas are called, following Langmuir, *divergent* and he provided the basic Griffith model as an example. The idea of Griffith essentially discusses the effect of stress fields on defects and was initially used as a model for the strength of brittle materials, as exemplified by glass. Subsequently, the model was applied to other brittle materials, first metals and later also inorganics. Still later the polymer community also accepted the idea. Other ideas are based on a broad basis of facts and existing models and converge to a certain conclusion for a specific problem. Obviously, these are called *convergent* and most of the daily used models in both science and engineering belong to this category. For example, for a particular fracture problem, where the question is to improve the strength of a certain material, one assumes that the Griffith concept is valid, measures the material properties and estimates the critical defect size. Combined with fractography, i.e. the art and science of inspecting fracture surfaces, one proposes actions to improve the processing so that better material properties result. Hence in the end a small set of divergent ideas

---

[e] Helmholtz, H. von (1847), *Über die Erhaltung der Kraft*, G. Reimer, Berlin. In this paper energy is called 'Kraft' (meaning force), indicating the confusion.

controls the flow of the many convergent ideas that are typically used to solve the problem at hand.

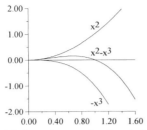

Mathematical similarities
Mathematics can be found throughout science and engineering and in quite a number of different phenomena similar mathematics is involved. Here we mention just one simple example although many more can be found. In a process often two energy terms are involved. For example, one energy term is dependent on the surface of an object and therefore scales quadratically with its size x while the other term is dependent on the volume of the object and therefore scales cubically with the linear dimension x. The process that comes directly to the mind is the fracture process, where a balance between the newly created surface and its associated energy (scaling with x^2) and the stress or strain energy (scaling with x^3) is present. The same mathematics applies e.g. to the heterogeneous nucleation of particles in a matrix. In that case the surface energy of the newly created particles balances the change in bulk Gibbs energy. In both cases a critical size appears (defect size and nucleus size), which, if exceeded, leads to fracture in the one case and growth of the nucleus in the other.

25.2   Materials versus design

Before we embark on the various areas of thermomechanics, it may be good to dwell a bit on the relationship between materials and design. To make the matter clear I provide an illuminating story of fracture, although a similar story about plasticity could have been used as well.

The story of brittle fracture has always been intimately connected to ships. Of course the wreckage of the RMS *Titanic* on 14 April, 1912 comes immediately to the mind. This loss was at those times largely attributed to reckless speed at the North Atlantic in an area with icebergs, although it was also speculated that the steel used was of inferior quality. Much later, only after the finding of the wreck in 1986 at a depth of 3 km by the deep sea vehicle *Alvin* (maximum diving depth 5 km) and the lifting of a piece of the Titanic of about 17 tons out of the ocean, guided by the deep sea vessel *Nautile* (maximum diving depth 6.6 km), it turned out that the steel used was of the current quality that could be made in those days, i.e. with a nil ductility temperature of about 0 °C. This high nil ductility temperature is mainly due to the size of the MnS inclusions. This is a much higher temperature than can be obtained nowadays, say –40 °C, due to a much reduced size of the inclusions that can be obtained by processing today. So indeed the hull was brittle given the circumstances but up to standards of that time[f]. A double hull as used in present day tanker design was evidently not even discussed.

---

[f] It must be said that there is a competing explanation, which blames the wrought iron rivets. Indeed it has been shown that the rivets of the sister ship RMS *Olympic*, built at the same shipyard, contain

Fracture of the S.S. *Schenectady* in port.

The very first attempts to describe failure were by Inglis, a ship engineer. He was inspired by the fracture of steel plates at certain important locations in the ship's hulls and introduced the idea of stress concentration. Griffith extended this work by discussing fracture as a quasi-static thermodynamic process in which the stress calculation of Inglis was incorporated. He did his experiments on glass and may be partially due to that fact his work, although now recognised as of the utmost importance[g], was only much later recognised. The problem of fracture was raised in an acute form following the occurrence of a large number of fractures of varying magnitude, including breaking-in-two and total loss, in some wartime built all-welded tankers and dry cargo ships. Before that time the failure of steel structures was only limitedly reported. The most well known is probably the all-welded Hasselt bridge in Belgium who failed in 1938. Therefore, although the problem was not new, it was only during World War II that one realised that it was a problem of great economic (and scientific) importance. In the UK the newly constituted Admiralty Ship Welding Committee asked in 1943 to professor J.F. Baker to investigate the fractures, which occurred in all-welded ships. The task of the metallurgical examination was given to his co-worker C.F. Tipper who later reported it extensively on her findings[h] and from which we draw most of the information given here. Similar information can be found in the book by W.D. Biggs[i].

The association between welding and brittle fracture was close as evidenced by the failure of the Hasselt bridge and from somewhat later accidents, e.g. the one of the S.S. *Schenectady* in 1943, who broke almost instantaneously into two apart from the bottom plates while in dock after trials at sea. For this reason it was only natural that the first investigations focused on welded structures. These investigations took two forms. In the first extensive measurements were made on the stresses in welded and riveted ships under different loading conditions. It was thought that riveted ships might have a greater capacity for yielding at points of high stress concentration due to slippage at the riveted seams. From these investigations no conclusive evidence was obtained. Nevertheless the presence of riveted seams seems to reduce the risk of failure by breaking the continuity of the structure and it was recommended to

---

banded silicate inclusions, so-called slag stringers, parallel to the rivet head. Obviously these inclusions render the rivets more brittle.

[g] The reference to his papers is one of most widely quoted ones in the history of science.

[h] Tipper, C.F. (1962), *The brittle fracture story*, Cambridge University Press, Cambridge.

[i] Biggs, W.D. (1960), *The brittle fracture of steel*, MacDonald and Evans, London.

introduce them as crack arresters. Design alterations were made to remove or mitigate stress concentrations, which proved to be the common source of fracture.

The second of these investigations considered the role of residual and reaction stresses introduced by the contraction and expansion of the weld metal and adjacent material following the joining process. The presence of Lüder's bands, only occurring above the yield strength, already indicated that considerable stress was present. Nevertheless the exact values of the stress were difficult to determine. The conclusion was reached that the residual welding stresses alone were unlikely to be the cause of failure but in conjunction with the external loads and metallurgical changes due to welding, they could initiate fracture in the presence of a stress raiser. These conclusions have later been verified. Consequently a great deal of effort was put on the material and the welding procedure.

This brief story illustrates that the focus of research shifted from the design of the structure to the material used for the construction. The work done was to improve the weld material so that it becomes more reliable and to minimise the effect of welding on the metal alloy used. Although in this case the insufficient material properties became clear, faulty designs, defective processing and exigencies of service conditions seriously enhanced the risks involved and both effects contributed to frequent failures. Having told this, the moral will be evident: *keep alert on both fronts*.

Let us briefly summarise the 'design problem'. If a structure is loaded in the elastic regime, the structure only shows *reversible deformation*. Even in this case the design may be inadequate if the deflections become too large with respect to the specifications. If a structure is loaded in the plastic regime, the structure shows *permanent* or *irreversible deformation* and the material behaviour is *dissipative*. This effect may be used as a safety device, preventing that the structure has to be designed too conservatively. Deformation by plastic processes is generally called *flow* or *yield*. Plastic deformation can be local or global. Visco-elasticity generally results in a permanent deformation but in this case the behaviour is *time dependent* and always dissipative. If a structure is loaded *until* it breaks one speaks of *fracture*. In many cases the fracture process is accompanied by crack tip plasticity and/or visco-elasticity. Flow and fracture are collectively known as *failure*.

Failure is complex and contains many aspects. If for a certain structure a failure problem occurs and has to be analysed, in most cases the following aspects are involved:

- *Loading configuration*. Knowledge of the loads applied to the structure and other operating conditions such as the temperature (distribution), a corrosive environment, etc. are the first parts of information to acquire.
- *Continuum mechanics*. The stress distribution in the structure is usually first calculated assuming only elastic deformation. Before a numerical calculation is done, usually simple estimates are made first. In the case of a numerical calculation it is advisable to start with a two-dimensional model.
- *Fracture mechanics*. For the structure at hand and with the given stress distribution knowledge on the compliance function of any relevant crack configuration is required. Although there are many data books on this issue, it is quite often that a specific assessment has to be made. This can be done by approximate methods (which often work quite well) or by numerical methods.
- *Plasticity*. In many structures dimensional stability requires the absence of global plastic deformation. Therefore small-scale yielding corrections are generally

sufficient, provided the intrinsic length scale of the material given by ER/Y^2 is much smaller than the typical defect size a.

- *Visco-elasticity.* In quite a number of cases the material behaviour is time-dependent, even if no threshold for deformation or fracture is involved. In general this behaviour complicates the matter considerably.
- *Material properties.* Material properties must be known for a proper assessment of any problem. This includes e.g. Young's modulus E, Poisson's ratio v, thermal expansion coefficient α, fracture energy R, yield strength Y and hardening modulus h. Their temperature dependence might be important as well. Stating that these data should be available is a non-trivial remark because in many cases insufficient data are available and additional measurements are necessary.
- *Statistical considerations.* All results of experiments have a certain confidence interval. For a proper assessment the reliability of the necessary data must be known. Furthermore, in the case of a small process zone, brittle behaviour dominates and fracture is, apart from the material properties E, R and Y, also controlled by intrinsic statistical behaviour associated with the distribution of the defect length a.

From this short list it will be clear that in may cases the problems cannot be solved or analysed exactly and that one has to be satisfied with an approximate answer. With all this in mind, let us now overview the various areas as discussed in the core of this book.

25.3   Elasticity

The description of elastic behaviour is the simplest within thermomechanics. Upon loading, where loading may include thermal effects, the material deforms but upon unloading the displacements disappear again. Therefore there is no dissipation. From a fundamental point of view it is clear that a basic understanding of the phenomenological theory of elasticity is well established. The basic theory has since long been known but occasionally new theorems have been added. Of course, analytical solutions to many particular problems are not (and likely will not become) available in view of their geometrical complexity but a numerical solution by finite element methods (FEM), boundary elements methods or alike in principle can always be obtained. Easy-to-use software is abundantly available for this purpose, in fact so easy to use that solutions of quite complex problems can be obtained without unduly effort by workers not at all formally educated in mechanics. Whether the required associated understanding for the interpretation of the results of the calculation is accompanying these efforts is another matter. In fact many researchers in thermomechanics find the topic of elasticity not any longer to their taste[j].

It has to be said that elasticity theory is widely used in engineering. The theory of simple beams and plates is frequently applied in the design and construction of houses. In the early days, when computers were still not widely available, and a full FEM calculation was impossible, the use of truss networks was widely used for larger structures, e.g. power relay line towers, bridges and oil rigs. Nowadays, of course, full 3D FEM calculations are frequently done, not only for large structures but also in the design cycles of many products.

---

[j] A notable exception is the work by Maugin, G.A. (1993), *Material inhomogeneities in elasticity*, Chapman and Hall, London.

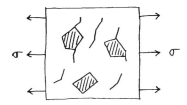

One of the problems in elasticity: cracked solids.

The basic explanation of elasticity in atomistic terms is also well established. The elastic moduli are in principle the second derivatives of an effective potential resulting from the joined effect of a static lattice potential and thermal motion and the influence of the various factors can be treated with a range of methods, varying in their degree of sophistication. Nowadays the elastic moduli of single crystals can be predicted within a few percent by techniques such as density functional theory. Also the influence of the microstructure and/or morphology is in principle well understood. The effect of random or non-random orientation of grains in polycrystalline materials can be taken into account although here exact solutions given certain boundary conditions are not available. The influence of microstructure is in principle nevertheless clear.

Still there are theoretical, not completely solved problems and they mostly can be found at the mesoscopic level. In particular in cases where the bonding between grains in a polycrystalline material or between inclusions with the matrix in a composite is incomplete or changes during service occur, the overall description of the elastic behaviour is a partially unsolved problem. In other words, a full description of the elastic behaviour of composites and cracked media is not available. Since a full description of the bonding for all grains or particles is impossible, and even if available impossible to handle, a sufficiently adequate 'equivalent' model has to be provided. Possibly internal variable theory can help here.

Summarising, in the realm of conservative deformation or elasticity, many problems have been solved but noteworthy problems remain. For example, the precise behaviour of composites, polycrystalline materials and (partially) micro-cracked bodies as a function of their components and their bonding is not satisfactorily solved. In these cases the partial de-bonding influences the elastic (and other) behaviour.

25.4   Plasticity

With plasticity we for the first time enter the realm of *structure-sensitive properties*, as Adolf Smekal (1895-1959) called properties dependent on the microstructure or morphology[k]. It appeared that after plastic deformation a permanent deformation remained the first appearance of which occurred at a smaller stress than theoretically expected. Deformed crystals showed an increased resistance to further deformation and the resulting stress-strain curves varied widely. The question arose why does slip occur on many planes (forming slip bands rather than leading to fracture on the first plane that slipped). These are the facts we have to account for. From the phenomenological point of view, the situation in plasticity is nowadays in principle

---

[k] Smekal, A. (1933), *Strukturempfindliche Eigenschaften der Kristalle*, Handbuch der Physik, ed. H. Geiger and K. Scheel, Springer, Berlin.

One of the problems in plasticity: collective dislocation behaviour.

clear, although to a much lesser extent as for elasticity. In phenomenology essentially two areas of attention exist. The first is the onset of yielding and the second is the occurrence of flow. For the onset of yielding in engineering situations almost invariably use is made of the equivalent stress, in particular the one as introduced by von Mises criterion. By the way, this might also be a divergent idea. It is, however, clear that the von Mises form of the criterion is not always the best suitable one and often only used because of the absence of other information. To describe flow still very often an empirical hardening law is used, which requires a significant amount of experimenting before it has been reliably established and that just for one particular material. More theoretically, the transfer of the description of thermo-elastic behaviour to more general, dissipative behaviour is not without pitfalls[1]. Several ways exist to do this but none of them is generally accepted.

In engineering plasticity theory is used for the description and prediction of shaping of metals and other materials and as a 'safety device' for loaded structures. Here also the use of simple structures and models has been largely superseded by FEM modelling. It is probably fair to say that the engineering applications of plasticity are largely due to the engineers themselves, implying that the understanding in molecular and crystallographic terms is useful but not crucial.

On the micro-scale one can say that for crystalline materials the individual carriers of plasticity, the dislocations, are well studied. The introduction of the concept of dislocation is another divergent idea. The continuum description of single dislocations has proven to be extremely fruitful and has been largely backed-up by molecular theory. The dynamics of a single dislocation and the interaction between two dislocations is well understood. These basic elastic models have been extended by computer calculations, which provided a great deal of trust in the classical models. The collective behaviour of dislocations, though necessary to predict macro-behaviour, that is to say the yield strength and hardening rules, is not so well developed and still intensively under study. This is even true for single crystals. Here use is made of large-scale computer calculations of dislocations in a network that interact via the continuum elasticity model[m]. These calculations certainly have led to improved insight. The presence of inclusions, either due to precipitation after annealing or incorporated as inert particles during processing, provides complications, which are not generally solved. An essential problem here is how to translate the collective behaviour of many dislocations to a set of manageable parameters that can describe their collective behaviour satisfactorily. In other words, what are 'good'

[1] For this see e.g. Woods, L.C. (1975), *The thermodynamics of fluid systems*, Clarendon, Oxford, and Lubliner, J. (1990), *Plasticity theory*, McMillan, New York.
[m] see e.g. Devincre, B., Kubin, L., Lemarchand, C. and Madec, R. (2001), Mater. Sci. Eng. A **309-310**, 211 and Devincre, B. and Kubin, L. (1997), Mater. Sci. Eng., A **234-236**, 8.

internal variables[n] to use in an internal variable theory (another divergent idea). For polycrystalline materials additional aspects such as the presence of grain boundaries and/or pores complicate the matter further. For point defects we also can calculate rather satisfactorily *ab-initio* their energy and this has largely been made possible by the increased computing possibilities. However, as soon as the collective behaviour of point defects is at stake, the situation is less than ideal. The degree of association of point defects in many materials is still not known. Nevertheless these processes are important in the thermo-mechanical but also the functional, i.e. the electrical, dielectrical, magnetical and optical behaviour of solids.

Comparable problems are present for the collective motion of long chain molecules in polymers. For these materials the problems of plasticity and visco-elasticity are often mixed-up so that a separate general assessment of the two effects is difficult, if not impossible.

In summary, with respect to dissipative deformation, many details are known about the individual behaviour of defects but the collective behaviour of defects in solids remains an important area of research. The combined energetics and dynamics of defects, in particular dislocations, and molecules, as e.g. in polymers can only be simulated using supercomputers and although this leads to improved insight, it has not provided so far a clue on how to define internal variables that can be effectively used in less computing time demanding theories.

25.5   Visco-elasticity

The more complex the material behaviour becomes, the less clear the situation is. For visco-elasticity time enters the picture of material deformation for the first time. On the level of phenomenology many basic things are pretty well clear. An extensive theory of linear visco-elasticity exists and attempts to enlarge the scope to non-linear visco-elasticity are present[o]. Nevertheless even on the phenomenological level several problems have not been solved. An example with practically useful implications is that of the receding contact of an indenter with a visco-elastic material. If this problem were completely solved, the indentation test could be done much more reliably to provide material data.

With respect to engineering a comparable situation exists as for plasticity but with more computational problems associated since not only the non-linearity of the problems has to be taken into account but also explicitly the time dependence. Of course, the correspondence principle provides solutions for those problems where an analogous elastic problem can be solved. However, in general only relatively simple problems from a geometrical point of view can be solved elastically in analytical way while in processes typically complex geometries are involved. Therefore one rapidly has to resort to numerical methods for the solution of visco-elastic problems. Further for lifetime predictions a reliable strain-time curve for the materials at hand is required. The conventional interpretation of creep in terms of molecular or microstructural mechanisms might not lead to the most suitable result. There are other

---

[n] The basic ideas provided by Ponter, A.R.S., Bataille, J. and Kestin, J. (1979), J. Méch. **18**, 511 are useful in this respect. See also Ziegler, H. (1983), *An introduction to thermomechanics*, North-Holland, Amsterdam, page 275-277.

[o] see e.g. Findlay, W.N., Lai, J.S. and Onaran, K. (1976) *Creep and relaxation of nonlinear viscoelastic materials*, North-Holland, Amsterdam; Christensen, R.M. (1971), *Theory of viscoelasticity*, Academic Press, New York.

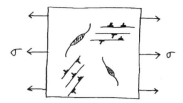

One of the problems in visco-elasticity: multiple mechanisms.

routes available of which in engineering terms rather successful attempts by Willshire[p] and his co-workers are the most notable. Whether their ideas are divergent remains to be seen.

With respect to molecular aspects also a reasonable basic understanding exists. For inorganics and metals creep and relaxation can be understood using diffusion via point defects or dislocation motion. For polymers the motion of individual chains can be described by reptation theory. Of course, many details are lacking but the overall picture is more or less clear.

The situation with respect to understanding becomes less clear when a link with the microstructural and/or morphological structure becomes important. The real problem is that the motion of individual molecules cannot be identified for each molecule separately, their interaction is complex and therefore one has to start seeking for 'equivalent' descriptors that can be used in thermomechanics. For inorganics and metals a strong microstructural dependence is present and this presents a problem in itself, namely the optimal statistical description of the microstructure suitable for the problem at hand.

Another problem is that of the presence of multiple operating mechanisms. These mechanisms possibly all contribute to the overall behaviour and moreover they may be interacting. As an example I mention creep of inorganics where interaction with in particular subcritical crack growth can be present. Both mechanisms are temperature dependent and contribute to the overall deformation. Other examples are a developing microstructure during the deformation process due to grain growth, microcrack development or void growth, and the crystallisation of second phases, which are less refractory.

Summarising, also for time-dependent dissipative deformation quite a number of details are well known but knowledge of the microstructural and morphological aspects is less than sufficient.

25.6 Fracture

As for plasticity and visco-elasticity, also for fracture a new factor is introduced. This is, of course, the disruption of the continuity of connectivity of the material, which may occur on the micro-, meso- and macro-scale.

For fracture even at the phenomenological level several problems are still present, in spite of the divergent idea of Griffith. First of all, it is not always clear whether a fracture experiment refers mainly to fracture initiation or to crack propagation and in fact quite different values might be obtained by different experiments for the same

---

[p] Evans, R.W. and Wilshire, B. (1985), *Creep of metals and alloys*, Institute of Metals, London. See also Evans, R.W. and Wilshire, B. (1992), Rev. Powder Metall. Phys. Ceram. **5**, 111.

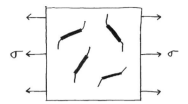

One of the problems in fracture: changing defect populations.

material. Evidently for the application to engineering problems, this situation is highly undesirable.

The use of a proper equivalent stress in multi-axial loading situations in fracture is another unsolved problem. Although frequently the straightforward Griffith criterion of the largest tensile stress is used, it has become clear that this criterion is often inadequate. Further the coupling effects between fracture and other phenomena, e.g. piezo-electricity, at present provide an active field of research. Finally, we mention that in particular for brittle materials the Weibull approach, another divergent idea, is often pursueded but that the conventional extrapolations necessary for large- or small-scale components are not always valid[q].

As before for the other types of material behaviour, most of the problems and successes of fracture theory occur at the meso-level. In this case let me deal with some successes first. An example of success is the use of ultra-high molecular weight polyethylene for hip joints instead of the normal low-density polyethylene[r]. The normal material shows a too high wear rate when in use and the debris is an important cause for necrosis. The smaller 'grain size' of high-density material as compared with the normal low-density polyethylene results in a much smaller wear rate and this is an important factor in its success. Another example is provided by the systematic removal of processing defects from ceramics, e.g. polycrystalline zirconia and zirconia-alumina composites, to result in a high strength, high toughness material[s]. This had led to a worldwide acceptance of the ceramic zirconia as a useful engineering material, which is now used for several applications, e.g. for drawing stones as used for wire drawing and as a material for valves parts in corrosive environments.

Of course, there are also meso-problem areas in fracture. One of these problematic areas is the effect of a changing defect population by corrosion and/or erosion and/or annealing upon strength and therewith upon lifetime in service. The quantification of a steady-state defect population is already not without pitfalls, to put it mildly, but a change of population due to the aforementioned processes is almost impossible to quantify. Nevertheless these processes happen concurrently and we have to deal with that.

Another problem is the handling of residual stress. In general the amount of work required for the quantification of residual stress both for metals and ceramics is quite large. Even larger is the effort that has to be put for assessing the influence of residual stress on the strength of inorganic materials. While a strengthening effect may occur

[q] See e.g. Scholten, H.F., Dortmans, L.J. and de With, G. (1994), page 192 in *Life prediction methodologies and data for ceramic materials*, ASTM-STP 1201, C.R. Brinkman and S.F. Duffy, eds. ASTM, Philadelphia.
[r] See e.g. Kurtz, M., Muratoglu, O.K., Evans, M. and Edidin, A. (1999), Biomaterials **20**, 1659.
[s] see e.g. Lange, F.F. (1989), J. Am. Ceram. Soc. **72**, 3.

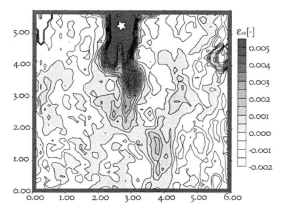

The process zone in a refractory composite, as represented by the in-plane strain ε_{xx}. The star indicates crack tip position, while the length scale is in mm.

due to compressive surface residual stress, the balancing interior tensile stress may affect the strength adversely by subcritical crack growth.

Still another research area of substantial interest is the domain of so-called *R*-curve effects. In conventional fracture theory, based on the Griffith formulation, the fracture energy is considered a material constant, independent of the length of the crack. In essence this is due to the fact that the crack surfaces are supposed to be traction-free. This, however, turns out to be not necessarily the case. It has been shown for several inorganic and metallic materials that the fracture energy *R* can rise substantially with increasing crack length. While for metals this effect is ascribed to plasticity effects, for ceramics one usually thinks of crack bridging phenomena, which dissipate energy. This energy dissipation increases until a steady-state situation is obtained. The length of the process zone, i.e. the zone where these effects are taking place, can be quite substantial. These effects may occur for fine-grained ceramics but are particularly evident in coarse-grained materials. Recently, for one of the first times the size of the process zone has been determined experimentally[t] in refractory material by imaging techniques.

This brings us to the value of the fracture energy itself. Apart from a few cases the fracture energy *R* has to be measured experimentally since it appeared impossible so far to predict the value of fracture energy *ab-initio*. This is true for metals, ceramics and polymers. Even a semi-quantitative model for the prediction of this property would be most welcomed.

Finally we have to say a few words on fatigue. Nowadays the study of fatigue of metals has a bit of the bad smell of being old-fashioned, although failure is in many cases due to fatigue and therefore it is still quite important. May be the most important engineering aspect is the calculation of reliable lifetime predictions. So far the accuracy of the predictions is relatively low and a large safety factor is usually employed. Moreover the data available are usually obtained for loading in full reversal while in practice other loading modes occur as well. The translation from the full reversal data to other loading modes, in particular semi-random loading, is not solved and probably will not be solved at all since the influence of the different loading

---

[t] Jiménez Piqué, E. (2002), *Fracture process zone for brittle materials: a model approach*, Thesis, Eindhoven University of Technology.

modes may affect the plastic zone in metals in a way not covered by full reversal loading. To tackle the problem nowadays extensive testing programs with simulated loading patterns are used to predict the development of damage and the associated lifetime.

For ceramics fatigue is often called subcritical crack growth. For ceramics with a 'normal' microstructure the behaviour for cyclic and random loading can be estimated from static loading. In this context 'normal' refers to a microstructure in which the process zone and *R*-curve effects do not play a significant role in fracture. For ceramics for which this is not true, e.g. zirconia, it is in all reality impossible to make such an estimate. There remain also other problems in this area. The problem of definition of the proper equivalent stress again pops up. Also interaction between fatigue and processes such as corrosion, to a significant extent controlled by the environmental conditions, is important but not fully understood.

Quite naturally within fracture research a great deal of attention has been paid to the crack tip. The nature of the singularity, so to speak, at the crack tip has led to many refinements of the classical continuum theory. These include non-local and/or Cosserat type theories. In this area intimate marriages between microscopic or atomic considerations at the crack tip with more mesoscopic or even macroscopic considerations at the further away loaded areas would be quite useful. At the micro-level the dynamics of atoms in solids is reasonably well described by lattice dynamics. At the meso-level the microstructure and/or morphology plays an important role. The connection, however, is largely absent. The situation is even worse if we also take into account macro-considerations. Leaving the last topic aside, i.e. considering only homogeneously loaded representative volume elements or meso-cells, a reliable theory connecting the atomistic effects with microstructural effects would be extremely useful from a pragmatic point of view as well as rewarding from a scientific point of view. The approach of local defect correction may be useful in this respect. Even a simple Einstein- or Debye-like dynamical behaviour was so far not linked with a continuum fracture process. Of course, large-scale quantum calculations have been done, but again the translation into practical internal variables is lacking. The incorporation of chemical effects is another complicating factor, which nevertheless is quite important for many solids. The use of an appropriate adsorption isotherm and the influence of structural relaxation or reconstruction are just two of the essential problems.

Summing up, for fracture the basic situation is again clear but also many questions remain. The interaction of mechanical effects with chemical, electric and magnetic effects, the interaction of a macrocrack with other features in the microstructure, such as microcracks, voids and inclusions, all deserve more attention.

25.7   The link, use and challenge

Having discussed the various fields of interest let us see what is the use of all that knowledge that is acquired and questions that remain. Superficially one may be inclined to think that thermomechanics in itself is interesting for those who take pleasure in this field of science. However, a moment of consideration will reveal that this discipline is not only interesting as such but also linked with respect to content to various other disciplines. Moreover, it is widely applied in various other disciplines to solve their problems more completely. Let us elaborate on both aspects a bit further and start with the link to other disciplines.

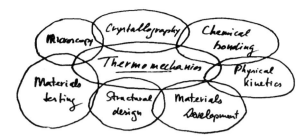

Thermomechanics using a wide range of scientific disciplines.

Thermomechanics of solid materials as defined in this book is related to a wide range of other topics from chemistry, physics and mechanical engineering. A schematic of these relations might be useful. We have encountered the following topics in arbitrary order and at various places: microscopy, crystallography, chemical bonding theory, physical kinetics, materials testing, materials development and structural design. At the heart of the circle of these (and probably other) disciplines, we find thermomechanics, the amalgamation of thermodynamics and mechanics. So thermomechanics is truly multidisciplinary, and not only requires input from many points of view but also provides input to several disciplines.

Elaborating a bit, I think the implications of thermomechanics for materials testing, structural design and materials development are rather straightforward. New materials are constantly being developed. For the assessment of their usefulness they have to be tested and, if of sufficient quality, they can be used in the design of structures. The need for a better identification of internal variables will stimulate microscopy as well as crystallography (in its widest sense). Macroscopic evolution equations may stimulate physical kinetics to provide a basic explanation for them.

From this relation to many disciplines it may come as no surprise that also the area of applications is wide. Making a similar schematic as before, we find thermomechanics again in the middle. At the circumference we find area such as catalysis, reactor design, fuel cells, solar cells, displays, coatings and surface technology. Again many more could be mentioned.

Starting with catalysis we realise that the usual industrial set-up contains many small-sized catalyst carriers in a vertical tube. The reactants are led in on one side and the products are removed on the other side. To have sufficient permeability combined with reactive surface area in the reactor the catalyst carriers have a more or less cylindrical shapes and they are porous themselves. Obviously, the higher the porosity, the higher the surface area and therefore the yield of the reactor. However, higher

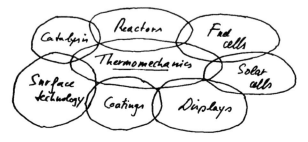

Thermomechanics in the middle of a wide range of scientific disciplines.

porosity implies lower strength, which might lead to crushing of the carriers to small fragments, which in its turn leads to a lower permeability of the reactor. There is therefore an optimum in porosity. This strength issue in a varying thermal and chemical environment is a thermomechanical problem.

In fuel and solar cells as well as displays many materials are combined to a device that has to endure severe thermal and mechanical loading. Differences in thermal expansion coefficients, thermal conductivity coefficients and elastic moduli between the various materials play a role and also their mutual adherence. These aspects together constitute a thermomechanical problem. In the aforementioned devices also coatings are frequently applied but their use is much more widespread. Examples from other application areas are hard coatings (e.g. TiN, TiC or SiC) on cutting and drawing tools, microporous coatings on membranes, optical coatings on lenses, etc. Finally, in chemical reactors a harsh chemical and thermal environment is usually present leading again to thermomechanical loading and the associated problems. By the way, also in reactors coatings are frequently used.

Turning back to science, what then, from a scientific point of view, is the challenge common to many of these application areas? Bringing back into remembrance one of the main lines of thought of this book, i.e. the micro-meso-macro connection, I postulate that many micro-aspects have been solved and that the same is true for the macro-aspects. The situation for the meso-level, on the other hand, seems to be quite different. The averaging of micro-aspects to meso-entities as well as the interpretation of macro-aspects in meso-entitites is insufficiently addressed. All too often changes at the micro-level are directly translated to macro-conclusions without taking into account the effect on or via the meso-level. Similarly, effects on the macro-level are directly translated to microscopic changes, again without considering the meso-level. A typical example is encountered in the paint industry where changes in chemical constitution of the components are directly translated in paint properties such as ageing and wheatherability, forgetting about the effects that such a change might have on the meso-level. As a schematic to represent this situation we could sketch it as a *meso-valley* between the micro- and macro-hill.

Of course, the reaching out of the micro-level to the macro- level and vice versa via the meso-level has met with different degrees of success in different areas. Relatively large success has been obtained in elastic problems. Briefly reiterating and slightly extending, in elasticity the route from micro via meso to macro comprises averaging of molecular behaviour over lattices and grains to yield macro-behaviour. Aspects related to partial de-bonding need further attention. For other areas, however,

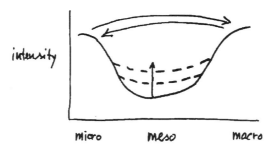

The meso-valley between the micro- and macro-hills.

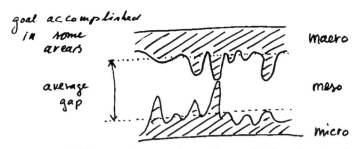

goal accomplished in some areas

average gap

macro

meso

micro

The variation in the bridging gap between micro- and macro-level as a function of topics considered.

the gap is not so small; in fact, it can be considered as wide. Although the individual behaviour of the most important carrier of plasticity in inorganics and metals, the dislocation, and in polymers, the molecule, is well understood, their collective behaviour is only clear in a qualitative (but not a quantitative) sense. For visco-elasticity in particular the connection between molecular considerations and the phenomenological behaviour is not completely well understood. For fracture the coupling with other phenomena, be it chemical (e.g. a reactive environment) or physical (e.g. piezo-electric phenomena), remains an important area of attention. Mapping the depth of the meso-valley as a function of the various topics illustrates the degree of proximity. The ideal situation would be a completely filled field between macro and micro, but from my point of view this situation is not met by far.

Although covering already a wide and important area of applications, in this book and the preceding outlook I have addressed only those problems that probably can be handled within the local state approximation in which the required size of the representative volume or meso-cell, necessary to be able to apply local equilibrium conditions, can be considered as small with respect to the macroscopic dimensions of the sample. For several problems it is not sure whether the local state approximation can be applied at all. It will be clear that too strong gradients make the local state approximation invalid. The dynamics of grain boundaries in inorganics and metals and spherulites or other morphological characteristics in polymers are examples of a strong gradient in a structure. An even stronger example might be the surface of materials. At the surface the density changes from typically about 1 to 10 g/cm^3 to almost zero in only a nanometre or so. Hence there is a strong gradient in density and in dynamics. It remains to be established whether these problems can be handled fully by the local state approximation. The situation becomes even worse if fields with strong gradients are applied and there effects studied. In that case the required size for meso-cell to describe effects in a local-equilibrium way may be so large that the gradient is no longer negligible over the meso-cell. This area is essentially completely open and its exploration will undoubtedly indicate possible routes that could lead to further progress.

Summary of thermodynamics

I. The first law says that you cannot win; the best you can do is break even.
II. The second law says you can break even only at the absolute zero.
III. The third law says you can never reach the absolute zero.
From the preface of Walter J. Moore (1972), *Physical chemistry*, 5th ed., Longman.

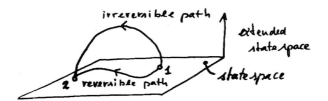

Schematic of the role of internal variables in thermomechanics. The transition from state 1, embedded in the equilibrium state space and characterised by a set of state variables, to state 2, also embedded in the equilibrium state space and characterised by another set of state variables, that is described via a path in extended state space.

Overall it thus appears that within the local state approximation the identification of the proper internal variables and their evolution behaviour is probably the largest obstacle for further applications of thermomechanics, in particular if one has the desire to cross the length scales so that a complete description becomes feasible. Exploring the coupling between different effects and the limits of the local state approximation will result in a better understanding of the history effect in materials. In the words of Muschik[u]: "One can imagine the concept of internal variables is flexible enough to describe a lot of phenomena easily. Therefore not to abuse the notion of internal variables they need physical background".

With this in mind the most important areas to invest in for the near future from my point of view will be clear: *To attempt to realise the further identification and interpretation of internal variables for all kinds of processes so that internal variable theory for individual processes and their coupling gets its microscopic foundation.*

25.8   Epilogue or how hot it will be and how far it is

In the Dutch literature there is a book written in 1839 by Hildebrand (pseudonym for the preacher and later professor Nicolaas Beets, 1814-1903) entitled *Camera Obscura* and which portrays daily life histories. One of the stories is called *How hot it was and how far* and contains the story of a gentleman on the way to his destination. It is a hot, sunny day and the distance to travel is much further than he anticipated. During the travel he more or less complains about the heat and the distance, although he is fact in a favourable position. After all, it is nice weather and he can travel. Imagine what the problems in travelling could be in those days in bad weather! Besides that, many people were not in such a position so that they could travel.

I think this story is rather well applicable to modern science and scientists. We increasingly can use different experimental techniques to unravel scientific problems. Just remember the relatively recent discovery and development of scanning probe microscopy with its members such as atomic force microscopy and scanning tunnelling microscopy. It might be appropriate in this connection to recall the words of Sir Humphry Davy (1778-1829): "Nothing tends to the advancement of knowledge as the application of a new instrument". Also theoretical developments have not stand still. Existing theories have been extended, e.g. thermodynamics with the inclusion of internal variables, and new theories have been developed, e.g. density functional theory. These are just two examples that passed during reading of this book. Moreover, the intensive growth of computer possibilities has enlarged the domain of

---

[u] Muschik, W. (1990), J. Non-Equilib. Thermodyn. **15**, 127.

Look… it's one of those new microscopes that uses light instead of electrons.

solvable problems tremendously. Nevertheless researchers always complain, generally about funds and quarrels with other scientists.

With respect to funding the situation is not new although the structure of research funding has been drastically changed in the last century. In ancient days scientists were dependent on a patron who used to provide them a position and money. The famous astronomer Johannes Kepler (1571-1630), vividly described in the book by Arthur Koestler (*The sleepwalkers*), received so little money in general from his patrons that often he could not pay his debts. Others were luckier and obtained more or less steady positions due to a prince or king, see e.g. the panels about Euler and Lagrange. Still a few others had independent means so that external funding was not really required, e.g. Josiah Willard Gibbs inherited a small fortune from his mother so that, after his studies, he could work quietly in his parental house on his breakthrough work. In general though patrons, princes and kings provided the money. Various agencies, institutions and companies have now largely overtaken the funding and scientists are paid as in any other profession. With that the somewhat elite attitude of in particular nineteenth century scientists such as Lord Rayleigh and Lord Kelvin has largely disappeared but the danger of compromising (if not prostituting) oneself is there in both situations of financing. Kepler not only did his research on the planetary motion but he also had to make regularly horoscopes for his patrons, a job on which he had mixed feelings[v]. It might well be that the patrons found this activity more important than his real research. Nowadays results of research sometimes cannot be published because it does not suit the financier or, even worse, data are manipulated to please the financier.

With respect to quarrels the situation has probably even less changed. Kepler needed the many empirical astronomical data that were obtained by the Danish nobleman Tycho Brahe at his observatory at his island Uranienborg located in the Sont in support of his ideas on planetary motion. Brahe had his own model for the planetary motion, now known to be wrong, and was rather hesitating to provide Kepler his data. The chance existed, of course, that his data would support Kepler's ideas! An example from present times can be found within the field of irreversible thermodynamics. Nowadays there are several schools on irreversible thermodynamics that claim to be general. In reality their results are applied to different fields of science. While rational mechanics[w] has largely focused on the dissipative behaviour of materials, (standard) thermodynamics of irreversible processes is largely paying

---

[v] On the one hand he disliked the idea of nourishing 'the superstition of fat heads', but on the other hand he sincerely believed that the planets have a subtle influence on man and nature.

[w] As R.W. Cahn remarks, it is not clear why the adjective 'rational' is thought to be necessary to denote this branch of mathematics; one would have thought it tautological.

attention to chemical processes and hardly any reference is given in papers from the one sect to the papers of the other[x]. Of course, many other examples of stories about funding and quarrels can be found, both in the past and in the present.

Be all that as it is, I hope I made clear in the previous sections that significant and interesting problems in the area of thermomechanics remain. Further it will be clear, I hope, that I think that if one can be active in research he or she is a lucky person in spite of the occasional problems. I expect progress in most of the areas discussed but I also expect that the solution of problems encountered will be more difficult and further away than initially anticipated. I guess that a comparison with the development in e.g. jazz music may be illuminating. Like in science, in jazz music there are several styles, which originated in various time periods, e.g. swing, bebop and cool. After a style is in existence for some time, a new style is born. However, that does not mean that no further contributions are made to the earlier style. On the contrary, the new style inspires the old style to new efforts (phrases, harmonies, ...) that are subsequently incorporated in the new style. By now and then, also a completely new style is born, e.g. free jazz, which resents most of the 'old stuff'. Nevertheless, in the end both the old and the new style use elements from either of them, the sharp edges are removed (certainly in the free style) and in this way the musical scenery as a whole is enriched. Therefore, returning to science, arguing is fruitful, reworking and keeping what is good and deleting what is fault as well, but dogmatic behaviour (as in the case of quaternions, see the panel on Hamilton) will lead to a framework that will disappear into oblivion but of course, keeping human nature in mind, only after its last defenders have deceased[y].

Of course, my discussion is along the line of the statement of the composer Ludwig van Beethoven (1770-1827) that 'everything should be new and expected'. There is a great deal of truth in this statement. To make this a little clearer, some have divided knowledge into the 'known', the 'known unknown' and the 'unknown unknown'. Since, as Mark Twain (1835-1910) remarked, 'it is difficult to make predictions, particularly about the future', we have discussed obviously the 'known unknown'.

Coming to the end of this chapter I have to state that I am neither pessimistic nor optimistic about progress in thermomechanics but realise that in many cases the problems are larger than originally anticipated. You might call that realistic. Finally, I cannot resist quoting what Kondepudi and Prigogine[z] wrote in the preface and epilogue of their recent textbook on thermodynamics, namely that science has no final formulation and is moving from a static geometrical picture to a description in which evolution and history play essential roles. Their message is that for this new description of nature thermodynamics is basic. It will be clear that my message and belief is essentially the same.

---

[x] For a much deeper critique on this matter, see Lavenda, B.H. (1978), *Thermodynamics of irreversible processes*, MacMillan, London (see also Dover, 1997).

[y] This statement is sometimes referred to as Planck's principle, see Hull, D.L., Tessner, P.D. and Diamond, A.M. (1978), Science **202**, 217.

[z] Kondepudi, D. and Prigogine, Y. (1990), *Modern thermodynamics*, Wiley, Chichester.

Appendix A

Units, physical constants and conversion factors

Basic and derived SI units

Quantity	Unit	Symbol
length	metre	m
mass	kilogram	kg
time	second	s
electric current	ampere	A
temperature	kelvin	K
	°C	$t/°C = T/K - 273.15$
amount of substance	mole	mol
force	newton	$N = kg\ m/s^2$
work, energy, heat	joule	$J = N\ m$
power	watt	$W = J/s$
pressure	pascal	$Pa = N/m^2$
frequency	hertz	$Hz = s^{-1}$
electrical charge	coulomb	$C = A\ s$
electrical potential	volt	$V = J/C$
electrical resistance	ohm	$\Omega = V/A$

Physical constants

Constant	Symbol	Value
Avogadro's number	N_A	$6.022 \times 10^{23}\ mol^{-1}$
elementary charge	e	$1.602 \times 10^{-19}\ C$
electron rest mass	m_e	$9.109 \times 10^{-31}\ kg$
proton rest mass	m_p	$1.673 \times 10^{-27}\ kg$
neutron rest mass	m_n	$1.675 \times 10^{-27}\ kg$
atomic mass unit (dalton)	$amu\ (Da)$	$1.661 \times 10^{-27}\ kg$
gas constant	R	$8.315\ J/mol\ K$
Boltzmann's constant	$k = R/N_A$	$1.381 \times 10^{-23}\ J/K$
Planck's constant	h	$6.626 \times 10^{-34}\ J\ s$
	$\hbar = h/2\pi$	$1.055 \times 10^{-34}\ J\ s$
standard acceleration of gravity	g	$9.807\ m/s^2$
speed of light	c_0	$2.998 \times 10^8\ m/s$
Faraday constant	F	$9.649 \times 10^4\ C/mol$
permeability of the vacuum	μ_0	$4\pi \times 10^{-7}\ N/A^2$ (exact)
permittivity of the vacuum	$\varepsilon_0 = 1/\mu_0\ c_0^2$	$8.854 \times 10^{-12}\ F/m$

Conversion factors for non-SI units

1 dyne	$= 10^{-5}$ N	1 eV	$= 96.48$ kJ/mol
1 bar	$= 10^5$ Pa	1 atm	$= 1.013$ bar
1 int. cal	$= 4.187$ J	1 torr	$= 1/760$ atm
1 erg	$= 10^{-7}$ J	1 l	$= 1$ dm$^3 = 10^{-3}$ m^3
1 eV	$= 1.602 \times 10^{-19}$ J	1 psi	$= 6.895 \times 10^3$ Pa

Prefixes

nano n	$= 10^{-9}$	kilo k	$= 10^3$
micro μ	$= 10^{-6}$	mega M	$= 10^6$
milli m	$= 10^{-3}$	giga G	$= 10^9$

Greek alphabet

A, α	alpha		N, ν	nu
B, β	beta		Ξ, ξ	xi
Γ, γ	gamma		O, o	omicron
Δ, δ	delta		Π, π	pi
E, ε	epsilon		P, ρ	rho
Z, ς	zeta		Σ, σ	sigma
H, η	eta		T, τ	tau
Θ, θ, ϑ	theta		Y, υ	upsilon
I, ι	iota		Φ, φ, ϕ	phi
K, κ	kappa		X, χ	chi
Λ, λ	lambda		Ψ, ψ	psi
M, μ	mu		Ω, ω	omega

Appendix B

Properties of structural materials*

This table lists typical values for various properties of structural materials. A wide range of values for nominally the same materials can be exhibited for nearly all properties. The data should be taken therefore as an indication only and not for specific design purposes. The symbols have their usual meaning while UTS and k denote the ultimate tensile strength and thermal conductivity, respectively.

Material	ρ (Mg/m³)	E (GPa)	ν	G (GPa)	Y (MPa)	UTS (MPa)	K_{Ic} (MPa m$^{1/2}$)	α (10^{-6} K^{-1})	k (J/m K)
Alumina	4.00	380	0.26	125	4800	320	4.4	8.1	30
Aluminium	2.72	70	0.34	28	330	550	41	33.1	237
Beryllium	2.88	345	0.12	110	360	500	4.2	13.7	—
Bone	1.95	14	0.43	3.5	100	100	4.9	20.0	—
CFRP	1.55	1.5	0.28	53	200	550	37	12.5	—
Cermet	11.7	520	0.3	200	650	1200	13	5.8	—
Concrete	2.5	48	0.2	20	25	3	0.7	11	1.0-2.0
Copper	8.33	125	0.35	50	510	720	94	18.4	400
Cork	0.18	0.032	0.25	0.005	1.4	1.5	0.1	180	0.03
GFRP	1.75	26	0.28	10	125	530	30	18.5	—
Glass	2.4	63	0.23	26	1500	75	0.7	8.8	1.0
Granite	2.65	66	0.25	26	2500	2500	1.5	6.5	—
Ice	0.92	9.1	0.28	3.6	88	6.5	0.1	55	—

Appendix B: Properties of structural materials

Material	ρ (Mg/m^3)	E (GPa)	ν	G (GPa)	Y (MPa)	UTS (MPa)	K_{Ic} (MPa m$^{1/2}$)	α (10^{-6} K^{-1})	k (J/m K)
Lead	11.10	16	0.45	5.5	33	42	40	29	34
Nickel	8.49	180	0.31	70	900	1200	94	13	142
Polyamide	1.15	2	0.42	0.76	40	55	3.0	103	–
Polybutadiene	0.91	0.0016	0.5	0.0005	2.1	2.1	0.1	140	0.25
Polycarbonate	1.20	2.7	0.42	0.97	70	77	2.6	70	0.20
Polyethylene	0.95	1	0.42	0.31	25	33	3.5	225	0.38
Polypropylene	0.90	1.2	0.42	0.42	35	45	3.0	86	0.12
Polyurethane	1.17	0.025	0.5	0.0086	30	30	0.3	125	–
PVC	1.37	1.5	0.42	0.6	53	60	0.5	75	0.15-0.20
Silicon	2.32	110	0.24	44	3200	2500	1.5	6.0	142
Silicon carbide	2.85	430	0.15	190	9800	630	4.2	4.2	350
Spruce	0.40	9	0.3	0.8	48	50	2.5	4.0	0.15
C-Steel	7.85	210	0.29	76	590	1200	50	13.5	52
Stainless steel	7.85	210	0.28	786	870	1200	50	16.6	14-27
Titanium	4.65	100	0.36	39	70	850	87	9.4	6.0-16
WC	15.5	640	0.21	270	6800	450	3.7	5.8	–

* Adapted from a similar appendix in D. Roylance (1996), *Mechanics of materials*, Wiley, New York.

Appendix C

Properties of plane areas[a]

In this appendix some properties of plane areas, such as the centroid of a single area, the centroid of composite areas and the various moments of inertia are presented.

C.1   Centroid of an area

In order to define the co-ordinates of the centroid of an area, let us consider Fig. C.1. The total area A is calculated from

$$A = \int dA \tag{C.1}$$

The co-ordinates of the *centroid* C are then given by

$$\bar{x} = \int x dA \Big/ \int dA = Q_x / A \quad \text{and}$$
$$\bar{y} = \int y dA \Big/ \int dA = Q_y / A \tag{C.2}$$

The numerators Q_x and Q_y are known as the *first moments* of the area with respect to the x- and y-axis, respectively. Whenever the boundaries of the area A are given by simple analytical expressions the integrals can be evaluated in closed form to calculate the co-ordinates of the centroid.

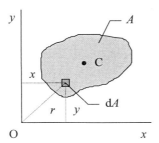

Fig. C.1: Plane area A with centroid C.

Example C.1

For a parabola given by $y = h[1-(x/b)^2]$ the elements of area dA and total area A are given by

$$dA = y dx = h\left(1 - \frac{x^2}{b^2}\right) dx \quad \text{and} \quad A = \int dA = \int_0^b h\left(1 - \frac{x^2}{b^2}\right) dx = \frac{2bh}{3}$$

---

[a]Taken with modification from a similar appendix in *Mechanics of Materials* by S.P. Timoshenko and J.M. Gere, Van Nostrand, New York, 1973.

respectively. The first moments are given by

$$Q_x = \int \frac{y}{2}\, dA = \int \frac{y^2}{2}\, dx = \int_0^b \frac{1}{2}\left[h\left(1-\frac{x^2}{b^2}\right)\right]^2 dx = \frac{4bh^2}{15} \qquad \text{and}$$

$$Q_y = \int x\, dA = \int xy\, dx = \int_0^b xh\left(1-\frac{x^2}{b^2}\right) dx = \frac{b^2 h}{4}$$

In the expression for Q_x, one has to substitute $y/2$ since the y co-ordinate of the centroid of the area element is located at $y/2$ (Fig. C.2). The co-ordinates of the centroid consequently are

$$\bar{x} = \frac{Q_y}{A} = \frac{3b}{8} \qquad \text{and} \qquad \bar{y} = \frac{Q_x}{A} = \frac{2h}{5}$$

In case the shape cannot be described by a simple function, one has to integrate numerically.

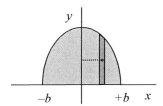

Fig. C.2: Calculation of Q_x.

In many cases the position of the centroid can be obtained by inspection. For example (Fig. C.3), if the area has two axes of symmetry, the centroid lies at their intersection. When the area has one symmetry axis, the centroid lies somewhere on that axis. Finally, if the area has inversion symmetry, the inversion point is the centroid.

In practice, it frequently happens that an area is composed of several parts, each part having a familiar geometric shape for which the co-ordinates of the centroid are already known. In determining the centroids of such areas, we need to divide the area into suitable parts and use an area weighted summation of the co-ordinates of the parts to arrive at the centroid of the area. For example, if the area is composed of two rectangular parts, as in the fourth area from the left in Fig. C.3, where $(\bar{x}_1, \bar{y}_1)$ and $(\bar{x}_2, \bar{y}_2)$ are the co-ordinates of the centroids of the parts, it holds that

$$\bar{x} = \frac{\int_A x\, dA}{\int_A dA} = \frac{\int_{A_1} x\, dA + \int_{A_2} x\, dA}{\int_{A_1} dA + \int_{A_2} dA} = \frac{\bar{x}_1 A_1 + \bar{x}_2 A_2}{A_1 + A_2} \quad \text{and} \quad \bar{y} = \frac{\bar{y}_1 A_1 + \bar{y}_2 A_2}{A_1 + A_2} \tag{C.3}$$

This is easily generalised to

$$A = \sum_i A_i \qquad Q_x = \sum_i \bar{y}_i A_i \qquad Q_y = \sum_i \bar{x}_i A_i \tag{C.4}$$

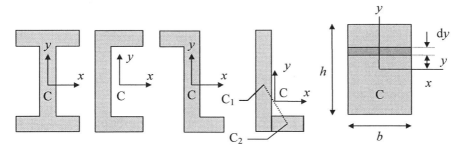

Fig. C.3: Various areas.

C.2   Moments of inertia of an area

The *second moments* of area, also denoted as *moments of inertia*, are defined with respect to the *x*- and *y*-axis, respectively, as

$$I_x = \int y^2 \mathrm{d}A \qquad \text{and} \qquad I_y = \int x^2 \mathrm{d}A \tag{C.5}$$

For simple geometrical shapes the integrals can be calculated analytically. Related to the second moment of area is the *polar moment of inertia*. It is defined as

$$I_p = \int r^2 \mathrm{d}A = \int \left(x^2 + y^2\right) \mathrm{d}A = I_x + I_y \tag{C.6}$$

Example C.2

For a rectangular section with a width b and a height h (Fig. C.3), one easily calculates

$$I_x = \int_A y^2 \mathrm{d}A = \int_{-b/2}^{+b/2} \mathrm{d}x \int_{-h/2}^{+h/2} y^2 \mathrm{d}y = bh^3/12$$

Example C.3

For a circular section with a radius r one easily calculates

$$I_p = \int_0^r r^2 \mathrm{d}A = \frac{\pi r^4}{2} = \frac{\pi d^4}{32} \qquad \text{where} \quad d = 2r$$

This also results in an easy method for calculating the second moment for a circular sections or parts of them. Because the moment is the same for all diameters, we have

$$I_x = I_y = I_p/2$$

For a quarter circle the polar moment is 1/4 of that of a full circle. Therefore the moments of inertia with respect to the *x*- and *y*-axis are

$$I_x = I_y = \pi r^4/16$$

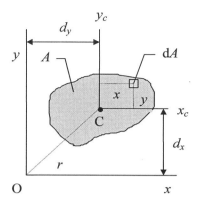

Fig. C.4: Parallel axis theorem.

The calculation of the moment of inertia with respect to any axis in the plane is done with the *parallel axis theorem* (Steiner's rule). To obtain this theorem consider Fig. C.4. The point C is the centroid and co-ordinate axes through them are denoted by x_c and y_c, respectively. The co-ordinate axes with respect to the point O are indicated by x and y, respectively. Then by definition the moment of inertia with respect to the x-axis is given by

$$I_x = \int (y+d_x)^2 \, dA = \int y^2 \, dA + 2d_x \int y \, dA + d_x^2 \int dA \tag{C.7}$$

The first integral is the moment of inertia I_{xc} about the x_c-axis. The second vanishes because the x_c axis passes through the centroid and the third is just the area. Therefore the expression reduces to

$$I_x = I_{xc} + Ad_x^2 \tag{C.8}$$

and similarly for I_y one obtains

$$I_y = I_{yc} + Ad_y^2 \tag{C.9}$$

The above equations represent the parallel axis theorem. They show that the moment of inertia with respect to any axis is equal to the moment of inertia with respect to a parallel axis through the centroid plus the product of the area and the square of the distance between the two axes. For the polar moment obviously we obtain

$$I_p = I_x + I_y = I_{xc} + I_{yc} + A(d_x^2 + d_y^2)$$

$$= I_{pc} + Ar^2$$

where r is the shift of the origin. Using this theorem, it is easy to calculate the moment of inertia with respect to any axis.

Example C.4

Consider the cross-section as shown in Fig. C.5. In this case the moment of inertia is

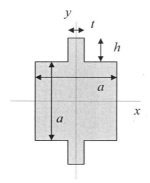

Fig. C.5: Square cross-section with two extra ribs.

$$I_x = \frac{a^4}{12} + \frac{2th^3}{12} + 2th\left(\frac{a+h}{2}\right)^2$$

If the size of the rib is small as compared with the square cross-section (t, $h <<$ a), then it holds that

$$I_x = \frac{a^4}{12} + 2th\left(\frac{a}{2}\right)^2$$

The maximum distance to the x-axis, y_{max} is equal to

$$y_{max} = a/2 + h$$

so that the section modulus for a beam with this cross-section is given by

$$Z_{ela} = \frac{I_x}{y_{max}} = \frac{\frac{a^4}{12} + 2th\left(\frac{a}{2}\right)^2}{\frac{a}{2} + h} = \frac{a^3}{6} \frac{1 + \left(\frac{2h}{a}\right)\left(\frac{3t}{a}\right)}{1 + \frac{2h}{a}} \qquad (C.10)$$

The term $a^3/6$ is the section modulus of a square beam with side a, so that the extra ribs introduce a correction factor. If $3t/a < 1$ the correction factor is smaller than 1. Hence, since the outer fibre stress is inversely proportional to the section modulus, for a fixed applied moment the outer fibre stress is larger in the beam with rib than in the beam without rib.

Appendix D

Statistics

In this appendix, some basic aspects of statistics, which are used in these notes, are mentioned, mainly for setting the terminology. More details can be found in the textbooks quoted in the bibliography.

D.1   Moments and measures

A *stochastic* or *random variable* is defined by the set of possible values, the *range*, and a probability distribution over this set. The *probability distribution* (or *density*) *function* (pdf) is denoted by f. The *cumulative distribution function* (cdf) $F(x)$ is the integral of f up to a certain value x (Fig. D.1)

$$F(x) = \int_{-\infty}^{x} f(x')\, dx' \tag{D.1}$$

The *expectation value* for any variable y, possibly a function of x, is denoted by $E(y)$ and defined by

$$E(y) = \int_{-\infty}^{\infty} y(x) f(x)\, dx \tag{D.2}$$

If $y = x$, the resulting expectation value is usually called the *mean* and is denoted by μ. It is a measure for the centre of the distribution. More generally, the pdf may be characterised by its expansion in *moments* m' of order n given by

$$m_n' = E(x^n) = \int_{-\infty}^{\infty} x^n f(x)\, dx \tag{D.3}$$

or by its expansion in *central moments* m given by

$$m_n = E((x - \mu)^n) = \int_{-\infty}^{\infty} (x - \mu)^n f(x)\, dx \tag{D.4}$$

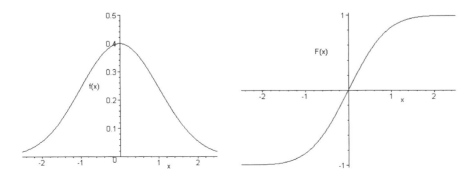

Fig. D.1: The pdf $y = f(x)$ and cdf $y = F(x)$.

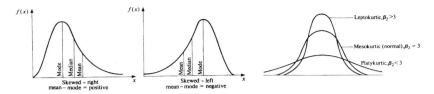

Fig. D.2: Characteristics of a (skew) pdf.

The first moment m_1' is equal to the mean μ. The first central moment m_1 is zero. The second central moment m_2 is called the *variance* of the distribution, V. Its square root is denoted as *standard deviation*, σ. It is a measure for the width of the distribution. The odd central moments of a symmetrical pdf are zero. Consequently the third central moment is a simple measure for the asymmetry of the distribution. It is usually used in its nondimensional form m_{ske}, generally called coefficient of *skewness* (Fig. D.2).

$$m_{ske} = m_3^2 / m_2^3 \tag{D.5}$$

The fourth central moment measures the degree of peakedness of a pdf. This moment is also normally used in its normalised form m_{kur}, generally denoted as coefficient of *kurtosis* (Fig. D.2).

$$m_{kur} = m_4 / m_2^2 \tag{D.6}$$

For many distributions a *moment generating function* (mgf) $M_x(t)$ may be defined by the convenient expression

$$M_x(t) = E[\exp(tx)] = \int_{-\infty}^{\infty} \exp(tx) f(x)\, dx \tag{D.7}$$

where t is a dummy variable. This function is unique in the sense that variables with the same mgf have the same pdf. The moment m_n is calculated from

$$m_n = \frac{d^n M_x(t)}{dt^n}\bigg|_{t=0} \tag{D.8}$$

that is, the derivative evaluated at $t = 0$. Apart from the mean, other measures exist for the central tendency of the variable x (Fig. D.2). The most relevant ones are
- The *median* x_{50} defined by the value of x which divides the area beneath the $f(x)$ curve into half. The median is the so-called 50% *percentile*. For percentile other percentages z may be used leading to $F(x_z)+F(x_{100-z}) = 1$,

$$\int_{-\infty}^{x_{50}} f(x)\, dx = \int_{x_{50}}^{\infty} f(x)\, dx \qquad \text{or} \qquad F(x_{50}) = 1/2 \tag{D.9}$$

- The *mode* x_{mod} defined as the most frequently occurring value of x,

$$df(x)/dx\big|_{x=x_m} = 0 \tag{D.10}$$

- The logarithmic or *geometrical mean*, x_{geo}, for functions with $x > 0$ defined by

$$\ln x_{geo} = E(\ln x) = \int_0^{\infty} (\ln x) f(x)\, dx \tag{D.11}$$

D.2   Distributions

Many forms of distribution functions are known. The *Gaussian* (also called *normal*) pdf is particularly important in statistics. This is due to the *central–limit theorem*. This theorem states that if we have several independent stochastic variables, distributed according to some (not necessarily Gaussian) pdf with finite mean and variance, the pdf of the sum is still approximately Gaussian for sufficiently large number of variables. Since an experimental estimate is usually influenced by a number of, sometimes unknown, factors, the pdf for an experimental estimate is approximately Gaussian. Let us denote the mean and standard deviation of the Gauss distribution by μ and σ, respectively. Introducing a standardised variable z by

$$z = (x - \mu)/\sigma \tag{D.12}$$

the Gauss pdf, $g(x)$, and cdf, $G(x)$, are given by

$$g(z) = (2\pi)^{-1/2} \exp(-z^2/2) \tag{D.13}$$

$$G(z) = (2\pi)^{-1/2} \int_{-\infty}^{z} \exp(-t^2/2)\, dt \tag{D.14}$$

The mgf for a standardised Gauss distribution is

$$M_z(t) = \exp(t^2/2) \tag{D.15}$$

where t is a dummy variable. For the original Gauss distribution the mgf is

$$M_x(t) = \exp(\mu t + \sigma^2 t^2/2) \tag{D.16}$$

All odd moments equal zero and the even central moments are given by

$$m_n = n!\sigma^n / [(n/2)! 2^{n/2}] \tag{D.17}$$

Consequently for a Gauss distribution the value of the kurtosis $m_k = 3$ (Fig. D.2). When $m_k > 3$ ($m_k < 3$), the pdf is more (less) peaked than for a Gauss pdf.

In some cases the mechanisms determining an experimental estimate are multiplicative in nature rather than additive e.g. in the case of crushing of particles. In that case the parameter $y = \ln x$ is distributed according to a Gauss distribution and x is said to be distributed log-normally. The *log-normal* pdf is often used to describe skew distributions. Obviously the log-normal distribution can be used only when the variable x is always positive. The standardised parameter

$$v = (\ln x - \mu')/\sigma' \tag{D.18}$$

is introduced where μ' and σ' represent the mean and standard deviation of the distribution in the variable y. The pdf for the log-normal distribution in terms of v is given by $g(v)$. Therefore

$$E(y) = E(\ln x) = \mu' \quad \text{and} \quad V(y) = V(\ln x) = \sigma'^2 \tag{D.19}$$

A mgf is not useful for the log-normal distribution but the moments m_n' are given by

$$m_n' = \exp(n\mu' + n^2\sigma'^2/2) \tag{D.20}$$

Henceforth the mean and variance are given by

$$E(x) = \exp(\mu' + \sigma'^2/2) \tag{D.21}$$

$$V(x) = E^2(y)\exp(\sigma'^2 - 1) = \exp(2\mu' + 2\sigma'^2) - \exp(2\mu' + \sigma'^2) \tag{D.22}$$

Because the log-normal distribution is symmetrical in y, the median y_{50}, the mode y_{mod} and mean y_{ave} all have the same value μ'. Furthermore, since the 50% percentile of the transformed variable equals the transformed percentile of the original value $\mu' = \ln y_{geo}$ so that we have altogether

$$\ln x_{geo} = \ln x_{50} = y_{50} = y_{mod} = y_{ave} = \mu' \tag{D.23}$$

There exist also relations between the mode x_{mod} and the mean x_{ave} for the variable x in the log-normal distribution:

$$\ln x_{mod} = \mu' - \sigma'^2 \quad \text{and} \quad \ln x_{ave} = \mu' + \sigma'^2/2 \tag{D.24}$$

Another frequently encountered distribution in materials science is the *Weibull* distribution. The Weibull pdf $w(x)$ and cdf $W(x)$ are given by

$$w(x) = abx^{b-1}\exp(-ax^b) \quad \text{and} \tag{D.25}$$

$$W(x) = 1 - \exp(-ax^b) \tag{D.26}$$

respectively. The mgf for the Weibull distribution is given by

$$M_x(t) = a^{-t/b}\Gamma\left(1 + \frac{t}{b}\right) \tag{D.27}$$

Therefore the moments m_n' are given by

$$m_n' = a^{-n/b}\Gamma\left(1 + \frac{n}{b}\right) \tag{D.28}$$

where $\Gamma(t)$ denotes the gamma function defined by

$$\Gamma(t) = \int_0^\infty x^{t-1}e^{-x}\,dx \tag{D.29}$$

For this function generally holds $\Gamma(t+1) = t\Gamma(t)$. For integer n it is connected to the more familiar factorial function $n!$ by $\Gamma(n+1) = n!$. Consequently the mean μ and variance V for the Weibull distribution are given by

$$\mu = a^{-1/b}\Gamma\left(1 + \frac{1}{b}\right) \quad \text{and} \quad V = a^{-2/b}\left[\Gamma\left(1 + \frac{2}{b}\right) - \Gamma^2\left(1 + \frac{1}{b}\right)\right] \tag{D.30}$$

The shape parameter b is in materials science frequently called the *Weibull modulus* and denoted by m, while the location is frequently given by the *characteristic value* $x_0 = a^{-1/b}$, representing the x-value corresponding to 63% probability.

D.3 Testing hypotheses

One must distinguish between the measures for a variable and its experimental estimates or, equivalently, between a *distribution parameter* and a *sample parameter*. A sample parameter is an estimator for a distribution parameter. From the outcome of n measurements of the variable x, one obtains the *average* $\bar{x}$ defined by

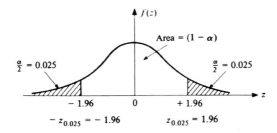

Fig. D.3: Confidence interval of 95% for a Gauss distribution.

$$\bar{x} = \frac{1}{n}\sum_i x_i \qquad\qquad (D.31)$$

The average is an unbiased estimate for the mean μ:

$$E(\bar{x}) = \mu \qquad\qquad (D.32)$$

In other words, the average is the expectation value or an (experimental) estimate for the mean. An experimental estimate for the logarithmic mean is given by

$$x_g = \left(\prod_i x_i\right)^{1/n} \qquad\qquad (D.33)$$

Estimates for the mode and median are made directly according to their definition.

From the outcome of n measurements, one can also calculate the standard deviation of the sample, s, defined by

$$s^2 = \frac{1}{n-1}\sum_i (x_i - \bar{x})^2 \qquad\qquad (D.34)$$

The number $v = n-1$ is called the (number of) degrees of freedom. The parameter s^2 is an unbiased estimate for σ^2

$$E(s^2) = \sigma^2 \qquad\qquad (D.35)$$

Frequently used is the coefficient of variation given by the ratio $s/\bar{x}$. Since the sample mean $\bar{x}$ is in itself a random variable, the standard deviation for $\bar{x}$ is denoted by $s_{\bar{x}}$ and is related to s by

$$s_{\bar{x}} = s/n^{1/2} \qquad\qquad (D.36)$$

After an experimental estimate for the parameter x is made, we normally want to make some statement about its reliability. When we have many measurements at our

Table D.1: Values for the parameter $z(\alpha/2)$ for two-tailed testing for various values of the confidence level $(1-\alpha)$.

$1-\alpha$	$z(\alpha/2)$	$z(\alpha/2)$	$1-\alpha$
0.90	1.64	1.00	0.682
0.95	1.96	2.00	0.955
0.99	2.56	3.00	0.997

Table D.2: Values for Student's $t(q,v)$ for a two-tailed test for various values of the confidence level $(1-\alpha)$ and number of degrees of freedom v. The parameter q denotes either $\alpha/2$ or $1-\alpha/2$.

$1-\alpha$	$t(q,1)$	$t(q,2)$	$t(q,3)$	$t(q,4)$	$t(q,5)$	$t(q,10)$	$t(q,20)$	$z(q)$
0.90	6.31	2.92	2.35	2.13	2.02	1.81	1.73	1.64
0.95	12.7	4.30	3.18	2.78	2.57	2.23	2.09	1.96
0.99	63.6	9.93	5.84	4.60	4.03	3.17	2.85	2.58

disposal (that is large n) we are sampling from a Gauss distribution for x with unknown μ but known σ because σ can be replaced by s with sufficient accuracy. In that case we can say that with a confidence of $100(1-\alpha)\%$ the value of x is within the interval (Fig. D.3)

$$\bar{x} - z(\alpha/2)s_{\bar{x}} < x < \bar{x} + z(\alpha/2)s_{\bar{x}} \tag{D.37}$$

The values of z are tabulated for various values of $(1-\alpha)$ in standard statistical tables (Table D.1). The distribution for z is symmetrical, that is $z(\alpha/2) = z(1-\alpha/2)$.

In general we are sampling, however, from a Gauss distribution for x with both μ and σ unknown (that is for small n). It can be shown that the estimate for the mean in this case is distributed according to the so-called *t-distribution*. The parameter $z(\alpha/2)$ must be replaced by $t(\alpha/2, v)$ (Student's t) which is, apart from the confidence level $(1-\alpha)$, also dependent on the number of measurements n through v. This parameter is tabulated again for various values of $(1-\alpha)$ and v (Table D.2). It reaches the z-values for a Gauss distribution quite rapidly as the number of measurements increases. If we predict e.g. by theory, that the outcome for a variable x is x_{the} and the experimental value for x after n measurements is x_{exp}, the 'zero hypothesis' $x_{\text{exp}} = x_{\text{the}}$ is said to be true with $100(1-\alpha)\%$ confidence if the value of x is within the interval

$$x_{\text{exp}} - t(\alpha/2, v)s_{\bar{x}} < x_{\text{the}} < x_{\text{exp}} + t(\alpha/2, v)s_{\bar{x}} \tag{D.38}$$

The distribution for t is also symmetrical.

Further, it can be shown that for samples of size n with sample variance s^2 for a Gauss distribution with variance σ^2, the variable $(n-1)s^2/\sigma^2$ has a so-called '*chi-square*' distribution (χ^2) with $n-1$ degrees of freedom. With a confidence of $(1-\alpha)$, σ^2 is in the interval

$$(n-1)s^2/\chi^2(\alpha/2, v) < \sigma^2 < (n-1)s^2/\chi^2(1-\alpha/2, v) \tag{D.39}$$

Values for χ^2 are again tabulated for various values of $(1-\alpha)$ and v (Table D.3). Note that the distribution for χ^2 is not symmetrical.

Finally, if we want to compare the outcome of two different experiments for the same parameter x, statistical theory offers no exact answer. To a good approximation,

Table D.3: Values for $\chi^2(q,v)$ for various values of the confidence level $(1-\alpha)$ and number of degrees of freedom v. The parameter q denotes either $\alpha/2$ or $1-\alpha/2$.

$1-\alpha$	$\chi^2(q,v)$ $\alpha/2$	$1-\alpha/2$	$\chi^2(q,v)$ $\alpha/2$	$1-\alpha/2$	$\chi^2(q,v)$ $\alpha/2$	$1-\alpha/2$	$\chi^2(q,v)$ $\alpha/2$	$1-\alpha/2$
0.90	0.352	7.81	0.711	9.49	1.15	11.1	3.94	18.3
0.95	0.216	9.35	0.484	11.1	0.831	12.8	3.25	20.5
0.99	0.0717	12.8	0.207	14.9	0.412	16.7	2.16	25.2

however, the *Welch-Aspin T-test* can be used. This test is a 'generalised' *t*-test. The parameter *T* is defined by

$$T = \left((\bar{x}_1 - \bar{x}_2) - (\mu_1 - \mu_2)\right) / \left(s_1^2/n_1 + s_2^2/n_2\right)^{1/2} \qquad \text{(D.40)}$$

and is approximately distributed as *t* with v degrees of freedom where v is given by

$$1/\upsilon = (s_1^2/S)^2/\upsilon_1 + (s_2^2/S)^2/\upsilon_2 \qquad S = s_1^2/n_1 + s_2^2/n_2 \qquad \text{(D.41)}$$

The parameters *n*, *v*, *x* and *s* have the same meaning as before. The 'zero hypothesis'

$$\bar{x}_1 - \bar{x}_2 = \bar{\mu}_1 - \bar{\mu}_2 \qquad \text{(D.42)}$$

is said to be true with (1–α) confidence if the inequality

$$(\bar{x}_1 - \bar{x}_2) < \left(s_1^2/n_1 + s_2^2/n_2\right)^{1/2} t(1 - \alpha/2, \upsilon) \qquad \text{(D.43)}$$

is satisfied.

D.4   Extreme value statistics

Apart from statistics of mean values also the statistics of extreme values, that is maxima and minina, are relevant in materials science, e.g. in mechanical, dielectric or electric breakdown. We recall that the central-limit theorem gives an asymptotic distribution for the average of *x*, namely the Gauss distribution, which does not depend on the parent distribution. The key question is: is there any limit distribution (or family of distributions) for maxima and minima which does not depend on the parent cdf? The answer to this question is given by the following theorems:

*   The only three types of (limit) distributions for maxima are

    Frechet: $F(z) = \exp(-z^{-\gamma})$ if $z > 0$ and $F(z) = 0$ otherwise

    Weibull: $W(z) = 1$ if $z \geq 0$ and $W(z) = \exp[-(-z)^{-\gamma}]$ otherwise

    Gumbel: $G(z) = \exp[-\exp(-z)]$, $-\infty < z < \infty$

*   The only three types of (limit) distributions for minima are

    Frechet: $F(z) = 1 - \exp[-(-z)^{-\gamma}]$ if $z < 0$ and $F(z) = 1$ otherwise

    Weibull: $W(z) = 1 - \exp(-z^{\gamma})$ if $z > 0$ and $W(z) = 0$ otherwise

    Gumbel: $G(z) = \exp[-\exp(z)]$, $-\infty < z < \infty$

In all these cases the independent variable *z* is scaled as $z = (x-\lambda)/\delta$, $\delta > 0$, where *x* is the original independent variable. It also holds that the exponent $\gamma > 0$. Rules for the domain of attraction can be given explicitly so that for any parent distribution the limit distribution can be known. A parent distribution with a non-finite end-point in the tail of interest cannot lie in the Weibull domain of attraction. A parent distribution with a finite end-point in the tail of interest cannot lie in the Frechet domain of attraction. For example for defects with size *a* with $0 < a < L$, where *L* is a length scale defining the macroscopic size of a structure, the only two options for the limit distributions are the Frechet and Gumbel type.

D.5   Change of variable

Frequently it occurs that a change of variable has to be made. That is, for a given random variable x we seek the distribution of $y = g(x)$ for a given function g. If we denote with F_x the cdf with respect to x, it holds that

$$F_y(y) = F_x[g^{-1}(y)] \qquad \text{if } g' > 0 \quad \text{and} \tag{D.44}$$

$$F_y(y) = 1 - F_x(g^{-1}(y)) \qquad \text{if } g' < 0 \tag{D.45}$$

For the pdf it holds that $f_y(y) = |dg^{-1}(y)/dy|f_x[g^{-1}(y)]$. For example, let x be a random variable with uniform distribution $F(x)$ over the interval $(0, 1)$ and let $y = g(x) = x^2$. Here $g'(x) > 0$ and thus $F_y(y) = F_x(\sqrt{y}) = \sqrt{y}$, while $f_y(y) = 1/(2\sqrt{y})$.

D.6   Basic reliability equations

The *reliability function* $R(t)$ is defined as the (cumulative) probability of survival as a function of time t under operating conditions. The function $R(t)$ is thus continuously decreasing from $R = 1$ at $t = 0$ to $R = 0$ for $t \to \infty$. The *failure function* is given by $F(t) = 1 - R(t)$. The *(instantaneous) hazard function* $h(t)$ is defined such that $h(t)\,dt$ is the probability of failure in an infinitesimal time interval dt, conditional to the constraint that at time t no failure has occurred yet. Hence

$$h(t) = f(t)/R(t) \qquad \text{where, as before,} \tag{D.46}$$

$$f(t) = dF(t)/dt \qquad \text{or equivalently,} \qquad F(t) = \int_0^t f(t')\,dt' \tag{D.47}$$

It follows that

$$R(t) = 1 - F(t) = \exp[-H(t)] \tag{D.48}$$

Only for the exponential distribution $R(t) = \exp(-\lambda t)$ the hazard function $h(t)$ is constant and given by λ. For all other distributions a non-constant $h(t)$ arises. In practical situations the hazard function $h(t)$ frequently shows the so-called bath-tub behaviour: an 'infant' mortality region with decreasing $h(t)$ in the beginning of the lifetime followed by a more or less constant region for $h(t)$ and a wear-out region with increasing $h(t)$ at the end of the lifetime. Finally, it should be stated that a similar reasoning can be made if time t is replaced by another independent variable, e.g. stress σ, provided a single-valued relationship exists between R and σ.

D.7   Bibliography

Castillo, E., (1988), *Extreme value theory in engineering*, Academic Press, New York.

Green, J.R., Margerison, D. (1978), *Statistical treatment of experimental data*, Elsevier, Amsterdam.

Hamilton, W.C. (1964), *Statistics in physical science*, Ronald Press, New York.

Mood, A.M., Graybill, F.A., Boes, D.C. (1974), *Introduction to the theory of statistics*, 3rd ed., McGraw-Hill Kogakusha, Tokyo.

Appendix E

Contact mechanics

The mechanics of surfaces in contact is called contact mechanics (Johnson, 1985). From this branch of science, we discuss here only the some simple contact situations. The most basic calculation in the field of contact mechanics is the calculation of the elastic strains and stresses of point loads and of line loads. These are briefly discussed first. After that the spherical contact situation is sketched followed by a description of the effects of sharp indenters. This includes their elastic and inelastic deformation as well as crack formation. More details can be found in the textbooks and papers quoted in the reference list. In particular we refer to Johnson (1985). As usual, Young's modulus, Poisson's ratio and the shear modulus are represented by E, v and $G = E/2(1+v)$, respectively.

E.1   Line loading

Flamant[a] treated the line loading of a semi-infinite solid in 1892. Consider a single normal line load n per unit length in the plane xz and applied at a point O' at coordinates $(\varepsilon,0)$ defined with respect to the origin O on the surface $z = 0$ of semi-infinite solid (Fig. E.1). This is a two-dimensional situation where the solid is in a plain strain mode.

Using the stress function $\phi(r,\theta) = Ar\theta \sin \theta$ and the associated expressions for the stresses, the radial stress σ_{rr} is given by

$$\sigma_{rr} = -(2n \cos \theta)/\pi r \qquad (E.1)$$

where r and θ denote the distance and angle with respect to O' respectively. Since the tangential stress $\sigma_{\theta\theta}$ and the shear stress $\sigma_{r\theta}$ are equal to zero, σ_{rr} a principal stress. In Cartesian co-ordinates the stresses with respect to O' are

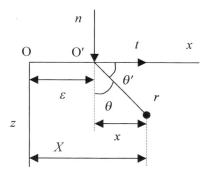

Fig. E.1: A line load acting on a semi-infinite body with a normal component n and tangential component t.

---

[a] Alfred-Aimé Flamant (1839-1914). French scientist who was mainly interested in the theory of granulated masses.

$$\sigma_{xx} = \sigma_{rr}\sin^2\theta = -\frac{2n}{\pi}\frac{x^2 z}{\left(x^2 + z^2\right)^2}$$ (E.2)

$$\sigma_{zz} = \sigma_{rr}\cos^2\theta = -\frac{2n}{\pi}\frac{z^3}{\left(x^2 + z^2\right)^2}$$ (E.3)

$$\sigma_{xz} = \sigma_{rr}\cos\theta\,\sin\theta = -\frac{2n}{\pi}\frac{xz^2}{\left(x^2 + z^2\right)^2}$$ (E.4)

A co-ordinate transformation $x = X - \varepsilon$ can be applied to obtain the expressions for the stresses with respect to O.

Using Hooke's law for plane strain the strains become

$$\varepsilon_{rr} = \frac{\partial u_r}{\partial r} = -\frac{\left(1 - v^2\right)}{E}\frac{2n\cos\theta}{\pi r}$$ (E.5)

$$\varepsilon_{\theta\theta} = \frac{u_r}{r} + \frac{1}{r}\frac{\partial u_\theta}{\partial\theta} = \frac{v\left(1 + v\right)}{E}\frac{2n\cos\theta}{\pi r}$$ (E.6)

$$\varepsilon_{r\theta} = \frac{1}{r}\frac{\partial u_r}{\partial\theta} + \frac{\partial u_\theta}{\partial r} - \frac{u_\theta}{r} = \frac{\sigma_{r\theta}}{G} = 0$$ (E.7)

Solving for the displacements u_r and u_z yields the general expressions

$$u_r = \frac{(1-v^2)}{\pi E}2n\cos\theta\ln r - \frac{(1-2v)(1+v)n}{\pi E}n\theta\sin\theta + C_1\sin\theta + C_2\cos\theta$$

$$u_\theta = \frac{(1-v^2)}{\pi E}2n\sin\theta\ln r + \frac{v(1+v)}{\pi E}2n\sin\theta$$

$$-\frac{(1-2v)(1+v)}{\pi E}2n\theta\cos\theta + \frac{(1-2v)(1+v)}{\pi E}n\sin\theta + C_1\cos\theta - C_3\sin\theta + C_3 r$$

If the solid does not tilt for points on the z-axis ($\theta = 0$) no lateral displacements are allowed and $C_1 = C_2 = 0$. By putting $\theta = \pm\pi/2$ in the above equations, the horizontal displacement $\bar{u}_r$ and vertical displacement $\bar{u}_\theta$ at the surface can be obtained and are given by

$$\bar{u}_r = u_r\big|_{\theta=\pi/2} = u_r\big|_{\theta=-\pi/2} = -\frac{(1-2v)(1+v)n}{2E} \quad \text{and}$$ (E.8)

$$\bar{u}_\theta = u_\theta\big|_{\theta=\pi/2} = -u_\theta\big|_{\theta=-\pi/2} = -\frac{(1-v^2)}{E}2n\ln(r_0/r)$$ (E.9)

where r_0 is taken as a reference point for the normal displacements. Thus for all points on the boundary of the solid, there is a constant displacement towards the origin. At the point of load application $\bar{u}_\theta$ becomes infinite.

Similarly for a tangential line load t (Fig. E.1), the situation is similar to before if we use the angle θ' instead of θ and the stresses are given by

$$\sigma_{rr} = -(2t\cos\theta')/\pi r \quad \text{and} \quad \sigma_{\theta\theta} = \sigma_{r\theta} = 0$$ (E.10)

which in Cartesian co-ordinates read

$$\sigma_{xx} = -\frac{2t}{\pi} \frac{x^3}{\left(x^2 + z^2\right)^2} \qquad \sigma_{zz} = -\frac{2t}{\pi} \frac{xz^2}{\left(x^2 + z^2\right)^2} \qquad \sigma_{xz} = -\frac{2t}{\pi} \frac{x^2 z}{\left(x^2 + y^2\right)^2} \qquad (E.11)$$

Making the appropriate medications for the angle, the strains are still given by Eqs. (E.5), (E.6) and (E.7) while the horizontal displacement $\bar{u}_r = -u_r|_{\theta'=0} = u_r|_{\theta'=\pi}$ and vertical displacement $\bar{u}_\theta = -u_\theta|_{\theta'=0} = -u_\theta|_{\theta'=\pi}$ at the surface are given by Eq. (E.9) and Eq. (E.8), respectively.

A frictional contact situation thus in principle can be described by $t = \mu n$ where μ is the appropriate friction coefficient. Since both the normal and tangential load situations lead to an infinite stress at the point O', this description is inadmissible and we have to allow for a finite area of contact.

For a distributed load $n(x)$ and $t(x)$ loaded over a strip $(-b < x < a)$, the integral over the load distribution has to be taken to obtain the total normal load N and tangential load T leading to

$$N = \int_{-b}^{a} n(x)\, dx \qquad \text{and} \qquad T = \int_{-b}^{a} t(x)\, dx$$

For uniform loads the above expressions result in $N = na$ and $T = ta$. For an arbitrary distribution of normal loads, the stress σ_{xx} at any point (x,z) is then given by

$$\sigma_{xx} = -\frac{2}{\pi} \int_{-b}^{a} n(s) \frac{z(x-s)^2}{\left[(x-s)^2 + z^2\right]^2}\, ds$$

Similar expressions hold for σ_{yy} and σ_{xy}. For the case of tangential loads, replacing $n(x)$ by $t(x)$ results in the proper expressions. The overall frictional contact situation is now described by the expression $T = \mu N$ and the stresses within the solid are given by the superposition of the stresses due to normal and tangential loading. For example, for an elastic half-space loaded over a strip $(-b < x < a)$ by a normal pressure $n(x)$ and a tangential traction $t(x)$, the normal component σ_{xx}

$$\sigma_{xx} = -\frac{2}{\pi} \int_{-b}^{a} n(s) \frac{z(x-s)^2}{\left[(x-s)^2 + z^2\right]^2}\, ds - \frac{2}{\pi} \int_{-b}^{a} t(s) \frac{(x-s)^3}{\left[(x-s)^2 + z^2\right]^2}\, ds$$

and similar expressions for σ_{zz} and σ_{xz}. For the displacements again the integral has to be taken over the displacement distribution due to the distributed load.

The maximum shear stress according to the Tresca criterion for a line load is given by $\tau = \sigma_{rr}/2 = -(n \cos \theta)/\pi r$. If we consider a circle of diameter b (Fig. E.2), we have $r = b \cos \theta$ and $\tau = -n/\pi b$ results. Consequently the shear stress remains constant at all points of that circle. The lines of constant shear stress are called *isochromatics*. For a continuous distribution of normal loads, integrals over the load distribution have to be

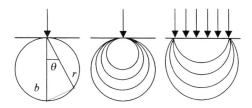

Fig. E.2: Patterns for isochromatics for point and distributed normal load.

calculated. For a point load as well as a uniformly distributed load, the shear stress reaches its maximum at the surface of the material (Fig. E.2).

E.2   Point loading

For point loading the situation is similar to that of line loading, but somewhat more complex. The calculation of the elastic strains and stresses of point loads was analysed first by Cerruti in 1882 and later by Boussinesq[b] in 1885. From the appropriate stress function for the case of a normal point load n acting on the surface the stresses are given by

$$\sigma_{rr} = \frac{n}{2\pi}\left[(1-2v)\left(\frac{1}{r^2} - \frac{z}{\rho r^2}\right) - \frac{3zr^2}{\rho^5}\right] \tag{E.12}$$

$$\sigma_{\theta\theta} = -\frac{n}{2\pi}(1-2v)\left(\frac{1}{r^2} - \frac{z}{\rho r^2} - \frac{z}{\rho^3}\right) \tag{E.13}$$

$$\sigma_{zz} = -\frac{3n}{2\pi}\frac{z^3}{\rho^5} \quad \text{and} \quad \sigma_{rz} = -\frac{3n}{2\pi}\frac{rz^2}{\rho^5} \tag{E.14}$$

where ρ is given by $\rho^2 = x^2 + y^2 + z^2$ and $r^2 = x^2 + y^2$. At any point the trace of the stress tensor is given by $\sigma_{rr}+\sigma_{\theta\theta}+\sigma_{zz} = -n(1+v)z/\pi\rho^3$. The displacements are given by

$$u_r = \frac{n}{4\pi G}\left[\frac{rz}{\rho^3} - (1-2v)\frac{\rho-z}{\rho r}\right] \quad \text{and} \quad u_z = \frac{n}{4\pi G}\left[\frac{z^2}{\rho^3} + \frac{2(1-v)}{\rho}\right] \tag{E.15}$$

Similar equations can be derived for a tangential point load t applied in the y-direction on the surface. The stresses are given by (with $\rho^2 = x^2 + y^2 + z^2$)

$$\frac{2\pi\sigma_{xx}}{t} = -\frac{3x^3}{\rho^5} + (1-2v)\left[\frac{x}{\rho^3} - \frac{3x}{\rho(\rho+z)^2} + \frac{x^3}{\rho^3(\rho+z)^2} + \frac{2x^3}{\rho^2(\rho+z)^3}\right]$$

$$\frac{2\pi\sigma_{yy}}{t} = -\frac{3xy^2}{\rho^5} + (1-2v)\left[\frac{x}{\rho^3} - \frac{x}{\rho(\rho+z)^2} + \frac{xy^2}{\rho^3(\rho+z)^2} + \frac{2xy^2}{\rho^2(\rho+z)^3}\right]$$

$$\frac{2\pi\sigma_{zz}}{t} = -\frac{3xz^2}{\rho^5}$$

$$\frac{2\pi\sigma_{xy}}{t} = -\frac{3x^2y}{\rho^5} + (1-2v)\left[-\frac{y}{\rho(\rho+z)^2} + \frac{x^2y}{\rho^3(\rho+z)^2} + \frac{2x^2y}{\rho^2(\rho+z)^3}\right]$$

$$\frac{2\pi\sigma_{xz}}{t} = -\frac{3x^2z}{\rho^5} \quad \text{and} \quad \frac{2\pi\sigma_{yz}}{t} = -\frac{3xyz}{\rho^5}$$

At any point the trace of the stress tensor is given by $\sigma_{rr}+\sigma_{\theta\theta}+\sigma_{zz} = -t(1+v)x/2\pi\rho^3$. The displacements are given by

---

[b] Joseph Valentin Boussinesq (1842-1929). French scientist, pupil of St.-Venant, who contributed to the theory of elasticity but also to hydrodynamics, optics and thermodynamics.

$$u_x = \frac{t}{4\pi G}\left\{\frac{1}{\rho} + \frac{x^2}{\rho^3} + (1-2v)\left[\frac{1}{\rho+z} - \frac{x^2}{\rho(\rho+z)^2}\right]\right\}$$

$$u_y = \frac{t}{4\pi G}\left[\frac{xy}{\rho^3} - (1-2v)\frac{xy}{\rho(\rho+z)^2}\right]$$

$$u_z = \frac{t}{4\pi G}\left[\frac{xz}{\rho^3} - (1-2v)\frac{x}{\rho(\rho+z)}\right]$$

E.3 General loading

It will be clear that by using these basic load situations in principle the stress distribution for any type of load distribution over the contact region can be obtained. In practice, however, the number of loading situations that can be solved analytically is rather restricted. One important loading situation is the axi-symmetric case of the form

$$p = p_0(1 - r^2/a^2)^\alpha \qquad (E.16)$$

where α is a parameter and a the radius of loaded area. For $\alpha = 0$ the expression represents uniform surface pressure while for $\alpha = -\frac{1}{2}$ it represents uniform normal surface displacements. For $\alpha = \frac{1}{2}$ it represents the Hertz contact situation (see the next paragraph). The basic solutions for point and line load can be used conveniently though to calculate the stress and displacement distribution numerically through the superposition principle.

E.4 Contact of cylindrical and spherical surfaces

Hertz started the analysis of curved surfaces in contact in 1882 and we will present the main results for the case of cylindrical surfaces in this section before we proceed to the spherical contact situation in the next section. The radii of curvature and denoted by R while a subscript denotes the material.

For the contact between two identical cylinders of radius R under a normal load P, the pressure distribution is given by Eq. (E.16) with $\alpha = \frac{1}{2}$ and thus

$$p(x) = \frac{2P}{\pi a}\left(1 - \frac{x^2}{a^2}\right)^{1/2} = p_0\left(1 - \frac{x^2}{a^2}\right)^{1/2}$$

where $2a$ is the width of the contact zone and p_0 is the maximum pressure. The mean pressure p_m equals $p_m = P/2a = \pi p_0/4$. On the basis of dimensional arguments we can say that the stress σ should be proportional to P/a, while the strain ε should be proportional to a/R. Therefore $P/a \sim Ea/R$ or $a^2 \sim PR/E$. The actual solution is

$$a^2 = \frac{4(1-v^2)RP}{\pi E}$$

For non-identical cylinders the solution remains approximately true provided the angle subtended by the contact width at the centre of the cylinders is less than $30°$ by using the effective modulus $E*$ and the radius $R*$ given by

$$\frac{1}{E*} = \frac{1-v_1^2}{E_1} + \frac{1-v_2^2}{E_2} \qquad \text{and} \qquad \frac{1}{R*} = \frac{1}{R_1} + \frac{1}{R_2}$$

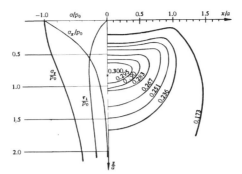

Fig. E.3: Hertzian contact situation. Left half: σ_{xx}, σ_{zz} and $\sigma_{xz,max}$. Right half: contours of principal shear stress.

For a load P the decrease in distance between the two centres of gravity δ is given by

$$\delta = a^2/2R^*$$

where the width of the contact area a is given by

$$a^2 = 4PR^*/\pi E^*$$

The stress in the material is described by a somewhat complicated expression and shown in Fig. E.3. The maximum pressure is p_0. At the surface the stress outside the contact zone is zero, inside it is given by $p(x)$. The stress along the z-axis is given by

$$\sigma_{xx} = -\frac{p_0}{a}\left[\left(a^2 + 2z^2\right)\left(a^2 + z^2\right)^{-1/2} - 2z\right] \quad \text{and} \quad \sigma_{zz} = -p_0 a\left(a^2 + z^2\right)^{-1/2}$$

These are the principal stresses so that the principal shear stress is given by

$$\sigma_{xz} = p_0 a\left[z - z^2\left(a^2 - z^2\right)^{-1/2}\right]$$

The maximum shear stress is about $0.30p_0$ and occurs at a depth of about $0.78a$. The isochromatics for the contact of a cylinder and a plane are shown in Fig. E.3. When tangential loading is also applied, the stress field can be obtained by the method mentioned before. As can be seen in Fig. E.4 for the particular case of a friction coefficient $\mu = 0.2$, the location of the maximum shear stress is now much nearer to the surface.

Similar results are obtained for contact of spherical bodies. The effective modulus

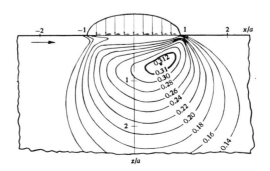

Fig. E.4: Isochromatics for the contact of a cylinder and a plane for normal and tangential loading with a friction coefficient $\mu = 0.2$.

and the radius of curvature are again denoted by E^* and R^*, respectively. The stress distribution due to load P is

$$p(x) = \frac{3P}{2\pi a^2}\left(1 - \frac{r^2}{a^2}\right)^{1/2} = p_0\left(1 - \frac{x^2}{a^2}\right)^{1/2}$$

where p_0 is the maximum pressure. The mean pressure p_m is given by $p_m = P/\pi a^2 = 2p_0/3$. The radius of the contact circle a is in this case given by

$$a = \frac{\pi p_0 R^*}{2E^*} = \left(\frac{3PR^*}{4E^*}\right)^{1/3}$$

while the mutual approach of the centres of gravity $\delta = a^2/R^*$ is represented by

$$\delta = \frac{\pi a p_0}{2E^*} = \left(\frac{9P^2}{16R^*E^{*2}}\right)^{1/3}$$

The stress distribution for the Hertzian contact is given in Fig. E.5 and compared with that for uniform pressure. Apart from the surface region the stress distributions are highly similar. The stress at the surface inside the contact zone is everywhere compressive except at the very edge where a radial tensile stress with maximum $(1 - 2\nu)p_0/3$ occurs. This is the largest tensile stress present anywhere and responsible for the Herztian 'cone' cracks. The principal stresses along the z-axis are given by

$$\frac{\sigma_{zz}}{p_0} = -\left(1 - \frac{z^2}{a^2}\right)^{-1} \quad \text{and} \quad \frac{\sigma_{rr}}{p_0} = \frac{\sigma_{\theta\theta}}{p_0} = -(1+\nu)\left[1 - \frac{z}{a}\tan^{-1}\left(\frac{a}{z}\right)\right] + \frac{1}{2}\left(1 - \frac{z^2}{a^2}\right)^{-1}$$

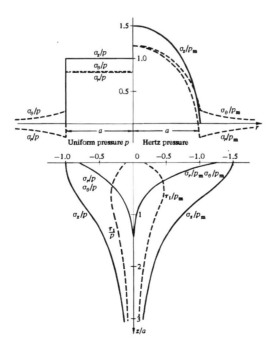

Fig. E.5: Stress distribution due to a Hertz contact (right) and due to uniform pressure (left) acting on a circular area with a radius a.

The principal shear stress

$$\tau = |\sigma_{rr} - \sigma_{zz}|/2$$

has a maximum value of $\cong 0.31p_0$ at a depth of approximately $0.48a$ (for $v = 0.3$). For comparison, for the uniform pressure distribution the maximum shear stress $\tau = -0.33p$ at a depth of $\cong 0.64a$. Again, for simultaneous tangential loading the maximum shear stress location occurs at a point closer to the surface. Also the tensile stress at the surface is enhanced significantly to (Halling, 1975)

$$\sigma = [(1-2v)p_0/3][1 + C'\mu]$$

where $C' = 3\pi(4+v)/8(1-2v)$ and μ is the friction coefficient. Initiation of cracks is therefore much easier as compared with the static case. The effect on the in-depth stress distribution is extensively discussed by Frank and Lawn[c], Lawn[d] and extended by Zeng et al[e].

Heinrich Rudolf Hertz (1857-1894)

Born in Hamburg, Germany after a year of study in engineering at the Polytechnical Institute in Munich, he moved to Berlin to study with Helmholtz and Kirchoff. He received his doctorate degree in 1880 on a topic from electrodynamics. In 1880 he also became an assistant to Helmholtz, working on mechanical problems. His famous work on the theory of compression of elastic bodies was done and published in 1881 at the age of 24. He not only offered the general solution but also the application to particular situations. The paper not only attracted the attention of physicists but also of engineers for whom he prepared the paper published in 1882 and which also contains the experimental verification of the theory. In 1883 he became a lecturer at the University of Kiel while in 1885 he was elected professor of physics at the Karlsruhe Polytechnical Institute, where he did his famous work on electrodynamics. In 1889 Hertz was elected to the chair of physics at the University of Bonn, where he worked on the discharge of electricity in rarified gases and wrote his treatise on *Die Prinzipien der Mechanik* published in 1894.

E.5   Blunt wedges and cones

Although the Hertzian contact situation is frequently encountered, the contact situation with sharp corners is at least as important. The most important examples of this situation are due to wedge and cone indenters. The stress distribution for these situations can be derived from the basic line and point contact situations by the superposition principle. We quote only the most important results.

---

[c] Frank, F.C. and Lawn, B. (1967), Proc. Roy. Soc. A **229**, 291.
[d] Lawn, B. (1968), J. Appl. Phys. **39**, 4828.
[e] Zeng, K., Breder, K. and Rowcliffe, D.J. (1992), Acta Metall. Mater. **10**, 2595

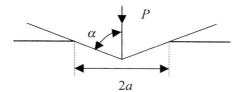

Fig. E.6: Indentation by a wedge with apex semi-angle α and a flat plane.

Consider first a two-dimensional wedge indenting a flat surface in such a way that the width of the contact area $2a$ is small as compared with the size of the solids. The semi-angle of the wedge is indicated by α (Fig. E.6). For such a wedge the pressure distribution is

$$p(r) = \frac{E * \cot\alpha}{\pi} \cosh^{-1}\left(\frac{a}{x}\right) = \frac{P}{a\pi} \cosh^{-1}\left(\frac{a}{x}\right)$$

where $P = E * a \cot\alpha$ is the total load. The pressure at the apex is infinite but the principal shear stress $\tau = |\sigma_{xx} - \sigma_{zz}|/2$ is finite. Along the z-axis τ is given by

$$\tau = \frac{E * a \cot\alpha}{\pi}\left(a^2 + z^2\right)^{-1/2}$$

which has a maximum of $(E * \cot\alpha)/\pi$ below the apex.

For a cone with a semi-angle α, similar results arise. In this case the pressure distribution is

$$p(r) = \frac{E * \cot\alpha}{2} \cosh^{-1}\left(\frac{a}{r}\right) = \frac{P}{\pi a^2} \cosh^{-1}\left(\frac{a}{r}\right)$$

with the total load $P = \frac{1}{2}\pi a^2 E * \cot\alpha$. Again at the apex, the pressure is infinite but the principal shear stress $\tau = |\sigma_{rr} - \sigma_{zz}|/2$ remains finite. Along the z-axis for $v = 0.5$, τ is given by

$$\tau = \frac{E * a^2 \cot\alpha}{2}\left(a^2 + z^2\right)^{-1}$$

which has a maximum of $(E * \cot\alpha)/2$ at the apex.

E.6 The effect of adhesion

In the standard Hertzian contact situation adhesive forces between the indenter and the half space are neglected. We discuss the effect of this adhesion here. The general solution for the pressure distribution for a contact of two axi-symmetric bodies with a contact radius a is given by

$$p(r) = p_0\left(1 - \frac{r^2}{a^2}\right)^{1/2} + p_0'\left(1 - \frac{r^2}{a^2}\right)^{-1/2}$$

where $p_0 = 2aE*/\pi R*$. In the adhesion-free situation p_0' is taken zero but in the presence of adhesive forces p_0' can be taken negative. The elastic energy E_{ela} stored by the two bodies by the compressive forces is given by

$$E_{\text{ela}} = \frac{\pi^2 a^3}{E^*}\left(\frac{2}{15}p_0^2 + \frac{2}{3}p_0 p_0' + p_0'^2\right)^2$$ (E.17)

while the compression δ is found to be

$$\delta = \frac{\pi a}{2E^*}(p_0 + 2p_0')$$ (E.18)

Since $(\partial U_{\text{ela}}/\partial a)_\delta = \pi^2 a^2 p_0'^2/E^* = 0$ in equilibrium, $p_0' = 0$ in the adhesion-free situation.

In the case of adhesion, the adhesive forces introduce a surface energy U_{sur} given by

$$U_{\text{sur}} = -2\pi\gamma a^2$$

and hence the total energy is $U = U_{\text{ela}} + U_{\text{sur}}$. In equilibrium $(\partial U_e/\partial a)_\delta = 0$ resulting in

$$p_0' = -(4\gamma E^*/\pi a)^{1/2}$$

The net force P is given by

$$P = \int_0^a p(r)\,2\pi r\,dr = \left(\frac{2}{3}p_0 + 2p_0'\right)\pi a^2$$

Substituting p_0 and p_0' yields

$$(P - 4E^* a^3/3R^*)^2 = 16\pi\,\gamma\,E^* a^3$$

This relationship is shown in Fig. E.7. From this figure, it is clear that upon unloading at point B, where $P = P_c = -3\pi\gamma R^*$ and $a = a_c = (9\gamma R^{*2}/4E^*)^{1/3}$, the situation becomes unstable and the contact recedes. These predictions have been verified experimentally.

E.7   Inelastic contact

So far we limited the discussion to elastic deformation. However, at a certain stress level yielding occurs. The onset is usually described by one of the two criteria:

$$\sqrt{J_2} = k = Y/\sqrt{3}   \quad\text{(von Mises' criterion)}$$

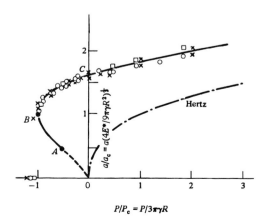

$$P/P_c = P/3\pi\gamma R$$

Fig. E.7: Variation of the contact radius with load in the case of adhesion for gelatine spheres with perspex. Radius $R \circ = 24.5$ mm, $\times = 79$ mm and $\square = 255$ mm.

$$\max |\sigma_I - \sigma_{III}| = 2k = Y \quad \text{(Tresca's criterion)}$$

where k and Y represent the yield strength in simple shear and in uniaxial tension, respectively.

For a contact of cylinders σ_{xx}, σ_{yy} and σ_{zz} are the principal stresses. The principal shear stress $\tau = |\sigma_{xx} - \sigma_{zz}|/2$ is $0.30p_0$ at a depth of $0.78a$. Tresca's criterion thus reads $2k = 0.60p_0 = Y$, independent of Poisson's ratio v. Hence yield commences when $p_0 = (p_0)_Y = 4p_m/\pi = 3.3k = 1.67Y$. Von Mises' criterion reaches a maximum of $0.322p_0$ at a depth of $0.70a$ for $v = 0.3$. Hence for this criterion yield starts when $(p_0)_Y = 3.1k = 1.79Y$. The value of the contact load for initial yield, as given by $P = \pi R^*(p_0)_Y^2/E^*$, is thus only slightly dependent on the choice of the yield criterion.

For a contact of spherical surfaces σ_{rr}, $\sigma_{\theta\theta}$ and σ_{zz} are the principal stresses. The maximum shear stress $\tau = |\sigma_{zz} - \sigma_{rr}|$ for $v = 0.3$ is $0.62p_0$ at a depth of $0.48a$. The Tresca criterion thus reads $p = (p_0)_Y = 3p_m/2 = 3.2k = 1.60Y$, while von Mises' criterion is $p = (p_0) = 2.8k = 1.60Y$. The load for initial yield is thus the same for both criteria and reads $P = \pi^3 R^{*2}(p_0)_Y^3/6E^{*2}$.

Elliptical contact situations are in between the spherical and cylindrical situations. Consequently, while the value of pressure for initial yield is approximately constant at about $1.7Y$, the depth changes from $0.5a$ for a spherical situation to about $0.86a$ for a cylindrical situation. The location of initial yield for a Hertzian contact is thus always below the surface using the Tresca criterion. Similar results are obtained for the von Mises criterion.

For wedges and cones, the situation is quite different. It can be proved that during two-dimensional frictionless indentation by a wedge, the tangential stress σ_{xx} is equal to the normal pressure p. If $v = 0.5$, the axial stress σ_{zz} is also equal to p. Moreover, in this case von Mises' and Tresca's criteria are identical if expressed in terms of k. The maximum shear stress $(E^* \cot \alpha)/\pi$ is at the apex as indicated before. Hence for a wedge yield will start when $\cot \alpha \geq \pi k E^*$.

For a cone the principal stress difference $|\sigma_{zz} - \sigma_{rr}|$ has also a maximum of $(E^* \cot \alpha)/2$ the apex. In this case two principal stresses are equal and von Mises' and Tresca's criterion are equal when expressed in terms of Y. Hence yield will commence when $\cot \alpha \geq Y/E^*$. It can be argued that for compressible materials ($v < 0.5$) these relations remain approximately true. Finally it should be remarked that yield starts at the interface contrary to the situation in Hertzian contact.

E.8 The pressurised cavity model

A permanent deformation due to an indentation occurs if the maximum shear stress exceeds the yield strength Y. The question is now to describe such an indentation. From the discussion on the simple contact situations given above, it should be clear that for the general contact situation simple but reliable expressions for stresses, strains and displacements are difficult to obtain. One necessarily has to resort to approximations. Fortunately the (far) stress fields of various blunt indenters are all approximately radial from the first point of contact with roughly hemi-spherical contours of strain, as confirmed by observations on the subsurface displacements by blunt indenters (Fig. E.8). A frequently adopted model for a sharp indentation is the *pressurised cavity model* (Hill, 1950) In this model one considers the indentation to be equivalent to a pressurised cavity (or hydrostatic core) within the material. The basic parts are the core with the radius a and the pressure p, the plastic zone surrounding the core with the radius c and an elastic field surrounding the plastic zone.

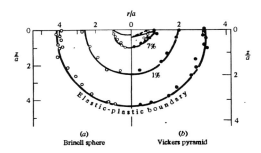

Fig. E.8: A comparison between the displacements due to a spherical indenter (left) and of a Vickers indenter (right) and the pressurized cavity model.

The stresses in the plastic zone ($a \le r \le c$) are

$$\sigma_{rr}/Y = -2 \ln(c/r) - 2/3 \qquad \text{and} \qquad \sigma_{\theta\theta}/Y = -2 \ln(c/r) + 1/3$$

while in the elastic zone ($r \ge c$)

$$\sigma_{rr}/Y = -2(c/r)^3/3 \qquad \text{and} \qquad \sigma_{\theta\theta}/Y = (c/r)^3/3$$

At the boundary of the core the pressure p is given by

$$p/Y = -[\sigma_{rr}/Y]_{r=a} = 2/3 + 2 \ln(c/a)$$

while the radial displacements are given by

$$du(r)/dr = (Y/E) [3(1-v)(c^2/r^2) - 2(1-2v)(r/c)] \tag{E.19}$$

Finally we require conservation of volume resulting in

$$2\pi a^2 \, du(a) = \pi a^2 \, dh = (\pi a^2 \tan \beta) \, da \tag{E.20}$$

where $\beta = \pi/2 - \alpha$, and the similarity of the strain field with progressing indentation expressed by

$$dc/da = c/a = \text{constant}$$

Combining yields

$$(E \tan \beta)/Y = 6(1-v)(c/a)^3 - 4(1-2v)$$

Substitution of c/a in the pressure equation yields p. For an incompressible material a simple equation is obtained

$$p/Y = (2/3)[1 + \ln(E \tan \beta/3Y)] \tag{E.21}$$

The pressure is thus dependent on the non-dimensional variable $E \tan \beta/Y$, which may be interpreted as the ratio of the strain $\tan \beta$ imposed by the indenter to the elastic strain capacity of the material Y/E. Replacing E by E^* accounts for the elasticity of the indenter. A graph of the non-dimensional variable p_m/Y versus $E^* \tan \beta/Y$ reveals the overall behaviour (Fig. E.9). For a Vickers indenter $\beta = 19.7°$ and first yield occurs at $p_m \cong 0.5Y$ while full plastic deformation sets an upper limit of about $3Y$, reached for $E \tan \beta/Y \cong 40$. For a sphere we may take $\tan \beta \cong \sin \beta \cong a/R$, which varies during indentation. Integration of Eqs. (E.19) and (E.20) with $c/a = 1$ at the point of first yield leads to Eq. (E.21) with an additional constant of about 0.19 on the right-hand side. First yield occurs for a spherical indenter at $p_m = 1.1Y$ while full plastic deformation is reached for $E \tan \beta/Y \cong 30$.

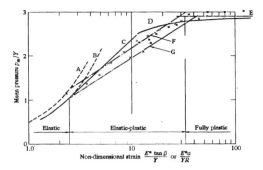

Fig. E.9: Indentation of an elastic-plastic material by spherical and conical indenters. Elastic: A cone, B sphere. Cavity model: F cone, G sphere. The lines C and D represent finite element results while the circles and crosses indicate experiment with pyramids and spheres, respectively. The line E denotes the full plastic situation.

So far we have discussed only elastic-perfectly plastic materials. Tabor (1951) has shown that for strain-hardening materials the above results may be applied if Y is replaced by a representative flow stress Y_R, measured in simple compression at a representative strain $\varepsilon_R \cong 0.2 \tan \beta$. For a Vickers indenter $\varepsilon_R \cong 0.07$ while for a spherical indenter $\varepsilon_R \cong 0.2a/R$. For strain-hardening materials obeying a power law with exponent n Matthews[f] obtained

$$\varepsilon_R = 0.28(1+1/n)^{-n}(a/R)$$

which varies from $0.188a/R$ for $n = 1$ to $0.171a/R$ for $n = \infty$, in reasonable agreement with Tabor's result. Fig. E.9 thus shows the mean indentation pressure by an axi-symmetrical indenter of arbitrary profile pressed into any elastic-plastic material whose stress-strain curve in compression is known. In particular an estimate may be made of the flow stress from the hardness. For a Vickers indenter the hardness $H_V = 0.93p_m \cong 2.8Y_R$.

E.9   Indentation in visco-elastic materials

Indentation of visco-elastic materials is an active field of research. We recall the superposition principle, which describes the response of a linear visco-elastic material in creep and relaxation. Denoting the stress by s and the strain by e we have

$$s(t) = \int_0^t \Psi(t-t')\frac{\partial e(t')}{\partial t'}\,dt' \qquad \text{or} \qquad e(t) = \int_0^t \Phi(t-t')\frac{\partial s(t')}{\partial t'}\,dt'$$

where Ψ and Φ represent the relaxation function and creep compliance, respectively. These functions are in general rather complex and we give only two examples. The first example represents delayed elasticity and is described by an analogue model consisting of a spring with spring constant k_1 in series with a Kelvin element with viscosity η and spring constant k_2. It is sometimes addressed as the *standard model*. The creep response to a step change in stress s_0 is given by

$$e(t) = \Phi(t)s_0 = \left\{\frac{1}{k_1} + \frac{1}{k_2}\left[1 - \exp(-t/t_1)\right]\right\}s_0$$

---

[f] Matthews, J.R. (1980), Acta Met. **28**, 311.

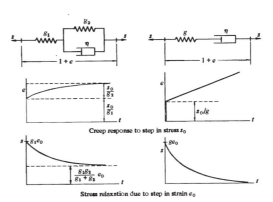

Fig. E.10: Visco-elastic standard and Maxwell models.

where $t_1 = \eta/k_2$. The relaxation function describes the response to a step in the strain e and is given by

$$s(t) = \Psi(t)e_0 = \frac{k_1}{k_1 + k_2}\left[k_2 + k_1 \exp(-t/t_2)\right]e_0$$

where $t_2 = \eta/(k_1+k_2)$. The second example represents unrestricted creep and is represented by the well-known *Maxwell model* consisting of a spring with spring constant k in series with a dashpot with the viscosity η. In this case the creep function is linearly related to time and reads

$$e(t) = \Phi(t)s_0 = \left(\frac{1}{k} + \frac{1}{\eta}t\right)s_0$$

The relaxation function is a simple exponential and reads

$$s(t) = \Psi(t)e_0 = \left[k\exp(-t/t_0)\right]e_0 \tag{E.22}$$

where $t_0 = \eta/k$. These relations are illustrated in Fig. E.10.

   In order to find a solution for the indentation problem, use is made of the correspondence principle. If we assume that the indenter is rigid and the material is incompressible, the expressions for the contact radius a and the pressure distribution $p(r)$ are

$$a^3 = (R\delta)^{3/2} = \frac{3}{8}\frac{RP}{2G} \qquad \text{and} \qquad p(r) = \frac{4}{\pi R}2G(a^2 - r^2)^{1/2}$$

For the visco-elastic material $2G$ is replaced by the relaxation operator so that for $r < a(t')$ we obtain for the pressure distribution

$$p(r,t) = \frac{4}{\pi R}\int_0^t \Psi(t-t')\frac{d}{dt'}\left[a^2(t') - r^2\right]^{1/2}dt' \tag{E.23}$$

This distribution produces normal displacements of the surface, which conform to the profile of the indenter. For the total force the result is

$$P(t) = \frac{8}{3R}\int_0^t \Psi(t-t')\frac{d}{dt'}a^3(t')dt'$$

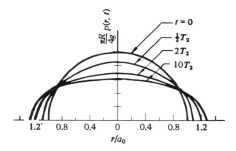

Fig. E.11: Variation of the pressure distribution with a step load for the standard model.

If the total displacement $\delta(t)$ is prescribed we directly have $a^3(t) = R\delta(t)$ which can be substituted in Eq. (E.23). If the total load is $P(t)$ prescribed, we replace $1/2G$ by the creep operator and obtain

$$a^3(t) = \frac{3}{8}R\int_0^t \Phi(t-t')\frac{d}{dt'}P(t')\,dt'$$

For a single step, $P(t) = 0$ for $t < 0$ and $P(t) = P_0$ for $t > 0$, the contact radius becomes

$$a^3(t) = \frac{3}{8}R\Phi(t)P_0 \tag{E.24}$$

In this case the calculation of the time dependence of the contact radius is simple.
     For the standard model with a single-load step we thus obtain

$$a^3(t) = \frac{3}{8}RP_0\left\{\frac{1}{k_1} + \frac{1}{k_2}[1-\exp(-t/t_1)]\right\} \tag{E.25}$$

Directly after application of the load, the elastic response leads to a contact radius

$$a_0 = (3RP_0/8k_1)^{1/3}$$

and then the contact radius grows with time and eventually becomes

$$a_1 = [3RP_0(1/k_1 + 1/k_2)/8]^{1/3}$$

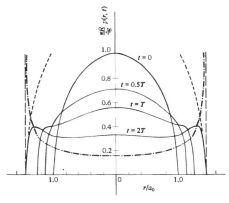

Fig. E.12: Variation of the pressure distribution with a step load for the Maxwell model. —— =
Maxwell model, — · — · = viscous fluid, - - - = elastic solid.

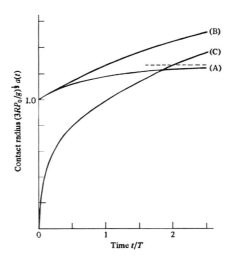

Fig. E.13: Growth of the contact radius $a(t)$ due to a step load. A = standard model, B = Maxwell model and C = viscous fluid.

The initial Hertz pressure distribution is characterised by $2G = k_1$ and $a = a_0$, while the final pressure distribution is again of the Hertz type and is characterised by $2G = k_1 k_2/(k_1+k_2)$ and $a = a_1$. At intermediate times the pressure distribution can be obtained by substituting $\Psi(t)$ and $a(t)$ in the $p(r,t)$ and performing the integrations. For $k_1 = k_2$ the results are shown in Fig. E.11. It appears that at all times the distribution is close to the Hertz type.

For the Maxwell model the creep behaviour can be obtained by the creep compliance $\Phi(t)$ in the expression for $a(t)$. Directly after application of the load, the elastic response leads to the contact radius again

$$a_0 = (3RP_0/8k_1)^{1/3}$$

and thereafter grows according to Eq. (E.24). The substitution of the relaxation function $\Psi(t) = k \exp(-t/t_0)$ together with Eq. (E.24) in $p(r,t)$ yield the pressure distribution. The numerically evaluated result is shown in Fig. E.12. The initial distribution is of the Hertz type but the distribution changes drastically with time and approaches the result for a viscous fluid.

In conclusion, the effect of visco-elasticity on the indentation behaviour can be quite profound as shown by the last example. For comparison the increase of the contact radius with time for the standard model, the Maxwell model and a viscous fluid is shown in Fig. E.13. Since the boundary conditions change during the indentation process, the direct application of the correspondence principle is not straightforward and more complex procedures have to be used. These procedures include full numerical approaches as well as attempts to solve the problem analytically by integral transforms. The problem is rather complex though and easily workable solutions have not been found yet.

E.10   Cracking

Apart from the deformation, another aspect that has to be considered is the appearance of cracks upon indentation. The appearance of cracks used to be seen as a nuisance, but their presence has been taken to advantage lately to characterise the

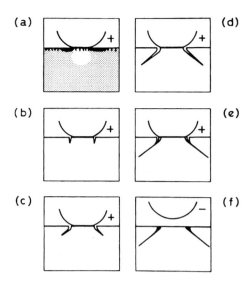

Fig. E.14: Evolution of a Hertzian cone crack showing loading (+) and unloading (–). In part (a), the black area represents the tensile and the grey area represents the compressive stress field.

fracture behaviour of brittle materials. In general the near stress field of indenters is complex so that the initiation of cracks is heavily dependent on the precise type of indenter. However, the far stress field of indenters is very similar and therefore the growth of indentation cracks can be considered in general terms. However, the fracture mode is different for sharp and blunt indenters. For a blunt indenter a Hertzian cone crack develops. Fig. E.14 illustrates the process. For a sharp indenter cracks appear at the corners of the indentation and they can be distinguished in median/radial cracks that occur during loading and lateral (or Palmquist) cracks that occur during unloading. The various steps are represented in Fig. E.15. In view of the complexities it is not surprising that many equations have been proposed. Ponton and Rawlings[g] provide a useful review for the Vickers indentation. Apart from general considerations about the surface quality and testing procedures, they show that almost all existing equations can be used for qualitative ranking. However, for quantitative ranking only three are able to rank various brittle materials without significant bias and reasonable accuracy using the radial-median crack system. The next paragraph discusses the background of these equations.

By carrying out a dimensional analysis of the elastic-plastic indentation process using the elastic stress field solutions modified by the presence of the indentation plastic zone, it can be shown that the indentation crack length c should be related to the indentation half-diagonal a by

$$K_{\text{ind}}/Y\sigma\sqrt{a} = F_1(c/a)F_2(r_\text{p}/a)F_3(v)F_4(\mu) \tag{E.26}$$

where K_{ind} is the fracture toughness as measured by indentation, Y is the uniaxial yield strength, r_p the indentation plastic zone radius and the Fs are empirical functions. As usual, v is Poisson's ratio and μ is the shear modulus. Often the mean indentation pressure $H = P/a^2$ with the load P is used as the measure for hardness and in this case

---

[g] Ponton, C.B. and Rawlings, R.D. (1989), Mater. Sci. Tech. **5**, 865 and **5**, 961.

the Vickers hardness $H_V = 0.9272H$. We then can write $\phi = H/Y$ with $\phi = \phi(E/Y, \nu)$ with the Young modulus E. From fitting on various data, it was found that for Vickers indentation data, the value of c/a correlated with $K_{ind}\phi/(H_V a^{1/2})$ and increased with decreasing hardness. It was further assumed that the influence of ν on c/a was insignificant and that a dependence of μ on H was unlikely. The hardness dependence of c/a was assumed to be due to r_p/a decreasing with hardness and to follow a power function of E/Y (or $E\phi/H$). Using these assumptions the above expression reduces to

$$K_{ind}\phi/H\sqrt{a} = F_1(c/a)F_2(E\phi/H) \tag{E.27}$$

From data fitting $F_2 = (E\phi/H)^{2/5}$ was obtained. Plotting $\log\{[K_{Ic}/(H_V a^{1/2})](H_V/E\phi)^{2/5}$ versus c/a for many ceramic materials, a slope of $-3/2$ for high values of c/a was obtained, agreeing with the slope for a penny-shaped crack loaded by P^* at its centre $K = P^*(\pi c)^{-3/2}$. With $P^* = P/(2\tan 74°)$ and $P = H_V a^2/0.4636$, one obtains

$$K = H_V a^2/0.4636(2\tan 74°)(\pi c)^{3/2}$$

Using the Tabor value $\phi = 2.7$, one can arrive at

$$K_{ind}\phi/H_V\sqrt{a} = k\left[2.7/(0.4636\pi^{3/2}\, 2\tan 74°)\right](c/a)^{-3/2} = 0.15k(c/a)^{-3/2}$$

where k is a correction factor for the presence of a free surface, empirically found to be 3.2 for large values of c/a. Finally substituting $H_V = 0.4636P/a^2$ the result as obtained by Evans and Charles[h] is

$$K_{ind} = 0.1777H_V a^2/c^{3/2} = 0.0824P/c^{3/2} \tag{E.28}$$

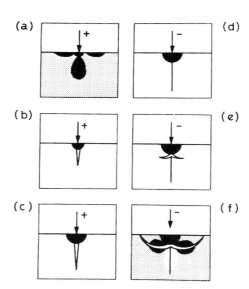

Fig. E.15: Evolution of the median, radial and lateral crack systems during sharp indenter loading (+) and unloading (–). The black area represents the plastic zone. In parts (a) and (f), the residual stress is shown (black = tensile and white = compressive).

[h] Evans, A.G. and Charles, E.A. (1976), J. Am. Ceram. Soc. **59**, 371.

Another attempt[i] to describe the indentation fracture process starts with Eq. (E.27) but omitting the constraint factor ϕ. Again using $F_1 = (c/a)^{-3/2}$ and using a fit on various data to obtain $F_2 = (E/H)^{2/5}$, a plot of $\log\{[K_{Ic}/(H_V a^{1/2})](H_V/E)^{2/5}$ versus c/a produced a good correlation in spite of the absence of ϕ. A polynomial fit to this plot yielded

$$\log\{[K_{Ic}/(H_V a^{1/2})](H_V/E)^{2/5} \equiv F \quad \text{with}$$

$$F = -1.59 - 0.34B - 2.02B^2 + 11.23B^3 - 24.97B^4 + 16.32B^5 \quad \text{where} \quad B = \log(c/a)$$

The final result becomes

$$K_{ind} = H_V a^{1/2} (E/H_V)^{2/5} 10^F = 0.6305 E^{0.4} P^{0.6} 10^F a^{-0.7} \tag{E.29}$$

However, the slope of fit using data from $c/a = 2$ to $c/a = 7$ resulted in -1.32 instead of -1.5.

Blendell[j] fitted the data of Evans and Charles and obtained

$$\left[K_{ind}\phi/(H_V a^{1/2})\right](H_V/E\phi)^{2/5} = 0.055 \log(8.4a/c)$$

which, again using $\phi = 2.7$, results in

$$K_{ind} = 0.0303 (H_V a^{1/2})(E/H_V)^{2/5} \log(8.4a/c) \tag{E.30}$$

The above-indicated equations were capable of ranking data of 16 different materials according to

Eq. (E.28): $K_{ind}/K_{Ic} = 0.97 \pm 0.23$

Eq. (E.29): $K_{ind}/K_{Ic} = 1.07 \pm 0.13$

Eq. (E.30): $K_{ind}/K_{Ic} = 1.14 \pm 0.15$

where '$\pm$' indicates the sample standard deviation. As judged from this analysis, Eq. (E.29) therefore produces the smallest standard deviation but a slightly larger bias, while Eq. (E.28) hardly shows any bias but a larger standard deviation.

The above three equations are all based ultimately on fitting and from this study and others[k] it appears that the indiscriminate use of an indentation equation could be highly misleading.

There have been also attempts to determine the fracture toughness using Hertzian cracks. Although somewhat more complicated, they seem also capable of a measurement[l] of the toughness with an accuracy of about 10%. By the way, also attempts have been made to determine the elastic modulus from Hertzian indentation[m] with reasonable success.

It should be mentioned that there are at least three other aspects that have to be considered with indentation. The first is the effect of residual stress. Attempts have been made to analyse the residual stress in a brittle surface via analysis of the slope of K_{ind} versus $c^{1/2}$. According to Ponton and Rawlings, the values derived should be considered carefully but in a number of cases good agreement with the value as

[i] Evans, A.G. (1979), page 112 in *Fracture mechanics applied to brittle materials*, ASTM-STP 678, S.W. Freiman, ed., ASTM, Philadelphia.
[j] Blendell, J.E. (1979), Ph.D. thesis, MIT.
[k] Ray, K.K. and Dutta, A.K. (1999), Brit. Ceram. Trans. **98**,165.
[l] Zeng, K. Breder, K. and Rowcliffe, D.J. (1992), Acta Metall. Mater. **10**, 2601.
[m] Zeng, K. Breder, K., Rowcliffe, D.J. and Herrström, C. (1992), J. Mater. Sci. **27**, 3789.

determined with the X-ray $\sin^2\psi$ technique is obtained[n]. It appears though that the externally applied load influences the residual stress term[o]. Second, it can be shown that elastic recovery[p] occurs after indentation and the effect has been used to estimate Young's modulus. The recovery is strongest for the Knoop indenter but still the accuracy is rather low and other methods are to be preferred. Third, during indentation mechanical energy is dissipated via plastic deformation and cracking. Although not straightforward, it appears possible to analyse the various contributions and attribute them to individual processes. In particular for the case of layered materials such as coatings, this is a helpful new development. Ultimately one tries to determine the stress-strain curve via indentation but so far it is not a standard procedure. Much more can be said and a general survey of problems associated with contact mechanics is given by Barber and Ciavarella[q].

Finally it must be remarked that the indentation process is crucial for the understanding of abrasive machining. In this process material removal occurs by scratching, sharp indenters, primarily by lateral crack formation. At the same time radial and median cracks develop which act as defects for overall fracture so that with increasing material removal rate, implying increasing loads, also leads to a decreasing strength.

E.11   Bibliography

Halling, J, ed. (1975), *Principles of tribology*, MacMillan, London.

Hill, R. (1950), *The mathematical theory of plasticity*, Oxford University Press, Oxford.

Johnson, K. L. (1985), *Contact mechanics*, Cambridge University Press, Cambridge.

Lawn, B. and Wilshaw, R. (1975), J. Mater. Sci. **10**, 1049.

Tabor, D. (1951), *The hardness of metals*, Oxford University Press, Oxford.

---

[n] De With, G. and Sweegers, N. (1995), Wear **188**, 142.
[o] Fett, T. (1995), Eng. Fract. Mech. **52**, 773.
[p] Lawn, B. and Howes, V.R. (1981), J. Mater. Sci. **16**, 2745.
[q] Barber, J.R. and Ciavarella, M. (2000), Int. J. Sol. Struct. **37**, 29.

Index